Modern Trends in Applied Aquatic Ecology

Modern Trends in Applied Aquatic Ecology

Edited by

R. S. Ambasht and
Navin K. Ambasht
Banaras Hindu University
Varanasi, India

Kluwer Academic/Plenum Publishers
New York, Boston, Dordrecht, London, Moscow

ISBN 0-306-47334-8

© 2003 Kluwer Academic/Plenum Publishers
233 Spring Street, New York, NY 10013

http://www.wkap.nl/

10 9 8 7 6 5 4 3 2 1

All rights reserved.

No part of this book may be reproduced, stored in a retrieval system, or transmitted in any form or by any means, electronic, mechanical, photocopying, microfilming, recording, or otherwise, without written permission from the Publisher, with the exception of any material supplied specifically for the purpose of being entered and executed on a computer system, for exclusive use by the purchaser of the work.

Printed in the United States of America

CONTRIBUTORS LIST

1. **Jacob John**
   Postgraduate Coordinator School of Environmental Biology
   Curtin University
   GPO Box U1987, Perth, WA 6845
   AUSTRALIA
   E-Mail: J.John@curtin.edu.au

2. **Paul A. Keddy**
   Department of Biological Sciences
   Southeastern Louisiana University
   Hammond, Louisiana, USA
   E-Mail: pkeddy@selu.edu

3. **Lauchlan H. Fraser**
   Department of Biology
   University of Akron
   Akron, OH 44325-3908, USA
   E-Mail: lfraser@uakron.edu

4. **Brian Whitton**
   School of Biological and Biomedical Sciences
   University of Durham
   Durham DH1 3LE, UNITED KINGDOM
   E-Mail: b.a.whitton@durham.ac.uk

5. **Alain Vanderpoorten**
   Department of Botany
   Duke University
   P.O. Box 90339
   Durham, NC 27708, USA
   E-Mail: alain@duke.edu
   vanderpoorten@ecol.ucl.ac.be

6. **Lisandro Benedetti-Cecchi**
   Dipartimento di Scienze dell'Uomo e dell'Ambiente
   Università di Pisa
   Via A Volta 6, I-56126 Pisa
   ITALY
   E-Mail: bencecc@discat.unipi.it

7. **Donat-P. Häder**
   Friedrich-Alexander Universität Erlangen-Nürnberg
   Institut für Botanik und Pharmazeutische Biologie
   Staudstr. 5, D-91058 Erlangen
   GERMANY
   E-Mail: dphaeder@biologie.uni-erlangen.de

8. **H. E. Zagarese**
   Centro Regional Universitario Bariloche
   Universidad Nacional del Comahue
   8400 - San Carlos de Bariloche
   ARGENTINA
   E-Mail: zagarese@crub.uncoma.edu.ar

9. **Barbara Tartarotti**
   Institute of Zoology and Limnology
   University of Innsbruck
   Technikerstrasse 25, 6020 Innsbruck
   AUSTRIA
   E-Mail: Barbara.Tartarotti@uibk.ac.at

10. **D. A. Añón Suárez**
    Centro Regional Universitario Bariloche
    Universidad Nacional del Comahue
    8400 - San Carlos de Bariloche
    ARGENTINA
    E-Mail: danon@crub.uncoma.edu.ar

11. **Rajeshwar P. Sinha**
    Friedrich-Alexander Universität Erlangen - Nürnberg
    Institut für Botanik und Pharmazeutische Biologie
    Staudstr. 5, D-91058 Erlangen
    GERMANY
    E-Mail: r.p.sinha@gmx.net

12. **Nándor Oertel**
    Hungarian Danube Research Station
    Hungarian Academy of Science
    Göd, HUNGARY

13. **János Salánki**
    Balaton Limnological Research Institute
    Hungarian Academy of Science
    Tihany, Pf35, 8237
    HUNGARY
    E-Mail: salanki@tres.blki.hu

14. **Arthur J. McComb**
    School of Environmental Science
    Murdoch University
    Perth, WA 6150, AUSTRALIA
    E-Mail: mccomb@essun1.murdoch.edu.au

15. **Jane M. Chambers**
    School of Environmental Science
    Murdoch University
    Perth, WA 6150, AUSTRALIA
    E-Mail: J.Chambers@murdoch.edu.au

16. **R. S. Ambasht**
    Honorary Scientist
    Department of Botany
    Banaras Hindu University
    Varanasi - 221 005, INDIA
    E-Mail: rambasht@banaras.ernet.in

17. **Navin K. Ambasht**
    Ecology Research Laboratory
    Department of Botany
    Banaras Hindu University
    Varanasi - 221 005, INDIA
    E-Mail: asnambasht@yahoo.com

18. **Robert Zwahlen**
    Environment and Social Development Specialist
    Electrowatt Engineering Ltd (EWE)
    P.O. Box 8037
    Hardturmstr. 161, Zurich
    SWITZERLAND
    E-Mail: robert.zwahlen@ewe.ch

PREFACE

Organisms and environment have evolved through modifying each other over millions of years. Humans appeared very late in this evolutionary time scale. With their superior brain attributes, humans emerged as the most dominating influence on the earth. Over the millennia, from simple hunter–food gatherers, humans developed the art of agriculture, domestication of animals, identification of medicinal plants, devising hunting and fishing techniques, house building, and making clothes. All these have been for better adjustment, growth, and survival in otherwise harsh and hostile surroundings and climate cycles of winter and summer, and dry and wet seasons. So humankind started experimenting and acting on ecological lines much before the art of reading, writing, or arithmetic had developed. Application of ecological knowledge led to development of agriculture, animal husbandry, medicines, fisheries, and so on. Modern ecology is a relatively young science and, unfortunately, there are so few books on applied ecology. The purpose of ecology is to discover the principles that govern relationships among plants, animals, microbes, and their total living and nonliving environmental components. Ecology, however, had remained mainly rooted in botany and zoology. It did not permeate hard sciences, engineering, or industrial technologies leading to widespread environmental degradation, pollution, and frequent episodes leading to mass deaths and diseases. Awareness of the dimensions of environmental problems threatening not only to the fast-deteriorating quality of human life but also survival has been generated by the United Nations Conference on Environment and Development (UNCED; Rio de Janeiro, 1992) and subsequent efforts of governments and nongovernment organizations. In this direction, there is an acute need for books on applied ecology. With the experience of writing our highly popular book, *A Textbook of Plant Ecology* (14th edition in press), we undertook the task of preparing this multiauthored volume. In recent years, wetlands have drawn world attention, particularly through the Ramsar Convention, IUCN, International Association of Ecology Wetland Group, and UNEP. Water is vitally required by all organisms; it is the medium of life processes and nutrient transports in plants. Water is the ultimate source of atmospheric oxygen. Surface water flows have built extensive, productive alluvial plains all over the world. Therefore, application of ecology in managing water quality and aquatic organisms, and impact of UV-B enhancement responses to other stresses in wetlands form the main themes of this book. Specific topics and subject specialists were identified. These world authorities have produced different chapters.

Jacob John, at the outset, has explained the concept of aquatic system health and attributes of biomonitors. Ubiquitous diatoms have been found to be an ideal tool in preparing predictive models for the health of aquatic systems. Case studies of Western Australia and South Western Australia have greatly helped to validate the conclusions. Keddy and Fraser have developed four principles involved in the management of biodiversity in wetlands: (1) water level fluctuations (depth), (2) fertility range from distinctly infertile to highly fertile zones, (3) competitive hierarchies, and (4) centrifugal organizations of species from fertile, sheltered sites to infertile sites. In managing aquatic ecosystems, we must expand from traditional natural systems to constructed wetlands for treatment of wastewater. Lotic bodies such as rivers with highly dynamic seasonal phases and fluxes are important types of aquatic bodies. Brian Whitton has highlighted the importance of indicator plants to monitor heavy metals in rivers. Aquatic bryophytes are sensitive to changes in water chemistry. Vanderpoorten has shown how, through calibration techniques, aquatic bryophytes may be used for predicting water chemistry with least prediction errors. The value of the chapter is enhanced by explaining the limits of transposing regional models. Improved knowledge of ecophysiology is needed. The concept of succession tells about the ever-changing nature of biotic communities and environmental complexes, and has widely applied implications. Bendetti-Cecchi has contributed a comprehensive discussion on current understanding of plant succession in littoral habitats. There are large gaps in understanding of variable patterns of recovery that preclude accurate predictions. Most of the lakes and rivers have become the dumping place of wastes. They are presently experiencing symptoms of eutrophication, siltation, and contamination by toxic chemicals, heavy metals, and secondary salinization, which put great stress on aquatic vertebrates. Artificial wetlands created by extensive mining in the form of mine-voids or pit wetlands can be utilized for aquaculture and biodiversity conservation. Phytoremediation, that is, environmental remediation of contaminated soil and water by hyperaccumulator plants, is assuming great applied ecological importance. Jacob John has elaborated this energy-friendly or "green" tool of use of algae for remediation of mine-voids and other degraded wetlands. With regard to enhancing UV-B levels at the surface of aquatic bodies, Häder has elaborated UV-B radiation impacts on primary producers found in freshwater and marine ecosystems. Bacterioplankton are more susceptible to UV-B than larger, eukaryotic organisms, and bacteria are the major degraders of organic material. Phytoplankton is the most important biomass producer of aquatic ecosystems. Protective mechanisms against UV-B by bacteria, cyanobacteria, and macroalgae are explained. Zagarese, Tartarotti, and Añón Suárez have regarded solar UV radiation as a strong and ubiquitous force in aquatic ecosystems. It causes impairment of essential biomolecules (nucleic acids and proteins), damage to exposed parts by way of sunburns and cataracts, and depression of immune systems. Indirectly, it photoactivates chemical pollutants. All these aspects of UV-B impacts are reviewed by Zagarese et al. and Sinha has reviewed responses of cyanobacteria to other kinds of stresses.

Aquatic ecosystems, the main receiving body of the runoff material from excessively pesticide-loaded agricultural fields and toxic materials released into the environment by modern anthropogenic activities, need to be studied for life forms that have biomonitoring and bioindicator values. Oertel and Salánki have elaborately defined biomonitoring and its relationship to toxicology. Origin of types of toxic chemicals and their effects on river and lake ecosystems, accumulation and synergism, action mechanisms, advantages, and limitations of biological versus chemical monitoring are nicely reviewed by McComb and Chambers. Biological early warning systems (BEWSs) such as fish and multisensor monitors of the Rhine Action Program are described, together with advantages and shortages.

Ambasht and Ambasht have reviewed work on the conservation role of ecotonal vegetation in checking soil erosion, water runoff, and nutrient losses across embankment slopes of river corridors and lake margins. Proper application of research results in selection of plants, naturally growing and artificially planted on sloping margins of wetlands, that can prevent eutrophication, siltation, upwelling of rivers, lakes, dam water storages, and floods considerably. Finally, Zwahlen has taken up the entirely man-made aquatic ecosystems created by dam projects for "identification," assessment, and mitigation of environmental impacts. The chapter is of immense value to students as well as to policymakers and executors of high dams at a very early stage of the planning process. Otherwise, if the project is contested at the half-way point, not for engineering but environmental and social reasons, then delay causes not only loss of time but also colossal loss of extra money. Zwahlen's chapter provides necessary information required at the early stage of planning of Dam projects and watershed management programs.

Acknowledgements

During the course of preparation and publication we received help from many people and thank them all. We are particularly grateful to all the chapter authors, who very readily agreed to take the trouble of writing these chapters, with the purpose of introducing the topics and then discussing the latest findings, ideas, and concepts from applied ecology viewpoints. We are indeed so grateful to our editor, Andrea Macaluso, and others at Kluwer Academic/Plenum Publishers for taking great interest and more than normal pains in giving a proper shape to the volume and publishing it so nicely and in such a short time. For his help during the preparatory phase, we thank Dr. R. Prasad. We also thank our family members Mrs. Annpurna, Dr. Pravin, Mrs. Sandhya, Mrs. Anupama, Ms. Prakriti, Ms Sukriti, and Ms. Soumya for their help and forbearance.

We are thankful to the Council of Scientific and Industrial Research, New Delhi, for awarding Emeritus Scientistship to RSA and Senior Research Associateship to NKA, and the authorities of the Banaras Hindu University for providing facilities. Thanks are also due to the Indian National Science Academy, New Delhi for the award of Honorary Scietistship to RSA.

R. S. Ambasht
Navin K. Ambasht
Banaras Hindu University, India

CONTENTS

1. **Jacob John** (Australia) Curtin University	*Bioassessment of Health of Aquatic Systems by the Use of Diatoms*	1
2. **Paul A. Keddy & Lauchlan H. Fraser** (USA) Univ. Louisiana & Univ. of Akron	*The Management of Wetlands for Biological Diversity: Four Principles*	21
3. **Brian A. Whitton** (UK) Durham University	*Use of Plants in Monitoring Heavy Metals in Freshwaters*	43
4. **Alain Vanderpoorten** (USA) Duke Univ., Durham NC	*Hydrochemical Determinism, Ecological Polymorphism, and Indicator Values of Aquatic Bryophytes for Water Quality*	65
5. **Lisandro Benedetti-Cecchi** (Italy) Università di Pisa	*Plant Succession in Littoral Habitats: Observations, Explanations, and Empirical Evidence*	97
6. **Jacob John** (Australia) Curtin University	*Phycoremediation: Algae as Tools for Remediation of Mine-Void Wetlands*	133
7. **Donat-P. Häder** (Germany) Erlangen Universitat	*UV-B Impact on the Life of Aquatic Plants*	149
8. **H. E. Zagarese, Barbara Tartarotti** (Austria) **and D. A. Añón Suárez** (Argentina)	*The Significance of Ultraviolet Radiation for Aquatic Animals*	173
9. **Rajeshwar P. Sinha** (Germany) Erlangen Universitat	*Stress Responses in Cyanobacteria*	201
10. **Nándor Oertel and János Salánki** (Hungary)	*Biomonitoring and Bioindicators in Aquatic Ecosystems*	219
11. **Arthur J. McComb and** **Jane M. Chambers** (Australia) Murdoch University	*The Ecology of Wetlands Created in Mining-Affected Landscapes*	247
12. **R. S. Ambasht & Navin K. Ambasht** (India) Banaras Hindu University	*Conservation of Soil and Nutrients through Plant Cover on Wetland Margins*	269

13. Robert Zwahlen (Switzerland)   *Identification, Assessment and Mitigation of Environmental Impacts of Dam Projects*   281

Index   371

Modern Trends in Applied
Aquatic Ecology

1

Bioassessment of Health of Aquatic Systems by the Use of Diatoms

Jacob John

Introduction

Physical and chemical monitoring of water quality has been practiced for a long time. There are standard techniques and guidelines for measuring light penetration, turbidity, conductivity, total dissolved salts, dissolved oxygen, biological oxygen demand, and nutrients: phosphates, nitrates, nitrites, ammonia, and so on (Chapman, 1992). Many regulatory agencies routinely monitor the physical and chemical properties of rivers, streams, and lakes. Measurements of these properties provide us with simple values, determined at a given time, but do not provide an overview of the health of the ecosystems. Water quality and habitats are integral components of an aquatic system. As water resources on regional, national, and international levels are undergoing rapid degradation, assessment of their health is considered urgent (Norris & Norris, 1995). Such assessment will enable us to take preventive as well as restorative measures. This chapter deals with a group of ubiquitous, easily quantifiable organisms as diagnostic tools in assessing aquatic system health—the diatoms.

Concept of Aquatic System Health

Forests are often described as the "lungs" of the earth and wetlands the "kidneys." A healthy ecosystem is composed of biotic communities and abiotic characteristics that form a self-regulating and self-sustaining unit. Any aquatic ecosystem has some capacity to assimilate stresses, such as effluent discharge or excessive nutrients. When the capacity to absorb stress has been exceeded, the ecosystem develops symptoms of degradation and becomes unhealthy; the biodiversity decreases (Loeb, 1994). Just as a physician diagnoses

---

Jacob John • Department of Environmental Biology, Curtin University, GPO Box U1987, Perth WA 6845 Australia

sickness by gathering lots of information (e.g., heart rate, pulse, breathing, weight, temperature, blood tests) an aquatic ecologist has to collect information on the physical, chemical, and biological characteristics of a system to assess its health. The dynamic interactions (physical, chemical, and biological) that characterize the ecosystems are best reflected by aquatic biota.

Biological Monitoring

Chemicals in water undergo large fluctuations in a relatively short time. Measuring water chemistry, even at frequent intervals, can miss biologically significant peak concentrations (Abel, 1989). Chemical analyses fail to account for the rates of transformation of a chemical by biological organisms and its possible relocation within an aquatic ecosystem (Loeb, 1994). A static chemical concentration represents a residual of that chemical after the biota have ingested, absorbed, stored, or transformed it. Moreover, chemical analyses alone do not account for the rates of transformation of a chemical by biota. Chemical attributes alone lack the responsiveness necessary to evaluate the health of an aquatic system.

Certain chemicals present in an aquatic system may be too small to detect. There are biota that concentrate such chemicals (e.g., mercury and radionuclides by diatoms). Biological indicators integrate the responses to the environmental changes on a short- as well as long-term basis.

A healthy ecosystem is maintained by a variety of healthy aquatic biota. Studying these aquatic biota directly provides the most information on changing conditions. Any surveillance program needs to protect the components of the aquatic system. For example, invasion of exotic species may destroy the habitat of native biota, or a native species may disappear as pollution increases. Again, these changes affecting biota may not be detected by chemical analyses (Loeb, 1994; Dixit et al., 1992; Cairns & Pratt, 1992; Lowe & Pan, 1996).

Attributes of Ideal Biomonitors

Virtually all biota tell us something about their environment. Some biota are more useful as biomonitors than others; specific biota may be suitable for specific monitoring. On the one hand, the presence or absence of a certain organism may not in itself be significant. However, a change from dominance to gradual disappearance of a species is of ecological significance.

The organisms selected as a biomonitor should be close to the transfer of nutrients and energy in the food web, simple, not complicated by life-cycle stages, and identifiable to the species or even the morphotype level. They should be sensitive even to fine changes in the environment and ideally should have a well-demarcated range of tolerance and preference to environmental variables, so that a change in the environment may be reflected in a shift in species dominance. The biota should be distributed widely throughout a water body, along the length of a river, for example, and preferably be universal in distribution or have a high degree of universality in distribution (Round, 1991).

Biomonitors should consist of groups of species, each group with well-defined habitats, so that they may reflect changes in a variety of habitats. They should be present in abundance and easily quantifiable. The data derived from monitoring then become amenable to multivariate statistical analysis, predictive models, transfer functions, and other statistical tools (Lowe & Pan, 1996). The bioindicators selected should be suitable for demonstrating causal relationships with the environmental change. They should not be selected purely on the basis of association between their abundance and environmental variables (Cox, 1991). The ideal biomonitors should be useful for both long- and short-term monitoring. Current conditions may be linked to the past conditions very effectively, if the same biomonitors are used for both short- and long-term monitoring (Dixit et al., 1992).

Ubiquity and Diversity of Diatoms

Diatoms are single-cell algae found in all lakes, rivers, and marine waters in abundance. Their cell walls are made up of silica with characteristic patterns useful in the identification of species. Accurate identification is desirable for their use as bioindicators. There are mainly two groups of diatoms: centric and pennate (Figure 1.1). Centric diatoms are radially symmetrical; characteristic of deep lakes, oceans, estuaries, and low velocity rivers; and generally float as phytoplankton. Pennate diatoms are rectangular or linear and grow attached to macrophytes, benthic substrata such as rocks and mud, and form the bulk of the periphyton (attached algae).

In flowing streams, rivers, and shallow wetlands, the periphyton is dominated by diatoms (Lowe, 1996). Diatoms differ in their microhabitats: epiphytic (growing on plants), episamic (on sand), epipelic (on mud), and epizoic (on animals). There are characteristic species assemblages in lentic (stagnant), lotic (flowing), brackish, saline, and marine waters.

Diatoms as Biomonitors

Diatoms are more specific in their preference and tolerance of environmental conditions than most other aquatic biota. Periphyton were the first group of biota ever suggested by scientists for use to detect organic pollution (e.g., the saprobien system by Kolkwitz and Marsson in 1909 (cited in Cairns & Pratt, 1992)). The classic systems to characterize organic pollution in streams and rivers (oligosaprobic, beta-mesosaprobic, alpha-mesosaprobic, and polysaprobic) are best reflected by the distribution pattern of benthic and periphyton diatom communities.

No other single group of biota has been studied as extensively as diatoms regarding their individual tolerance and preference to environmental conditions at the species level. The literature on the autecology of diatoms is vast and covers a wide range of geographical areas. It was Patrick's work (United States) in the 1950s that provided the impetus for river health assessment in America. Patrick (nicknamed "River Doctor" by the popular press) and her team worked on several groups of biota. However, her greatest contribution has been in the field of diatom studies, and river monitoring based on diatoms, which continues to gather momentum throughout the United States (Patrick, 1984, 1994; Patrick et al., 1968; Charles, 1996). In Canada, the United States, and Europe,

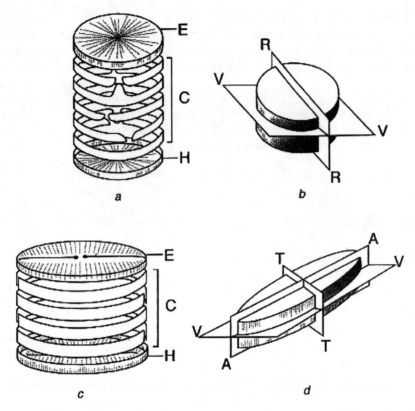

Figure 1.1. Structure of cell walls (frustule) and symmetry of diatoms. (a) Centric diatom: E = epivalve, C = cingulum, H = hypovalve. The cingulum (girdle) consists of copulae (girdle bands). The copula adjacent to the valve is *valvocopula*. (b) Centric diatom symmetry: VV = valvar plane, RR = radial plane. (c) Pennate diatom: E = epivalve, C = cingulum with cingular elements (girdle bands), H = hypovalve. *Valve face* is the flat surface of the valve up to the margin; *mantle* is the steep edge (side) of the valve. (d) Pennate diatom symmetry: VV = valvar plane, AA = apical plane, TT = transapical plane. *Apical axis* connects the two poles through the median line. *Transapical axis* connects the center of the epivalve to that of the hypovalve. The markings on the cell walls (STRIAE), size and shape, and the RAPHE (a longitudinal slit seen in most pennate diatoms) are criteria for classification.

diatoms have been used for several years to determine the effects of acidification on lakes and rivers, under the project known as PIRLA (Paleoecological Investigation of Recent Lake Acidification). Changes in pH have been inferred by analyzing the remnants of diatoms in the sediment of lakes (Charles et al., 1994).

Diatoms can be identified at the species level by microscopic examination, and the relative frequency of individual species can be easily enumerated. Based on the known environmental preference of species, the assemblages of diatoms may be used as diagnostic tools to assess water quality. The community structure of diatom assemblages from a pristine site may be used to compare the water quality of test sites.

Data on species and their relative abundance can be plotted to obtain a truncated curve model. A natural diatom community corresponds to a truncated normal curve in structure, and pollution of various types affects the structure of the community in various ways. For example, with high nutrient pollution, the curve lengthens; that is, certain

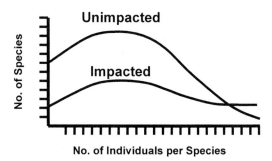

Figure 1.2. Distribution curve of species and individuals in a pristine (unimpacted) and an impacted aquatic system. (In a pristine system, species richness is high, with individuals in each species being low compared to an impacted system in which species richness is low but populations of each species may be high.) Adapted from Cairns & Pratt (1992).

species become much more common than others (Figure 1.2). Toxic chemicals often cause the survival of a few species in large numbers. Pristine sites can be compared to impacted sites using this method. However, this model may not be suitable for all the different types of pollution impacts (Patrick, 1994).

Round (1991) has shown that many rivers in the United Kingdom have distinct diatom assemblages in the pristine sites near their sources and the impacted sites passing through farmlands in both groups of sites dominated by different species. The first step in using diatoms as indicators is to obtain quantitative data on their ecological characteristics along the environmental gradients. The applications of diatoms, including bioassessment of aquatic systems, are treated in detail by Stoermer and Smol (1999).

Ecological Preferences

Based on their preference and tolerance of various environmental variables (e.g., pH, salinity, nutrients), diatoms can be grouped into various categories. A diatom community can thus be analyzed according to the dominance or relative frequency of species (John, 1993). Saline waters are dominated by species that tolerate or even prefer high salinity. In eutrophic systems, diatoms that tolerate or prefer high amounts of nutrients dominate. The association of pH and diatom assemblages has been well documented (Huttunen & Merilianen, 1983; Kwandrans, 1992; Coring, 1996), with the development of various models (Fabri & Leclerq, 1984; Ten Caté et al., 1993). Nutrient enrichment has also been shown to cause easily observed variations in diatom assemblages at the species level (Marcus, 1980; Chessman, 1985; Neiderhauser & Schanz, 1993; Ollikainen et al., 1993). The effects of changes in the levels of nitrogen, phosphorus, and carbon on diatom assemblages have been investigated extensively (Hoagland, 1983; Fairchild & Lowe, 1984; Fairchild et al., 1985; Kiss & Genkal, 1993).

The specific effects of organic pollution on diatom communities have been explored by several researchers (e.g., Watanabe et al., 1986; Hoffman, 1996). Similarly, nutrient enrichment and diatom community interactions have been investigated in many studies (Uomo, 1996; Kelly et al., 1996; Prygiel & Coste, 1996). The effects of salinity on diatoms have also been well investigated (Blinn, 1983). Diatom species have various salinity tolerances, with assemblages varying in accordance with the salinity regimen present (Ehrlich & Ortal, 1979; Miller, 1984; Oppenheim, 1991; Juggins, 1992). Predictive models have been developed to relate diatom species to salinity (Cummings & Smol, 1993).

Predictive Models

Another approach developed recently in the United Kingdom is to sample large numbers of pristine sites to obtain data on the distribution of biota and then compare these data with test sites (both pristine and impacted) to determine the degree to which the test sites differ from the pristine sites. Computer packages are currently available on these methods.

Canonical Correspondence Analysis (CCA), a direct gradient analysis, is widely used. Taxa and samples can be directly related to measured environmental variables. From the CCA biplot, it is possible to approximate the indicator value of the most common taxon. Once the dominant environmental variables that determine the species distribution have been identified using CCA techniques, transfer functions can be generated to infer environmental characters from diatom assemblage data (Dixit et al., 1992; Lowe & Pan, 1996). Given the assemblages of diatoms, the environmental conditions of sites can be predicted by this method.

Although there is no shortage of methods available for the use of bioindicators to assess water quality, basic work needs to be done on several geographic regions to develop suitable assessment systems on a regional level. Regional models have to be developed for each geographical region for appropriate water quality biomonitoring. Research activities should start at a regional level to solve regional problems, incorporating the findings of research done elsewhere: "Think globally, act locally."

Multivariate analyses are increasingly being used to relate diatom distribution patterns to environmental factors. Most of the predictive models are based on cluster analysis, which arranges sampling sites into a small number of groups or clusters and generates a hierarchical diagram of relationships based on taxonomic structural similarity among the samples (Lowe & Pan, 1996), called a dendrogram.

CCA is a powerful tool in detecting patterns of species distribution related to associated physical and chemical parameters. Since the introduction of CCA, the computer program CANOCO (ter Braak 1987–1992) has been extensively used in developing diatom-based predictive models (John & Helleren, 1998). The weighted average models applied to specific habitat conditions are used not only in paleoecological studies but also in assessment of the current ecological health of wetlands and rivers (Stevenson et al., 1985). Another approach is similar to the RIVPACS (River Invertebrate Prediction and Classification System) program developed by the Institute of Freshwater Biology in the United Kingdom using macroinvertebrates (Wright, 1995). The RIVPACS program is used to generate site-specific predictions of macroinvertebrate fauna expected in rivers and streams. A similar predictive model has been developed for urban streams in Australia using diatoms (John, 2000a).

Ecological profiles have been generated for hundreds of periphytic diatoms. For example, van Dam et al., (1994) generated a checklist of 948 taxa with ecological indicator values from the Netherlands. In addition, information is available on ecological preferences of several hundred diatoms (Descy, 1979; Lowe, 1974; Schoeman, 1976; Lange-Bertalot, 1979; Lange-Bertalot & Metzelin, 1996). From the diatom profiles, indicator species or indicator values of species can be generated. Diatom indices are useful analytical tools derived from diatom profiles and indicator values (Watanabe et al., 1990; Watanabe & Asai, 1992; Sumita & Watanabe, 1983; Kelly et al., 1996; Lenior & Coste, 1996; John, 2000a).

Diatom Indices

Generally, a healthy system tends to have a high diversity of species, each represented by small numbers. In a system under stress, species richness changes from high to low. Certain sensitive species disappear as pollution increases. Several diversity indices reduce the community structure of biota into single numbers for comparative purposes. The Shannon–Weaver index may be used to represent changes in water quality in a system (Archibald, 1983; van Dam, 1982).

$$H = -\sum_{i=1}^{S} \frac{n_i}{N} \operatorname{Log}_2 \frac{n_i}{N}$$

where $N =$ the number of individuals in a sample: $n_i =$ the number of individuals of a species i in a sample, and $S =$ the total number of species.

It is not necessary to identify individual species for the Shannon–Weaver diversity index, as long as one recognizes each species as different from the others. However, the diversity indices may not be a valid tool for assessment of health for some aquatic systems. For example, in some of the pristine rivers and lakes in thick forests, species richness may be low due to the presence of humic acids and tannins (John, 1998). In general, the more severe the stress, the greater the reduction in species diversity, but in some instances, disturbances may increase species diversity (Stevenson, 1984). Therefore, a diversity index alone may not give a true picture of health, at least in some systems. What is of significance is the change in diversity index experienced by a particular system.

Several other indices (e.g., biotic and pollution indices) developed by research workers in the United Kingdom, United States, Europe, Japan, and Australia may be used to compare pristine and polluted sites, and there is a vast amount of literature available on the topic.

Sampling

Uniform samples of periphyton are essential for comparisons of control sites (unpolluted or nonimpacted) and polluted or impacted sites. Absolute pristine sites of aquatic systems are becoming sparser day by day, due to ever-increasing urbanization and development. However, attempts should be made to select as many relatively pristine sites as possible in any biomonitoring project.

Natural substrates (e.g., stones, pebbles, benthic substrates such as mud, or macrophytes) are ideal for sampling periphyton. But in reality, the main problem is the heterogeneity of natural substrates with regard to texture, shape, size, and depth at which they are found (Austin et al., 1981). Currently, collection of uniform pebbles, stones, macrophytes, and a constant area of benthic sediments (mud or sand) are being practiced. Artificial substrates do guarantee uniformity of sampling. The term *artificial*

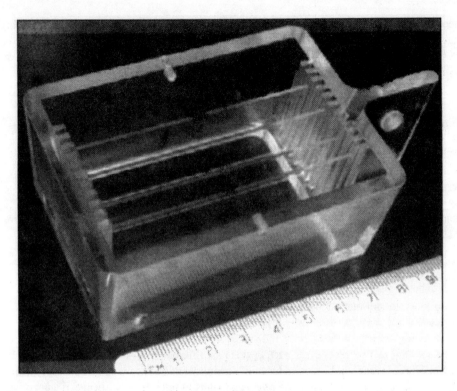

Plate 1.1. The JJ Periphytometer—an artificial substrate collector that holds 10 microscope slides for colonization by periphyton; it is ideal for uniform sampling of a large number of sites. It can be installed at any desired depth in a water body and should be secured by strong strings attached to fixed objects such as nearby stones or wood.

substrate has been defined as a device placed in an aquatic ecosystem to study colonization by indigenous organisms (Cairns, 1982). The advantages of artificial substrates include fixed habitats, so that pollution cannot be avoided, relatively quick colonization after changes in water quality or flow, ease of sample preparation for analysis, and standardization between sites (Biggs, 1989). A periphyton sampler (the JJ Periphytometer) designed by John (2000a) has been used successfully in monitoring metal pollution in streams, the impact of aquaculture on water quality, and in health assessment of streams, rivers, and wetlands in Australia (Plate 1.1). Adaptability to a variety of aquatic situations and robustness are additional attributes of this substrate. However, vandalism and severe storm events can easily result in the loss of the samplers. The artificial substrates can be placed at varying depth, attached to a wooden stake or to a nearby heavy object (stone or wood) by strong strings hidden from curious trespassers.

Identification and Enumeration of Diatoms

The samples collected either by natural or artificial substrates are converted into permanent preparations for identification and enumeration. Samples from a constant area of benthic sediment (collected by thrusting an inverted open vial or tube into the sediment and pulling it up) should contain living diatoms in the top 1–3 cm. Stones and pebbles collected from streams, rivers, and wetlands normally have a biofilm consisting of diatoms and other periphytic algae. Replicate samples from each site minimize sampling errors. The diatoms from the stone or pebble surfaces can be easily removed by a brush. The top 1–3 cm layer of the benthic samples of mud or sand can be treated with a strong clearing agent (60% nitric acid, 30% hydrogen peroxide) to extract and clean the diatoms. The diatoms from glass slides in the artificial substrates should be scraped and treated in acids or hydrogen peroxide to clear them of all organic materials (John, 1983; 2000b). The cleared diatom samples are mounted in a medium such as Naphrax®, whose refractive index is higher than that of water or glass, to make permanent mounts of the diatom frustules. It is advisable to consult specialized literature before trying out the techniques. Making neat slides with uniformly distributed diatoms is essential for accurate identification, enumeration, and subsequent analysis of the community structure. The permanent slides essentially become an invaluable record of the diatom assemblages from a site and can therefore be the basis of a "snapshot" of the health of a water body for future references. At least two permanent slides from each site should be examined under a high power light microscope ($\times 1,000$) with oil immersion for identification of the species. Literature on local diatom flora should be consulted for identification to species level.

Diatom taxonomy has been undergoing rapid revision due to increased understanding of the structure of the frustule (the siliceous cell wall of the diatom), especially due to ultrastructural studies with electron microscopy. However, many of the diatom species are universal in distribution, and with the use of most common literature on the diatom flora of Europe or the United States, at least most genera could be identified. For the Indian subcontinent, the contribution of Gandhi (1999) is a valuable reference. Examples of useful modern references for freshwater diatoms are Patrick & Reimer (1966, 1975), Round (1991), and John 1983, 2000b).

Enumeration of the relative frequency of species is done by scanning the slide from side to side and scoring the species counted. A standard total count of 300 to 500 valves per site is the normal practice, although 100, 200, 300, 400, 500, or even 1,000 valves are counted by researchers, depending on the specific objective of the study.

From the data on relative frequency, often expressed as a percentage composition of the diatom assemblages, various statistical values can be derived to assess the health of a system.

Case Histories

Three case histories from Australia are described to illustrate the practical use of diatoms as tools for biomonitoring. The first illustrates the use of diatoms to classify the natural wetlands of the Swan Coastal Plain in Western Australia. The second involves the use of diatoms in assessing the development of artificial wetlands created after sand mining. The third case history involves the development of a predictive model for assessing the health of inland rivers and streams in Western Australia.

Wetlands of the Swan Coastal Plain, Western Australia

I have been studying the taxonomy and distribution patterns of diatoms of Western Australia for the past 25 years. Salinity was demonstrated as the most important factor responsible for the distribution patterns of diatoms in the Swan River estuary in Western Australia (John, 1983, 1988, 1994).

Recently, in a study of the diatom assemblages of the Swan Coastal Plain wetlands in Western Australia, it was found that the vast majority of the 72 wetlands investigated were alkaline, shallow, and mesotrophic or eutrophic, with high conductivity. These included one acidic lake, a few saltlakes, and several brackish lakes. A total of 227 diatom taxa were identified. Out of these, 94% were periphytic or benthic in habitat, with only 6% being euplanktonic, reflecting the shallow nature of the wetlands. The majority of the species ($>65\%$) were reported as occurring in oligohalobous water bodies; the rest were tolerant to mesohalobous, euhalobous, or euryhalobous conditions, according to the information compiled from the literature (see Table 1.1 for an explanation of these terms).

The wetlands of the Swan Coastal Plain are located within 1 to 25 km of the coastline. Not surprisingly, the same ions dominate the freshwater lakes as in the sea water, although in reduced concentration. Many have been impacted by eutrophication.

A species–environment relationship was inferred from the diatom community structure by using CCA (Ter Braak, 1987–1992; Lowe & Pan, 1996). CCA grouped species and sites in relation to the selected environmental variables. All the wetlands were positioned along the axes representing the environmental variables of pH and salinity. When all species of diatoms were included in the analysis, the sites were clustered into 12 groups. When rare species were down-weighted (species of less than 1% abundance), the sites were clustered into 10 groups (Figure 1.3).

Group 1 was represented by the only acidic lake (pH 3.5) in the coastal plain—Lake Gnangara; the largest number of lakes (31 lakes) was in Group 3. This group represents mostly mesotrophic lakes. Groups 8, 9, and 10 were saline lakes dominated by alkaliphilous, mesohalobous to euhalobous diatom species. These three lakes were very similar in the composition of the diatom species and water chemistry. All the lakes were grouped according to their water chemistry and salinity. This study demonstrated that pH and electrical conductivity (salinity) had an overriding influence on diatom distribution over other factors.

Electrical conductivity and pH are important parameters of the water quality of wetlands. This case history illustrates clearly the potential for the use of diatoms to classify wetlands based on the distribution patterns related to these environmental factors. CCA is

Table 1.1. Environmental Preference of Diatoms

	pH preference
Acidobiontic	<5.5 (found only in acidic waters)
Acidophilous	<7
Indifferent	Around 7
Alkaliphilous	>7
Alkabiontic	Found only in alkaline waters
	Halobien preference (Salinity preference)
Oligohalobous	<0.5 ppt
Mesohalobous	0.5–0.30 ppt
Euhalobous	20–40 ppt
Euryhalobous	Wide range of salinity

Modified from Lowe (1974).

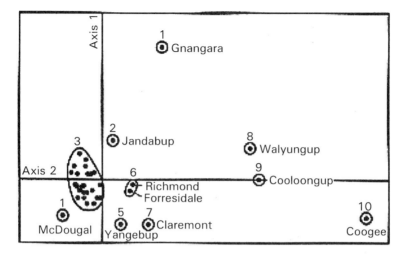

Figure 1.3. CANOCO biplot of wetland groupings for all diatom species with rare species down-weighted. Axes 1 and 2 represent pH and salinity, respectively.

extremely powerful in detecting patterns of species distribution related to associated physical and chemical parameters (Lowe & Pan, 1996).

Monitoring the Development of Created Wetlands in Mine-Voids

Since European settlement, there has been a considerable decline of natural wetlands in Western Australia. In recent years, there has been some attempt to offset this balance. Several of the mine pits left after the extraction of coal and sand minerals are being rehabilitated into productive wetlands. Biomonitoring the development of these artificial wetlands has now been recognized as integral to their management.

Former sand mining pits at Capel, 200 km south of Perth in Western Australia, have been rehabilitated into self-sustaining wetlands since 1975. A chain of freshwater lakes was created as a potential waterbird refuge and for passive recreation. These wetlands occupy an area of 44 hectares now known as the RGC (Rennison Goldfields Consolidated) Wetlands Centre. The chain of lakes is mostly interconnected, receives treated effluent water from the sand-mining processing plant at one end of the chain, and there is an outlet and a discharge point at the other end.

Initial studies in 1984 showed that the lakes were deficient in primary production, which is required to attract ample bird life. The factors limiting primary production were probably low pH; high ammonia, sulphate, and manganese, and low levels of phosphorus. The effluent water discharging into the lakes was buffered to high pH in 1986 by the sand mining company. Within 6 months, the concentration of iron and manganese in the water declined considerably, and the pH increased to 7. Reshaping steep shorelines, creation of islands, general landscaping, and intensive planting of emergent vegetation were the other steps taken to develop the wetlands into sustainable ecosystems.

Diatom assemblages in these lakes were monitored from 1986 to 1992 by a multidisciplinary team studying the development of the system (John, 1993). Sampling of diatoms involved hand collection of benthic, planktonic, and epiphytic diatoms. The

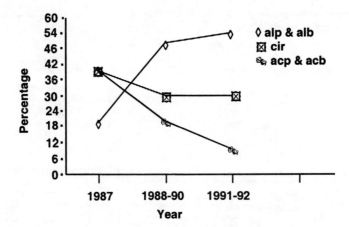

Figure 1.4. Mean percentage of species according to their pH preference in RGC wetlands, Capel, Western Australia; alp & alb = alkaliphilous and alkalibiontic; cir = circumneutral; acp & acb = acidophilous and acidobiontic.

species richness of diatom assemblages increased from 19, in 1987, to 34, in 1990, and further to 61, in 1992, reflecting the development of the system. All of the 61 species of diatoms encountered in this study in 1992 could be classified according to their pH preference from the information gathered from both the literature and the field data collected. There was a remarkable increase in the percentage of alkaliphilous and alkalibiontic species. The percentage of circumneutral species declined slightly. However, the most dramatic change was the decline in the percentage of acidophilous and acidobiontic species after the improvement of pH in the lakes in 1987 (Figure 1.4). Species diversity measured with the Shannon–Weaver index increased from 0.28 to 0.62 over the same period. A species list of the diatom assemblages, with relative abundance of each species and autoecological information, provides a vast amount of information compared to any single number, such as the diversity or saprobien index (Round, 1991; Schoeman, 1976).

This study clearly showed that it was the shift in dominance of certain species, as shown by the relative abundances, that reflected the water quality. Following a change in pH and improvement in water quality, there was a sharp increase in abundance and dominance of the diatoms *Cymbella minuta* and *Achnanthes minutissima*. These two species were identified as marker–indicator species appearing in large numbers after "liming" to neutralize an acidified Welsh lake (Round, 1990). The decline in numbers of several *Pinnularia* species, except *P. sub-capitata* and *P. viridis*, clearly reflected the change from acidic to neutral conditions. *Pinnularia* species in general are well recognized as acidophilous or acidobiontic diatoms. As the diatom communities increased in diversity with the increase in pH in the RGC wetlands, macroinvertebrate diversity and bird life also increased. Only 27 species of birds were recorded from the lakes in 1987. In 1992, this number had increased to 104 (Doyle, 1993). The rapid response by diatoms to change in water quality was clearly demonstrated in this study.

These case-histories illustrate the major attributes of ideal biomonitors presented earlier. The first case history involved 227 species and the second 61 species, most of which are universal in distribution. Sensitivity to pH and electrical conductivity (salinity), wide distribution, a simple life cycle, ease of sampling, and collection of quantitative data suitable for statistical analysis are points demonstrated in these case histories.

A Snapshot of River Health in Southwest Australia

Study Sites

The southwest of Western Australia was used in this study, which required for the development of a predictive model a large number of relatively pristine and impacted sites. The southwest offers an array of diverse, even paradoxical environmental variables. It is in this region of Western Australia that we have the largest area of forests, with the highest number of streams and rivers in the state. Extensive stands of Jarrah (*Eucalyptus marginata*) and Karri (*Eucalyptus diversicolor*)—two of the tallest majestic native *Eucalyptus* species—exist here. Highly significant conservation areas of native fauna and flora characterize this region. Yet this land includes some of the worst examples of human exploitation of nature for timber and agricultural practices, resulting in severe degradation of streams, rivers, and land. Loss of habitat diversity, riparian degradation, erosion, siltation, segmentation, and invasion of exotic weeds are among the consequences. The land use in this region varies from intense agriculture to "nature reserve," with the average rainfall ranging from 1,400 mm per annum in the southwest to 500 mm per annum toward the north and east, with a corresponding decline in native vegetation density.

Methods

Extensive historical and contemporary information was collected at the outset from catchment groups, the Water Authority of Western Australia, the Water and Rivers Commission, and the Department of Conservation and Land Management, on the streams and rivers of the region before selecting the sampling sites. The vast majority of the sites used were the same as those used by the state macroinvertebrate monitoring group. This certainly provides an opportunity for comparing the relative merits of using macroinvertebrates or diatoms for the bioassessment of stream and river health.

More than 100 sites were chosen as reference sites—relatively pristine sites—and 30 sites as impacted (monitoring) sites. The statistical package Pattern Analysis Package (PATN) (Belbin, 1993) was used to determine the species–site relationships and investigate the distribution pattern of diatom assemblages in relation to the environmental factors. Some 20 sites randomly selected from the reference sites were used as test sites to determine the internal consistency of the reference sites.

Results and Discussion

Accurate identification of diatoms to the species level and the use of the relative abundance of each species rather than mere presence or absence adds to the strength of this research, providing a sound ecological basis for the possible use of diatoms as biomonitors. Diatoms were found to be very sensitive to environmental changes, and a shift in dominance was found to be closely associated with the environmental parameters.

Out of 200 species of diatoms, 170 were represented by more than 1% in their relative abundance. Several multivariate statistical tools such as multidimensional scaling (MDS) for ordination and principal axis correlation (PCC) were used to investigate the relative abundance of these diatom species in relation to the environmental variables.

Riparian vegetation, electrical conductivity (salinity), alkalinity, pH, depth, flow rate, temperature, nitrate, and total nitrogen were the most significant environmental variables separating the monitoring sites from the reference sites, and the test sites were grouped with the reference sites. Riparian degradation, alkalinity, pH, temperature, electrical conductivity (salinity), width, shallowness, and low flow rate were positively correlated with the monitoring sites. The multidimensionality of the complex data was reduced by ordination. Hierarchical classification of diatom distribution patterns clearly grouped diatom species with similar ecological preferences together. The ordination technique revealed highly significant correlation values between monitoring and reference sites, and the specific assemblages of diatom species. The assemblages of the most pristine sites were totally different from those of the most impacted sites. Between these two extremes were mixed assemblages of diatoms reflecting varying degrees of degradation.

A list of indicator species was generated with a highly significant correlation coefficient, characteristic of impacted and reference sites. The autecological information gathered from the literature on individual species of diatoms, validated by this research, was used not only to assess the current status of the rivers and streams in the southwest, but also to diagnose the type of degradation present. Salinization could be detected at a very early stage, because diatoms are most sensitive to salinity (e.g., species of *Rhopalodia*, *Amphora*, and *Mastogloia* (Plate 1.2) increased in abundance as salinity increased). The potential use of diatoms for monitoring the effectiveness of restoration is immense. As a "pre-warning" system, diatoms appear most promising.

Species of *Eunotia*, *Cymbella*, *Tabellaria*, *Frustulia*, and *Pinnularia* displayed maximum diversity in the most pristine sites (Plate 1.3). Diatom species *Cyclotella meneghiniana*, *Ctenophora pulchella*, and *Bacillaria paxillifer* were characteristic of mesotrophic sites with higher electrical conductivities.

Two-way tables (Belbin, 1993) of species and site data, with sites constrained into a few groups, provide the most valuable information on the health of the rivers and streams in the southwest. Any newly sampled sites can be tested to see where they fit in, and these tables can be used as diagnostic tools.

A panoramic view of the state of the health of the river systems in the southwest of Australia from this study clearly shows that the most impacted rivers are in the inland Blackwood Catchment area. The most pristine streams and rivers are in the remnant state forests in the Darling Scarp, and the Donnelly and Shannon River drainage division, where Jarrah and Karri forests occur. However, a large area starting from the Moore River in the north to the Kalgan River in the south in Albany show symptoms of degradation crying for restoration measures—"prewarning degradation."

The scope of the current research extends far beyond Australia given that diatoms tend to be more universal in distribution than any other comparable biomonitors.

Concluding Remarks

Fish, macroinvertebrates, macrophytes, and diatoms are the most commonly used bioindicators of aquatic system health, although there are specific advantages and disadvantages associated with the use of each of these groups. Fish are large enough for contaminant residue analysis and are increasingly being used for ecotoxicological tests

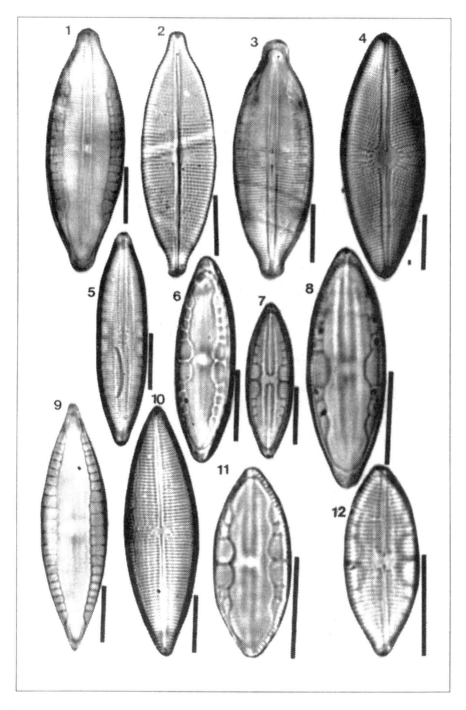

Plate 1.2. Diatom species characteristic of streams affected by salinization in southwest Australia: Figures 1–3, *Mastogloia balticum*, Grun; Figure 4, *Mastogloia elliptica* (Ag.) Cl.; Figures 5–8, *Mastogloia pumila* (Grun.) Cl.; Figures 9–10, *Mastogloia reimerii*, John; Figures 11–12, *Mastogloia braunii*, Grun.

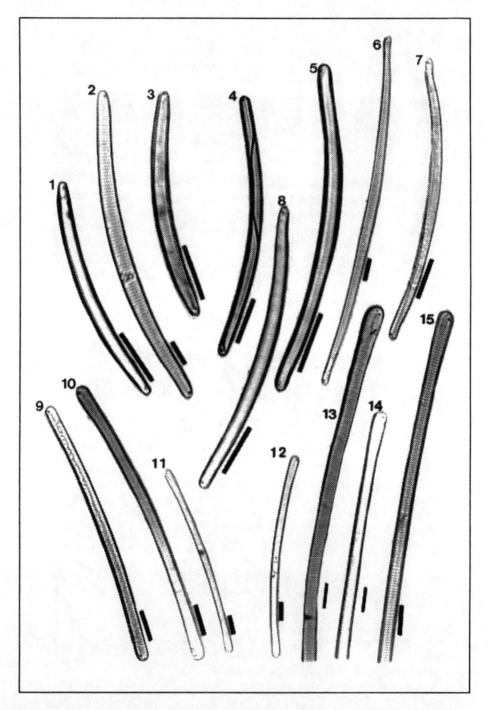

Plate 1.3. Species characteristic of pristine (unimpacted) streams in southwest Australia. Figures 1–8, *Eunotia curvata*, Kütz Lagerst; Figures 9–12, *Eunotia flexuosa* (Breb), Kütz; Figures 13–15, *Eunotia eurycephaloides*, Närpel-Sch & Lange-Bertalot.

to monitor the effects of toxins in aquatic systems. Decline in native fish species may be an obvious symptom of deterioration of water quality, invasion of exotic species, habitat loss, or overfishing. By the time fish species disappear totally, as has happened in many streams in the world, it may be too late for any recovery measures.

Macrophytes often may not be present throughout the length of a stream or river for continuous monitoring. However, presence or absence, relative density, overgrowth, and types of species may be valuable in assessing the health of aquatic systems.

Macroinvertebrates do share many desirable attributes of ideal biomonitors with diatoms but are more sensitive to habitats than to water quality (personal observations). Many of them may only spend part of their life cycle in water, and seasonality is an important factor affecting their distribution. Diatoms, on the other hand, have the advantage of being abundant in all aquatic systems throughout the year. Although there are seasonal changes in species composition, a single-season collection is sufficient for their use as biomonitors (John, 2000a). The sensitivity of diatoms to changes in salinity, acidity, trophic status, organic pollution, and habitats has been well researched (Stoermer & Smol, 1999). However, taxonomy remains problematic, and expert help is required to ensure taxonomic precision. Protocol on sampling, processing, and identification and statistical analysis should be well established before diatom monitoring becomes widespread.

There are numerous examples of diatom studies and uses other than those mentioned in this chapter. The impact on aquatic life of point source of specific pollutants such as heavy metals and toxic chemicals discharged can be monitored by the use of periphyton. Artificial substrate collectors (e.g., JJ periphytometer) are ideal for this purpose, because the cumulative effect of even small concentrations of pollutants in receiving waters can be monitored. Water quality in fish farms can be monitored with the use of diatoms. Aquaculture in both inland ponds and shallow marine embayments is on the increase. The effects of ammonia generated by waste products of fish and crustaceans on water quality need to be monitored, and, again, diatoms are most suitable for this purpose (personal observations in Western Australia). The impact of acid mine drainage (AMD) on lakes and streams is of great environmental concern in mining countries. Several projects are currently involved in the rehabilitation and restoration of acid mine voids and degraded wetlands in Australia. Passive treatment involving anoxic limestone drain (ALD) compost pond wetlands for heavy metal stripping is one such example. The effectiveness of each component of the treatment system can be easily monitored by analyzing the diatom community (John, 1999; unpublished data). Within 6 months of neutralization of acidic waters, an increase in species richness of diatoms has been observed.

Decline in aquatic biodiversity is a worldwide problem. It is expensive and often unrealistic to conduct surveys on biodiversity on every group of organisms. The concept of "surrogate biodiversity surveys" focusing on selected organisms is becoming a practical alternative. Diatoms are ideal as biomonitoring tools for gaining some insight into aquatic biodiversity. Generally, it is safe to assume that a high species diversity of diatoms reflects a high diversity of other forms of life.

As environmental protection agencies, conservation societies, and water resource managers become more aware of the need to implement appropriate management strategies for conserving aquatic systems, it is hoped that biomonitoring with diatoms will become universal as part of the management tools.

References

Abel, P. D. (1989). *Water pollution biology*. Ellis Howard.
Archibald, R. E. M. (1983). The diatoms of the Sundays and Great Fish Rivers in the eastern cape province of South Africa. In J. Cramer (Ed.), *Bibliotheca diatomologica*. Vaduz.
Austin, A., Lang, S., & Pomeroy, M. (1981). Simple methods for sampling periphyton with observations on sampler design criteria. *Hydrobiologia, 85*, 33–47.
Belbin, L. (1993). PATN: *Pattern analysis package*. Canberra: CSIRO, Division of Wildlife and Rangelands Research.
Biggs, B. J. F. (1989). Biomonitoring of organic pollution using periphyton, South Branch, Canterbury, New Zealand. *New Zeal. J. Mar. Freshw. Res., 23*, 263–274.
Blinn, D. W. (1983). Diatom community structure along physiochemical gradients in saline lakes. *Ecology 74*(4); 1256–1263.
Cairns, J. Jr. (Ed.) (1982). *Artificial substrates*. Ann Arbor: Ann Arbor Science.
Cairns, J., Jr., & Pratt, J. R. (1992). A history of biological monitoring using benthic macroinvertebrates. In: D. M. Rosenberg & V. H. Resh (Eds.), *Freshwater biomonitoriing and macroinvertebrates*.
Chapman, D. (Ed.). (1992). *Water quality assessment—a guide to the use of biota, sediments and water in environmental monitoring*. Chapman & Hall.
Charles, D. F. (1996) Use of algae for monitoring rivers in the United States: Some examples. In: B. A. Whitton & E. Rott (Eds.), *Use of algae for monitoring rivers II* (pp. 109–118). Innsbruck: Institut für Botanik, Universitat Innsbruck.
Charles, D. F., Smol, J. P. & Engstrom, D. R. (1994). Palaeolimnological approaches to biological monitoring. In: S. L. Loeb & A. Spacie (Eds.), *Biological monitoring of aquatic systems* (pp. 233–293). Boca Raton, FL: Lewis publishers.
Chessman, B. C. (1985). Artificial-substratum periphyton and water quality in the lower La Trobe River, Victoria. *Aust. J. Mar. Freshw. Res., 36*, 855–871.
Coring, E. (1996). Use of diatoms in monitoring acidification in small mountain rivers in Germany with special emphasis on diatom assemblage type analysis (DATA). In: B. A. Whitton & E. Rott (Eds.), *Use of algae for monitoring rivers II* (pp. 7–16). Innsbruck: Institut für Botanik, Universitat Innsbruck.
Cox, E. J. (1991). What is the basis for using diatoms as monitors for river quality? In: B. A. Whitton, E. R. Rott, & Friedrich (Eds.), *Use of algae for monitoring rivers* (pp. 33–40). Innsbruck: Institut für Botanik, Universitat Innsbruck.
Cummings, B. F., & Smol, J. P. (1993). Development of diatom based salinity models for paleoclimatic research from lakes in British Columbia (Canada). *Hydrobiologia, 269/270*, 179–196.
Descy, J. P. (1979). A new approach to water quality estimation using diatoms. *Nova Hedwigia, 64*, 305–323.
Dixit, S. S., Smol, J. P., Kingston, J. C., & Charles, D. F. (1992). Diatoms: powerful indicators of environmental change. *Environmental Science and Technology, 26*(1): 23–33.
Doyle, F. W. (1993). *Waterbird usage of the lakes at the RGC Wetlands Centre, Capel*. Western Australia: RGC Mineral Sands Ltd.
Ehrlich, A., & Ortal, R. (1979). The influence of salinity on the benthic diatom communities of the Lower Jordan River. *Nova Hedwigia, 64*, 325–334.
Fabri, R., & Leclerq, L. (1984). Diatom communities in the rivers of Ardenne (Belgium): Natural types and impact of pollutions. In: M. Ricard (Ed.) *Proceedings of the Eighth Diatom Symposium* (pp. 337–346). Koenigstein: Koeltz Scientific Books.
Fairchild, W. G. & Lowe, R. L. (1984). Artificial substrates which release nutrients: Effects on periphyton and invertebrate succession. *Hydrobiologia, 114*: 29–37.
Fairchild, W. G., Lowe, R. L., & Richardson, W. B. (1985). Algal periphyton growth on nutrient-diffusing substrates: An in situ bioassay. *Ecology, 66*, 465–472.
Gandhi, H. P. (1999). *Freshwater diatoms of central Gujarat*. India: Bishen Singh Mahendra Pal Singh Dehra Dun.
Hoagland, K. D. (1983). Short-term standing crop and diversity of periphytic diatoms in a eutrophic reservoir. *J. Phycol, 19*, 30–38.
Hoffman, G. (1996). Recent developments in the use of benthic diatoms for monitoring eutrophication and organic pollution in Germany and Austria. In: B. A. Whitton & E. Rott (Eds.) *Use of algae for monitoring rivers II*. (pp. 73–77). Innsbruck: Institut für Botanik, Universitat Innsbruck.
Huttunen, P., & Merilainen, J. (1983). Interpretation of lake quality from contemporary diatom assemblages. *Hydrobiologia, 103*: 91–97.
John, J. (1983). The diatom flora of the Swan River estuary. *Bibliotheca Phycologia*. J. Cramer.

John, J. (1988) Observations on the morphology and ecology of Navicula elegans from Western Australia. In: F. E. Round (Ed.), (pp. 406–417). *Proceedings of the 9th International Diatom Symposium*. UK: Bio Press Ltd.

John, J. (1993). The use of diatoms in monitoring the development of artificial wetlands at Capel. *Hydrobiologia, 269/270*, 427–439.

John, J. (1994). Eutrophication of the Swan River estuary and the management strategy. In: W. J. Mitsch (Ed.), *Global wetlands: Old world and new* (pp. 749–757), Amsterdam: Elsevier.

John, J. (1998). Diatoms: Tools for bioassessment of river health. Land and Water Resources Research and Development Corporation, Canberra. 388pp.

John, J. (1999). Biological monitoring of water quality of rivers: The ideal bioindicators. In *Aquatic Science 302 lecture notes* (pp. 1–10). Perth: Curtin University of Technology.

John, J. (2000a). *Diatom prediction and classification system for urban streams*. Canberra: Land and Water Resources Research and Development Corporation.

John, J. (2000b). *A guide to diatoms as indicators of urban stream health, National River Health Programme: The urban sub-programme*. Canberra: Land and Water Resources Research and Development Corporation.

John, J., & Helleren, S. (1998). Diatom assemblages—ideal biomonitors of wetlands: Two case histories from Western Australia. In: A. J. McComb & J. A. Davis (Eds.), *Wetlands for the future* (pp. 529–538), Adelaide: Gleneagles Publishing.

Juggins, S. (1992). Diatoms in the Thames Estuary, England: Ecology, palaeoecology and salinity transfer function. *Bibl. Diatom., 25*, 1–216.

Kelly, M. G., Whitton, B. A. & Lewis, A. (1996). Use of diatoms to monitor eutrophication in UK rivers. In: B. A. Whitton & E. Rott (Eds.), *Use of algae for monitoring rivers II* (pp. 79–86). Innsbruck: Institut für Botanik, Universitat Innsbruck.

Kiss, K. T., & Genkal, S. I. (1993). Winter blooms of centric diatoms in the River Danube and in its side-arms near Budapest (Hungary). *Hydrobiologia, 269/270*, 317–325.

Kwandrans, J. (1992). Diatom communities of the acidic mountain streams in Poland. *Hydrobiologia, 269/270*, 335–342.

Lange-Bertalot, H. (1979). Pollution tolerance of diatoms as a criterion for water quality estimation. *Beih. Nova Hedwigia, 64*, 285–304.

Lange-Bertalot, H., & Metzeltin, D. (1996). Indicators of oligotrophy: 800 taxa representative of three ecological distinct lake types—carbonate buffered, oligodysrophic, weakly buffered soft water. *Iconographia diatomologica* (Vol. 2). Koenigstein: Koeltz Scientific Books.

Lenior, A., & Coste, C. (1996). Development of a practical diatom index of overall water quality applicable to the French National Water Quality Board network. In: B. A. Whitton & E. Rott (Eds.), *Use of algae for monitoring rivers II* (pp. 29–43). Innsbruck: Institut für Botanik, Universitat Innsbruck.

Loeb, S. L. (1994). An ecological context for biological monitoring. In S. L. Loeb & A. Spacie (Eds.), *Biological monitoring of aquatic systems* (pp. 3–7). Boca Raton, FL: Lewis.

Lowe, R. L. (1974). *Environmental requirements and pollution tolerance of freshwater diatoms*. Cincinnati: National Environmental Research Centre, Office of Research and Development, U.S. Environmental Protection Agency.

Lowe, R. L. (1996). Periphyton patterns in lakes. In: R. J. Stevenson, M. L. Bothwell, & R. L. Lowe (Eds.), *Algal ecology fresh water ecosystems* (pp. 58–72). San Diego: Academic Press.

Lowe, R. L. & Pan, Y. (1996). Benthic algal communities as biological monitors. In: R. J. Stevenson, M. L. Bothwell, & R. L. Lowe (Eds.), *Algal ecology freshwater benthic ecosystems* (pp. 705–733). San Diego: Academic Press.

Marcus, M. D. (1980). Periphytic community response to chronic nutrient enrichment by a reservoir discharge. *Ecology, 61*, 387–399.

Miller, U. (1984). Ecology and palaeocology of brackish water diatoms with special reference to the Baltic Basin. In: M. Ricard (Ed.), *Proceedings of the Eighth Diatom Symposium* (pp. 601–612). Koenigstein: Koeltz Scientific Books.

Neiderhauser, P., & Schanz, F. (1993). Effect of nutrients (N, P, C) enrichment upon the littoral diatom community of an oligotrophic high mountain lake. *Hydrobiologia, 269/270*, 453–462.

Norris, R. H., & Norris, K. R. (1995). The need for biological assessment of water quality: Australian perspective. *Australian Journal of Ecology, 20*, 1–6.

Ollikainen, M., Simola, H., & Niinoija, R. (1993) Changes in diatom assemblages in the profundal sediments of two large oligohumic lakes in eastern Finland. *Hydrobiologia, 269/270*: 405–413.

Oppenheim, D. R. (1991). Seasonal changes in epipelic diatoms along intertidal shore, Berrow Flats, Somerset. *J. Mar. Biol. Ass. UK., 71*, 579–596.

Patrick, R. (1984). Diatoms as indicators of changes of water quality. In: M. Ricand (Ed.), *Proceedings of the Eighth Diatom Symposium* (pp. 759–766). Koenigstein: Koeltz Scientific Books.

Patrick, R. (1994). What are the requirements for an effective biomonitor? In: S. L. Loeb & A. Spacie (Eds.), *Biological monitoring of aquatic systems* (pp. 23–29). Boca Raton: Lewis.

Patrick, R., & Reimer, C. W. (1966). The diatoms of the United States (Vol. 1). *Monographs of the Academy of Natural Sciences of Philadelphia, 13*, 688 pp.

Patrick, R., & Reimer, C. W. (1975). The diatoms of the United States (Vol. 2, Part 1.) *Monographs of the Academy of Natural Sciences of Philadelphia, 13*, 213 pp.

Patrick, R., Roberts, N. A., & Davies, B. (1968). The effects of changes in pH on the structure of diatom communities. *Notulae Naturae, 416*: 1–13.

Prygiel, J., & Coste, M. (1996). Recent trends in monitoring French rivers using algae, especially diatoms. In: B. A. Whitton & E. Rott (Eds.), *Use of algae for monitoring rivers II* (pp. 7–16). Innsbruck: Institut für Botanik, Universitat Innsbruck.

Round, F. E. (1990). The effect of liming on the benthic diatom populations on three upland Welsh lakes. *Diatom Research, 5*: 129–140.

Round, F. E. (1991). Use of diatoms for monitoring rivers. In: B. A. Whitton, E. Rott, & G. Friedrich (Eds.), *Use of algae for monitoring rivers* (pp. 25–32). Innsbruck: Institut für Botanik, Universitat Innsbruck.

Schoeman, F. R. (1976). Diatom indicator groups in the assessment of water quality in the Jukskei-Crocodile River System (Transvaal Republic of South Africa). *J Limn. Soc. S.A., 2*(1), 21–24.

Stevenson, R. J. (1984). Epilthic and epipelic diatoms in the Sandusky River, with emphasis on species diversity and water pollution. *Hydrobiologia, 114*, 161–175.

Stevenson, R. J., Singer, R., Roberts, D. A., & Boylen, C. W. (1985). Patterns of epipelic algal abundance with trophic status and acidity in poorly buffered New Hamptonshire Lakes. *Can. J. Fish. Aquat. Sci., 42*, 1501–1512.

Stoermer, F., & Smol, J. P. (1999). *The diatoms: Applications for the environmental and earth sciences.* Cambridge University Press.

Sumita, N., & Watanabe, T. (1983). New general estimation of river pollution using new diatom community index (NDCI) as biological indicator based on specific composition of epilithic diatom communities, applied to the Asana-gawa and the Sai-gawa Rivers in Ishikawa Prefecture. *Jpn. J. Limnol., 44*: 329–340.

Ten Caté, J. H., Maasdam, R., & Roijackers, R. M. M. (1993). Perspectives for the use of diatom assemblages in the water management policy of Overijssel (the Netherlands). *Hydrobiologia, 269/270*, 351–359.

ter Braak, C. J. F. (1987–1992). *CANOCO*—a Fortran programme for canonical community ordination. Technical report LW-88-02. Ithaca, NY: Micro Computer Power.

Uomo, A. D. (1996). Assessment of water quality of an apennine river as a pilot study for diatom based monitoring of Italian watercourses. In: B. A. Whitton & R. Rott (Eds.), *Use of algae for monitoring rivers II* (pp. 65–72). Innsbruck: Institut für Botanik, Universitat Innsbruck.

van Dam, H. (1982). On the use of measures of structure and diversity in applied diatom ecology. *Beih. Nova Hedwigia, 73*: 97–113.

van Dam, H., Mertens, A., & Simkeldam, J. (1994). A coded checklist and ecological values of fresh water diatoms from the Netherlands. *Netherlands Journal of Aquatic Ecology, 28*(1): 117–133.

Watanabe, T., & Asai, T. (1992). Simulation of organic water pollution using highly prevailing diatom taxa (4). Diatom assemblage in which the leading taxon belongs to *Nitzchia, Pinnularia, Swinella* or *Synedra. Diatom, 7, 37–42.*

Watanabe, T., Asai, K. & Houki, A. (1986). Epilithic diatom assemblage index to organic pollution (DAIpo) and its ecological significance. *Ann. Rep. Graduate Division of Human Culture, Nara Women's Univ., 1*, 76–94.

Watanabe, T., Asai, K., Houkia, & Yamada, T. (1990). Pollution spectrum by dominant diatom taxa in flowing and standing waters. In: H. Simola (Ed.), *Proceedings of the 10th International Diatoms Symposium 1988*. (pp. 563–572), Koenigstein: Koeltz Scientific Books.

Wright, J. F. (1995). Development and use of a system for predicting the macroinvertebrate fauna in flowing waters. *Australian Journal of Ecology, 20*, 181–197.

2

The Management of Wetlands for Biological Diversity: Four Principles

Paul A. Keddy and Lauchlan H. Fraser

Wetlands and Gradients

There are vast areas of major kinds of wetlands: swamps, marshes, fens, and bogs (Table 2.1). Because plant and animal species, vegetation, and wetland types are so variable, it may seem difficult to treat all of these together. Far too often, one encounters specialist publications on the plants or animals of a particular bog, fen, marsh, mire, reed swamp or aquatic community; these balkanized treatments detract from the general principles involved in managing wetlands. Furthermore, because so much focus in wetland management is placed on fish and wildlife production this too often takes precedence over other ecological objectives. Large expanses of wetland vegetation are generally ignored or treated in passing as "aquatic plants." Our objective here is to try to pull together all these disparate vegetation types, species, and physiographic types, and present four general principles necessary for managing them to maintain and enhance biological diversity.

From one perspective, referring to wetlands as a whole may appear foolhardy; wetlands appear to have little in common with one another, because one can find so many types of wetlands, from a ombrotrophic peat bog to a wet meadow on a shoreline. However, all types of wetland are, in fact, controlled by only a short list of environmental factors: water levels, soil fertility, disturbance, salinity, grazing, and burial. This is true whether one is talking about the largest tropical floodplain complex in the world, the Amazon basin (Junk, 1983, 1986; Goulding, 1980; Lowe-McConnell, 1975, 1986), or the shorelines of small, temperate zone lakes (Pearsall, 1920; Spence, 1964; Bernatowiscz & Zachwieja, 1966; Keddy, 1981, 1983, 1984). Regardless of location, wetlands have

---

Paul Keddy • Department of Biological Sciences, Southeastern Louisiana University, Hammond, Louisiana 70402. **Lauchlan H. Fraser** • Department of Biology, University of Akron, Akron, Ohio, 44325-3908.

Table 2.1. A preliminary list of the world's largest wetlands. At the lower limit of 30,000 km² a number of other candidate areas appear possible (after Keddy, 2000).

Rank	Continent	Wetland	Description	Area (km²)*
1/2	Eurasia	West Siberian Lowlands	Peat bogs, boggy forests, meadows	780,000 (up to 1,000,000)
1/2	South America	Amazon River	Large river floodplain 300,000	> 800,000
			Small river floodplain > 500,000	> 90,000
3	North America	Hudson Bay Lowlands	Peatlands	> 200,000 320,000
4	South America	Pantanal	Marsh, swamp, floodplain	120,000 140,000 200,000
5/6/7	Africa	Upper Nile Swamps (Sudd)	Swamps and floodplains	90,000 + 50,000
5/6/7	Africa	Chari-Logone (drains into L. Chad)	Seasonal floodplain	90,000
5/6/7	North America	Mississippi River floodplain	Bottomland hardwoods	86,000
8/9	Eurasia	Papua-New Guinea	Swamp, bog	69,000
8/9	Africa	Zaire-Congo system	Riverine swamps and floodplain	40,000 + 80,000
10	North America	Upper Mackenzie River	Marsh, fen, floodplain	60,000
11	South America	Chilean Fjordlands	No published description available	55,000
12	North America	Prairie potholes	Marsh	40,000
13	South America	Orinoco River delta	Floodplain, swamp, marsh	30,000

*different areas for the same wetland reflect different data sources

gradients of nutrients, soil depth and flooding duration that act like a prism to subdivide them into regions with different abiotic characteristics and corresponding species adundances.

From another perspective, wetlands show differences because each is subject to particular influences such as rates of water level change, waves, fire, and ice scour. In general, these kinds of natural disturbances lead to high biological diversity and many unusual or rare species. Well-studied examples include the wet pine savannas of southeastern North America (Christensen, 1988) and the rich Atlantic coastal plain communities of the Great Lakes and the eastern seaboard of North America.

In this chapter, we want to combine these two perspectives in order to explore how to manage aquatic ecosystems to maintain the different types of wetlands found in them and, in particular, to protect or create the unusual features that promote high species diversity. A commonsense application of a few general principles provides practical guidelines for the management of aquatic ecosystems.

Four Principles

Water Level Fluctuations

Whereas all wetlands are associated with flooded soils, the duration of flooding is largely responsible for different vegetation types. This leads to conspicuous vegetation zonation because different species tolerate different degrees of flooding (Figure 2.1). The

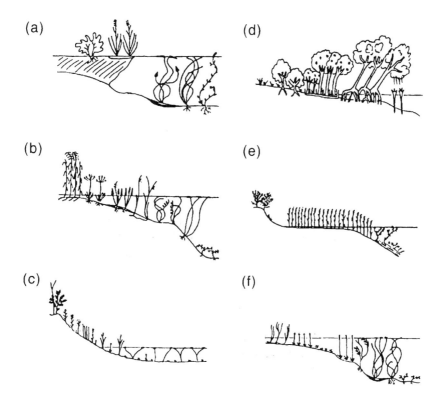

Figure 2.1. Some examples of plant zonation: (a) a bog (after Dansereau, 1959); (b) the St. Lawrence River (after Dansereau, 1959); (c) Wilson's Lake, Nova Scotia (after Wisheu & Keddy, 1989); (d) a mangrove swamp of the Caribbean (after Bacon, 1978); (e) the eastern shore of Lake Kisajno, northeastern Poland, a typical small lake phyto littoral (after Bernatowicz & Zachwieja, 1966); (e) a sandy shoreline (after Dansereau, 1959).

relationship between water level and plant species diversity for two types of shoreline wetland are shown in Figure 2.2. This first source of biological variation is so conspicuous that far too many ecologists seem content to profile the vegetation and conclude that they have completed their study of the wetland.

Zonation is a dynamic, not static, property of wetlands. Let us consider two extreme (and limiting) cases to frame the discussion and clarify thinking. If water levels were entirely stable, the result would be a two-zone system (Figure 2.3, bottom). There would be aquatic communities with some emergent species in the water, whereas woody plants would occur above the waterline. If water levels fluctuated widely and wildly, they would frequently exceed the natural tolerance limits of most species, producing habitats devoid of plants, or dominated by only a few weedy species.

Somewhere within these two extremes lies the regimen, then, that promotes maximum ecological diversity. Year-to-year fluctuations are an important factor generating plant diversity. High water periods kill shrubs that dominate the upper zone, and low water periods allow many other species to regenerate from buried seeds. If mean water levels change from one year to the next, one can then transform the two-phase system to a four-phase system (Figure 2.3, top). In this case, the simple practice of changing water levels from one year to the next doubles the number of vegetation types. It more than doubles the number of plant species, because the new vegetation types, emergent marsh and wet

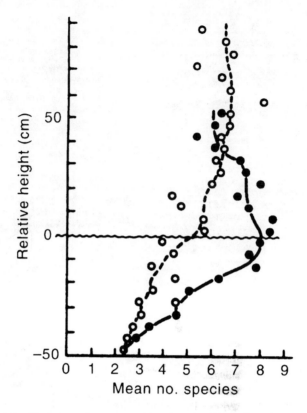

Figure 2.2. The relationship between the mean number of species (0.025 m^2) and the August water level for wetlands on sheltered (solid circles) and exposed (open circles) shores in a Nova Scotia lake (after Keddy, 1984).

meadow, generally support large numbers of plant species. Such diverse plant communities will lead to animal diversity too (Figure 2.4). Furthermore, fluctuations in water level can generate additional levels of biological diversity (Figure 2.5).

How much year-to-year change is enough? This probably varies with climate, but wetland research in the Great Lakes suggests that changes over many meters are required; in smaller inland lakes, this probably drops to less than a meter. Superimposed upon year-to-year variation is variation within a year. This is probably less important for producing rich wetland plant communities, but some seasonal decline in water level is natural and is generally found in species-rich shoreline wetlands. Dropping water levels by roughly one-half meter during the growing season is probably a good first approximation.

A Predictive Model for Changes in Wetland Extent

Water levels of the Great Lakes have changed over both geological and historical time scales (Figure 2.6), so there is now a rich array of shoreline wetland types. Rich wet meadow flora are particularly well developed on gently sloping sandy shorelines (Reznicek & Catling, 1989). Great Lakes wetlands provide important habitats for fish, waterfowl, and rare plant species (Smith et al., 1991). Large areas of these wetlands have been drained, and humans have also already reduced the amplitude of water level fluctuations. More

Management of Wetlands for Biological Diversity

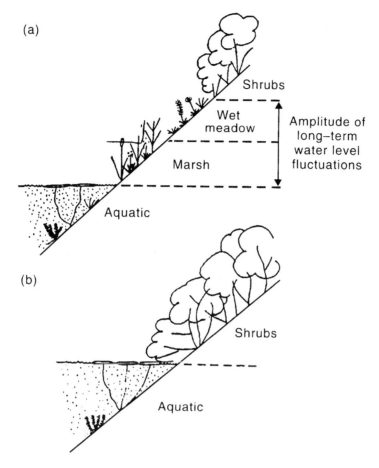

Figure 2.3. The constriction of water level fluctuations reduces wetland habitats from four zones (a) to two zones (b) (from Keddy, 1991).

recently, there has been added pressure to control fluctuations further. Quinlan and Mulamoottil (1987) report that around the shore of Lake Ontario, wet meadows accounted for a staggering 65.4% of the wetlands in 1927, and only 22.1% in 1978. Figure 2.7 provides preliminary estimates of the probable effects of water level fluctuation on wetland area based on a model that predicted the upper boundary of wet meadows and the lower boundary of the marsh.

To model the landward edge of the wet meadow, it was necessary to consider the dieback and recolonization by woody plants. Two assumptions were made. The first was that the dieback of woody plants was directly related to high water levels during the growing season; secondly, woody plants reinvaded according to an exponential model. This allowed predictions of the lower limit of woody plants from projected water levels (Figure 2.7, top line). Lag times of 15, 18, or 20 years, made little difference.

To model the lower boundary of the marsh required the assumption that marsh plants move downslope the same year that water levels fall; this would most likely be the result of germination from buried seeds. As water levels rose, the marsh plants would die back over several years (Figure 2.7, bottom line). Lag times of 2, 3, or 4 years made little difference.

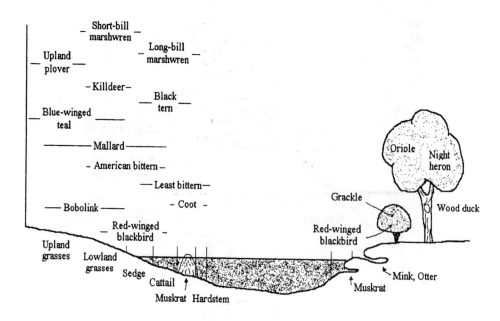

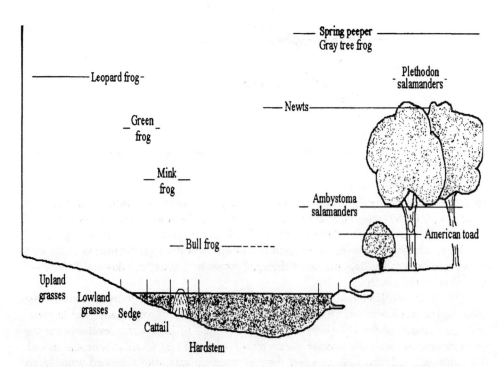

Figure 2.4. Zonation in some birds and mammals (top) and amphibians (bottom) in relation to water level and vegetation (after Weller, 1994).

Management of Wetlands for Biological Diversity 27

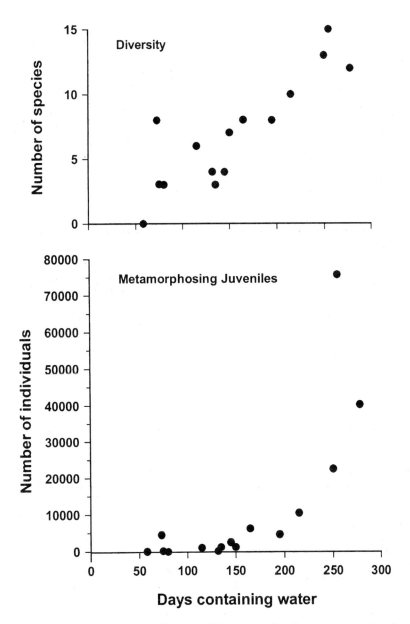

Figure 2.5. Relationship between water level and amphibian species diversity (top) and number of metamorphosing juvenile amphibians (bottom) (after Pechmann et al., 1989).

The area between these lines is then the area of wet meadow and marsh as a function of time. Figure 2.7 shows, for example, the great areas of wetland that occurred during the low water period of the mid-1930s. This model was then used on projected water level scenarios to forecast the effects on wet meadow/marsh area in the Great Lakes. If further reductions in amplitude occurred as opposed the model predicted losses approximating 30% of the wetlands in Lake Ontario alone.

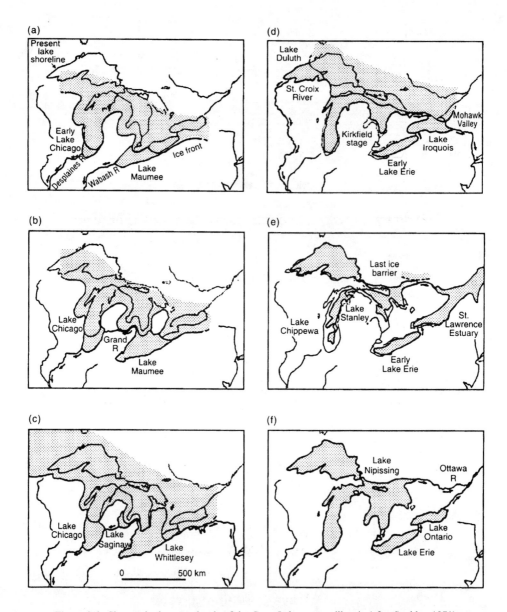

Figure 2.6. Changes in the water levels of the Great Lakes over millennia (after Strahler, 1971).

A Descriptive Model for Frequency and Intensity of Flooding

Two of the most important components of flooding, frequency and intensity, can be plotted on orthogonal axes to represent all possible pairwise combinations. We can then plot, for many sites, the frequency of flooding against the depth of flooding, or perhaps more conveniently, the frequency of flooding and amplitude of water level changes. There are several important properties that could then be plotted for sites located in relation to these axes. One could plot the many reservoirs or wetlands of the world in order to explore patterns. Are there, for example, certain combinations that are rare and others that are

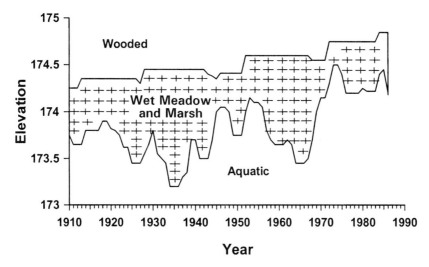

Figure 2.7. A simple simulation model showing how wetland vegetation changes with water levels in the Great Lakes (after Painter & Keddy, 1992). The upper line is the woody plant boundary (18 yr lag), whereas the lower line is the lower marsh boundary (3 yr lag). Note that the area of wet meadow and marsh varies with water level history.

common? One could plot important properties such as productivity or wildlife diversity, and explore how they are related to water levels. Unfortunately, the required data on water levels are scattered through a broad literature describing individual cases, and buried in reams of unpublished reports. Often the data are not quite comparable. As a first step in this direction, Figure 2.8 shows such a plot for a few lakes, and identifies a corridor of high plant species richness. This is based upon a set of lakes in eastern North America, and there is currently no way to know how well we can extrapolate from this geographic region or to other properties.

Fertility

Variation in fertility occurs among and within wetlands. In wetlands exposed to waves and ice scour, silt and clay are constantly eroded and exported, so these areas tend to be infertile and support distinctive plant species. In contrast, silt and clay are deposited in bays, and these wetlands tend to be dominated by few large clonal species with dense canopies. Fertility gradients are therefore an important feature in producing different plant communities in all wetlands. The greater the array of fertility levels, the greater will be the array of vegetation types and plant species (Pearsall, 1920; Spence, 1964; Bernatowiscz & Zachwieja, 1966; Auclair et al., 1972). This is most easily illustrated by use of the biomass of plant communities as a measure of their location along the fertility gradient. Figure 2.9 plots the total number of plant species and rare plant species against biomass for wetlands. The diversity of plant species is highest at intermediate levels of biomass (fertility), and the number of rare species is greatest at the lowest levels of biomass (lowest fertility).

Eutrophication will reduce the length of this fertility gradient and slowly cause more areas of a wetland to converge on the high biomass vegetation type. In one experiment, 12 different habitat types were created in plastic containers (Weiher & Keddy, 1995; Weiher et

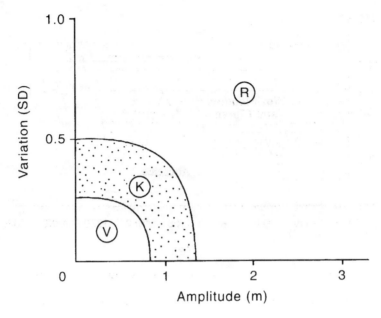

Figure 2.8. A corridor of high species richness (stippled region) is associated with yearly water level amplitudes of roughly 1 m and standard deviations of roughly 0.5 m. The circles indicate three representative lakes from Nova Scotia: V, Vaughan; K, Kejimkujik; R, Rossignal (from Hill et al., 1998).

al., 1996). These habitats included sand, gravel, cobbles, and stable and fluctuating water levels, each replicated 10 times. Half of these wetland types received additional NPK fertilizer and, as shown in Figure 2.10, in every case, the number of plant species was lower in the fertilized plant communities.

A Mathematical Model for the Prediction of Maximum Potential Species Richness

The relationship between diversity and productivity is well documented and applies to both plant and animal distributions (see Tilman & Pacala, 1993; Rosenzweig & Abramsky, 1993; Keddy & Fraser, 1999, for overviews). One promising general model for the prediction of herbaceous plant diversity is the standing crop–species richness relationship ("hump-backed model") first proposed by Grime (1973a, 1973b, 1979), which states that species richness reaches a maximum at intermediate standing crop. Grime suggests that species richness is limited at low standing crop by minimal availability of mineral nutrients and, in some cases, high levels of disturbance, whereas at high standing crop, interspecific competition limits the number of species. We investigated the general applicability of the "hump-backed model" in herbaceous wetlands by searching the literature for all studies that presented species richness as a function of biomass. Our objective was to explore both the generality and the potential limitations of the model. We also wanted to predict quantitatively the decline in species richness with increasing biomass using known biological parameters and equations. In particular, we wanted to determine the maximum species richness we would expect to find at any given standing crop in a herbaceous wetland.

Management of Wetlands for Biological Diversity   31

(a)

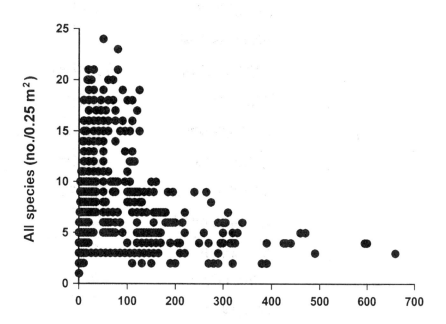

(b)

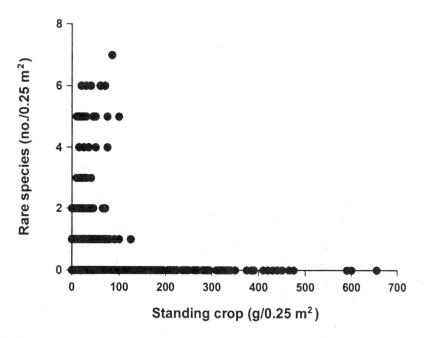

Figure 2.9. Plant species richness along gradients of standing crop (a) for 401 0.25 m² quadrats in eastern North America and (b) same quadrats, but nationally rare species only (from Moore et al., 1989).

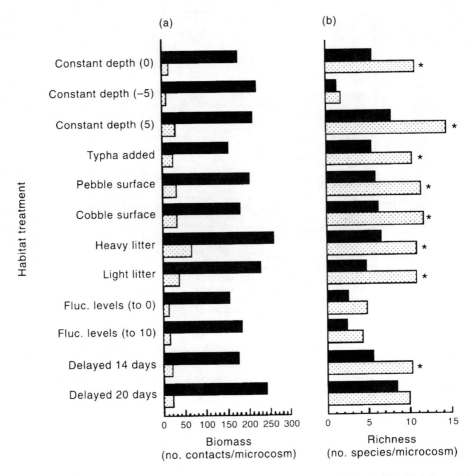

Figure 2.10. The effect of fertilization on biomass and species richness of wetland plants in 12 artificially created habitats (from Wisheu et al., 1991). Stippling represents the infertile treatment, whereas solid bars are the fertilized plants. *$P < .001$.

We found 22 published studies that measured both standing crop and species richness, totalling 1,367 data points, from fresh- and saltwater marshes, bogs, fens, and wet meadows around the world, varying in sample size from 0.20 to 50 m². In order to standardize the scale of sample size to 1 m², we generated a species richness versus log area linear regression from the data. This regression line was: $S^A = 3.42 \ln A + 13.28$, where S^A = species richness per unit area of measurement (A), and A = area (m²) (F ratio = 77.42; $df = 1,909$); $P < .001$) (see Connor & McCoy, 1979; Palmer & White, 1994, for further discussion of species-area relationships). We then standardized the scale to 1 m² by applying the equation: $S = S^o / S^A$, where S = species richness m⁻², and S^o = observed species richness.

The data display the classic "hump-backed" shape, with richness reaching a maximum at about 500 g m⁻² (Figure 2.11). There have been previous attempts to fit polynomial regressions through biomass-richness data sets with varying degrees of success (Wisheu & Keddy, 1989; Puerto et al., 1990, Garcia et al., 1993; Muotka & Virtanen,

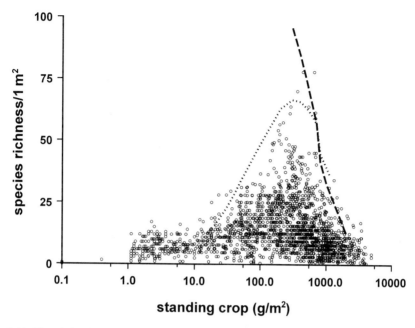

Figure 2.11. The relationship between species richness and aboveground biomass in herbaceous wetlands. The lines shown, equations 2 (dots) and 3 (dashes), represent the maximum species richness possible as a function of biomass and are explained in the text. The data represent 1,367 points from 22 published studies (Auclair et al., 1972; Day et al., 1988; Forrest & Smith, 1975; Garcia et al., 1993; Gough et al., 1994; Grace & Pugesek, 1997; Klinkhamer & de Jong, 1985; Lambert, 1976; Moore & Keddy, 1989; Muotka & Virtanen, 1995; Partridge & Wilson, 1987; Puerto et al., 1990; Rejmankova et al., 1995; Vermeer, 1986; Vermeer & Verhoeven, 1987; Vince & Snow, 1984; Vitt, 1990; Wheeler & Giller, 1982; Wheeler & Shaw, 1991; Wilson & Keddy, 1986a; Wilson & Keddy, 1988; Zobel & Liira, 1997).

1995). We found a significant nonlinear parabolic relationship between species richness and standing crop, but it explained only 13.5% of the variance in richness. Many other factors have been shown to influence plant species richness in wetlands, including climate (Vitt, 1990), pH (Gough et al., 1994), soil nutrients (Willis, 1963; Verhoeven et al., 1996), salinity (Partridge & Wilson, 1987; Grace & Pugesek, 1997; Keogh et al., 1999), abiotic disturbance (Keddy, 1983), and herbivory (Lubchenko, 1978), which may explain the large variance.

Despite a statistically significant relationship, the biomass–richness relationship did not follow a specific statistical pattern that could be described by any one line. The data could certainly be hiding families of lines and curves. Our intention, though, was to look for emergent patterns from the 1,367 data points in order to make general predictions (e.g., Brown, 1995). We found that richness varied from 0 to $75\,m^{-2}$, with a clear maximum occurring in the range of 300–$600\,g\,m^{-2}$. Although collected from a wide array of vegetation types and geographic locales, and corrected for differences in sample area, the points are clearly consistent with Grime's model.

Boundaries (as opposed to means) are difficult, if not impossible, to define statistically (but see Scharf et al., 1998). Nevertheless, Brown and Maurer (1987, 1989) have suggested that many relationships between ecological variables are not characterized by a regression line through the data, but fall instead within a region to which lines set boundaries. For example, the range of trees is limited along an altitudinal gradient by frost, which forms a boundary condition. Applying a similar approach to the standing crop–richness relationship, traditional regression lines may simply not be appropriate; indeed, as originally formulated (Grime, 1973a, 1973b), the relationship is a boundary condition (maximum rather than mean richness) that is specified as the dependent variable (see also Marrs et al., 1996). It is clear that the data follow a pattern, and that the greatest species richness occurs between 300 and 600 g/m^{-2}. Here, we examine a quantitative method of estimating an upper boundary to species richness.

To estimate this boundary, the first step was to calculate N (the estimated number of shoots in a quadrat), and then to calculate S (species number) for the given N. Because we used the same procedure to calculate S from N in both the upslope and downslope equations, we begin by describing the technique to calculate S. We used the so-called "collector's curve" (Pielou, 1977) to predict species richness from the number of individuals (shoots, ramets, rosettes, etc.): $S = S^*[1 - (1 + N/kS^*)^{-k}]$, where S is species number, S^* is species pool, N is number of individuals (shoots), and k is a constant describing the slope (generally, the larger the k value, the greater number of rare species in the community). The collector's curve describes a relationship in which S approaches S^* asymptotically. The model has two parameters: S^* and k.

In nature, the species pool, S^*, varies between communities. Species pool is a measure of the total set of species that are potentially capable of coexisting in a particular community (Partel et al., 1996). There is evidence of a positive correlation between the species pool of a particular area and species richness measurements at the smaller scale (Wisheu & Keddy, 1996; Partel et al., 1996; reviewed by Zobel, 1997). Therefore, if the species pool is large, all things considered, there will be a greater chance that richness will be higher in, say, a 1 m^2 area. In this case, we used data from Wisheu and Keddy (1996) to determine, via polynomial regression, that S^* could be expressed as a function of biomass between 12 and 1680 g m^{-2}, such that $S^* = -9.56 \ln B^2 + 86.21 \ln B - 76.49$, where $B =$ standing crop (m^{-2}) ($F = 67.356$; $df = 25$; $P < .0001$).

We acknowledge that k, like S^*, also varies between communities. We have given k a value of 1, but if k is varied, the shape of the slope stays approximately constant. Most of the difference, though slight, occurs in the tail of the curve. Given S^* and k, our provisional formula for the calculation of S for both the upslope and the downslope is as follows:

$$S = (-9.56 \ln B^2 + 86.21 \ln B - 76.49) \\ \times [1 - (1 + N/(-9.56 \ln B^2 + 86.21 \ln B - 76.49)^{-1}] \quad (1)$$

The calculation of the number of individual shoots, N, required different methods and data for the upslope and downslope.

Upslope. At any standing crop, N can be calculated by dividing standing crop (B) by the mean individual weight (C). The problem is that C does not remain constant with increasing biomass. Consider, for example, the small rosette species such as *Isoetes* or *Lobellia* on gravel shores compared with large shoots of *Typha* or *Phragmites*. One way we can estimate C requires us to know total biomass, B, and the number of shoots. Consequently, Auclair et al. (1972) measured both numbers of individuals and total aboveground biomass. Using the field data from Auclair et al., we established through polynomial regression that mean individual weight could be expressed as a function of standing crop between 20 and $1100\,\mathrm{g\,m^{-2}}$, such that $C = 3.54 \times 10^{-6}B^2 + 6.81 \times 10^{-4}B + 1.10$ (F = 9.64; $df = 104$; $P < .001$). Therefore, using our "working formula" (equation 1), the final equation for the upslope (over the range of 20–1100 g m^{-2}) can be written as follows:

$$S = (-9.56\ \ln B^2 + 86.21\ \ln B - 76.49)$$
$$\times [1 - (1 + B/(-9.56\ \ln B^2 + 86.21\ \ln B - 76.49)$$
$$\times (3.54 \times 10^6 B^2 + 6.81 \times 10^4 B + 1.10)^{-1}]. \quad (2)$$

Downslope. N was calculated by applying the self-thinning rule for plants (Yoda et al., 1963), which has been shown to apply for almost any fully occupied plant stand dominated by a single species such that: $\omega = 9670 d^{1.49}$, where ω = shoot dry weight (g), and d = density (m^{-2}) (Gorham, 1979). Since $\omega d = B$, and $d = N$, we can rewrite the self-thinning equation as $N = (9{,}670/B)^{2.04}$. Therefore, by using equation 1, the final equation for the downslope (between 12 and 1,680 g m^{-2}) is as follows:

$$S = (-9.56\ \ln B^2 + 86.21\ \ln B - 76.49)$$
$$\times [1 - (1 + (9{,}670/B)^{2.04}/(-9.56\ \ln B^2 + 86.21\ \ln B - 76.49)^{-1}]. \quad (3)$$

Figure 2.11 shows that these two simple equations, one relating plant size to standing crop, and one relating shoot density to standing crop, when combined with the standard collector's curve, can set upper limits to species richness. Furthermore, when superimposed on our data from 1,367 samples in 22 published studies, a close fit is apparent. Similar to Oksanen's (1996) "no-interaction" model, the number of shoots determines the upper limits to species richness. Points may, of course, fall well below these lines, but our objective was to define upper limits rather than account for diversity on a quadrat by quadrat basis.

Competitive Hierarchies

Over the last few decades, it has become apparent that most plant communities are organized by competitive hierarchies (Keddy, 2001); a small number of strong competitors tend to make up most of the biomass in a wetland, whereas larger numbers of weak competitors use the remaining space. The stronger competitors occupy the fertile sites and exclude the weaker competitors to infertile areas. Furthermore, increasing soil fertility increases the ability of these few species to dominate a wetland and exclude neighbors (Keddy et al., 2000). The dominance achieved by *Typha* or *Phragmites* in fertile soils can be compared to the rich array of plant types in interdunal meadows, wet prairies, or fens. The competitive ability of seven wetland plant species measured in an experiment was able to predict their field distributions along the shoreline; the large leafy plants occupied the fertile sites, and small rosette species were excluded to the sandy shores (Wilson & Keddy, 1986a, b). This experiment was repeated with a much larger sample of species and the same result was found (Gaudet & Keddy, 1995). This merges with the work on fertility, because it is well established that fertilization enhances the competitive performance of the large leafy species. The more eutrophic a site, the greater the likelihood of dominance by a few aggressive wetland species such as *Typha, Phragmites,* or *Phalaris*.

Centrifugal Organization

The same species usually prosper in the fertile and sheltered sites within a wetland, but different species can occur on other sites depending on the constraints. Shallowly sloping sands may develop fens, gravel shorelines may have isoetid plants, wet prairies may occur where fire or water level fluctuations kill woody plants, pannes may develop between alkaline sand dunes, and so on. The benign ends of many gradients are similar enough that we can describe them as a "core" habitat that can be dominated by the same species. At the peripheral end of each gradient, however, species with specific adaptations to particular sources of adversity occur. This pattern is termed *centrifugal organization* (Figure 2.12). Many peripheral habitats radiate outwards from the single, central core habitat.

The core habitat has low disturbance and high fertility, and is dominated by large leafy species capable of forming dense canopies. In northeastern North America, *Typha* dominates the core region. In other climatic regions, herbaceous perennials in the genera *Papyrus*, *Phragmites*, *Phalaris*, *Calamagrostis*, or *Rhynchospora* may play a similar role. Outside the core, different constraints create radiating axes along which different groups of species and vegetation types are arrayed. Along an axis of soil phosphorus, for example, the community composition changes from a high biomass *Typha*-dominated wetland to sparse vegetation dominated by isoetid and insectivorous species (Moore et al., 1989), two

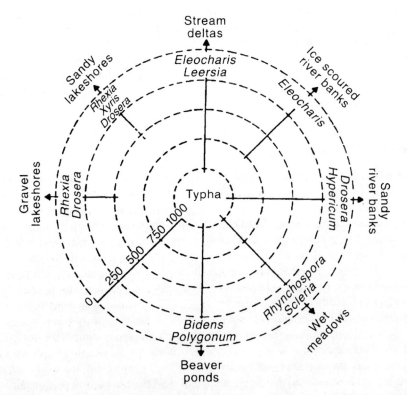

Figure 2.12. When many gradients radiate outward from a shared core habitat, the pattern is termed *centrifugal organization* (from Wisheu & Keddy, 1992).

groups of plants indicative of infertile conditions (Boston, 1986; Givnish, 1988). Furthermore, the shortage of phosphorus, as opposed to nitrogen, creates rather different plant communities (Verhoeven et al., 1993, 1996).

Nutrient concentrations, however, are only some of the many gradients that occur in wetlands. Another gradient is disturbance, and the species found along these gradients differ from those found along gradients of fertility. *Typha* would again occupy fertile, protected areas, but where ice scour or severe flooding occur, either reeds or annual species would be abundant (Day et al., 1988; Moore et al., 1989). The deeply buried rhizomes of reeds protect them from moderate ice scour, while fast-growing annuals are able to set seed between periods of mud deposition (Grubb, 1985; Day et al., 1988). Peripheral habitats formed by different kinds and combinations of infertility and disturbance support distinctive floras that reflect differing environmental conditions [e.g., shoreline fens (Charlton & Hilts, 1989; Yabe, 1993; Yabe & Onimaru, 1997), interdunal swales and sand spits (Willis, 1963; Reznicek & Catling, 1989), coastal plain wetlands (Keddy & Wisheu, 1989), river banks (Brunton & Di Labio, 1989; Nilsson et al., 1989), bottomland swamps (Penfound & Hathaway, 1938; White, 1983), and floodplains (Salo et al., 1986; Duncan, 1993)].

Our objective here is to introduce this model and stress its value in creating biologically diverse wetlands. The postulated mechanisms that produce this arrangement of vegetation, and tests of these postulates, are discussed elsewhere (Keddy, 1990, 2000, 2001; Wisheu & Keddy, 1992; Gaudet & Keddy, 1995).

Management Guidelines

The following guidelines apply to all wetland types. The numbers given are drawn from temperate zone lakes and will have to be calibrated for other wetland types. Tropical floodplains, for example, may have much larger seasonal variation than temperate lakes.

1. Water levels should change from year to year. A 10-year cycle with changes of 1–4 m is probably typical of smaller lakes, with greater fluctuations in larger lakes. The highest water levels will determine the area of herbaceous wetlands by setting the lower limit of trees and shrubs.
2. Within years, high water levels in spring will further retard invasion by shrubs and trees.
3. During the growing season, water levels should fall approximately 0.5 to 1.0 m.
4. The gradient(s) from infertile to fertile sites greatly increase the number of ecological communities that can arise.
5. The more kinds of infertile and otherwise constrained habitats available, the more kinds of plants that can coexist in a lake, reservoir, or wetland.
6. Eutrophication usually reduces both the number of species at individual sites and the total number found in a lake, reservoir, or wetland.

Wetland Management for Water Resources

The supply of renewable freshwater for human use is limited. Postel et al. (1996) estimated that by the year 2025, humans will appropriate over 70% of the total renewable runoff of freshwater. Many regions of the globe are already experiencing severe water shortages or unsanitary water (Gleick, 2000). Due to their location between terrestrial and aquatic communities, wetlands form an important component of the hydrological cycle. Most surface waters originating from terrestrial systems must pass through wetlands in order to reach their aquatic outlet. Because these waters can contain an array of dissolved substances, wetlands receive and collect many different compounds. Consequently, wetlands can potentially play an important role in the transformation and reduction of these compounds. Constructed wetlands for the purpose of treating wastewater may therefore play a very important role in cleaning freshwater and water recycling for human use. These so-called "treatment" wetlands (Kadlec & Knight, 1996) can be applied to a number of different management situations, including agricultural runoff, human waste, and acid–mine drainage. Treatment wetlands vary in size from a few square meters to many acres. Some treatment wetlands are designed with water flowing above the soil surface, whereas others have a subsurface flow of water (Kadlec & Knight, 1996). It is very difficult to prescribe general design criteria for treatment wetlands because of the vast diversity of types that are found and the conditions under which they are applied. The listed management suggestions are for natural wetland systems and do not necessarily apply to treatment wetlands. Treatment wetlands are not designed to maximize diversity, but to treat wastewater. The high nutrient conditions commonly found in treatment wetlands for agricultural runoff would tend to reduce plant species diversity by increasing biomass and interspecific competition.

Nevertheless, the four major principles should not be ignored when considering the management of treatment wetlands. Water level fluctuations are important for all wetlands and therefore should be considered for treatment wetlands. It may be possible to include nutrient gradients within treatment wetlands by increasing the size of the wetland and introducing heterogeneity of the substrate, for example, soil depth and the size of soil particles. It may also be useful to consider constructing treatment wetlands at the head of threatened natural wetlands to act as preliminary sieves, thereby reducing the detrimental effects of eutrophication.

References

Auclair, A. N., Bouchard, A., & Pajaczkowski, J. (1972). Plant composition and species relations on the Huntingdon Marsh, Quebec. *Canadian Journal of Botany*, *51*, 1231–1247.
Bacon, P. R. (1978). *Flora and fauna of the Caribbean*. Trinidad: Key Caribbean Publications.
Bernatowiscz, S., & Zachwieja, J. (1966). Types of littoral found in the lakes of the Masurian and Suwalki Lakelands. *Komitet Ekolgiezny–Polska Akademia Nauk XIV*, 519–545.
Boston, H. L. (1986). A discussion of the adaptation for carbon acquisition in relation to the growth strategy of aquatic isoetids. *Aquatic Botany*, *26*: 259–270.
Brown, J. H. (1995). *Macroecology*. Chicago: University of Chicago Press.
Brown, J. H., & Maurer, B. A. (1987). Evolution of species assemblages: Effects of energetic constraints and species dynamics on the diversification of North American avifauna. *American Naturalist*, *130*, 1–17.
Brown, J. H., & Maurer, B. A. (1989). Macroecology: The division of food and space among species on continents. *Science*, *243*, 1145–1150.
Brunton, D. F., & Di Labio, B. M. (1989). Diversity and ecological characteristics of emergent beach flora along the Ottawa River in the Ottawa-Hull region, Quebec and Ontario. *Naturaliste Canadien*, *116*: 179–191.
Charlton, D. L., & Hilts, S. (1989). Quantitative evaluation of fen ecosystems on the Bruce Peninsula. In: M. J. Bardecki & N. Patterson (Eds.), *Ontario wetlands: Inertia or momentum* (pp. 339–354). Proceedings of conference, Ryerson Polytechical Institute, Toronto, October 21–22, 1988.

Christensen, N. L. (1988). Vegetation of the southeastern coastal plain. In: M. G. Barbour, & W. D. Billings (Eds.), *North American terrestrial vegetation* (pp. 317–363) Cambridge, UK: Cambridge University Press.

Connor, E. F. & McCoy, E. D. (1979). The statistics and biology of the species–area relationship. *American Naturalist, 113*, 791–833.

Czaya, E. (1983). *Rivers of the world*. Cambridge, UK: Cambridge University Press.

Dansereau, P. (1959). Vascular aquatic plant communities of southern Quebec. A preliminary analysis. *Transactions of the Northeast Wildlife Conference, 10*: 27–54.

Day, R. T., Keddy, P. A., McNeill, J. & Carleton, T. (1988). Fertility and disturbance gradients: A summary model for riverine marsh vegetation. *Ecology, 69*: 1044–1054.

Duncan, R. P. (1993). Flood disturbance and the coexistence of species in a lowland podocarp forest, south Westland, New Zealand. *Journal of Ecology, 81*: 403–416.

Forrest, G. I., & Smith, R. A. H. (1975). The productivity of a range of blanket bog vegetation types in the northern Pennines. *Journal of Ecology, 63*, 173–202.

Garcia, L. V., Maranon, T., Moreno, A., & Clemente, L. (1993). Above-ground biomass and species richness in a Mediterranean salt marsh. *Journal of Vegetation Science, 4*, 417–424.

Gaudet, C. L., & Keddy, P. A. (1995). Competitive performance and species distribution in shoreline plant communities: A comparative approach. *Ecology, 76*, 280–291.

Givnish, T. J. (1988). Ecology and evolution of carnivorous plants. In: W. B. Abrahamson (Ed.), *Plant–animal interactions*. New York: McGraw-Hill.

Gleick, P. H. (2000). *The world's water 2000–2001: The biennial report on freshwater resources*. Washington, DC: Island Press.

Gorham, E. (1979). Shoot height, weight and standing crop in relation to density of monospecific plant stands. *Nature, 279*, 148–150.

Gough, L., Grace, J. B., & Taylor, K. L. (1994). The relationship between species richness and community biomass: the importance of environmental variables. *Oikos, 70*, 271–279.

Goulding, M. (1980). *The fishes and the forest: Explorations in Amazonian natural history*. Berkeley: University of California Press.

Grace, J. B., & Pugesek, B. H. (1997). A structural equation model of plant species richness and its application to a coastal wetland. *American Naturalist, 149*, 436–460.

Grime, J. P. (1973a). Control of species density in herbaceous vegetation. *Journal of Environmental Management, 1*, 151–167.

Grime, J. P. (1973b). Competitive exclusion in herbaceous vegetation. *Nature, 242*, 344–347.

Grime, J. P. (1979). *Plant strategies and vegetation processes*. Chichester, UK: Wiley.

Grubb, P. J. (1985). Plant populations and vegetation in relation to habitat disturbance and competition: problems of generalizations. In: J. White, (Ed.), *The population structure of vegetation* (pp. 595–621). The Hague: Junk.

Hill, N. M., Keddy, P. A., & Wisheu, I. C. (1998). A hydrological model for predicting the effects of dams on shoreline vegetation of lakes and reservoirs. *Environmental Management, 22*: 723–736.

Junk, W. J. (1983). Ecology of swamps on the Middle Amazon. In: D. W. Goodall (Ed.), *Ecosystems of the world 4B: Mires: Swamp, bog, fen and moor*, (pp. 269–294) Amsterdam: Elsevier Science.

Junk, W. J. (1986). Aquatic plants of the Amazon system. In: B. R. Davies & K. F. Walker (Eds.), *The ecology of river systems*, (pp. 319–337) Dordrecht, The Netherlands: Junk.

Kadlec, R. H., & Knight, R. L. (1996). *Treatment wetlands*. New York: Lewis.

Keddy, P. A. (1981). Vegetation with coastal plain affinities in Axe Lake, near Georgian Bay, Ontario. *Canadian Field Naturalist 95*: 241–248.

Keddy, P. A. (1983). Shoreline vegetation in Axe Lake, Ontario: Effects of exposure on zonation patterns. *Ecology, 64*: 331–344.

Keddy, P. A. (1984). Plant zonation on lakeshores in Nova Scotia: a test of the resource specialization hypothesis. *Journal of Ecology 72*: 797–808.

Keddy, P. A. (1990). Competitive hierarchies and centrifugal organization in plant communities. In: J. B. Grace, & D. Tilman (Eds.), *Perspectives on plant competition* (pp. 265–290). San Diego: Academic Press.

Keddy, P. A. (1991). Water level fluctuations and wetland conservation. In: J. Kusler, & R. Smardon (Eds.), *Wetlands of the Great Lakes: Protection and restoration policies, Status of the science* (pp. 79–91). New York: Managers Inc.

Keddy, P. A. (2000). *Wetland ecology: Principles and conservation*. Cambridge, UK: Cambridge University Press.

Keddy, P. A. (2001). *Competition* (2nd ed.). London: Chapman & Hall.

Keddy, P. A. & Fraser, L. H. (1999). On the diversity of land plants. *EcoScience, 6*, 366–380.

Keddy, P. A., Gaudet, C. & Fraser, L. H. (2000). Effects of low and high nutrients on the competitive hierarchy of 26 shoreline plants. *Journal of Ecology, 88*, 413–423.

Keddy, P. A., & Wisheu, I. C. (1989). Ecology, biogeography, and conservation of coastal plain plants: Some general principles from the study of Nova Scotian wetlands. *Rhodora, 91*: 72–94.

Keogh, T. M., Keddy, P. A., & Fraser, L. H. (1999). Patterns of tree species richness in forested wetlands. *Wetlands, 19*: 639–647.

Klinkhamer, P. G. L., & de Jong, T. J. (1985). Shoot biomass and species richness in relation to some environmental factors in a coastal dune area in The Netherlands. *Vegetatio, 63*, 129–132.

Lambert, J. D. H. (1976). Plant succession on an active tundra mud slump, Garry Island, Mackenzie River Delta, Northwest Territories. *Canadian Journal of Botany, 54*, 1750–1758.

Lowe-McConnell, R. H. (1975). *Fish communities in tropical freshwaters: Their distribution, ecology and evolution.* London: Longman.

Lowe–McConnell, R. H. (1986). Fish of the Amazon system. In: B. R. Davies & K. F. Walker (Eds.), *The ecology of river systems* (pp. 339–351). Dordrecht, The Netherlands: Junk.

Lubchenko, J. (1978). Plant species diversity in a marine intertidal community: importance of herbivore food preference and algal competitive abilities. *American Naturalist, 112*, 23–39.

Marrs, R. H., Grace, J. B., & Gough, L. (1996). On the relationship between plant species diversity and biomass: A comment on a paper by Gough, Grace and Taylor. *Oikos, 75*, 323–326.

Moore, D. R. J., & Keddy, P. A. (1989). The relationship between species richness and standing crop in wetlands: the importance of scale. *Vegetatio, 79*: 99–106.

Moore, D. R. J., Keddy, P. A., Gaudet, C. L., & Wisheu, I. C. (1989). Conservation of wetlands: Do infertile wetlands deserve a higher priority? *Biological Conservation, 47*: 203–217.

Muotka, T. & Virtanen, R. (1995). The stream as a habitat templet for bryophytes: Species' distributions along gradients in disturbance and substratum heterogeneity. *Freshwater Biology, 33*, 141–160.

Nilsson, C., Grelsson, G., Johansson, M., & Sperens, U. (1989). Patterns of plant species richness along riverbanks. *Ecology, 70*, 77–84.

Oksanen, J. (1996). Is the humped relation between species richness and biomass an artefact due to plot size? *Journal of Ecology, 84*, 293–295.

Painter, S., & Keddy, P. A. (1992). Effects of water level regulation on shoreline marshes: A predictive model applied to the Great Lakes. National Water Research Institute, Environment Canada, Burlington.

Palmer, M. W., & White, P. S. (1994). Scale dependence and the species–area relationship. *American Naturalist, 144*, 717–740.

Partel, M., Zobel, M., Zobel, K., & Van der Maarel, E. (1996). The species pool and its relation to species richness—evidence from Estonian plant communities. *Oikos, 75*, 111–117.

Partridge, T. R., & Wilson, J. B. (1987). Salt tolerance of salt marsh plants of Otago, New Zealand. *New Zealand Journal of Botany, 25*, 559–566.

Pearsall, W. H. (1920). The aquatic vegetation of the English Lakes. *Journal of Ecology, 8*, 163–201.

Pechmann, J. H. K., Scott, D. E., Whitfield, J., & Semlitsch, R. D. (1989). Influence of wetland hydroperiod on diversity and abundance of metamorphosing juvenile amphibians. *Wetlands Ecology and Management, 1*: 3–11.

Penfound, W. T., & Hathaway, E. S. (1938). Plant communities in the marshlands of southeastern Louisiana. *Ecological Monographs, 8*: 1–56.

Pielou, E. C. (1977). *Mathematical ecology.* New York: Wiley.

Postel, S. L., Daily, G. C., & Ehrlich, P. R. (1996). Human appropriation of renewable fresh water. *Science, 271*: 785–788.

Puerto, A., Rico, M., Matias, M. D., & Garcia, J. A. (1990). Variation in structure and diversity in Mediterranean grasslands related to trophic status and grazing intensity. *Journal of Vegetation Science, 1*, 445–452.

Quinlan, C., & Mulamoottil, G. (1987). The effects of water level fluctuation on three Lake Ontario shoreline marshes. *Canadian Water Resources Journal, 12*, 64–77.

Rejmankova, E., Pope, K. O., Pohl, M. D., & Rey-Benayas, J. M. (1995). Freshwater wetland plant communities of northern Belize: implications for paleoecological studies of Maya wetland agriculture. *Biotropica, 27*, 28–35.

Reznicek, A. A., & Catling, P. M. (1989). Flora of Long Point. *Michigan Botanist, 28*, 99–175.

Rosenzweig, M. L., & Abramsky, Z. (1993). How are diversity and productivity related? In: R. E. Ricklefs, & D. Schluter (Eds.), *Species diversity in ecological communities* (pp. 52–65). Chicago: The University of Chicago Press.

Salo, J., Kalliola, R., Hakkinen, I., Makinen, Y., Niemela, P., Puhakka, M., & Coley, P. D. (1986). River dynamics and the diversity of Amazon lowland forest. *Nature, 322*, 254–258.

Scharf, F. S., Juanes, F., & Sutherland, M. (1998). Inferring ecological relationships from the edges of scatter diagrams: Comparison of regression techniques. *Ecology*, *79*, 448–460.

Smith, P. G. R., Glooschenko, V., & Hagen, D. A. (1991). Coastal wetlands of three Canadian Great Lakes: inventory, current conservation initiatives, and patterns of variation. *Canadian Journal of Fisheries and Aquatic Sciences*, *48*, 1581–1594.

Spence, D. H. N. (1964). The macrophytic vegetation of freshwater lochs, swamps and associated fens. In: J. H. Burnett (ed.), *The Vegetation of Scotland* (pp. 306–425). Edinburgh: Oliver & Boyd.

Strahler, A. N. (1971). *The earth sciences* (2nd ed.). New York: Harper & Row.

Tilman, D. & Pacala, S. (1993). The maintenance of species richness in plant communities. In: R. E. Ricklefs & D. Schluter (Eds.), *Species diversity in ecological communities* (pp. 13–25). Chicago: University of Chicago Press.

Verhoeven, J. T. A., Kemmers, R. H., & Koerselman, W. (1993). Nutrient enrichment of freshwater wetlands. In: C. C. Vos, & P. Opdam (Eds.), *Landscape ecology of a stressed environment* (pp. 33–59). London: Chapman & Hall.

Verhoeven, J. T. A., Koerselman, W., & Meuleman, A. F. M. (1996). Nitrogen- or phosphorus-limited growth in herbaceous, wet vegetation: Relations with atmospheric inputs and management regimes. *Trends in Ecology and Evolution*, *11*: 493–497.

Vermeer, H. J. G. (1986). The effect of nutrients on shoot biomass and species composition of wetland and hayfield communities. *Acta Oecologica/Oecologia Plantarum*, *7*, 31–41.

Vermeer, J. G., & Verhoeven, J. T. A. (1987). Species composition and biomass production of mesotrophic fens in relation to the nutrient status of the organic soil. *Acta Oecologica/Oecologia Plantarum*, *8*, 321–330.

Vince, S. W., & Snow, A. A. (1984). Plant zonation in an Alaskan salt marsh. *Journal of Ecology*, *72*, 651–667.

Vitt, D. H. (1990). Growth and production dynamics of boreal mosses over climatic, chemical and topographic gradients. *Botanical Journal of the Linnean Society*, *104*, 35–59.

Weiher, E., & Keddy, P. A. (1995). The assembly of experimental wetland plant communities. *Oikos*, *73*: 323–335.

Weiher, E., Wisheu, I. C., Keddy, P. A., & Moore, D. R. J. (1996). Establishment, persistence, and management implications of experimental wetland plant communities. *Wetlands*, *16*, 208–218.

Weller, M. W. (1994). *Freshwater Marshes: Ecology and Wildlife Management*. 3rd edn. Minneapolis: University of Minnesota.

Wheeler, B. D., & Giller, K. E. (1982). Species richness of herbaceous fen vegetation in Broadland, Morfolk in relation to the quantity of above-ground plant material. *Journal of Ecology*, *70*, 179–200.

Wheeler, B. D., & Shaw, S. C. (1991). Above-ground crop mass and species richness of the principal types of herbaceous rich-fen vegetation of lowland England and Wales. *Journal of Ecology*, *79*, 285–301.

White, D. A. 1983. Plant communities of the lower Pearl River basin, Louisiana. *American Midland Naturalist*, *110*: 381–396.

Willis, A. J. (1963). Braunton Burrows: The effects on the vegetation of the addition of mineral nutrients to the dune soils. *Journal of Ecology*, *51*, 353–374.

Wilson, S. D., & Keddy, P. A. (1986a). Species competitive ability and position along a natural stress/disturbance gradient. *Ecology*, *67*: 1236–1242.

Wilson, S. D., & Keddy, P. A. (1986b). Measuring diffuse competition along an environmental gradient: Results from a shoreline plant community. *American Naturalist*, *127*, 862–869.

Wilson, S. D., & Keddy, P. A. (1988). Species richness, survivorship, and biomass accumulation along an environmental gradient. *Oikos*, *53*, 375–380.

Wisheu, I. C., & Keddy, P. A. (1989). The conservation and management of a threatened coastal plain plant community in eastern North America (Nova Scotia, Canada). *Biological Conservation*, *48*, 229–238.

Wisheu, I. C., & Keddy, P. A. (1992). Competition and centrifugal organization of plant communities: Theory and tests. *Journal of Vegetation Science*, *3*, 147–156.

Wisheu, I. C., & Keddy, P. A. (1996). Three competing models for predicting the size of species pools: A test using eastern North American wetlands. *Oikos*, *76*, 253–258.

Wisheu, I. C., Keddy, P. A., Moore, D. J., McCanny, S. J., & Gaudet, C. L. (1991). Effects of eutrophication on wetland vegetation. In: J. Kuslor, & R. Smardon (Eds.), *Wetlands of the Great Lakes: Protection and restoration policies, Status of the science* (pp. 112–121). New York: Managers Inc.

Yabe, K. (1993). Wetlands of Hokkaido. In S. Higashi, A. Osawa, & K. Kanagawa (Eds.), *Biodiversity and ecology in the northernmost Japan* (pp. 38–49). Hokkaido University Press.

Yabe, K., & Onimaru, K. (1997). Key variables controlling the vegetation of a cool–temperate mire in northern Japan. *Journal of Vegetation Science*, *8*, 29–36.

Yoda, K., Kira, T., Ogawa, H., & Hozumi, K. (1963). Self-thinning in overcrowded pure stands under cultivated and natural conditions. *Journal of Biology Osaka City University, 14*, 107–129.

Zobel, M. (1997). The relative role of species pools in determining plant species richness: An alternative explanation of species coexistence. *Trends in Ecology and Evolution, 12*, 266–269.

Zobel, K., & Liira, J. A. (1997). A scale–independent approach to the richness vs biomass relationship in ground-layer plant communities. *Oikos, 80*, 325–332.

3

Use of Plants for Monitoring Heavy Metals in Freshwaters

Brian A. Whitton

Introduction

There have been observations on the effects of heavy metals on aquatic plants for three-quarters of a century, but awareness of their potential for monitoring received a boost when Whitehead and Brooks (1969) reported on the use of stream bryophytes in New Zealand to locate uranium enrichment. Prior to this, there had of course been studies on metal pollution and plant communities in the field and the laboratory, though interpretation of field results was often complicated by the simultaneous presence of organic wastes or phosphate-rich water. Early experimental studies dealt with toxicity or with the concentrations of metals in aquatic plants such as *Cladophora*, often to assess whether these plants could be used to remove contamination. Government-sponsored surveys on *in situ* levels of radionuclides in aquatic bryophytes were carried out in several countries during the 1960s, but such studies were apparently always treated as restricted information. The early literature is not listed here, because it was reviewed nearer the time by the author (Whitton, 1970b, 1980). In addition to aquatic studies, there are many reports on the use of terrestrial plants for monitoring heavy metals, especially mosses in relation to atmospheric pollution (see Onianwa, 2001). Aquatic studies have sometimes been carried out without much awareness of the literature on terrestrial studies, and the converse is even more true.

Since the survey by Whitehead and Brooks (1969), there have been numerous studies on heavy metals and aquatic plants, many dealing entirely or in part with monitoring. (For convenience, all phototrophs are termed plants here.) A variety of methods have been proposed and assessed using field and/or laboratory data, and sometimes put to practical use for broad surveys or the investigation of point–source contamination. The aims of such studies have, however, not always been expressed clearly, leaving the reader to guess what the researcher expected to achieve. Most frequently, it is the ability of plants to provide an

---

Brian A. Whitton • School of Biological and Biomedical Sciences, University of Durham, Durham DH1 3LE, United Kingdom.

integrated response to exposure that is important in monitoring, but other situations require assessment of variability or the maximum levels in the environment.

Some of the more recent papers provide fresh insight into methodology, whereas others do little more than report on routine monitoring at a new site. However, before reading about individual accounts, the reader is encouraged to consider a number of general points, because the interpretation of the results in published studies often seems open to question (at least to me). This is a subject in which it is especially important to read the full account and not just rely on the abstract.

Reports on monitoring differ a lot in the amount of background information on sites and organisms. However, without adequate information, it is hard to select the best method or interpret the data. Has metal contamination already started and, if so, for how long? Is contamination due to recent human activity, or has it extended for periods on the geological time scale? In the case of rivers, are there hot spots of contamination within the catchment that might provide periodic inoculation of highly metal-tolerant strains to a downstream site? Is the site subject to wide variation in aqueous metal levels and/or other features of the water? What is the relationship between metal levels in the water and those in the sediments? Such questions seem obvious, yet are often ignored.

There are many features of the plants themselves that influence their suitability for monitoring or subsequent interpretation of data. The most obvious is whether the plant is exposed only to the water or also to sediments. Contact with sediments is clear-cut in most rooted plants, although some species also have roots exposed to the water column. Algal and bryophyte species differ in the extent to which they are in contact with both water and sediment, but organisms used for monitoring have mostly been taken from microhabitats where contact is largely with the water. Other features of plants that influence the information obtained include the growth rate, the length of the growth cycle, and whether there are growth phases especially sensitive or resistant to heavy metals.

It is also important to know whether the plants are living or dead. In the case of field results, this applies especially at sites with highly variable levels of metals, including the assessment of a catastrophic event such as a mine spill. Laboratory studies on metal binding by aquatic plants have frequently used levels of metals that are high compared with those likely to be encountered in nature, yet all too often, there is no mention of the condition of the organism. Dead or highly stressed material can sometimes be used for monitoring, but information from plants in the process of dying is difficult to interpret.

Finally, it should be pointed out that some accounts are enthusiastic about the successes of using aquatic plants in monitoring heavy metals, whereas others are less convinced. The purpose of this chapter is to put the diverse literature into perspective.

Metal Accumulation

Relationship between Plant and Aqueous Environments

Numerous studies have reported that aquatic plants often accumulate heavy metals in concentrations much higher than in their aqueous environment, even when those metals are not essential for metabolism or are potentially toxic. Studies in the 1960s and 1970s emphasized the value of analyzing plants rather than water for monitoring purposes, because of the increased sensitivity of detection, but this has become less important as analytical methods have become more sensitive and the necessary equipment more widely available. Other considerations have come to the fore, such as the fact that the metal accumulated by a plant gives a better indication of the fraction of the metal in the environment likely to affect an aquatic ecosystem than do most types of direct chemical analysis (Empain et al., 1980).

The simplest way to use information on metal concentrations for monitoring is to compare values for specimens of a particular species at different sites, such as in a river downstream from an effluent or at a range of sites within a particular catchment or geographical region. The moss *Rhynchostegium riparioides* was used, for instance, to produce striking color maps of metal contamination in rivers of the Meuse catchment (Descy et al., 1981). Usually, such an approach has been conducted at the same time as a survey of water or sediment chemistry, but the biological information tends to be especially convincing to those involved in water management. Since the study by Whitehead and Brooks (1969), bryophytes have been used a number of times in geochemical surveys and prospecting (e.g., Schacklette & Erdman, 1982).

A more demanding but potentially much more useful approach is to establish a database about accumulation in a species independently from the use of that species to monitor contamination in one particular study. Any further sample, such as stored plant material lacking information about ambient water chemistry at the time of sampling, can then be compared with the database and the probability estimated that the plant was taken from water within a certain metal range. So far, this has only been done for a few species of river algae and bryophytes (see Kelly & Whitton, 1989a). Where such a database has been established for particular metals and a species solely in contact with the water (e.g., *Cladophora glomerata*; Whitton et al., 1989), there is a linear relationship between the logarithm of metal concentration in the plant and that in the water (Figures 3.1 and 3.2). At least in the case of aquatic mosses, the relationship is higher if tissue analysis is confined to the apical part of the shoot rather than the whole plant (e.g., *Rhynchostegium riparioides*;

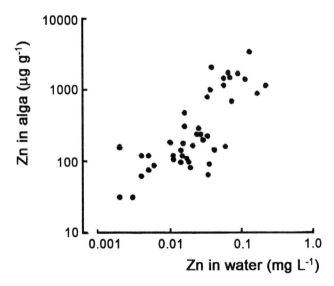

Figure 3.1. Zn concentration in *Cladophora glomerata* versus Zn concentration in water (passing through 0.2 μm filter) at 60 sites in northern England. Redrawn from Whitton et al., 1989, with permission of Kluwer Academic Publishers.

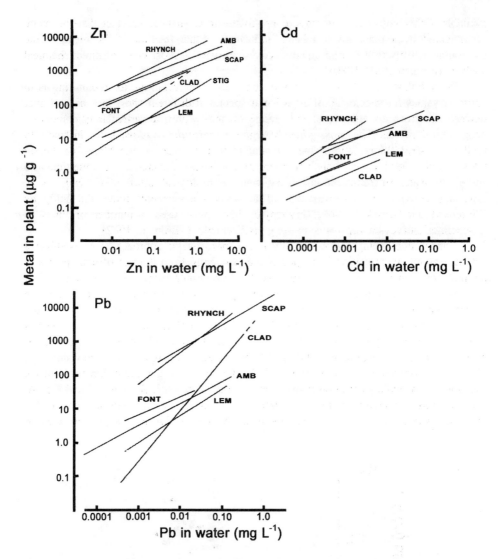

Figure 3.2. Relationships between metal concentrations in plants and water (passing through 0.2 μm filter). AMB, *Amblystegium riparium*; CLAD, *Cladophora glomerata*; FONT, *Fontinalis antipyretica*; LEM, *Lemanea fluviatilis*; RHYNCH, *Rhynchostegium riparioides*; SCAP, *Scapania undulata*; STIG, *Stigeoclonium*. Redrawn from Kelly and Whitton, 1989a, with permission of Kluwer Academic Publishers.

Wehr & Whitton, 1983b), even though metal concentrations are almost always higher in the older part of the shoot (Wehr et al., 1983; Siebert et al., 1996). The extent to which metals can move from older to younger parts of moss plants, or vice versa, has received some attention in terrestrial mosses (e.g., Brown & Brümelis, 1996) but apparently not in aquatic mosses.

As there is usually a broad scatter of points around the regression line relating metal in an aquatic alga or moss to that in its environment, it is important to obtain a relatively large database from as chemically diverse a range of environments as tolerated by that species. In order to provide a database for routine monitoring of rivers, ideally there should be at least 100 sites, as far as possible, each from a different river. A comparison of the bivariate relations between Zn, Cd, and Pb in plant and water with large databases for three algae, one liverwort and three mosses (Figure 3.2) sampled from rivers in northwestern Europe (though mainly the United Kingdom) showed marked differences in this relationship for both metal and species (Kelly & Whitton, 1989a). The data were based on analysis of the youngest part of the plant, in the case of mosses, the apical 2 cm of shoot.

These studies showed a significant positive relationship between the concentrations of the metal in the plant and the water in almost every case. However, the bryophytes generally accumulated the metal in a much higher concentration than the algae. With an ambient Zn concentration of 0.01 mg L^{-1}, the difference in Zn concentration between the bryophyte with the highest value (*R. riparioides*) and the alga with the lowest value (*Stigeoclonium*) was almost two orders of magnitude. There was a similar contrast at 0.001 mg L^{-1} Pb between the liverwort *Scapania undulata* and *C. glomerata*. However, the slope was in each case much steeper for the alga, indicating a proportionately greater change in tissue concentration in response to change in ambient concentration. *Cladophora* is therefore especially useful for detecting changes in heavy metal concentration (especially Pb), such as might occur in a short-term pollution event.

A study in northwestern Spain (López & Carballeira, 1993) was based on three of the same species as reported by Kelly and Whitton (1989a), *Scapania undulata*, *Fontinalis antipyretica*, and *R. riparioides*, together with two other mosses, *Fissidens polyphyllus* and *Brachythecium rivulare*. The 2-cm apical parts of shoots were used for analysis. There was a statistically significant bivariate relationship for most metals and species, although the value was usually low. The Spanish study agreed with the UK study that found a high accumulation of Cd and Pb by *Scapania undulata*, but differed for Zn. In the Spanish study, this liverwort also accumulated Zn to the highest level, whereas Kelly and Whitton reported considerably higher accumulation in *Rhynchostegium* (Figure 3.2). Among the bryophytes, *Rhynchostegium* tended to show the widest range of tissue metal concentrations for a particular range of aqueous metal concentrations; in the Spanish study, this was a feature shared with *Brachythecium rivulare*.

Samecka-Cymerman et al. (1991) introduced the concept of a background level of any metal in a species, using *Scapania undulata* as an example, while Carballeira and López (1997) tried to test this on a more rigorous basis by comparing four different methods (see section on pigment ratios). However, the concept seems difficult to interpret. There is presumably a minimum level of any heavy metal essential for metabolism, but otherwise, the level in the plant may be expected to vary according to the environment. Carballeira and López found the highest "background" levels of almost all the nine metals studied to be higher in *Scapania undulata* and *Rhynchostegium riparioides* than in *Fontinalis antipyretica* and *F. polyphyllus*, corresponding to the results for these three species included in Figure 3.2.

Although a number of studies on the monitoring of mercury pollution include data on accumulation by aquatic plants, in most cases, these have not added much more insight than that available from analysis of water and sediments. However, a study by Stokes et al. (1983) on 11 lakes in Ontario, Canada, shows the potential for developing the methodology. Using filamentous green algae grown on artificial substrates to monitor low-level methyl mercury contamination, the authors found that intrasite variation in the mercury content of the algae was low, whereas intersite differences were significant. This mercury was almost entirely methyl mercury, whereas algae from natural substrates showed more variation in total mercury content and a lower proportion of this as methyl mercury. The mercury content of the algae on artificial substrates showed no relationship to mercury in water or sediments but did show a significant correlation with mercury in perch. The authors suggested that the uptake of mercury by both algae and fish may be primarily determined by the availability of methyl mercury in the water.

There are no reports of aquatic plants acting as hyperaccumulators of one particular metal, as known for a number of terrestrial plants. However, it might be argued that the typically much higher level of metal accumulated by bryophytes than algae is a form of hyperaccumulation. Even if this is simply an inevitable effect of bryophyte metabolism and wall structure, it still means that potential grazers have to withstand these high levels of metals.

Influence of Environment on Metal Accumulation

Among the possible reasons for differences in the relationship between the concentration of a metal in a plant and that in the water is the influence of other environmental factors. In the case of a database for field results, each set of values is usually based on material collected at one time, so the concentration in the plant reflects some sort of integrated response to past events in the environment, whereas that in the water is simply a spot measurement. This is a particular problem when a species occurs in habitats that vary markedly with respect to a factor influencing metal uptake, such as is often the case with *S. undulata* and pH (Whitton et al., 1982; Vincent et al., 2001). As other environmental factors may influence the uptake of a particular metal, this affects the concentration in the plant at the time of sampling for both the database and for routine monitoring.

The problem of assessing the influence of other environmental factors on accumulation of a particular metal can be partially overcome by applying multivariate approaches to the database, such as multiple stepwise regression, as has been done for *F. antipyretica* (Say & Whitton, 1983) and *R. riparioides* (Wehr & Whitton, 1983b). Such an approach is optimized when values for variables in both the environment and the plant tissue are included in the statistical analysis. This means analyzing the plant tissue not only for the heavy metal being monitored but also other elements, especially those known to influence accumulation of the metal. In the case of the databases brought together by Kelly and Whitton (1989a), aqueous Ca is the principal non-heavy-metal factor reducing the accumulation of Zn and Cd, though not Pb, in plant tissue. Aqueous phosphate markedly reduces Zn and Pb, but not Cd, accumulation in *Rhynchostegium* (Wehr & Whitton, 1983b). This general approach is more reliable when experimental studies have confirmed the influence of those factors suggested by statistical analysis. In the case of the alga *Stigeoclonium*, a combination of experimental and statistical analysis showed (Kelly & Whitton, 1989b) that Fe and Ca and/or Mg were the most important factors influencing both Zn toxicity and accumulation in populations taken from field sites with a wide range of aqueous Zn ($0.002-2$ mg L^{-1}).

The ambient pH is an important factor influencing the accumulation of almost all metals, but unfortunately, it is difficult to assess this just from the plant sample alone. Metal accumulation is in most cases higher at higher pH values, as shown for *F. antipyretica* (Vázquez et al., 2000). A combined field and laboratory study of *S. undulata* showed that Cu, Zn, Cd, and Pb, but not Al, all showed increased binding at higher pH (Vincent et al., 2001). In the case of Zn and Cd, it was suggested that this was due to decreased proton competition for surface sites and probably also less Al competition. Although decreased proton competition applied also to Cu, in this case, the effect on accumulation was reduced because of a marked decrease in metal aquo ion concentration with increased pH due to complexation with fulvic and carbonate species. Due to the problems of ion competition, solution speciation, and pH effects, the authors concluded that bryophytes are not especially good chemical analytical devices, though they are useful indicators of bioavailability. In view of the many previous studies in which bryophytes have successfully monitored heavy metals, their conclusion seems somewhat biased. Factors that may have led to this include the facts that they studied one species from streams where fulvic acids played an especially important role and that this species occurs across a wide pH range. However, their comments do emphasize the need to take care in planning surveys of rivers and streams with a high fulvic acid content, or where there are wide temporal or spatial differences in pH.

Uptake and Loss

Aquatic plants seldom exist under steady-state conditions with respect to ambient heavy metal levels or the environmental factors likely to influence their accumulation. Even if contamination arises from a relatively constant source, such as drainage from a deep mine where the annual temperature range is low, there is still the seasonal change in light flux influencing growth rate (Patterson & Whitton, 1981; Wehr & Whitton, 1983c). However, a number of studies on metal uptake and loss have helped to interpret values for metal analysis obtained from sites with highly variable metal concentrations. Both field transplant (see sections on genetic aspects and procedures in the field) and laboratory experiments have been used.

Uptake and loss of six heavy metals were compared by López et al. (1994) in *F. antipyretica* transplants. Uptake over a 28-day period was predicted well by a two-compartment kinetic model: Uptake velocity was initially high and gradually declined. Mean uptake rate, time to reach equilibrium, and metal concentration in the moss at equilibrium all increased with increasing metal concentration in the water. Bioconcentration constants were in the order Zn < Ni < Co < Cu < Pb < Cd. When the metal-enriched moss was returned to a clean site, it was possible to distinguish two phases of loss: initially rapid and subsequently slow. Uptake and release rate constants for copper and the same moss were reported by Gonçalves and Boaventura (1998). The test concentration ranged from 0.09 to 0.75 mg L^{-1} Cu, so it is unclear whether the results were influenced by toxicity at the higher concentrations. Further study by the same research group (Vázquez et al., 1999) with high concentrations (1–200 mg L^{-1}) of the same six metals with shoot tips of *Fontinalis antipyretica*, *Scapania undulata*, and *Fissidens polyphyllus* showed that there were considerable differences in the cation-binding capacity of extracellular sites, being high in *Scapania* and relatively low in *Fissidens*. Uptake of heavy metals led to considerable losses of K, especially in *Scapania*, probably due to effects on the plasma membrane.

Quantifying the loss of Cd from *Fontinalis antipyretica* was proposed by Carballeira et al. (2001) as a way to monitor intermittent acidification of rivers, though their study to assess this was restricted to the laboratory. Cd-free moss was incubated in Cd solutions such that saturation conditions were obtained in the extracellular (presumably cell wall) compartment and near-saturation concentrations in the extracellular compartment. The concentration used to load the plant with metal (up to 100 mg L^{-1} Cd) was sufficient high that, presumably, the plant was killed. This seems an interesting approach to follow further, but with much lower concentrations of metal to preload the moss.

Genetic Aspects

Another factor that may influence the information in a database is the extent to which differences exist between populations in their ability to accumulate a metal. Such differences are most likely in species with a marked ability to develop genetically tolerant strains, as known for the algae *Klebormidium* (formerly *Hormidium*; Say et al., 1977), *Stigeoclonium* (Harding & Whitton, 1976), *Microthamnion* (Whitton, 1980), and *Mougeotia* (Patterson & Whitton, 1982), all of which can occur at much higher concentrations of Zn, Cd, and Pb than tolerated by nonadapted strains. In the case of *Stigeoclonium*, the more tolerant the strain is to Zn, the greater is the relative influence of Mg, as opposed to Ca, in reducing Zn toxicity (Harding & Whitton, 1977). Unfortunately, there appear to have been no studies to compare accumulation by tolerant and sensitive strains over a range of Zn concentrations, which would establish whether or not the tolerant strains have acquired the ability to reduce Zn accumulation.

Little is known about the extent to which populations of otherwise widely distributed bryophytes occurring at sites with high concentrations of a heavy metal are ones that have undergone genetic changes in response to that metal. Unlike the situation with most filamentous green algae, transplant experiments with *Rhynchostegium riparioides* attached to boulders suggest that populations from sites free of Zn enrichment can tolerate a considerably higher concentration of ambient Zn than that occurring at their original site (Whitton, unpublished data). This is apparently also the situation with other bryophytes that occur over a wide range of Zn concentrations. However, this does not rule out the possibility of a population becoming adapted to the metal in some way that increases the success of the population at a high metal concentration. For instance, one possible explanation for the lower level of Zn, but not Cd and Pb, accumulation in U.K. than Spanish populations of *Scapania undulata* (see above) is that some U.K. populations from sites likely to have been contaminated by Zn for at least six centuries may have acquired an ability to reduce Zn uptake.

Rooted Plants

Higher plants pose problems for monitoring metals, because they usually have roots in contact with sediments. Based on both large-scale empirical studies and direct measurements, it is generally concluded that sediments are the source of most metals for rooted aquatic plants (e.g., Jackson, 1998). Jackson assembled data to show that the concentration of an element in a rooted macrophyte is usually proportional to the metal's concentration in the underlying sediments. However, plants such as *Ceratophyllum*, *Elodea* (e.g., *E. nuttallii*; Nakada et al., 1979), and the various genera of duckweed, which have few, if any, roots penetrating the sediments, must obtain most of their metal content from the water column. It has also been suggested that the water column is the main source at sites where metal levels in the water column are high, but aqueous levels in the sediments are low (Denny, 1980; Guilizzoni, 1991).

Interpretation of data from rooted plants is complicated, because metals are transported between various parts of the plant, and this may vary seasonally. Higher plants have therefore been used in monitoring studies largely at sites where they are easier to collect than filamentous algae or bryophytes, or where sediments are more likely to be enriched in heavy metals than the water. However, there are a number of reports of monitoring with both predominantly submerged rooted species (e.g., *Myriophyllum exalbescens*; Franzin &

McFarlane, 1980) and predominantly emergent species (e.g., wetland *Equisetum arvense*; Ray & White, 1979).

Practical Matters

Procedures in the Field

The simplest way to obtain plant material for analysis subsequent to its exposure to a particular environment is to sample an *in situ* population. This was the approach adopted in the majority of studies listed in Table 3.1. However, if the species is absent or there is a need to quantify metal contamination over a defined period, then transplants may be useful. This has usually been done with mosses attached to boulders (Mersch & Johansson, 1993; Mouvet, 1985). There have also been several attempts to use this approach with *Cladophora* attached to boulders, but this alga is especially sensitive to heavy metals (Whitton, 1970c), and it is difficult to interpret the data if the alga has been subjected to toxic concentrations of metal.

If suitable material of moss attached to movable boulders is not available for transplant, a modification of the approach is to use detached shoots in a mesh bag. This worked well in U.K. rivers using *Fontinalis antipyretica* or *Rhynchostegium riparioides* (Kelly et al., 1987). Pieces of plant were put in a bag with 0.7 or 0.9 mesh cm^{-1}, taking care to ensure that the shoots were packed loosely, and the bag was staked in the river such that it was suspended in a fast-flowing stretch. Brightly colored supermarket bags for oranges proved to be a temptation for passersby, so they were replaced by less colorful ones. After the bag had been recovered from the river, the 2-cm apical parts of the shoots were removed for analysis. In one series of experiments done in summer, shoots of

Table 3.1. Examples of Aquatic Studies Using Accumulation to Monitor Metal Contamination

Country	Location	Taxon	Reference
Belgium	R. Amblève	7	Mouvet, 1985
	R. Meuse, Sambre	4, 6, 7, 12	Empain, 1976b
Canada	Ontario lakes	14	
France	R. Cance, Bienne	7, 12	Mouvet, 1985
	Jura streams	2, 7, 12	Mouvet et al., 1986
	Somme	7, 12	Empain, 1976a
	Rhône tributaries	1, 2, 3, 4, 7, 12	André & Lascombe, 1987
	Moselle	7	Mersch & Kass, 1994
Germanuy	R. Elbe	7	Siebert et al., 1996
Greece	Macedonian rivers	5, 14	Sawidis et al., 1995
			Sawidis, 1996
Portugal	Ave Basin (Selho, Este)	7	Gonçalves et al., 1992
	Cávado Basin	7	Gonçalves et al., 1994
UK	Welsh streams	8, 13	Burton & Peterson, 1979
	Derwent Reservoir	11	Harding & Whitton, 1978
	Northeast England rivers	9	Harding & Whitton, 1981b
	Mine drainage	10	Patterson & Whitton, 1981
	R. Etherow	8, 12	Say et al., 1981
	R. Roding	5	McHardy & George, 1985

Key to taxa: 1, *Amblystegium riparium*; 2, *Cinclidotus aquaticus*; 3, *Cinclidotus danubicus*; 4, *Cinclidotus nigricans*; 5, *Cladophora glomerata*; 6, *Fissidens crassipes*; 7, *Fontinalis antipyretica*; 8, *Fontinalis squamosa*; 9, *Lemanea fluviatilis*; 10, *Mougeotia*; 11, *Nitella flexilis*; 12, *Rhynchostegium riparioides*; 13, *Solenostoma crenulata*; 14, other species.

R. riparioides transplanted between rivers from 0.01 to 0.3 mg L^{-1} Zn reached a new saturation concentration (ca. 2000 µg g^{-1} Zn) after about 6 hours, whereas it took three times as long to reach this concentration in winter. (All tissue concentrations in this review refer to dry weight.) Providing that care was taken in placing the moss into the mesh bag, the rate of Zn uptake by *Rhynchostegium* in the bag was the same as that by moss still attached to a boulder.

Mesh bags have also used in other studies on *F. antipyretica*, including the kinetics of metal uptake and loss (López et al., 1994) and a comparison of different types of response by plants transplanted to polluted sites on the Elbe River, Germany (Bruns et al., 1997).

Choice of Organism

Some of the features influencing the choice of organism have already been mentioned: the presence or absence of roots, ease of carrying out transplant experiments, enrichment ratio for a particular metal, and steepness of the regression line relating metal in plant and water. Epiphytes are especially frequent on *Cladophora* and *Myriophyllum*, making preparation of samples and interpretation of data more difficult. Bryophytes are in general more robust than algae at every sampling and preparation stage. If anything more than spot measurements is required, then it is important to have detailed information about accumulation of a particular metal in a particular species (Figure 3.2). This means that monitoring should, as far as possible, be confined to species such as *Fontinalis antipyretica* and *Rhynchostegium riparioides*, for which there is already a lot known, or to species that have sufficient monitoring potential to make the necessary background research worthwhile.

Ten species of aquatic plant were recommended for monitoring heavy metals in rivers in a standard methods book for the United Kingdom (Whitton et al., 1991): 4 algae, 1 liverwort, 3 mosses, 2 flowering plants. One of the flowering plants, *Potamogeton pectinatus*, grows rooted in sediments but occurs in downstream stretches of river where other species suitable for monitoring metals may be absent. In addition, it is one of the most widespread submerged vascular plants in the world, growing typically, but not always, in nutrient-rich rivers, so further data on metal accumulation by this species would be of value in many countries.

Several species of duckweed have been used in studies of heavy metals, mostly various types of bioassay or their potential for stripping metals from contaminated water, but also a few studies on concentrations in field populations. Prasad et al. (2001), who studied some physiological and biochemical effects of Cu and Cd accumulation by *Lemna trisulca*, provide a recent guide to the literature.

Most studies on metal accumulation have been conducted in temperate regions, and the only taxa that have received much study in the subtropics and tropics are the floating-leaved vascular plants, *Salvinia*, *Pistia stratiotes*, *Eichhornia crassipes*, and several species of duckweed. Although some researchers of these organisms mention monitoring, most have focused on their potential for metal removal, especially *Eichhornia* (Maine et al., 1999) and *Lemna* (Jain et al., 1989). However, in a comparison of Cd uptake by four species, Maine et al. (2001) concluded that *Pistia stratiotes* was the most useful, retaining the ability to accumulate Cd even after toxicity symptoms appeared. The sparse occurrence of bryophytes in most lowland tropical rivers means that they can seldom be used for sampling.

Preparation of Samples for Analysis

The adoption of metal analysis in plant tissue as a routine approach for monitoring requires that practical methods be standardized as much as possible at all stages. Marked differences exist in the older literature concerning the methods for sampling, treatment at the time of collection, pretreatment in the laboratory, and analysis. The influence of such differences was evaluated practically for *Fontinalis antipyretica* and *Rhynchostegium riparioides* by Wehr et al. (1983). Although these authors concluded that no single sequence of methods is ideal for all purposes, nevertheless the procedure in Table 3.2 is suitable for routine monitoring with these two mosses and probably also other submerged bryophytes.

The most important difference in procedure between various studies is whether or not the samples have been air-dried or kept fresh until the time they are washed. Air-dried samples of whole plants showed higher levels of all metals tested than those kept fresh (Wehr et al., 1983). Almost certainly this is because it is more difficult to wash the shoots thoroughly after they have been dried and rewetted, so the fresh material probably provides a better indication of metals linked to cell wall materials or accumulated inside cells, rather than loosely attached to particles at the surface. Material should therefore only be air-dried at the time of sampling if there is likely to be a long period before it can be washed and prepared for analysis. If information from air-dried samples is especially important, such as tests on old herbarium samples, then it is necessary to conduct studies on differences between metal uptake by air-dried and fresh material of that particular species.

Another difference in procedure that can have a marked influence on the results is the method of washing. Some authors have used automated procedures, whereas others have done washing by hand, using forceps and a series of containers with distilled water. Wehr et al. (1983) found that automated procedures led to lower values than hand washing for most elements in *R. riparioides*, but only for Fe and Pb in *Fontinalis antipyretica*. Although there is no firm evidence to decide which procedure is "correct," it seems possible that *Rhynchostegium* may suffer mechanical damage with automated washing, so washing by hand is recommended.

Provided material has been dried to steady state, the choice of temperature is less important, because it is simple to prepare a curve relating mass to temperature for each species and thus permit a fairly reliable conversion between results obtained with different drying temperatures. A value of 105°C has been used in the majority of studies, because it is easy to reach a constant value at this temperature, and there is a neglible change in mass with a further 10°C rise in temperature; it is also the temperature used most widely in the general ecological, though not the agricultural, literature.

Table 3.2. Key Steps in Preparing Material of Submerged Moss for Heavy Metal Analysis

A. If the survey is on a river, collect whole shoots from a fast-flowing stretch; throw away any heavily encrusted parts.
B. Wash material thoroughly in water from the sites.
C. Store moist material loosely in a plastic bottle, but without free water; keep in cool box or refrigerator.
D. Prepare material for analysis as soon as possible, but certainly within 1 day. Remove apical 2 cm of shoots and transfer to petri dish with distilled water.
E. Wash samples thoroughly, throwing away any shoots that are difficult to clean well due to encrustations.
F. Dry to constant weight at 105°C.
G. Typically, 3–10 shoot apices are required to provide sufficient material for accurate measurement of mass.

Further evaluation of methods for preparation and analysis of aquatic bryophytes, together with statistical analysis of the data, are given by Mouvet (1991).

Use of Dried Material

Although it is recommended to keep materials fresh until the final washing stages prior to drying, there are situations in which samples that have already been dried need to be analyzed. Reference material of *R. riparioides* supplied as a powder and with a certified content of a number of metals has been available at various times from several sources (e.g., BCR 61; see Commission of the European Communities, 1982). Material on herbarium sheets may provide information on the former condition of a river, as done with the moss *Cinclidotus fontinaloides* from the River Wear, England, collected in about the year 1800. Comparison with the same species collected at the same site in 1979 suggested that the levels of Zn and Pb in this river were at least twice as high in 1800 than at the later date (Wehr & Whitton, unpublished data), which fits with the known history of lead mining in the catchment. This species grows in intermittently submerged parts of rivers, so it can often be sampled dry. It has considerable potential for monitoring purposes but first requires study to establish the relationship between metal in the moss and its aquatic environment, and also whether loss of metal is quantitatively important when the moss is not submerged in river water. Another study on herbarium material is that reported by Sergio et al. (1992) on *Sphagnum auriculatum* growing near the unpolluted source of the River Ave, Portugal, with a comparison of the heavy metal contents of materials collected in 1924, 1981, and 1990.

Cell Location of Metal Accumulation

The cell sites where metals are accumulated in aquatic plants have been studied with a range of physical and chemical techniques. Only those species important in monitoring are discussed here. Such information is especially helpful when planning and interpreting the results of transplant experiments.

Although there have been many studies on the localization of heavy metals in algal cells, these have seldom been developed to the stage where they contribute much to monitoring. According to Almeida et al. (1999), M. J. Pereira has tested a combination of scanning electron microscopy and energy dispersive spectrometry with Euglenophyta and Chlorococcales for this purpose. However, information on metal localization is more useful with bryophytes, because it can help to interpret data on accumulation.

A number of authors have suggested the importance of surface deposits on aquatic bryophytes, where these are obvious Mn- and or Fe-rich crusts. The most detailed study is that by Sergio et al. (2000), who used electron microscopy on *Fontinalis antipyretica* transplanted to a stream in Portugal contaminated by mine effluent; the moss showed localized deposits on the surface of live cells but not dead ones. These deposits were strongly enriched with Mn but also included heavy metals. It was suggested that Mn deposition might be caused by oxygen released by photosynthesis. If this is the case and such deposition enhances binding by heavy metals, then passive as well as active accumulation may be influenced by light, a factor that needs to be considered in interpreting metal values in plants influenced by short-term pollution events.

The cell wall appears to be the most important site for metal binding in most bryophytes, but there are considerable differences in the extent to which the metals can be exchanged with other metals or H⁺ or other cations in their environment. In the case of Cu and *Rhynchostegium riparioides*, Mouvet and Claveri (1999) used a chemical approach to separate the copper accumulated into three fractions: the intercellular fraction located in the free-water space of the cells wall; the exchangeable fraction corresponding to metal trapped in cell wall exchange sites; and the residual fraction within the cell. It was suggested that the fraction of Cu not readily desorbed during a loss experiment was that which had reached the cytoplasm.

The metal present in the cell walls is sometimes particulate and not in an easily exchangeable form. For instance, Hg, which reached 0.1–1.2% dry weight in the liverworts *Jungermannia vulcanicola* and *Scapania undulata* sampled from an acidic stream in Japan (Satake et al., 1983), was found to occur in the walls as fine particles of insoluble HgS (Satake et al., 1990). However, although X-ray microanalysis of *S. undulata* from a mine site in the English Lake District showed the Pb to be present at very high concentrations in the walls and apparently not easily exchangeable, there was no evidence for the presence of PbS (Satake et al., 1989). Even if some of the Pb had been bound as the sulphate, this would not have exceeded 3% of the total Pb in the shoots. In spite of further study by Radecki et al. (1992), the exact form in which Pb is bound in the walls of this species is still uncertain.

Biochemical and Physiological Approaches

Pigment Ratios

Several studies have investigated the influence of heavy metals on the pigment composition of aquatic bryophytes, and there are a few scattered records for other aquatic plants. A decrease in the chlorophyll *a*/phaeophytin *a* was first reported (McLean & Jones, 1975) for several bryophytes in streams in western Wales polluted by Zn, Cd, and Pb from old mine workings. A more detailed use of this approach was applied (López and Carballeira, 1989) to the terminal 2–3 shoot apices of five species of bryophyte from a wide range of aquatic sites in northwestern Spain. The authors concluded that the pigment ratio is indicative of "stress" at a site, whether due to heavy metals or organic pollution. However, Bruns et al. (1997), in their study of moss transplanted to the River Elbe, found no correlation between pigment ratio and heavy metal concentrations of the moss or ambient water.

In their study to establish "background" levels of metals in aquatic bryophytes (see earlier section on the relationship between plant and aqueous environments) at a large number of sites, Carballeira and López (1997) compared the pigment ratio method with three statistical methods of analyzing data on metal composition. There were considerable differences in the results from the different methods, but the authors favored the pigment ratio method, because they suggested that this requires no assumptions about natural variability.

Although the pigment ratio provides an easily quantifiable approach to monitoring, doubts have been raised (Whitton & Kelly, 1995) about interpretation of the results, so the method should only be used as one of several methods for monitoring. The shift in pigment ratio may be influenced by the fact that more old plant material is likely to be included in samples from sites where the organism is stressed. Another unknown is the extent to which species that have acquired genetic tolerance to a heavy metal have pigment ratios different from populations at unpolluted sites. Bruns et al. (1997) did not speculate on the apparent lack of response of the pigment ratio of *F. antipyretica* transferred to the Elbe. However, their material originated from sites on the east of the Harz Mountains, a region with much former mining activity, so the possibility that they used a population with enhanced metal tolerance should be considered. Intuitively, pigment ratios would seem most likely to be of value for monitoring sites only recently subject to contamination or when material lacking tolerance is transferred to a polluted site.

Another biochemical approach is to characterize and quantify molecules, the occurrence of which reflects the presence of heavy metals. The terminology of such compounds found in algae has been reviewed by Robinson (1989). Enhanced levels of soluble ligands binding copper and cadmium, but not lead, were found in *R. riparioides* at sites with elevated concentrations of these metals (Jackson et al., 1991). Ding et al. (1994) also found a relationship between thiol group content and Cd in the floating vascular plant, *Eichhornia crassipes*. In a study of the River Elbe, Germany, Bruns et al. (1997) reported that values for thiol-containing peptides in *F. antipyretica* transported from the laboratory to polluted sites in the river showed an initial reduction in levels of these compounds but subsequently increased. Although it was concluded that phytochelatins were playing a role, the same authors subsequently (1999) disproved this, and the nature of the metal-binding ligands in this case is uncertain. In general, reports on the presence of phytochelatins in nonvascular aquatic plants exposed to heavy metals should be assessed with caution.

Toxicity Studies

The presence of metal-tolerant populations at metal-contaminated sites has been shown for a number of genera of freshwater algae, including blue-green algae (Shehata & Whitton, 1981), filamentous greens (see section on genetic aspects), and diatoms (Ivorra et al., 2002). Providing that the tolerance exceeds that of populations growing at sites lacking contamination and occurs in several different populations, this suggests that the metal in question has had a selective effect at the site, although the possible influence of inocula from upstream should be considered. In order to compare tolerance of populations or individual plants from different sites, they should as far as practicable be assayed under the same conditions. A culture method suitable for screening a range of taxa and metals was tested by Whitton (1970a) and a somewhat similar approach was applied by Foster (1982). Both studies showed evidence for populations tolerant to particular metals. Although the method is straightforward and gives reproducible results, it requires up to 10 days.

If the purpose of toxicity tests is to predict the quantitative effect of a metal under particular field conditions, then the tests need to be refined. The assay medium should be similar to water at the field site with respect to factors such as chelating agents that influence toxicity. Some algae differ markedly in their sensitivity to heavy metals according to whether the organism is in the light or dark at the time of exposure (e.g., Whitton, 1968). The results of growth-based assays can be influenced by whether the test organism has already been exposed to another heavy metal (Wong & Beaver, 1981).

Short-term assays have been described by a number of authors, usually based on some feature related to photosynthesis. Takamura et al. (1989) developed a methodology that they then applied to 118 algal strains isolated from field sites, with or without enrichment by Cu, Zn, and Pb. Carbon uptake was then measured over a 4-h period with the use of $Na^{13}HCO_3$. The relationship between tolerance levels in the field and those measured in the laboratory was more clear-cut for Cu than for Zn or Pb. Cu-tolerant isolates tended to be Zn-tolerant in the diatoms, whereas Cd-tolerant isolates tended to be Zn-tolerant in the green algae.

Pulse amplitude modulated fluorimetry was used by Ivorra et al. (2002) to measure chlorophyll fluorescence and photon yield in strains of the diatom *Gomphonema parvulum*. The Zn concentration causing a 50% reduction of the photon yield was significantly higher in an isolate from a Zn-polluted than a reference site. So far, too few studies have compared the quantitative effects of metal pollution assessed using short-term features of photosynthesis to those assessed using long-term growth to be sure how reliable are the short-term assays. Nevertheless, this is a very promising approach.

Use of Whole Communities

A development of the toxicological approach is to sample a whole community and test its tolerance to a potentially toxic chemical. Such a community-scale approach integrates the specific tolerances of all the taxa present in a given community. This was shown especially clearly in the case of marine periphyton and arsenate by Blanck and Wängberg (1988), and their study subsequently stimulated similar studies in freshwaters, including ones on heavy metals. Most studies have been based on the quantification of some feature of photosynthesis (e.g., Admiraal et al., 1999a). Ivorra et al. (2000) used the approach to show that microalgae present in young biofilms were more sensitive to Zn or Cd than microalgae in old biofilms, the difference being apparent even when both types of biofilm were taken from a clean stream.

Copper was one of two toxicants tested by Navarro et al. (2002) on microalgal communities at two different seasons (spring and summer) in the River Ter, northeastern Spain. Artificial substrata were left in the river at a number of sites and short-term (1–4 h) toxicity tests were subsequently carried out in the laboratory on these communities with photon yield to measure the ecotoxicological end point. Tolerance was lower in spring than summer.

Floristic Approaches

Ecological Surveys

Many studies of the impact of heavy metals on the floristic composition of aquatic communities have been reported in the literature, including the study of *in situ* populations (Table 3.3) and the effect of transplanting communities on artificial substrata (Admiraal et

Table 3.3. Examples of Ecological Studies of the Impact of Heavy Metals on the Floristic Composition of Aquatic Plants

Country	Location	Key topic	Reference
Canada	Kootenay Paint Pots	Algae in natural springs	Wehr & Whitton, 1983a
India	R. Tungabhadra	Paper mill metals on algae	Reddy & Venkateswarlu, 1985
UK	Gillgill Burn	Zn gradient	Say & Whitton, 1980
	R. Etherow	Compare plants and animals	Harding & Whitton, 1981a
USA	California stream	Copper and periphytic algae	Leland & Carter, 1984
	Missouri ponds and streams	High Zn and blue-green algae	Whitton et al., 1981

al., 1999b). In general, the higher the concentration of metal, the fewer the species present, as shown in a survey down a stream showing a gradient from high to low Zn (Say & Whitton, 1980) and in a comparison of data from a large number of individual sites differing in Zn concentrations by more than three orders of magnitude (Whitton & Diaz, 1980). However, at least in the case of Zn and Cd, the only sites found in which plants were completely absent are where contamination had occurred within the previous year (Whitton, unpublished data). A small stream in France with over 3 g L^{-1} Zn, but where contamination had probably existed for some years, had three species of green algae (Say & Whitton, 1982a).

Floristic studies can be especially helpful to assess the damage caused by a major but unexpected pollution event. Sabater (2000) used diatom communities for this purpose following a major mine tailings spill in April 1998, into the Guadiamar River in southwestern Spain, where the toxic material overflowed onto the floodplain in the vicinity of the Doñana National Park. Because no information on diatom communities was available from the river prior to the spill, the approach used by the author was to survey a reference site upstream of the spill and seven sites downstream, 7 and 14 months after the event. Statistically significant relationships were detected between a number of measures of the diatom community and various heavy metals in the environment. Comparison between the reference and affected sites showed a shift from a diatom community dominated by *Nitzschia palea*, *Achnanthes minutissima*, and *Amphora pediculus* to one dominated by *Nitzschia palea* and *Gomphonema parvulum*. Values for the Shannon–Wiener diversity index strongly decreased in the affected area. The author concluded that heavy metals in the water and sediments had a marked and persistent effect on the diatom communities. This study provides an excellent model for what can be done following a catastrophic event, though, ideally, the surveys should be commenced much sooner after the event.

Development of Indices

Although there have been a considerable number of indices developed to quantify the effects of pollution, their contribution to monitoring heavy metals has been minor. The earliest widely used indices, such as the quantified version of the saprobic index (Sládeček, 1973) and the diatom index of Descy (1979) did not attempt to differentiate the possible role of heavy metals in modifying the effects of organic pollution. More recently, the use of diatom indices has shifted toward monitoring distinct forms of pollution, especially the mineral nutrient status of the water. Although the effects of metal pollution have sometimes been noted during surveys based on diatom indices, so far no one has proposed an index aimed specifically at quantifying such effects. However, diversity indices have sometimes been used in studies of severe pollution in which diatoms are an important part of the flora, such as that described earlier for the Guadiamar River.

Perspective

In spite of the range of methods tested for monitoring heavy metals, they are by no means all those possible (see McCormick & Cairns, 1994). However, even among those tested, none has yet become routine for monitoring in the manner of benthic diatoms and macroinvertebrates for other aspects of pollution in rivers. In the case of heavy metals, aquatic plants have proved most useful in monitoring contamination from mining or industrial pollution, either as part of a routine program to minimize the effects of pollution or as an independent effort to increase public awareness (e.g., R. Vardar, Macedonia; S. Krstić and Z. Levkov, personal communication). In order to be successful, a method needs to be straightforward. Metal accumulation fits this well.

The value of the method based on accumulation is enhanced if the same species of aquatic plant can be used simultaneously for other purposes, such as monitoring insecticides and PCBs (Mouvet et al., 1985; Whitton et al., 1996). However, in spite of many reports on accumulation of organics, little basic research has been done to permit quantitative interpretation of the results. If such information were available, plants could be used within the same sampling program to monitor a variety of substances. Furthermore, the species of bryophyte used most widely for monitoring of heavy metals and, to a lesser extent, pesticides, based on accumulation, *Fontinalis antipyretica* and *Rhynchostegium riparioides*, have also proved useful for monitoring the phosphorus status of the environment (Christmas & Whitton, 1998).

Another aspect of monitoring based on accumulation, which has seldom been considered, is the possibility of standardizing the methods used for monitoring heavy metals with aquatic and terrestrial bryophytes. Most terrestrial studies have used the whole moss plant and in general a less thorough washing procedure, although Fernández and Carballeira (2001) used a broadly similar methodology with two terrestrial mosses as they have used in their various aquatic studies. It would be well worth trying to standardize aquatic and terrestrial procedures further, because this would encourage their use by environmental management organizations responsible for both types of environment.

References

Admiraal, W., Blanck, H., & De Jong, B. M., Guasch, H., Ivorra, N., Lehmann, V., Nystrom, B. A. H., Paulsson, M., & Sabater, S. (1999a). Short-term toxicity of zinc to microbenthic algae and bacteria in a metal polluted stream. *Water Research, 33,* 1989–1996.

Admiraal, W., Ivorra, N., Jonker, M, Bremer, S., Barranquet, C., & Guasch, H. (1999b). Distribution of idatom species in a metal-polluted Belgian–Dutch river: An experimental analysis. In: J. Prygiel, B. A. Whitton, & J. Bukowska (Eds.), *Use of algae for monitoring rivers III.* (pp. 240–244). Douai, France: Agence de l'Eau Artois–Picardie.

Almeida, S. F. P., Pereira, M. J., Gil, M. C., & Rino, J. M. (1999). Freshwater algae in Portugal and their use for environmental monitoring. In: J. Prygiel, B. A. Whitton, & J. Bukowska (Eds.), *Use of algae for monitoring rivers III* (pp. 10–16). Douai, France: Agence de l'Eau Artois-Picardie.

André, B., & Lascombe, C. (1987). Comparaison de deux traceurs de la pollution métallique des cours d'eau: Les sédiments et les bryophytes. *Sciences de l'Eau 6,* 225–247.

Blanck, H., and Wängberg, S.-A. (1988). Induced community tolerance in marine periphyton established under arsenate stress. *Can. J. Fish. Aquat. Sci., 45,* 1816–1819.

Brown, D. H., & Brümelis, G. (1996). A biomonitoring method using the cellular distribution of metals in moss. *Sci. Total Envir., 187,* 153–161.

Bruns, I., Friese, K., Markert, B., & Krauss, H.-J. (1997). The use of *Fontinalis antipyretica* L. ex Hedw. as a bioindicator for heavy metals: 2. Heavy metal accumulation and physiological reaction of *Fontinalis antipyretica* L. ex Hedw. in active biomonitoring in the River Elbe. *Sci. Total. Envir., 204,* 161–176.

Bruns, I., Friese, K., Markert, B., & Krauss, H.-J. (1999). Heavy metal inducible compounds from *Fontinalis antipyretica* reacting with Ellman's reagent are not phytochelatins. *Sci. Total Envir., 241,* 215–216.

Burton, M. A. S., & Peterson, P. J. (1979). Metal accumulation by aquatic bryophytes from polluted mine streams. *Envir. Pollut., 19,* 39–46.

Carballeira, A., & López, J. (1997). Physiological and statistical methods to identify background levels of metals in aquatic bryophytes: Dependence on lithology. *J. Envir. Quality, 26,* 980–988.

Carballeira, A., Vásquez, M. D., & López, J. (2001). Biomonitoring of sporadic acidification of rivers on the basis of release of preloaded cadmium from the aquatic bryophyte *Fontinalis antipyretica* Hedw. *Envir. Pollut, 111,* 95–106.

Christmas, M., & Whitton, B. A. (1998). Phosphorus and aquatic bryophytes in the Swale–Ouse river system, North-East England: 1. Relationship between ambient phosphate, internal N : P ratio and surface phosphatase activity. *Sci. Total Envir, 210/211,* 389–399.

Commission of the European Communities. (1982). The certification of the contents of cadmium, copper, manganese, mercury, lead and zinc in two plant materials of aquatic origin (BCR numbers 60 and 61) and in olive leaves (BCR number 62). Report EUR 8119 EN. Brussels and Luxembourg.

Denny, P. (1980). Solute movement in submerged angiosperms. *Biol. Rev., 55,* 65–92.

Descy, J.-P. (1979). A new approach to water quality estimation using diatoms. *Nova Hedwigia, 64,* 305–323.

Descy, J.-P., Empain, A., & Lambinon, J. (1981). La Qualité des Eaux Courantes en Wallonie—Bassin de la Meuse. Bruxelles, Belgium: Secretariat d'Etat à l'Environnement, à l'Aménagement du Territoire et à l'Eau pour la Wallonie.

Ding, X., Jiang, J., Wang, Y., Wang, W., & Ru, B. (1994). Bioconcentration of cadmium in water hyacinth (*Eichhornia crassipes*) in relation to thiol group content. *Envir. Pollut., 84,* 93–96.

Empain, A. (1976a). Estimation de la pollution par métaux lourds dans la Somme par l'analyse des bryophytes aquatiques. *Bull. Fr. Pisciculture, 48,* 138–142.

Empain, A. (1976b). Les bryophytes aquatique utilisés comme traceurs de la contamination en métaux lourds des eaux douces. *Mém. Soc. R. Bot. Belg, 7,* 141–156.

Empain, A., Lambinon, J., Mouvet, C., & Kirchmann, R. (1980). Utilisation des bryophytes aquatiques et subaquatiques comme indicateurs biologiques de la qualité des eaux courantes. In: P. Pesson (Ed.), *La Pollution des Eaux Continentales* (2nd ed.) (pp. 195–223). Paris: Gauthier-Villars.

Fernández, J. A., & Carballeira, A. (2001). A comparison of indigenous mosses and topsoils for use in monitoring atmospheric heavy metal deposition in Galicia (northwest Spain). *Environ. Pollut, 114,* 431–441.

Foster, P. L. (1982). Metal resistances of Chlorophyta from rivers polluted by heavy metals. *Freshwater Biology, 12,* 41–61.

Franzin, W. G., & McFarlane, G. A. (1980). An analysis of the aquatic macrophyte *Myriophyllum exalbescens*, as an indicator of metal contamination of aquatic ecosystems near a base smelter. *Bull. Envir. Contam. Toxicol., 24,* 597–605.

Gonçalves, E. P. R., & Boaventura, R. A. R. (1998). Uptake and release kinetics of copper by the aquatic moss *Fontinalis antipyretica*. *Water Research, 32,* 1305–1313.

Gonçalves, E. P. R., Boaventura, R. A. R., & Mouvet, C. (1992). Sediments and aquatic mosses as pollution indicators for heavy metals in the Ave river basin (Portugal). *Sci. Total Envir, 114,* 7–24.

Gonçalves, E. P. R., Soares, H. M. V. M., Boaventura, R. A. R., Machado, A. A. S. C., & Silva, J. C. G. E. (1994). Seasonal variations of heavy metals in sediments and aquatic mosses from the Cávado river basin (Portugal). *Sci. Total Envir., 142,* 143–156.

Guilizzoni, P. (1991). The role of heavy metals and toxic materials in the physiological ecology of submerged macrophtyes. *Aquatic Botany, 41,* 87–109.

Harding, J. P. C., & Whitton, B. A. (1976). Resistance to zinc of *Stigeoclonium tenue* in the field and the laboratory. *Br. Phycol. J., 11,* 417–426.

Harding, J. P. C., & Whitton, B. A. (1977). Environmental factors reducing the toxicity of zinc to *Stigeoclonium tenue*. *Br. Phycol. J., 12,* 17–21.

Harding, J. P. C., & Whitton, B. A. (1978). Zinc, cadmium and lead in water, sediments, and submerged plants of the Derwent Reservoir, Northern England. *Water Research, 12,* 307–316.

Harding, J. P. C., & Whitton, B. A. (1981a). R. Etherow: Plants and animals of a river recovering from pollution. *Naturalist, Hull 106,* 15–31.

Harding, J. P. C., & Whitton, B. A. (1981b). Accumulation of zinc, cadmium and lead by field populations of *Lemanea*. *Water Research, 15,* 301–319.

Hargreaves, J. W., & Whitton, B. A. (1976). Effect of pH on tolerance of *Hormidium rivulare* to zinc and copper. *Oecologia, 26,* 235–243.

Ivorra, N., Barranguet, C., Jonker, M., Kraak, M. H. S., & Admiraal, W. (2002). Metal-induced tolerance in the freshwater microbenthic diatom *Gomphonema parvulum*. *Envir. Pollut., 116,* 147–157.

Ivorra, N., Bremer, S., Guasch, H., Kraak, M. H. S., & Admiraal, W. (2000). Differences in the sensitivity of benthic microalgae to Zn and Cd regarding biofilm development and exposure history. *Environ. Toxicol. Chem., 19,* 1332–1339.

Jackson, L. J. (1998). Paradigms of metal accumulation in rooted aquatic vascular plants. *Sci. Total Envir., 219,* 223–231.

Jackson, P. P., Robinson, N. J., & Whitton, B. A. (1991). Low-molecular weight metal ligands in the freshwater moss *R. riparioides*: Analysis of plants exposed to elevated concentrations of Zn, Cu, Cd and Pb in the laboratory and field. *Environmental and Experimental Botany, 31*, 359–366.

Jain, S. K., Vasudevan, P., & Jha, N. K. (1989). Removal of some heavy metals from polluted waters by aquatic plants: Studies on duckweed and water velvet. *Biological Wastes, 28*, 115–126.

Kelly, M. G., & Whitton, B. A. (1987) Growth rate of the aquatic moss *Rhynchostegium riparioides*. *Freshwater Biology, 18*, 461–468

Kelly, M. G., Girton, C., & Whitton, B. A. (1987). Use of moss-bags for monitoring heavy metals in rivers. *Water Research, 11*, 1429–1435.

Kelly, M. G., & Whitton, B. A. (1989a). Interspecific differences in Zn, Cd and Pb accumulation by freshwater algae and bryophytes. *Hydrobiologia, 175*, 1–11.

Kelly, M. G., & Whitton, B. A. (1989b). Relationship between accumulation and toxicity of zinc in *Stigeoclonium* (Chaetophorales, Chlorophyta). *Phycologia, 28*, 512–517.

Leland, H. V., & Carter, J. L. (1984). Effects of copper on species composition of periphyton in a Sierra Nevada, California, stream. *Freshwater Biology, 14*, 281–296.

López, J., & Carballeira, A. (1989). A comparative sutudy of pigment contents and response to stress in five species of aquatic bryophytes. *Lindbergia, 15*, 188–194.

López, J., & Carballeira, A. (1993). Interspecific differences in metal bioaccumulation and plant–water concentration ratios in five aquatic bryophytes. *Hydrobiologia, 263*, 95–107.

López, J., Vázquez, M. D., & Carballeira, A. (1994). Stress responses and metal exchange kinetics following transplant of the aquatic moss *Fontinalis antipyretica*. *Freshwater Biology, 32*, 185–198.

Maine, M. A., Duarte, M. V., & Suñé, N. L. (2001). Cadmium uptake by floating macrophytes. *Water Research, 35*, 2629–2634.

Maine, M. A., Suñé, N. L., Panigatti, M. C., & Pizarro, M. J. (1999). Relationships between water chemistry and macrophyte chemistry in lotic and lentic environments. *Archives of Hydrobiology, 145*, 129–145.

McCormick, P. V., & Cairns, J., Jr. (1994). Algae as indicators of environmental change. *Journal of Applied Phycology, 6*, 509–526.

McHardy, B. M., & George, J. J. (1985). The uptake of selected heavy metals by the green alga *Cladophora glomerata*. In: J. Salanki (Ed.), *Heavy metals in water organisms*. Budapest: Akadémiai Kiado.

McLean, R. O., & Jones, A. K. (1975). Studies of tolerance to heavy metals in the flora of the rivers Ystwyth and Clarach, Wales. *Freshwater Biology, 5*, 431–444.

Mersch, J., & Johansson, L. (1993). Transplanted aquatic mosses and freshwater mussels to investigate the trace metal contamination in the rivers Meur and Plaine, France. *Envir. Technol., 14*, 1027–1036.

Mersch, J., & Kass, M. (1994). La mousse aquatique *Fontinalis antipyretica* comme traceur de la contamination radioactive de la Moselle en aval de la centrale nucléaire de Cattenom. *Bull. Soc. Nat. Luxemb., 95*, 109–117.

Mouvet, C. (1984). The use of aquatic bryophytes to monitor heavy metal pollution of freshwaters as illustrated by case studies. *Verh. Int. Verein. Limnol., 22*, 2420–2425.

Mouvet, C. (1985). Accumulation of chromium and copper by the aquatic moss *Fontinalis antipyretica* L. ex Hedw. transplanted in a metal-contaminated river. *Envir. Technol. Lett., 5*: 541–548.

Mouvet, C. (1991). *Métaux Lourds et Mousses Aquatiques: Standarisation des Aspects Analytiques*. R 32 744, PR 9304600317. BRGM, Belgium.

Mouvet, C., & Claveri, B. (1999). Localization of copper accumulated in *Rhynchostegium riparioides* using sequential chemical extraction. *Aquatic Botany, 63*, 1–10.

Mouvet, C., Galoux, M., & Bernes, A. (1985). Monitoring of polychlorinated biphenyls (PCB)s and hexachlorocyclohexanes (HCH) in freshwaters using the aquatic moss *Cinclidotus danubicus*. *Sci. Tot. Environ., 44*, 253–267.

Mouvet, C., Pattée, E., & Cordebar, P. (1986). Utilisation des mousses aquatiques pour l'identification et la localisation précise de sources de pollution métallique multiforme. *Acta Oecologica, 7*: 72–91.

Nakada, M., Fukaya, K., Takeshita, S., & Wade, Y. (1979). Accumulation of heavy metals in the submerged plant (*Elodea nuttallii*). *Bull. Envir. Contam. Toxicol., 22*, 21–27.

Navarro, E., Guasch, H., & Sabater, S. (2002). Use of microbenthic algal communities in ecotoxicological tests for the assessment of water quality: the Ter river case study. *Journal of Applied Phycology, 14*, 41–48.

Onianwa, P. C. (2001). Monitoring atmospheric metal pollution: A review of the use of mosses as indicators. *Environmental Monitoring and Assessment, 71*, 13–50.

Patterson, G., & Whitton, B. A. (1981). Chemistry of water, sediments and algal filaments in groundwater draining an old lead–zinc mine. In: P. J. Say, & B. A. Whitton (Eds.), *Heavy metals in Northern England: Environmental and biological aspects* (pp. 65–72). UK: Department of Botany, University of Durham.

Patterson, G., & Whitton, B. A. (1982). Influence of zinc on a zinc-tolerant strain of *Mougeotia. Br. Phycol. J., 17*, 237.

Prasad, M. N. V., Malec, P., Waloszek, A., Bojko, M. & Strzałka, K. (2001). Physiological responses of *Lemna trisulca* L. (duckweed) to cadmium and copper bioaccumulation. *Plant Science, 161*, 881–889.

Radecki, J. J., Soma, M., Satake, K., & Whitton, B. A. (1992). X-ray photoelectron spectroscopic characterization of lead in aquatic bryophytes and mineral phases of pebbles in stream passing through a lead mine. *Polish Journal of Experimental Studies, 1*(3), 39–41.

Ray, S. N., & White, W. J. (1979). Equisetum arvense—an aquatic vascular plant as a biological monitor for heavy metal pollution. *Chemosphere, 3*, 125–128.

Reddy, P. M., & Venkateswarlu, V. P. (1985). Ecological studies in the paper mill effluents and their impact on the river Tungadhadra: Heavy metals and algae. *Proceedings of the Indian Academy of Science (Plant Science), 95*, 139–146.

Robinson, N. J. (1989). Algal metallothioneins: Secondary metabolites and proteins. *Journal of Applied Phycology, 1*, 5–18.

Sabater, S. (2000). Diatom communities as indicators of environmental stress in the Guadiamar River, S-W. Spain, following a major mine spill. *Journal of Applied Phycology, 12*, 113–124.

Samecka-Cymerman, A., Kempers, A. J., & Bodelier, P. L. E. (1991). Preliminary investigations into the background levels of various metals and boron in the aquatic liverwort *Scapania uliginosa* (Sw.) Dum. *Aquatic Botany, 39*, 345–352.

Satake, K., Shibata, K., & Bando, Y. (1990). Mercury sulphide (HgS) crystals in the cell walls of the aquatic bryophytes, *Jungermannia vulcanicola* Steph. and *Scapania undulata* (L.) Dum. *Aquatic Botany, 36*, 325–241.

Satake, K., Soma, M., Seyama, H., & Ushiro, T. (1983). Accumulation of mercury in the liverwort *Jungermannia vulcanicola* Steph. in an acid stream Kashiranashigawa in Japan. *Archives of Hydrobiology, 99*, 89–92.

Satake, K. Takamatsu,T., Soma, M., Shibata, K, Nishikawa, M., Say, P. J., & Whitton, B. A. (1989). Lead accumulation and location in shoots of the aquatic liverwort *Scapania undulata* in stream water at Greenside Mine, England. *Aquatic Botany, 33*, 111–122.

Sawidis, T. (1996). Radioactive pollution in freshwater ecosystems from Macedonia, Greece. *Arch. Environ. Contam. Toxicol., 30*, 100–106.

Sawidis, T., Chettri, M. K., Zachariadis, G. A., & Stratis, J. A. (1995). Heavy metals in aquatic plants and sediments from water systems in Macedonia, Greece. *Ecotoxicol. Environ. Safety, 32*, 73–80.

Say, P. J., Diaz, B. M., & Whitton, B. A. (1977). Influence of zinc on lotic plants: I. Tolerance of *Hormidium* species to zinc. *Freshwater Biology, 7*, 357–376.

Say, P. J., Harding, J. P. C., & Whitton, B. A. (1981). Aquatic mosses as monitors of heavy metal concentration in the River Etherow, England. *Environ. Pollut. Ser. B, 2*, 295–307.

Say, P. J., & Whitton, B. A. (1977). Influence of zinc on lotic plants: II. Environmental effects on toxicity of zinc to *Hormidium rivulare. Freshwater Biology, 7*, 377–384.

Say, P. J., & Whitton, B. A. (1980). Changes in flora down a stream showing a zinc gradient. *Hydrobiologia, 76*, 255–262.

Say, P. J., & Whitton, B. A. (1982a). Chemie et écologie de la végétation de cours d'eau à forte teneur de zinc: 1. Massif Central. *Annals Limnologie, 18*, 19–31.

Say, P. J., & Whitton, B. A. (1982b). Chemistry and plant ecology of zinc-rich streams in France: 2. The Pyrenees. *Annals Limnologie, 18*, 19–31.

Say, P. J., & Whitton, B. A. (1983). Accumulation of heavy metals in aquatic mosses: 1. *Fontinalis antipyretica* Hedw. *Hydrobiologia, 100*, 245–260.

Schacklette. H. T., & Erdman, J. A., (1982). Uranium in spring water and bryophytes at Basin Creek in central Idaho. *J. Geochem. Exploration, 17*, 222–236.

Sergio, C., Figueira, R., & Crespo, A. M. V. (2000). Observations of heavy metal accumulation in the cell walls of *Fontinalis antipyretica*, in a Portuguese stream affected by mine effluent. *J. Bryol., 22*, 252–255.

Sergio, C., Seneca, A., Maguas, C., & Branquinho, C. (1992). Biological responses of *Sphagnum auriculatum* Schimp. to water pollution by heavy metals *Cryptogamie Bryologie Lichenologie, 13*, 155–163.

Shehata, F. H. A., & Whitton, B. A. (1981) Field and laboratory studies on blue-green algae from aquatic sites with high levels of zinc. *Verh. Int. Verein. Theor. Angew. Limnol., 21*, 1466–1471.

Siebert, A., Bruns, I., Krauss, G.-J., Miersch, J., & Markert, B. (1996). The use of the aquatic moss *Fontinalis antipyretica* L. ex Hedw. as a bioindicator for heavy metals. *Sci. Total Envir., 177*, 137–144.

Sládeček, V. (1973). System of water quality from the biological point of view. *Arch. Hydrobiol. Ergebn. Limnol., 7*, 1–218.

Stokes, P. M., Dreier, S. I., Frkas, M. O., & McLean, R. A. N. (1983). Mercury accumulation by filamentous algae: A promising biological monitoring system for methyl mercury in acid-stressed lakes. *Envir. Pollut. Ser. B*, 258–271.

Takamura, N., Kasai, F., & Watanabe, M. M. (1989). Effects of Cu, Cd and Zn on photosynthesis of freshwater benthic algae. *Journal of Applied Phycology*, *1*, 39–52.

Vázquez, M. D., López, J., & Carballeira, A. (1999). Uptake of heavy metals to the extracellular and intracellular compartments in three species of aquatic bryophyte. *Ecotoxicol. Envir. Safety*, *44*: 12–24.

Vzquez, M. D., Fernandez, J. A., López, J., & Carballeira, A. (2000). Effects of water acidity and metal concentration on accumulation and within-plant distribution of metals in the aquatic bryophyte *Fontinalis antipyretica*. *Water Air Soil Pollution*, *120*, 1–19.

Vincent, C. D., Lawlor, A. J., & Tipping, E. (2001). Accumulation of Al, Mn, Fe, Cu, Zn, Cd and Pb by the bryophyte *Scapania undulata* in three upland waters of different pH. *Environ. Pollut.*, *14*, 93–100.

Wehr, J. D., Empain, A., Mouvet, C., Say, P. J., & Whitton, B. A. (1983). Methods for processing aquatic mosses used as monitors of heavy metals. *Water Research*, *17*, 985–992.

Wehr, J. D., Kelly, M. G., & Whitton, B. A. (1987). Factors affecting accumulation and loss of zinc by the aquatic moss *Rhynchostegium riparioides*. *Aquatic Botany*, *29*, 261–274.

Wehr, J. D., Say, P. J., & Whitton, B. A. (1981). Heavy metals in an industrially polluted river, the Team. In: P. J. Say, & B. A. Whitton (Eds.). *Heavy metals in Northern England: Environmental and biological aspects* (pp. 99–108). UK: Department of Botany, University of Durham.

Wehr, J. D., & Whitton, B. A. (1983a). Aquatic cryptogams of natural acid springs enriched with heavy metals: The Kootenay Paint Pots, British Columbia. *Hydrobiologia*, *98*, 97–105.

Wehr, J. D., & Whitton, B. A. (1983b). Accumulation of heavy metals in aquatic mosses. 2. *Rhynchostegium riparioides*. *Hydrobiologia*, *100*, 261–284.

Wehr, J. D., & Whitton, B. A. (1983c). Accumulation of heavy metals in aquatic mosses: 3. Seasonal changes. *Hydrobiologia*, *100*, 285–291.

Whitehead, N. E., & Brooks, R. R. (1969). Aquatic bryophytes as indicators of uranium mineralization. *Bryologist*, *72*, 501–507.

Whitton, B. A. (1968). Effect of light on toxicity of various substances to *Anacystis nidulans*. *Plant Cell Physiology*, *9*, 23–26.

Whitton, B. A. (1970a). Toxicity of zinc, copper and lead to Chlorophyta from flowing waters. *Arch. Mikrobiol.*, *72*, 353–360.

Whitton, B. A. (1970b). Toxicity of heavy metals to freshwater algae: A review. *Phykos*, *9*, 116–125

Whitton, B. A. (1970c). Biology of *Cladophora* in freshwaters. *Water Research*, *4*, 457–476.

Whitton, B. A. (1980). Zinc and plants in rivers and streams. In: J. O. Nriagu (Ed.), *Zinc in the environment: Part II. Health effects* (pp. 364–400) New York, Chichester, UK: Wiley.

Whitton, B. A. (1984). Algae as monitors of heavy metals in freshwater. In: L. E. Shubert (Ed.), *Algae as ecological indicators* (pp. 257–280). London: Academic Press.

Whitton, B. A., Burrows, I. G., & Kelly, M. G. (1989). Use of *Cladophora* to monitor heavy metals in rivers. *Journal of Phycology*, *1*, 293–299.

Whitton, B. A., Darlington, S. T., & Say, P. J. (1996). Biological monitoring of insecticides and other low level contaminants in freshwaters using phototrophs. In: M. Moo-Young, W. A. Anderson, & A. M. Chakrabarty (Eds.), *Environmental biotechnology: Principles and practice*. International Symposium on Environmental Biotechnology, Waterloo, Canada, 4–8 July 1994. Dordrecht: Kluwer.

Whitton, B. A., & Diaz, B. M. (1980). Chemistry and plants of streams and rivers with elevated zinc. *Trace Substances in the Environment*, *14*, 457–463.

Whitton, B. A., Gale, N. L., & Wixson, B. G. (1981). Chemistry and plant ecology of zinc-rich wastes dominated by blue-green algae. *Hydrobiologia*, *83*, 331–341.

Whitton, B. A., Harding, J. P. C., Kelly, M. K., & Say, P. J. (1991). *Use of plants to monitor heavy metals in freshwaters*. Standing Committee of Analysts 1990. HMSO.

Whitton, B. A., & Kelly, M. G. (1995). Use of algae and other plants for monitoring rivers. *Australian Journal of Ecology*, *20*, 45–56.

Whitton, B. A., Say, P. J., & Jupp, B. P. (1982). Accumulation of zinc, cadmium and lead by the aquatic liverwort *Scapania*. *Environ. Pollut. Ser. B*, *3*, 299–316.

Wong, S. L., & Beaver, J. L. (1981). Metal interactions in algal toxicology: Conventional versus *in vivo* tests. *Hydrobiologia*, *85*, 67–71.

4

Hydrochemical Determinism, Ecological Polymorphism, and Indicator Values of Aquatic Bryophytes for Water Quality

Alain Vanderpoorten

Management of the large rivers has resulted in a rapid succession of modifications of their physicochemical (eutrophication, organic pollution, increase in the concentrations of pesticides and heavy metals) and hydrological characteristics. In most of the large rivers, the concentrations of pollutants have greatly varied across time. Indeed, these rivers have progressively been regulated to provide protection from flooding, to produce hydroelectricity, and to facilitate the movement of shipping. Their alluvial floodplains have been strongly industrialized. However, this industrialization was undertaken in total disregard of ecological consequences. As a result, water quality has decreased, giving rise to spectacular ecological disasters such as the pollution of the Rhine by xenobiotic factors following the fire in the chemical plant Sandoz in 1986. These ecological disasters revealed the vulnerability of large rivers. Several international agreements for improving water quality, such as the Rhine Action Program between Germany, France, the Netherlands, Luxembourg, and Switzerland, were signed. A large number of chemical components have consequently been regularly measured, showing the improvement of water quality for 15 years (Hellmann, 1994; Beurksens et al., 1994; Malle, 1996; Tittizer & Krebs, 1996).

Concern over the impact of nutrients in Europe has increased since the publication of the European Community's Wastewater Treatment Directive (Kelly, 1998). It was demonstrated that the natural geochemical content of P–PO_4^{3-} in the large European rivers is of about 10 µg L^{-1}, but that it is commonly 10 to 1,000 times greater in most of the rivers due to eutrophication (Carbiener et al., 1995). Despite efficient urban and industrial

Alain Vanderpoorten • Université Catholique de Louvain, Unité d'Ecologie & Biogéographie, 4–5 Place Croix du Sud, B-1348 Louvain-la-Neuve, Belgium.

discharge sanitation policies, and a great deal of campaign information on the dangers of abusive mineral fertilizer use, the waters of most of the rivers have not yet recovered the chemical quality they had at the end of the last century. This has been shown by long-term monitoring studies in rivers such as the Seine and the Marne (France), where the oxidation of ammonium into nitrates causes nitrogen concentrations to increase further (Cun et al., 1997). The great interest that developed with problems due to nutrient enrichment focused research on eutrophication and has involved a rapid increase in biological methods for monitoring nutrients (see Kelly & Whitton, 1998, for review). Living organisms have been more and more widely used in detecting the complex effect of environmental pollution. One objective of the Man and Biosphere program organized by the United Nations Educational, Scientific, and Cultural Organization (UNESCO) is the selection of bioindicators that may prove useful to detect environmental impact. It is declared in the program that bioindicators have to be used independent of the measurement of physical and chemical parameters of ecosystems.

It might be thought easier to measure environmental variables at a site rather than to infer their values from the species occurring there, but most of the time, it is not (ter Braak & Looman, 1987). Physical and chemical analyses themselves provide insufficient data on the environment and nothing on the relationship between environment and life (Kovacs & Podani, 1986). In addition, often the problem is that the more directly relevant a measurement, the slower and/or less precise it is. When total values over time are required, costly repeated measurements are needed. But the desired information can be inferred from faster or more precise measurements. The biological communities integrate the environmental conditions over time. This is one of the ideas behind biological evaluation of water quality and biomonitoring (e.g., Descy & Empain, 1981; Trémolières et al., 1994; Kelly, 1998). There are also situations in which it is impossible to measure environmental variables by direct means, whereas biological records exist. An example is the reconstruction of past changes in acidity in lakes from fossil diatoms from successive strata of the bottom sediment (Charles, 1985; Davis & Anderson, 1985; Davis & Smol, 1986; Charles & Smol, 1988).

Aquatic bryophytes are sensitive to water pollution, as shown by morphological damage (Benson Evans & Williams, 1976), reduced chlorophyll concentration and photosynthesis, and changes in pigment composition (Penulas, 1984; López & Carballeira, 1989; Martinez-Abaigar et al., 1993) after transplantation into polluted environments. They have been used in several direct methods for predicting water quality from species combinations (Frahm, 1974; Benson Evans & Williams, 1976; Vhrovsek et al., 1984; Arts, 1990; Tremp & Kohler, 1993; Haury et al., 1996). However, these methods only permitted the assessment of a global factor "pollution." Their validity and their precision were not evaluated. In addition, these methods were based on combinations of taxa whose ecology was poorly known. The species response curves have not yet been investigated. Few studies deal with the direct relationships between bryophyte communities and chemical factors. The chronic toxicity of heavy metals was recently investigated in a few terrestrial species (Brown & Whitehead, 1986; Brown & Wells, 1990; Wells & Brown, 1995; Sidhu & Brown, 1996). In aquatic bryophytes, conversely, the effects of pollutants are, except for experimental works such as those of Frahm (1975, 1976), Kosiba & Sarosiek (1995), Steinman (1994), and correlative studies on the relationships between aquatic bryophyte assemblages and water chemistry (Thiébaut et al., 1998; Vanderpoorten, 1999a, 1999b; Vanderpoorten & Klein, 1999a, 1999b; Vanderpoorten et al., 1999, 2000, 2001b), undersurveyed. The ecophysiological processes explaining their reaction to nutrients remain poorly understood (Bates, 1992). Hence, Düll (1992) and Glime (1992) gave no information concerning the trophic requirements of aquatic bryophytes in the ecological index system for central Europe and in a review of the effects of pollutants on aquatic species, respectively.

No investigations have yet been conducted to determine whether the populations of a same species exhibit similar reactions along a gradient of pollution. It had already been observed, however, that populations of species such as *Fontinalis antipyretica* and *Amblystegium tenax* present a contrasted ecology among different hydrographic networks, suggesting the possible existence of several ecotypes (Frahm & Abst, 1993; Vanderpoorten, 1999a, 1999b; Vanderpoorten & Durwael, 1999; Vanderpoorten & Empain, 1999; Vanderpoorten et al., 1999). The aim of this chapter is to improve our understanding of the hydrochemical determinism and the ecological polymorphism of aquatic bryophytes, in order to test their use as biomonitors of water quality.

How Do Aquatic Bryophytes Segregate along a Gradient of Water Pollution?

A first step in the understanding of the hydrochemical determinism of aquatic bryophytes is to determine the chemical components causing their segregation and to discover whether species segregate along a pollution gradient. There is thus a need for simultaneous vegetation and water chemistry records across a wide range of waters of varying chemical status (i.e., a need to seek gradients of chemical components that can be correlated to changes in species composition).

Vanderpoorten et al. (2001b) found that aquatic bryophytes, like vascular macrophytes (Robach et al., 1996), segregate first along a gradient of pH and buffering capacity, and second along a gradient of trophic level. This is shown by the dispersion of the species from a data set including a wide range of chemical conditions, from strongly acidic waters with low buffering capacity to alkaline waters rich in cations, and from oligotrophic to hypertrophic status (Figure 4.1). Species characteristic for acidic waters with low buffering capacity include *Marsupella emarginata*, *Jungermannia sphaerocarpa*, *Scapania undulata*, and *Racomitrium aciculare*. They are replaced by species of neutral waters, but with low buffering capacities, such as *Chiloscyphus polyanthos*, *Fissidens pusillus*, and *Dichodontium pellucidum*, and by species characteristic of hard waters, such as *Cratoneuron filicinum* and *Cinclidotus* spp., along a gradient in pH and cation content (Empain et al., 1980).

Himmler & Ness (1992) and Vanderpoorten & Klein (1999a) found a contrasting bryoflora, including species usually termed *acidophilous*, such as *Marsupella emarginata* and *Scapania undulata*, and species usually characteristic of neutral waters with good buffering capacities, such as *Chiloscyphus polyanthos*, in brooks with low buffering capacity in the Vosges (France) and the Black Forest (Germany). Species such as *Marsupella emarginata* and *Scapania undulata* tolerate neutral pH in these oligomineral waters. They are even present when pH is above 7.5 (Thiébaut et al., 1998; Vanderpoorten & Klein, 1999a). The occurrences of these species are not well correlated with pH: The correlation between pH and the presence of *Scapania undulata* was low ($r = .27, p < .01$) in a study of 125 brooks in Germany (Tremp & Kohler, 1993). These bryophytes are better correlated with the cation contents of the waters than with their pH. For example, in the Vosges and the Black Forest, *Hyocomium armoricum* showed a typical calcifuge behavior, disappearing when the concentrations of Ca^{++} exceeded $6\,mg\,L^{-1}$ (Himmler & Ness,

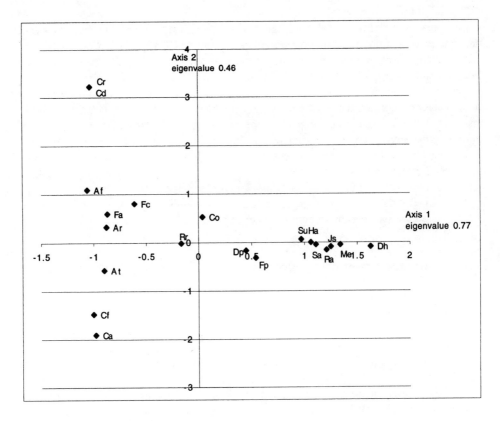

Figure 4.1. Correspondence analysis of 40 sampling sites ranging from strong acidity and low buffering capacity to alkalinity, and from oligotrophic to hypertrophic level, and 20 aquatic bryophyte species from northeast France (after Vanderpoorten et al., 2001b). Af = *Amblystegium fluviatile*; Ar = *A. riparium*; At = *A. tenax*; Ca = *Chiloscyphus pallescens*; Co = *C. polyanthos*; Cd = *Cinclidotus danubicus*; Cr = *C. riparius*; Cf = *Cratoneuron filicinum*; Dp = *Dichodontium pellucidum*; Dh = *Dicranella heteromalla*; Fc = *Fissidens crassipes*; Fp = *F. pusillus*; Fa = *Fontinalis antipyretica*; Ha = *Hyocomium armoricum*; Js = *Jungermannia sphaerocarpa*; Me = *Marsupella emarginata*; Ra = *Racomitrium aciculare*; Rr = *Rhynchostegium riparioides*; Su = *Scapania undulata*; Sa = *Sphagnum auriculatum* (with permission by Kluwer).

1992; Vanderpoorten & Klein, 1999a). Conversely, the species usually characteristic of waters with high buffering capacities, such as *Chiloscyphus polyanthos*, *Cratoneuron filicinum*, *Rhynchostegium riparioides*, and *Thamnobryum alopecurum*, are still present in very soft waters. *Cratoneuron filicinum* was found in waters with concentrations in Ca^{++} and Mg^{++} as low as 6 mg L^{-1} and 1.5 mg L^{-1}, respectively. *Chiloscyphus polyanthos* and *Thamnobryum alopecurum* were found in waters with less than 4 mg L^{-1} of Ca^{++} and 1.5 mg L^{-1} of Mg^{++}. High Ca^{++} and Mg^{++} concentrations are thus not required by these species. Conversely, these species are very acid-sensitive. A high correlation ($r = .97, p < .0001$) was found between the occurrence of *Chiloscyphus polyanthos* and pH in a study involving 125 brooks in Germany (Tremp & Kohler, 1993). Such species can tolerate waters with low cation concentrations if the concentrations of protons are low. The aquatic bryoflora thus includes a combination of species characteristic for waters with low concentrations of dissolved minerals on the one hand, and with low concentrations of protons on the other. This original combination relies on a fragile physicochemical balance, because the buffering capacity of the waters is very low. At such low concentrations, the chemical factors are limiting for the bryoflora, which may react rapidly to slight changes in ion concentration. Therefore, the aquatic assemblages of soft waters are strongly correlated to narrow gradients of pH, K^+, Mg^{++}, and N.

Sewage effluent discharges coming from villages or trout hatcheries lead to a simultaneous increase of water-dissolved mineral content (Mg^{++}, K^+) and trophic level. This change is shown by a decrease of the species characteristic for waters with low concentrations of dissolved minerals (e.g., *Brachythecium plumosum*, *Hygrohypnum duriusculum*, *Hyocomium armoricum*, and *Marsupella emarginata*). When disturbance increases, all these species and *Scapania undulata* disappear, whereas species such as *Cratoneuron filicinum* appear.

Conversely, acidic precipitation can cause sudden decreases of pH because of the very low buffering capacity of the waters (Probst et al., 1990). In this case, species diversity decreases (Yan et al., 1985). The absence of the acid-sensitive species (e.g., *Chiloscyphus polyanthos*, *Rhynchostegium riparioides*, and *Thamnobryum alopecurum*) and the presence only of species adapted to low pH, such as *Scapania undulata* and *Marsupella emarginata*, is indicative of chronic increases of protons, even though pH may remain most of the time close to neutrality. In the most acidic waters, *Sphagnum* species increase along a gradient of acidification to the detriment of the other aquatic macrophytes (Arts, 1990; Brandrud & Johansen, 1994; Stephenson et al., 1995).

In the hard waters of the Rhine floodplain in France, where waters range from a very oligotrophic status (less than $10\,\mu g\,L^{-1}$ of both $N-NH_4^+$ and $P-PO_4^{3-}$) to a eutrophic status (with about $1{,}000\,\mu g\,L^{-1}$ of $N-NH_4^+$ and $500\,\mu g\,L^{-1}$ of $P-PO_4^{3-}$), Vanderpoorten et al. (1999) found a fairly strong correlation between the main floristic gradient, derived from a principal components analysis of the floristic data set, and water trophic level (Figure 4.2). Trophic level thus seems to determine the segregation of the aquatic bryophyte species in hard waters. *Cinclidotus danubicus*, *C. riparius*, and *Fissidens crassipes* were the species presenting the most eutrophic status, whereas *Amblystegium tenax*, *Pellia endiviifolia*, and *Chiloscyphus pallescens* were restricted to the most oligotrophic part of the gradient. The occurrences of *Amblystegium riparium*, *Fontinalis antipyretica*, and *Rhynchostegium riparioides* were poorly correlated with the trophic gradient. They occurred in both oligotrophic and eutrophic waters, but showed a preference for eutrophic waters.

Aquatic bryophytes thus segregate along gradients of water pollution. The composition of their communities consequently fluctuated as a function of changes in water quality. The diversity of the aquatic bryophyte species has increased during the last decades in the Rhine (Frahm & Abst, 1993; Frahm, 1997) and in two large Belgian rivers, the Meuse and the Sambre (Vanderpoorten, 1999a), in which the changes in aquatic bryophyte assemblages between the 1970s and the 1990s, as assessed by the Jaccard's distance between the floristic assemblages, were best correlated with the changes in heavy metals concentrations and, to a lesser extent, with the changes of trophic level. The impact of an improvement of water quality regarding heavy metals and trophic level on the aquatic bryophyte assemblages is shown along the second axis of the correspondence analysis of the floristic data set, whose site scores are significantly negatively correlated with these factors [axis one with the micropollutants: atrazine ($r = .78$, $p < .05$), pyrene ($r = .93$, $p < .01$), cyanides ($r = .82$, $p < .01$), anionic detergents ($r = .77, p < .01$); axis 2 with the trophic level: NH_4^+ ($r = -.86$), $p < .001$), PO_4^{3-} ($r = .52, p < .05$); and heavy metals: Zn ($r = .73, p < .001$), Cu ($r = .76, p < .001$), Pb ($r = .81, p < .001$)] and along which the scores of the same sites increased between the surveys 1972–1973 and 1997 (Figure 4.3).

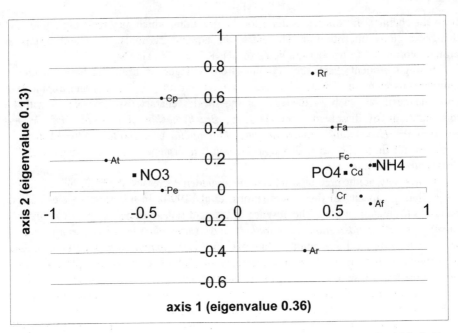

Figure 4.2. Distribution of aquatic bryophytes along a trophic gradient in the Rhine floodplain in France (redrawn from Vanderpoorten et al., 1999). Species and chemical variables are represented by their correlation coefficient on the principal components analysis of the floristic data set. Af = *Amblystegium fluviatile*, Ar = *A. riparium*, At = *A. tenax*, Cd = *Cinclidotus danubicus*, Cr = *C. riparius*, Cp = *Chiloscyphus pallescens*, Pe = *Pellia endiviifolia* (submersed), Rr = *Rhynchostegium riparioides*.

It thus seems that heavy metals and trophic factors, which heavily polluted the rivers in the 1970s, were limiting for the aquatic bryoflora that was able to spread because of the decrease in these factors. Hence, the study sites show more floristic affinities along a spatial than along a temporal gradient. In the Meuse, the floristic assemblages presented differences along a longitudinal gradient in the 1970s, with *Cinclidotus* spp. and *Amblystegium riparium* in the downstream part of the river, and *Cinclidotus* spp., *A. riparium*, *Fontinalis antipyretica*, *Fissidens crassipes*, and certain Amblystegiaceae such as *Hygrohypnum luridum* or *A. tenax* in the upstream part of the river. The decrease in heavy metals and trophic level, which is especially obvious in the downstream sites, led these two different clusters to converge into a single floristic cluster, including species such as *Amblystegium fluviatile*, *A. humile* and *Cratoneuron filicinum*, that were present in the river at the end of the last century and strongly decreased or even disappeared from the Meuse in the 1970s. Nowadays, the longitudinal floristic gradient is thus homogeneous. Hence, the first five sites recorded in 1997 present more floristic affinities with each other than with their homologues recorded in 1972–1973. Only one site, located downstream from the industrial area of Liège, showed no temporal shift of aquatic bryophyte assemblages, which is shown by the proximity of its site scores in 1972–1973 and 1997 along the first axes of the ordination, and which testifies to a still limiting pollution for its bryoflora. In the river Sambre, water pollution reached such a level in the 1970s that only *A. riparium*, which is well-known to be one of the most pollution-tolerant species (Hussey, 1982; Kelly & Huntley, 1987), or even no species occurred. The current floristic assemblages of the Sambre, characterized by *Rhynchostegium riparioides* and *Amblyste-*

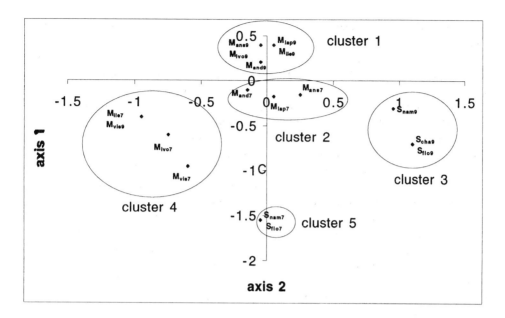

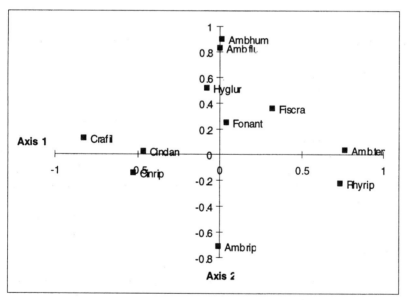

Figure 4.3. Spatiotemporal changes of aquatic bryophytes assemblages in the rivers Meuse and Sambre (Belgium) as a function of changes in water quality between the 1970s and the 1990s (reproduced with permission from Vanderpoorten, 1999a): (a) representation of the sampling sites in the first plane of the correspondence analysis of the floristic data set; (b) representation of the species. Ambflu = *Amblystegium fluviatile*, Ambhum = *A. humile*, Ambrip = *A. riparium*, Ambten = *A. tenax*, Cindan = *Cinclidotus danubicus*, Cinrip = *C. riparius*, Crafil = *Cratoneuron filicinum*, Fiscra = *Fissidens crassipes*, Fonant = *Fontinalis antipyretica*, Hyglur = *Hygrohypnum luridum*. M_{ans7}, M_{lap7}, M_{and7}, M_{ivo7}, M_{ile7}, M_{vis7}, S_{cha7}, S_{flo7}, S_{nam7}: collecting sites (upstream to downstream) in the rivers Meuse (M) and Sambre (S) in the 1970s; M_{ans9}, M_{lap9}, M_{and9}, M_{ivo9}, M_{ile9}, M_{vis9}, S_{cha9}, S_{flo9}, S_{nam9}: the same in the 1990s.

gium tenax, show consequently less similitude with their homologues of the 1970s than with the floristic assemblages of the river Meuse in the same year.

The site scores of the Sambre show a second temporal evolution trend, increasing between the surveys in 1972–1973 and 1997 along the first correspondence analysis (CA) axis. The site scores on axis 1 are significantly correlated with the mean concentrations of K^+ and with the maxima concentrations of $N-NO_2^-$ and $N-NO_3^-$. Exactly as in the river Rhine (Hellmann, 1994), these factors, as opposed to heavy metals and trophic level, have shown a continuous tendency to increase in the rivers Sambre and Meuse during the last decades. They currently reach much lower concentrations in the Meuse than in the Sambre, which is also much more loaded in N and micropollutants such as pyrene, anionic detergents, or atrazine. This might explain why the aquatic bryoflora of the two studied rivers has shown two different patterns of recolonization, that is, the spread of *Cratoneuron filicinum, Amblystegium fluviatile, A. humile, Hygrohypnum luridum,* and *Cinclidotus danubicus* in the river Meuse, and the spread of *Rhynchostegium riparioides* and *Amblystegium tenax* in the lower Sambre, where still no *Cinclidotus* species occur.

Thus, the upstream–downstream zonation patterns of aquatic bryophytes and the shift of their communities strongly correlate to temporal fluctuations of water chemistry, including micropollutants, nutrients, and heavy metals, and suggest their use as biomonitors of water quality.

How Can Aquatic Bryophytes Be Used as Biomonitors of Water Quality?

One of the most important applications of the study of the relationships between biological communities and their environment is the expression of an environmental variable as a function of species data. Calibration concerns how to determine the function $f(X)$ for a given type of variables (X, y) and a given type of sample (Martens & Naes, 1989). Ordinary least squares regression (OLS) is the most commonly used method to fit an equation of regression by minimizing the residual sum of squares. However, OLS can overfit the observed data when there are many predictors, that is, tailoring the model too much to the current data, to the detriment of future predictions. When the predictors show multicollinearity, the least squares estimators exhibit large variances and consequently are not stable (Martens & Naes, 1989; Palm & Iemma, 1995). In addition, the regression coefficients of multicollinear predictors may be not significant despite their significant correlation with the variable to be predicted (Tenenhaus, 1998). In ecology, other methods were thus developed. One of the most popular, thanks to its simplicity, is the weighted averaging of species indicator values (Ellenberg et al., 1991). New developments in the use of ordinations have more recently been used for calibration purposes, by using the ordination components as predictors (Joliffe, 1982). These methods differ by the algorithm used in the construction of the components, which may maximize the percent variance of the explanatory variables, without taking the variable to be predicted into account, as in principal components regression, symmetrically maximize the correlation between the two groups of variables, as in canonical correlation analysis or dissymmetrically maximize the covariance between the two groups of variables, as in partial least squares regression. The first applications of these methods were developed in chemistry by principal components analysis (Martens & Naes, 1989). The principal components regression, however, presents the disadvantage that it does not link the construction of the orthogonal components to the variables to be predicted. Further applications in the use of components derived from ordination techniques such as canonical analysis, interbattery factor analysis, redundancy analysis, and partial least squares regression were therefore developed (Tenenhaus, 1998). As opposed to methods such as canonical or redundancy analysis, in which the number of extractable components equals the number of variables of the smallest of the X and Y data sets, partial least squares regression is especially useful when one aims at describing a variable y thanks to several variables X strongly collinear or even linearly related, and when one does not wish to eliminate any explanatory variable (Tenenhaus et al., 1995).

Regression methods were already used to infer water chemistry from diatoms (Battarbee, 1984; Davis & Anderson, 1985; Oksanen et al., 1988, Holmes et al., 1989) or from aquatic bryophyte assemblages (Thiébaut et al., 1998). Following the results obtained after cross-validation, Birks et al. (1990) recommended simple weighted averaging regression as the easiest and most reliable pH reconstruction procedure currently available. Vanderpoorten & Palm (1998), however, showed that water trophic level can efficiently be inferred from a linear model using canonical variables of aquatic bryophyte assemblages at a regional scale. The aim of the present paragraph is to test the ability of aquatic bryophyte combinations to predict water chemistry, to review the current available calibration techniques used to infer water quality from aquatic bryophyte assemblages, and to compare the performance and reliability of these techniques as a function of the data set, sample size, and species response curve.

Data Sampling

The species are systematically searched at the water-quality measurement stations in at least 250-m-long river sections upstream and downstream. They are found submerged or, more often, at different levels on the banks, where they are only occasionally flooded. Presence–absence or degree of cover may be recorded. However, the degree of cover is closely related to vegetation dynamics due to physical events such as floods and thus does not necessarily mirror water chemistry (Tremp & Kohler, 1993) (Figure 4.4).

Chemical data usually come from seasonal measurements in the same sampling site, raising the problem of how to relate these multiple measurements to a single measurement concerning the bryophyte record. Most of the time, it is necessary to summarize the information included in the chemical data matrix with estimators such as the mean or the standard deviation. The extreme values of the chemical factors can markedly differ from their annual means. Concerning nutrients, however, very high correlation coefficients were found between the log-transformed seasonal averages of total P ($r = .97, p < .001$) and total N ($r = .87, p < .001$) and their variances (Faafeng & Fjeld, 1998), so that the choice of one of these estimators might not influence the analysis.

Data Analysis

Weighted Averaging of Species Indicator Values

Weighted averaging (WA) consists in using the average of the optima of species for a given environmental factor as a prediction of this factor (Ellenberg et al., 1991). Further developments in the use of this technique suggest the weighting of the species by their tolerance (ter Braak & Looman, 1986). Birks et al. (1990) found that weighting of the

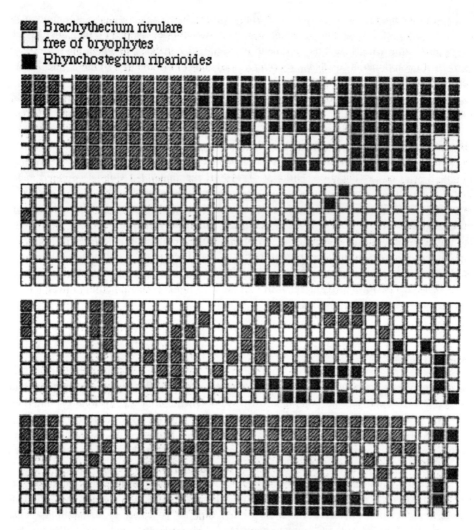

Figure 4.4. Variability of moss cover in water stretches submitted to flooding in the Black Forest (Germany) in Sept. 1991, Feb. 1992, Aug. 1992, and Jan. 1993, respectively (after Tremp & Kohler, 1993).

species indicator values by their tolerance did not improve the fit of the model after cross-validation, hence the use of the simple version of the method:

$$\hat{y}_i = \sum_{j=1}^{p} x_{ij} o_j / \sum_{j=1}^{p} x_{ij},$$

where x_{ij} is the species indicator variable for species j in site i ($x_{ij} = 0$ if species j is absent and $x_{ij} = 1$ if species j is present in site i).

Estimation of the species indicator values can be empirically derived or averaged over the values of the environmental variable for the samples in which a species occurs (e.g., Empain, 1977; Haury et al., 1996; Kelly, 1998). This may, however, give misleading results if the investigated species present a broad ecological range and the distribution of the environmental variable is not homogeneous over the whole range of occurrence of the species (ter Braak & Looman, 1986). Another alternative to define species indicator values is to estimate the species optimum by fitting a curve to the species data by regression. In the case of presence–absence data, the logistic regression fits a curve giving the probability of occurrence of a species as a function of a given environmental factor. A species presents a significant optimum if the factor b_2 of the relation between the probability p of occurrence of a species as a function of a factor y,

$$p = \exp(b_0 + b_1 y + b_2 y^2)]/[1 + \exp(b_0 + b_1 y + b_2 y^2)]$$ (ter Braak & Looman, 1987),

is significantly inferior from 0. The optimum o is then calculated as follows:

$$o = -b_1/2b_2$$

Multiple Regression on Species Indicator Variables

The prediction, $\hat{y}_i$, is derived from a linear combination of species indicator variables x_{ij}:

$$\hat{y}_i = b_0 + b_1 x_{i1} + \cdots + b_p x_{ip}.$$

The coefficients of the linear combination are calculated in order to minimize the sum of squares between the observed and predicted values. Two variants of the multiple regression can be used. In the first case (OLS), the prediction is inferred from the p available species indicator variables. In the second case (STEPWISE), the species indicator variables are selected in order to conserve only a subset of statistically significant variables. The selection of the subset of r variables can be performed by different methods, such as forward, backward, or STEPWISE selection [see Draper & Smith (1981) and Weisberg (1985) for further details].

Multiple Regression on Linear Combinations of Species Indicator Variables

The principle consists in calculating several linear combinations of all the species indicator variables,

$$t_{ik} = c_{k0} + c_{k1} x_{i1} + \cdots + c_{kp} x_{ip} (k = 1, \ldots, s, s \leq p),$$

and in calculating a linear regression using the first q linear combinations as explanatory variables:

$$\hat{y}_i = b_0 + b_1 t_{i1} + \cdots + b_q t_{iq}.$$

Different algorithms can be used to calculate the linear combinations t_{ik}. In the case of the principal components regression (PCR), the linear combinations are derived from the standardized variables

$$x_{ij}^o = (x_{ij} - \bar{x}_j)/\hat{\sigma}_j$$

as follows:

1. The linear combinations are computed in order to maximize their variance.
2. The linear combinations are uncorrelated.
3. The sum of the squared coefficients of each linear combination equals 1 [see Weisberg (1985) and Jackson (1991) for further details].

The linear combinations can also be derived from a correspondence analysis (CA). On a given component, the coordinate of a site is the arithmetic mean of species coordinates weighted by the site profile, multiplied by a constant, which is identical for all the sites and is a function of the order k of the component.

In PCR and in regression on CA components, the components are solely derived from the species indicator variables, without taking the variable to be explained into account. In partial least squares regression (PLS), conversely, the components are derived by taking the variable y into account. More precisely, the components t_{ik} are derived by maximizing the covariance between the standardized species indicator variables and the standardized variable y, by respecting the following constraints:

1. The components t_{ik} are uncorrelated.
2. The sum of the squared coefficients of each component equals 1 (Tenenhaus, 1998).

The maximum number of factors that can be computed equals the number of species indicator variables. The number of factors q to extract depends on the data. Basing the model on more extracted factors improves the model fit to the observed data, but extracting too many factors can cause overfitting, that is, tailoring the model too much to the current data to the detriment of future predictions. q can be determined either by selection (forward, backward...) or by cross-validation, using the predicted error sum of squares (PRESS) statistics:

$$\text{PRESS} = \sum_{i=1}^{n}(y_i - \hat{y}_{(i)})^2.$$

PRESS is calculated by cross-validation as follows:

1. The ith site is deleted from the data set.
2. The prediction model is derived from the $n - 1$ other sites.
3. The prediction for the deleted ith site, $\hat{y}_{(i)}$, is calculated from the model derived from the $n - 1$ other sites and the mean, $\bar{y}_{(i)}$, is calculated as the mean of the $n - 1$ other sites.
4. The steps 1, 2, and 3 are repeated for $i = 1, \ldots, n$.

The number of components q is thus that minimizing the PRESS.

Which Technique Performs Better with Which Data?

Although aquatic bryophytes have been used for biomonitoring water quality (e.g., Frahm, 1974; Benson Evans & Williams, 1976; Vhrovsek et al., 1984; Arts, 1990; Tremp & Kohler, 1993; Haury et al., 1996), the precision and the validity of these methods were not assessed. Reconstructing past pH from diatom assemblages has known further developments, and statistical methods, whose accuracy could be measured, were used (e.g., Charles, 1985). Some methods may produce lower apparent root mean squared errors. However, these error estimates are not based on rigorous error estimations such as cross-validation, but on regression statistics derived solely from training sets. As ter Braak & van Dam (1989) emphasized, root mean squared error based on training sets alone gives an overoptimistic idea of prediction. Vanderpoorten & Palm (2001) used the cross-validated r-squared Q^2 criterion to compare the performance of the models from four data sets.

For a data set of n sites, the Q^2 criterion is

$$Q^2 = 1 - \text{PRESS}/\text{SS},$$

where SS is the sum of squares,

$$\text{SS} = \sum_{i=1}^{n}(y_i - \bar{y}_{(i)})^2,$$

and where $\bar{y}_{(i)}$ is the "naive" prediction of y_i (i.e., the mean value of y_i after exclusion of site i).

In order to compare the performance exhibited by a general model including all the data and by local models, the Q^2 of the data set with the mean Q^2, $\bar{Q}^2$, of the local models is

$$\bar{Q}^2 = 1 - \sum_{a=1}^{4} \text{PRESS}_a / \sum_{a=1}^{4} \text{SS}_a,$$

where a refers to the local models.

In STEPWISE, the effects of a systematic reselection of variables for each of the n observations were compared. The selection of the variables n times on the one hand, and without reselecting each time the variables by using the variables selected once on the basis of n sites on the other, was carried out. In OLS, problems of multicollinearity between variables occurred when deleting one observation from the data set. In this case, one of the multicollinear variables was deleted, and the equation of regression was recalculated with the $p - 1$ remaining variables and used to predict y in the deleted site. In the case of the regression on components, the components were recalculated n times for the $n - 1$ sites. The number of components q included in the prediction model was, however, determined once on the basis of the n sites.

Table 4.1 summarizes the results obtained by each prediction method and for each data set. The investigated methods exhibited very contrasting performances. The STEP-

Table 4.1. Summary of the Performance of the Investigated Regression Methods (after Vanderpoorten & Palm, 2001).

Data set (number of sites)	WA Q^2	OLS p	Q^2	r	STEPWISE Q^2	q	PCR Q^2	q	CA Q^2	q	PLS Q^2
GEN (65)	−0.04	10	0.38	4.0	0.35 (0.47)	3	0.22	2	0.25	3	0.40
A (20)	—	8*	0.76	2.1	0.82 (0.91)	1	0.80	1	0.74	2	0.69
B (14)	—	9*	−1.09	2.0	0.62 (0.72)	2	0.24	3	0.54	1	0.42
C (19)	—	10	0.03	2.5	0.27 (0.65)	1	0.34	1	0.25	2	0.35
D (12)	—	8*	0.22	1.4	−0.07 (0.58)	2	0.72	2	0.40	1	0.68
mean Q^2			0.13		0.43		0.53		0.51		0.54

Q^2: Stone–Geisser criterion, r: mean number of bryophyte species selected by the stepwise procedure; the Q^2 obtained without reselecting the variables at each step of the cross-validation are given in parentheses; q: number of linear combinations selected for the prediction model; *one or two species have been suppressed, because present or absent in all sites. WA = weighted averaging, OLS = ordinary least squares regression without selection of variables, STEPWISE = stepwise least squares regression, PCR = principal components regression, CA = correspondence analysis regression, PLS = partial least squares regression.

WISE method exhibited the highest prediction ability for two of the five investigated data sets. However, terrible predictions for one data set were obtained with this method. It was thus impossible to select *a priori* the method producing the lowest prediction error for all the investigated data sets. Among the methods producing the lowest prediction errors, PLS and PCR provided acceptable predictions of the mean $N-NH_4^+$ concentrations for all the investigated data sets (i.e., Q^2 always positive). In reality, none of the PLS and related methods fit the observed data any better than OLS, because all the methods approach OLS as more factors are extracted. The crucial point is that when there are many predictors, OLS can overfit the observed data, and biased regression methods with fewer extracted factors can provide better predictability of future observations (SAS Institute, Inc., 1999). PLS and PCR, although not producing the best predictions in all the cases, could therefore be seen as the best compromise between prediction accuracy and constancy of the prediction quality. Figure 4.5 gives as an example the measured $N-NH_4^+$ concentrations versus the predicted ones after cross-validation for the five data sets in the case of PLS.

The ability of linear models such as PLS and PCR to provide the best prediction ability compared to other models stands in contrast with recent reviews on the use of ordination techniques in ecology (Okland, 1990). Birks et al. (1990) suggested that

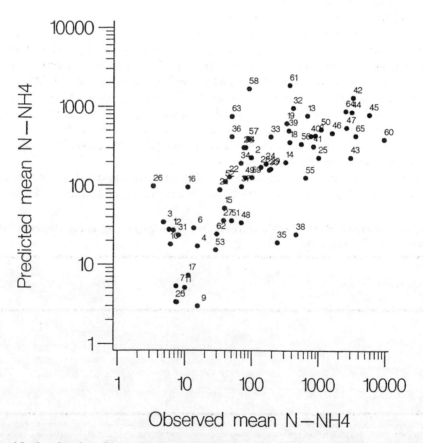

Figure 4.5a. Log–log plots of the predicted mean concentrations of $N-NH_4^+$ inferred by PLS from 10 aquatic bryophyte species after cross-validation as a function of the mean of the measured $N-NH_4^+$ concentrations: (a) large-scale model (n = 65) (with permission by Kluwer).

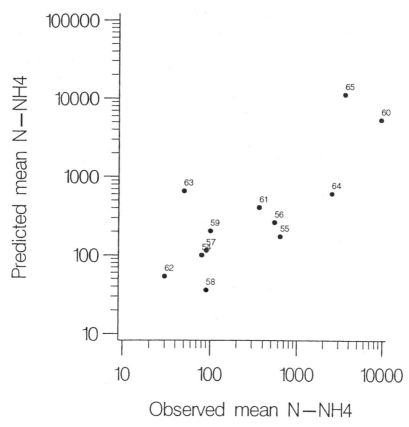

Figure 4.5b. Log–log plots of the predicted mean concentrations of N–NH$_4^+$ inferred by PLS from 10 aquatic bryophyte species after cross-validation as a function of the mean of the measured N–NH$_4^+$ concentrations: (b) regional model (Rhine floodplain in Eastern France, $n = 15$) (reproduced with permission, from Vanderpoorten & Palm, 2001).

methods relating environmental factors to bell-shaped species response curves sound theoretically better than other procedures. This is especially true in the case of data sets including large environmental gradients, along which the species response curves are commonly bell-shaped (Whittaker, 1967). Along a gradient of concentrations of N–NH$_4^+$ such as that investigated, which covers the whole range of the species, it could thus be expected that WA, which relies on a sum of species optima, and CA, which is related to a unimodal response model (ter Braak, 1985), give better results than the linear methods. Ter Braak & van Dam (1989) found that WA of diatom indicator values used to assess pH variations in water bodies gave better results than linear methods. The root mean squared error (RMSE) obtained in a training set and in a test set, respectively, were 0.46 and 0.75 with WA, 0.45 and 2.24 with OLS, and 0.51 and 2.74 with STEPWISE.

Two major characteristics of the data sets explain why WA and regression on CA components did not perform better. First, the diversity of aquatic and subaquatic bryophytes—between 1 and 10 species per sample—is very low compared to that of other taxonomic groups such as the diatoms. Indeed, although substitution habitats due to the regulation and/or the canalization of the large rivers have allowed the occurrence of allochtonous species in the downstream areas, increasing the number of species since the last century (Frahm, 1997; Vanderpoorten & Klein, 1999b), aquatic bryophytes are mainly characteristic for the upstream mountain parts of the streams and decrease in frequency and diversity in sluggish lowland rivers, where habitat conditions favor vascular macrophytes (Tremp & Kohler, 1995). Ter Braak & van Dam (1989) found that WA does not exhibit good performances when the taxa diversity is low (less than approximately 10 species). Second, although the gradients of $N-NH_4^+$ were large and covered the whole species spectra, only 5 of the 10 species exhibited significant optima. The five other species exhibited monotonic response curves and could consequently not be used as indicators in WA. The exclusion of species such as *Cratoneuron filicinum* or *Chiloscyphus pallescens*, which exhibit a typical oligotrophic status, leads to a drastic loss of information.

But the main reason explaining such differences in results relies on the method of validation itself: ter Braak & van Dam (1989) used a test set, whereas we used a cross-validation. The different ways to validate the results can lead to contrasting conclusions. In STEPWISE, the reselection of variables for each deleted observation during the procedure of cross-validation always led to an increase of the PRESS, compared to the single selection. The two methods even led to very contrasting conclusions with the data set D. In this case, STEPWISE was rejected because it produced a negative Q^2 statistics when reselecting the variables at each step of the cross-validation, whereas the Q^2 reached 0.58 when selecting the variables once on the basis of the data set including the n observations. In STEPWISE, thus, the ability of the regression models to predict future observations cannot be solely assessed by selecting once the variables of the complete data set but must be performed by reselecting the variables for each site. In order to compare our results with those obtained by the method of the test set used by ter Braak & van Dam (1989), we applied the model constructed with the data from the hydrographic network A to the data of the hydrographic networks C and D. When using this method of validation, we also obtained terrible predictions. The ability of the model A to predict $N-NH_4^+$ concentrations for the data set C and D was so bad that the RMSE were higher than the estimated standard deviations of these concentrations: For the hydrographic network C, the RMSE, expressed in percent of the standard deviation of y, reached 109 by OLS and 120 by STEPWISE. For the hydrographic network D, the RMSE reached 89 by OLS and 108 by STEPWISE.

These results show that regional models cannot be transposed from an area to another. Consequently, the mean Q^2 calculated on the Q^2 from the four hydrographic networks was much higher than that of the general model based on all the observations put together. Certain $N-NH_4^+$ values correctly predicted at a regional scale are ill-predicted at a large scale. Very similar results were obtained in models inferring past pH from diatom assemblages conserved in lake sediments, and many investigators have developed diatom-based pH inference equations for specific areas (Holmes et al., 1989).

Are Aquatic Bryophytes Good Indicators of Water Quality?

Figure 4.5 and Table 4.1 show that PLS and related methods allow the estimation of water chemistry from the presence–absence of aquatic bryophytes. However, the performance of the models, as assessed by the cross-validated r-squared ranging for log-transformed data, from about 0.25 to 0.80, was rather low. The fact that the aquatic vegetation does not directly reflect water quality is partly due to the ability of aquatic macrophytes to integrate the variability of water quality over time (Trémolières et al.,

1994). With their very slow population dynamics, aquatic bryophytes are especially relevant. Many sites currently exhibiting good-quality waters according to recent physicochemical measures and index of water quality based on organisms with faster population dynamics, such as diatoms, are considered polluted according to aquatic bryophytes (Descy & Empain, 1981). Vanderpoorten & Palm (1998) showed that in the brooks along the river Rhine, which have been totally disconnected from the main river since its canalization began in 1930, and therefore have only been fed by groundwater, the bryophytes gave estimations of the physicochemical variables greater than the measured variables. This is due to the presence of species with a large trophic range but with eutrophic tendencies, such as *Amblystegium riparium*. The presence of these species is evidence of the former eutrophic status of these streams when they were still connected to the Rhine. Their large ecological range allows them to develop in habitats that progressively became oligotrophic over about 30 years. However, besides this ability of aquatic bryophytes to testify to ancient disturbance of their environment, two main reasons explain their poor performance as bioindicators.

Aquatic Bryophytes Exhibit a High Tolerance to Pollutants

Aquatic bryophytes absorb chemical components through their surfaces. They differ from phanerogams in that they behave essentially as an ion exchange medium and lack the regulatory mechanisms of roots (Empain, 1985). Because the leaves are mostly one cell thick, they permit easy entry, hence a rapid absorption of the pollutants. Aquatic bryophytes therefore accumulate heavy metals, radionucleids, and organic pollutants (Empain, 1976; Say & Whitton, 1983; Mouvet, 1985; Gonçalves et al., 1994; Siebert et al., 1996). Frisque et al. (1983) found that the accumulation of polychlorobiphenyls (PCBs) in the mosses varies between 10,000 and 23,500. The liverwort *Jungermannia* accumulates the pesticide fenitrothion to over $100\times$ the water concentration (Eidt et al., 1984) and the moss *Fontinalis antipyretica*, $33\times$. As a comparison, phytoplankton accumulated only about $8\times$ the water concentration (Morrisson & Wells, 1981).

The increasing number of chemicals in the environments requires efficient protective mechanisms in affected organisms. Plants and animals are able to metabolize foreign compounds (xenobiotics) with different types of detoxifying enzymes systems. The detoxification process involves three steps: (1) transformation by oxidation, reduction, or hydrolyzation performed by an enzyme complex such as the cytochrom P-450, (2) conjugation on a substrate; and (3) accumulation under an inactive form in vacuole, lignin, or release (Sandermann, 1994). One of the best-investigated detoxification mechanisms is conjugation of xenobiotics with glutathione by glutathione S-transferase (GST). Nearly 100 GST isozymes have been characterized to date (Buetler & Eaton, 1992). Some investigations have demonstrated a change in GST activity in *Fontinalis antipyretica* exposed to benzo(a) anthracene and benzo(a) pyrene (Roy et al., 1995). Samples of the vascular macrophyte *Nuphar lutea* from two different lake areas exhibited increased GST activity in plants from the site contaminated with polyaromatic hydrocarbons (Schrenk et al., 1998). Aquatic macrophytes, including bryophytes, thus seem to be able to develop efficient enzyme systems to face the increase in pollutants in their environment. As a consequence, aquatic bryophyte species may present contrasting tolerance to pollutants, but most of the species seem to tolerate concentrations much higher than those encountered in nature.

Metals

The ability to sequester heavy metals in the cell wall and in vacuoles permits survival of bryophytes under conditions in which many vascular plants cannot survive. Sarosiek et al. (1987) reported lethal concentrations of 20 mg L^{-1}, 10 mg L^{-1}, and 30 mg L^{-1} for Zn, Cu, and Pb, respectively, on the liverwort *Ricciocarpos natans*. After 4 weeks of exposure, Frahm (1976) found no effects of Zn^{++}, Cd^{++}, Mn^{++}, and Pb^{++} at the concentrations of 0.8 mg L^{-1}, 0.02 μg L^{-1}, 0.3 mg L^{-1}, and 0.4 mg L^{-1}, respectively, on *Amblystegium riparium*, *Fissidens crassipes*, *Fontinalis antipyretica*, *Leskea polycarpa*, and *Schistidium alpicola*. As a comparison, these factors reached maximal concentrations of 0.5 mg L^{-1}, 0.01 μg L^{-1}, 0.2 mg L^{-1}, and 0.225 mg L^{-1} during the period of the heaviest pollution in the 1970s in the lower Rhine. Frahm (1975) found a tolerance to Fe^+ greater than 12 mg L^{-1} for *Amblystegium riparium*, *Leskea polycarpa*, and *Fissidens crassipes*, whereas *Schistidium alpicola* died at 4 mg L^{-1} and *Fontinalis antipyretica* at 2 mg L^{-1}. As a comparison, the highest concentrations in Fe, reached in the 1970s, were of about 2 mg L^{-1} in the Sambre, one of the most heavily polluted Belgian rivers (Descy & Empain, 1981). Thus, although metals can affect terrestrial bryophytes in heavily polluted sites of accumulation such as former mining areas, causing saturation of the available anionic exchange sites external to the cell cytoplasm, but also decline in photosynthesis in proportion to intracellular metals such as Cd (Rühling & Tyler, 1970; Brown & Whitehead, 1986; Brown & Wells, 1990; Wells & Brown, 1995; Sidhu & Brown, 1996), they are almost never found in rivers at concentrations approaching critical values for aquatic bryophytes.

Organic Pollutants

Kosiba & Sarosiek (1995) determined that two organic pollutants originating from textile industry in the Danube River, aniline and fuchsine, are toxic for *Fontinalis antipyretica* and *Rhynchostegium riparioides*. The LC_{50}, calculated by probit regression (see Ryan, 1997, for review), were 17.03 mg L^{-1} and 2.22 mg L^{-1} of aniline and fuchsine, respectively, for *Fontinalis antipyretica*, and 23.48 mg L^{-1} and 4.81 mg L^{-1} for *Rhynchostegium riparioides* (Figure 4.6). Fuchsine thus presents a higher degree of toxicity than aniline. The tolerance of the mosses toward these substances is different, *Fontinalis antipyretica* being more sensitive than *Rhynchostegium riparioides*. Kosiba and Sarosiek (1995), unfortunately, gave no measure of these pollutants in their study area. Frahm (1976) found a tolerance to phenol after 4 weeks of exposure of 0.02 mg L^{-1} for *Fontinalis antipyretica* and 0.08 mg L^{-1} for *Leskea polycarpa*, *Amblystegium riparium* and *Fissidens crassipes*. As a comparison, phenol concentrations up to 0.023–0.045 mg L^{-1} were measured in the lower Rhine in the 1970s. No effects of benzopyrene on the growth of four aquatic moss species were detected during a 1-month cultivation experiment at concentrations up to 1 mg L^{-1}, that is, 400 times their maximal values reached in 1996 in the running surface waters of southern Belgium (Vanderpoorten, 1999b).

Trophic level

Changes in aquatic bryophyte communities are strongly correlated to changes in the trophic level of the waters (Vanderpoorten, 1999a, Vanderpoorten et al., 1999). Karttunen and Toivonen (1995) attributed the spread of *Riccia fluitans* and *Ricciocarpos natans* in

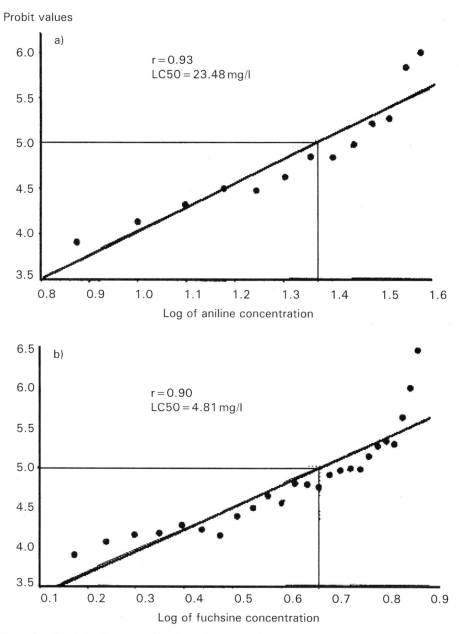

Figure 4.6. Correlation between probit values of mortality of *Rhynchostegium riparioides* and aniline (a) and fuchsine (b) concentrations (after Kosiba & Sarosiek, 1995).

southern Finland to eutrophication. However, experimental results on the effects of nutrient enrichment on aquatic bryophytes produced contradicting results. Baker & Boatman (1985) concluded that K, N, and P all affect the interfascicle length & vegetative reproduction (innovation formation) of *Sphagnum cuspidatum* within the supplied ranges of concentrations, but that the only element inducing a growth response in natural conditions is P. Conversely, no response to nutrient supply could be detected in a series of culture experiments of P enrichment on the moss supposed to be of eutrophic status, *Calliergonella cuspidata* (Kooijman & Bakker, 1993). Similarly, Steinman (1994) found that the leafy liverwort *Porella* is not phosphorus-limited from both primary production and tissue chemistry data after 5 weeks of phosphorus enrichments in two woodland streams in eastern Tennessee. Although tissue phosphorus concentrations increased after enrichment in *Porella*, corresponding increases in primary production did not occur, suggesting that the sorption of luxury consumption was more responsible for these phosphorus increases than limitation. If nutrients do not limit the growth of aquatic bryophytes, it would be of considerable interest to determine which mechanisms account for this phenomenon. Two possibilities that may warrant further attention are luxury consumption of potentially limiting nutrients and highly efficient cycling of nutrients, perhaps through sorption/desorption processes.

Increasing nutrient concentrations thus does not clearly cause an increase in the primary production of aquatic bryophytes. Conversely, factors such as NH_4^+/NH_3 can be toxic for bryophytes at high concentrations. Table 4.2 reviews the available lethal concentrations of NH_4^+ for six moss species. The tolerance ranged between 7.5 mg L^{-1} for *Fontinalis antipyretica* to 500 mg L^{-1} for *Amblystegium riparium*. The investigated species exhibit fairly high tolerance to NH_4^+. As a comparison, the maximal concentrations reached during the 1970s in one of the most polluted rivers of Belgium flowing in a heavily industrialized and urbanized area was 19.5 mg L^{-1}. Experimental data thus suggest that water trophic level does little to explain changes in species composition in most of the moderately polluted waters.

pH and Buffering Capacity

One of the main factors affecting aquatic bryophytes in culture experiments is pH and associated factors, such as the buffering capacity. One result of low pH is a decrease in aquatic bryophyte richness (Yan et al., 1985) and a shift of bryophyte communities, with an increase of *Sphagnum* species (Roelofs et al. 1984; Roberts et al., 1985). *Sphagnum* strong cation exchange capacity removes biologically important cations (Clymo, 1963) and replaces them with protons. Tremp & Kohler (1993) found a decrease in photosynthesis

Table 4.2. Lethal Concentrations for $N-NH_4^+$ in Aquatic Bryophytes (after Frahm, 1975, 1976; Vanderpoorten, 1999b).

Species	Lethal concentration (mg L^{-1})
Amblystegium riparium	500
Amblystegium tenax (1)	50
Amblystegium tenax (2)	100
Amblystegium tenax (3)	500
Fissidens crassipes	11.25
Fontinalis antipyretica	7.5
Rhynchostegium riparioides	50
Schistidium rivulare	11.25

Three populations of *Amblystegium tenax* coming from rivers with contrasting water qualities were investigated: (1) and (2) from unpolluted rivers and (3) from a heavily polluted area.

and development of protonema in *Chiloscyphus polyanthos* and *Rhynchostegium riparioides* when pH was decreased. Conversely, certain species, such as *Scapania undulata*, display a remarkable ability to tolerate acidity (Figure 4.7). *Jungermannia vulcanicola* was recorded in acid springs at a pH as low as 1.9 (Satake et al., 1989). Concentrations of K in shoots of *Jungermannia vulcanicola* and *Scapania undulata* were high in the acid streams, reaching 5.78% and 2.76% of dry weight, respectively. As a comparison, concentrations in *Rhynchostegium riparioides*, located further downstream when confluent water of pH 5.5 entered the river, were less than 1%. The additional H^+ in the water at low pH competes with other ions for the binding sites of bryophytes, causing nutrition troubles (Glime, 1992). Satake et al. (1989) suggested that acid-tolerant species possess a physiological system regulating the difference in H^+ concentration between the outside (acidic water) and inside (protoplast) to maintain activity for growth in acidic environments. An advantage of the ability to grow at low pH is the inverse relationship between pH and the ability to accumulate potentially toxic ions of Al, Mn, and Zn (Caines et al., 1985). pH

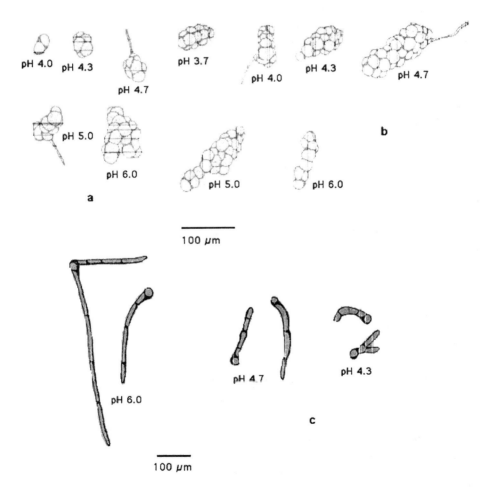

Figure 4.7. Development of the protonema of *Chiloscyphus polyanthos* (a), *Scapania undulata* (b), and *Rhynchostegium riparioides* (c) after 25 days of exposure as a function of pH (redrawn from Tremp & Kohler, 1993).

also has indirect effects on bryophyte nutrition. At low pH, all the CO_2 is present as free CO_2, making it possible for the bryophytes to obtain adequate CO_2 for photosynthesis, hence the tendency for normally terrestrial taxa, such as *Dicranella heteromalla*, *Polytrichum longisetum*, and *Bryum* spp. to grow submersed (Sand-Jensen & Rasmussen, 1978; Thiébaut et al., 1998). Indeed, unlike the strictly aquatic vascular macrophytes that can use the bicarbonates as a source of C, bryophytes can use only free CO_2 (Bain & Proctor, 1980; Allen & Spence, 1981). Another pH-mediated relationship is the change from NO_3^-- to NH_4^+-dominated N utilization. Bryophytic N nutrition seems to be NH_4^+-dominated. *Sphagnum flexuosum* exhibited no NO_3^- utilization in experiments. Likewise, *Fontinalis antipyretica* uses ammonium N (Schwoerbel & Tillmans, 1974), and *Riccia fluitans* occurs in waters that are low in NO_3^- but high in NH_4^+ (Roelofs, 1983). As a consequence, when NH_4NO_3 is supplied to acidified water, communities can shift, favoring *Sphagnum* spp. and *Drepanocladus fluitans* (Schuurkes et al., 1986).

There is thus strong experimental support for explaining aquatic bryophyte segregation as a function of pH, cation contents, and related factors. Even slight changes in ion concentrations in waters with low buffering capacities may cause drastic floristic changes. Conversely, culture experiments show that aquatic bryophytes present unexpected tolerances to trophic level and pollutants. Metals and micropollutants may induce contrasting response patterns among aquatic bryophyte species, but at concentrations much higher than those encountered in nature. NH_4^+ is toxic for certain species, such as *F. antipyretica* at concentrations of a few mg L^{-1}, which still may exist in the most heavily polluted rivers. Conversely, certain other species, such as *Amblystegium tenax*, exhibit such a tolerance that they should not be affected by the highest pollution peaks. Aquatic bryophytes, though, segregate even in brooks and rivers with NH_4^+ concentrations ranging between 10 and 1,000 µg L^{-1}, thus never experiencing high levels of NH_4^+ contamination. It cannot be excluded that combinations or interactions of factors may display effects at much lower concentrations than those of the factors considered individually. Examples of such interactions exist for ammonium and carbon dioxide, which are necessary for ammonium enrichment to increase the biomass of *Sphagnum cuspidatum* (Paffen & Roelofs, 1991). This is also true for mineral and nutrients, whose supply seem to compensate for the intolerance to high mineral levels in *S. fallax* (Kooijman & Kanne, 1993). However, the correlation between changes in aquatic bryophyte assemblages and water chemistry is not strongly supported by experimental data, and the spread of a number of species still remains poorly understood. In cryptogamic epiphytes, a similarly documented spread has largely been attributed to the improvement of air quality (Greven, 1992; Sjögren, 1995), but the link between the two phenomena has not been more successfully demonstrated than for aquatic bryophytes.

Aquatic Bryophytes Are Ecologically Polymorphic

The second most serious problems with the use of aquatic bryophytes as indicators of water quality is that a number of species display contradicting response curves in different areas. The different ecological profiles exhibited by such species, which may be linked to our failure to recognize different ecotypes, is very detrimental to the quality of the estimation (Battarbee, 1984). Vanderpoorten & Durwael (1999) compared the response curves of aquatic bryophytes in two different river basins in Wallonia (Belgium) and in Alsace (France). Most of the investigated aquatic bryophytes of calcareous streams showed unimodal response curves, modeled by logit regression (see ter Braak & Looman, 1987, and Ryan, 1997, for review), along large trophic gradients of N–NH$_4^+$ and P–PO$_4^{3-}$. This is in agreement with the general concept of the bell-shaped species response curves with respect to environmental gradients (Whittaker, 1967). The modeled response curves obtained in the Walloon hydrographic network, in the Alsatian hydrographic network and in both networks considered simultaneously showed a strong overlap in *Amblystegium riparium* (Figure 4.8). The populations of this species consequently belong to a single ecological phenotype regarding the water trophic level. *A. riparium* is the species exhibiting the broadest ecological range, occurring from oligotrophic streams in Alsace to hypertrophic rivers in the Walloon hydrographic network. *A. riparium* is in fact well known to be one of the most organic pollution–tolerant species (Frahm, 1974; Hussey, 1982; Kelly & Huntley, 1987). Its expected frequency in waters with N–NH$_4^+$ annual mean concentrations up to 20,000 µg L^{-1} reached 0.4. The frequency of occurrence of *Rhynchostegium riparioides* decreased much more quickly in eutrophic conditions than that of *Amblystegium riparium*. *R. riparioides* had an expected occurrence frequency of 0.4 at 5,000 µg L^{-1} whereas this frequency of occurrence was expected at 20,000 µg L^{-1} in *Amblystegium riparium*.

In *Amblystegium fluviatile*, conversely, the response curves did not overlap between the studied hydrographic networks. This was also the case for *Amblystegium tenax*, *Fissidens crassipes*, and *Fontinalis antipyretica*. *Amblystegium tenax* has a typical oligotrophic status in the Alsatian floodplain, but in the Walloon hydrographic network, it is still present in the most polluted waters, whose annual mean concentrations of N–NH$_4^+$ reach up to 9,000 µg L^{-1}. *Amblystegium fluviatile*, *Fissidens crassipes*, and *Fontinalis antipyretica* exhibited optima in eutrophic waters when data from the two hydrographic networks were simultaneously considered but reached their maximal occurrence in the most oligotrophic waters in the Walloon hydrographic network. Data from the literature also are contradictory. *Fontinalis antipyretica* is usually regarded as a species characteristic of good-quality waters but occurs in sites as heavily polluted as the lower Elbe (Germany) (Frahm & Abst, 1993). *Amblystegium fluviatile* is presented as a species of oligotrophic status in the Grand-Duchy of Luxembourg (Werner, 1996) but of eutrophic status in The Netherlands (Landwehr & Barkman, 1966). Such geographical differences in species niches have already been pointed out by Shaw (1985) on the basis of two ecological studies of *Tomentypnum falcifolium* and *T. nitens* performed in two different areas, producing contradictory results.

These differences in species ecology might be due to different intercorrelation between the environmental variables or the species in the study areas. For aquatic bryophytes, which are mostly relegated to the variable water level zone (Muotka & Virtanen, 1995), niche structure and habitat conditions are of prime importance (Vitt et al., 1986; Glime & Vitt, 1987) and might interfere with the suitability of water quality to explain the segregation of the species. For example, *Cinclidotus riparius* presents an ecological optimum in experimental conditions when submitted to variable water levels (Empain, 1977) and accordingly occurs along rivers with variable water discharges. In the studied basins, such conditions were found in the largest and thus moderately eutrophic rivers, so that *C. riparius* seems to have, as in the Netherlands (Landwehr & Barkman, 1966), a eutrophic status. *Cinclidotus* species, however, also occur abundantly in oligotrophic waters with variable levels, such as the temporary karstic groundwater-fed lakes of western Ireland (Reynolds, 1998). This suggests that significant correlation may occur between changes in water chemistry and aquatic bryophyte assemblages, which may be due to coevolution of the concentrations in certain pollutants, with little or no influence

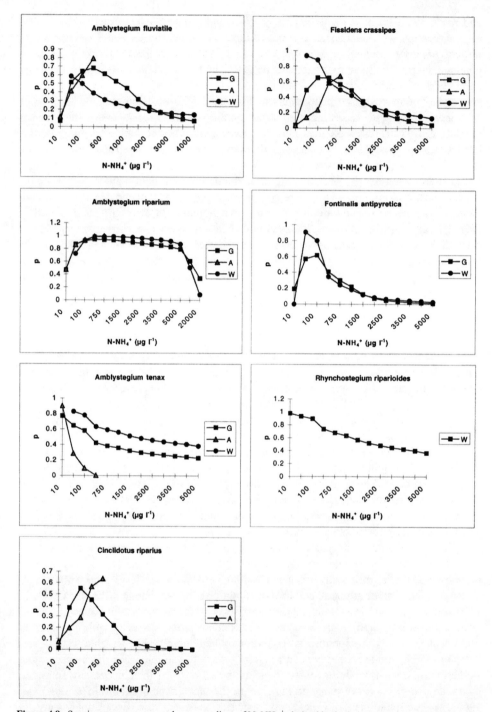

Figure 4.8. Species response curves along a gradient of $N-NH_4^+$ derived by logistic regression within the range of concentrations of two hydrographic networks: Alsace (A) and Wallonia (W). G = response curves calculated when considering the data from both hydrographic networks simultaneously (reproduced, with permission, from Vanderpoorten & Durwael, 1999).

on aquatic biocenoses with other environmental factors, such as niche structure and habitat conditions truly important for explaining changes in species composition.

A response curve is the result of the physiological response and competition between species (Fresco, 1982). A species can be outcompeted near its physiological optimum by more competitive species, whereas it may be able to cope with less favorable environmental conditions when competition is less. The floristic assemblages assessed in this study, however, were very similar in the investigated hydrographic networks. Moreover, the importance of the competition between bryophytes remains an area of controversy (Slack, 1990; Wilson et al., 1995). Consequently, the observed differences in species response curves cannot definitely be attributed to this factor.

The different response curves of a number of species in the investigated river basins may suggest that different genetic strains with contrasting physiological tolerances exist in these species. Experimental cultures revealed that populations of the widespread species *Fontinalis antipyretica*, *Rhynchostegium riparioides*, and *Amblystegium tenax* from areas with various pollution levels exhibited contrasting tolerances at very high pollution levels (Sarosiek & Samecka-Cymerman, 1987; Vanderpoorten, 1999b). The basins are indeed rarely interconnected, so that few diaspore exchanges occur between them. In addition, aquatic bryophytes are only rarely fertile. This may favor processes of regional microevolution without morphological variation, which might interfere with the suitability of aquatic bryophytes as biomonitors of water quality.

Molecular work on phylogenetic relationships among mosses has been progressing at a rapid pace over the last several years, but most such studies have focused on higher level relationships utilizing chloroplast and nuclear ribosomal DNA large subunit sequences (e.g., Hedderson et al., 1996; Goffinet et al., 1998; De Luna et al., 1999; Buck et al., 2000). At the population level, Shaw et al. (1987) found high isozyme differentiation between aquatic (var. *kindbergii*) and riparian populations of *Climacium americanum*. Amplified fragments length polymorphism used to determine the genetic distance between five populations of *Amblystegium tenax* compared to two populations of *A. fluviatile* suggested that populations of *A. tenax* were genetically clearly segregated from other populations to such an extent that this genetic differentiation was higher than that separating *A. tenax* from *A. fluviatile* (Vanderpoorten & Tignon, 2000) (Figure 4.9). There was thus evidence for a high genetic differentiation between the investigated strains of *A. tenax*. In higher plants, some authors reported significant correlation between habitat type and allozyme loci (Heywood & Levin, 1985; Van Rossum et al., 1996). However, investigations on allozyme frequency–environment correlation in outcrossing plants species have yielded contradictory results. Several studies failed to show allozymic differentiation among ecological races (Westerbergh & Saura, 1992; Freiley, 1993; Aitken & Libby, 1994) and concluded either that ecological separation was too recent for drift to have promoted allozymic divergence or that the latter had been prevented by sufficient gene flow. The level of genetic differentiation among strains found in *A. tenax* suggests that ancient genetic individualization occurred among the investigated strains and that these populations belonging to different hydrosystems with few connections between them are genetically isolated from each other.

Thus, taxonomically recognized species might not show the genetic differences that would be expected from their morphology. Confusing patterns of variation have led to highly variable assessments of specific and even generic delineation. Sporophytes are remarkably uniform within the most important families of pleurocarpous mosses (e.g., Amblystegiaceae, Brachytheciaceae, and Hypnaceae), so that both species and genera are

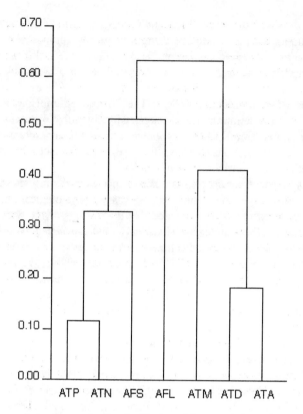

Figure 4.9. Summary cluster diagram for the classification of five populations of *Amblystegium tenax* (AT_x) compared to two populations of *A. fluviatile* (AF_x) according to the Jaccard's distances derived from the presence–absence of their amplified fragments of restriction and using average linkage analysis (after Vanderpoorten & Tignon, 2000). Dissimilarity, indicated on the vertical axis, is the root mean square distance between clusters.

distinguished by such gametophytic characters as plant size, leaf orientation, costa (midrib) prominence, presence or absence of multistratose regions on the leaves, and leaf cell. Some of these are precisely those characters that appear to be environmentally plastic. The observation that unrelated aquatic mosses, especially those growing in rheophytic habitats, often share a suite of morphological characters (Vitt & Glime, 1984) also raises the question of the extent species and genera are defined by convergent adaptations. Phylogenetic inferences gained from DNA sequence variation in the genus *Fontinalis* (Shaw & Allen, 2000) and the family Amblystegiaceae (Vanderpoorten et al., 2001a) suggest that these traits are evolutionarily labile, and that the species and genera of these aquatic mosses are defined by convergent similarities. A better understanding of aquatic bryophytic phylogeny is thus absolutely necessary before they can be used as biomonitors of water quality.

Acknowledgments

This chapter was written while I was a recipient of a fellowship of the Belgian American Educational Foundation. Many thanks are due to L. E. Anderson for reading a first draft of this chapter.

References

Aitken, S., & Libby, W. J. (1994). Evolution of the pygmy-forest edaphic subspecies of *Pinus contorta* across an ecological staircase. *Evolution, 48*: 1009–1019.

Allen, E. D., & Spence, D. H. N. (1981). The differential ability of aquatic plants to utilize the inorganic carbon supply in fresh waters. *New Phytologist, 87*, 269–283.

Arts, G. H. P. (1990). Aquatic Bryophyta as indicator of water quality in shallow pools and lakes in The Netherlands. *Annales Botanici Fennici, 27*, 19–32.

Bain, J. T., & Proctor, M. C. F. (1980). The requirement of aquatic bryophytes for free CO_2 as an inorganic carbon source: Some experimental evidence. *New Phytologist, 86*, 393–400.

Baker, R. G. E. (1992). The influence of copper and zinc on shoot length and dry weight of *Sphagnum palustre* and *Sphagnum cuspidatum* in aequous cultures. *Journal of the Hattori Botanical Laboratory, 72*, 89–96.

Baker, R. G. E., & Boatman, R. J. (1985). Some effects of nitrogen, phosphorus, potassium and carbon dioxide concentration on the morphology and vegetative reproduction of *Sphagnum cuspidatum* Ehrh. *New Phytologist, 116*, 605–611.

Bates, J. W. (1992). Mineral nutrient acquisition and retention by bryophytes. *Journal of Bryology, 17*, 223–240.

Battarbee, R. W. (1984). Diatom analysis and the acidification of lakes. *Philosophical Transactions of the Royal Society of London, B305*, 451–477.

Benson Evans, K., & Williams, P. F. (1976). Transplanting aquatic bryophytes to assess river pollution. *Journal of Bryology, 9*, 81–91.

Beurksens, J. E. M., Winkels, H. J., De Wolf, J., & Dekker, C. G. C. (1994). Trends of priority pollutants in the Rhine during the last fifty years. *Water Science Technology, 29*, 77–85.

Birks, H. J. B., Line, J. M., Juggins, S., Stevenson, A. C., & ter Braak, C. J. F. (1990). Diatoms and pH reconstruction. *Philosophical Transactions of the Royal Society of London, B327*, 268–278.

Brandrud, T. E., & Johansen, S. W. (1994). Effects of acification on macrophyte growth in the HUMEX lake Skjervatjern, with special emphasis on *Sphagnum auriculatum*. *Environment International, 20*, 329–342.

Brown, D. H., & Wells, J. M. (1990). Physiological effects of heavy metals on the moss *Rhytidiadelphus squarrosus*. *Annals of Botany, 66*, 641–647.

Brown, D. H., & Whitehead, A. (1986). The effects of mercury on the physiology of *Rhytidiadelphus squarrosus*. *Journal of Bryology, 14*, 367–374.

Buck, W. R., Goffinet, B., & Shaw, A. J. (2000). Testing morphological concepts of orders of pleurocarpous mosses (Bryophyta) using phylogenetic reconstructions based on *trn*L-*trn*F and rps4 sequences. *Molecular Phylogenetics & Evolution, 16*, 180–198.

Buetler, T. M., & Eaton, D. L. (1992). Glutathione S-transferases: Amino-acid sequence comparison, classification and phylogenetic relationship. *Environmental Carcinogenesis and Ecotoxicology Review—Part C, 10*, 181–203.

Caines, L. A., Watt, A. W., & Wells, D. E. (1985). The uptake and release of some trace metals by aquatic bryophytes in acidified waters in Scotland. *Environmental Pollution, 10*, 1–18.

Carbiener, R., Trémolières, M., & Muller, S. (1995). Végétation des eaux courantes et qualityé des eaux: Une thèse, des débats, des perspectives. *Acta Botanica Gallica, 142*, 489–531.

Charles, D. F. (1985). Relationships between surface sediment diatom assemblages and lake water characteristics in Adirondack lakes. *Ecology, 66*, 994–1011.

Charles, D. F., & Smol, J. P. (1988). New methods for using diatoms and chrysophytes to infer past pH of low-alkalinity lakes. *Limnology and Oceanography, 33*, 1451–1462.

Clymo, R. S. (1963). Ion exchange in *Sphagnum* and its relation to bog ecology. *Annals of Botany, 27*, 309–324.

Cun, C., Bousquet, G., & Vilagines, R. (1997). A 90-year record of water quality data of Paris Seine and Marne rivers. *Journal of Water SRT—Aqua, 46*, 150–164.

Davis, R., & Smol, J. (1986). The use of sedimentary remains of siliceous algae for inferring past chemistry of lake water problems, potential and research needs. *Development in Hydrobiology, 29*, 291–300.

Davis, R. B., & Anderson, D. S. (1985). Method of pH calibration of sedimentary diatom remains for reconstructing history of pH in lakes. *Hydrobiologia, 120*, 69–87.

De Luna, E., Newton, A. E., Withey, A., Gonzales, D., & Mishler, B. D. (1999). The transition to pleurocarpy: A phylogenetic analysis of the main diplolepidous lineages based on *rbc*L sequences and morphology. *Bryologist, 102*, 634–650.

Descy, J.-P., & Empain, A. (1981). Inventaire de la qualityé des eaux courantes en Wallonie (Bassin wallon de la Meuse).. Rapport de synthèse. Vol. I. Normes typologiques, cartographie des résultats et discussion générale. University of Liège, Department of Botany, unpublished report.

Draper, N. R., & Smith, H. (1981). *Applied regression analysis.* New York: Wiley.

Düll, R. (1992). Zeigerwerte von Laub- und Lebermoosen. *Scripta Geobotanica, 18,* 175–214.

Eidt, D. C., Sosick, A. J., & Mallet, V. N. (1984). Partitioning and short-term persistence of Fenitrothion in New Brunswick (Canada) headwater streams. *Archives of Environmental Contamination and Toxicology, 13,* 43–52.

Ellenberg, H., Weber, H. E., Düll, R., Wirth, V., Werner, W., & Paulllißen, D. (Eds.). (1991). Zeigerwerte von Pflanzen in Mitteleuropa. Göttingen: E. Goltze.

Empain, A. (1976). Les bryophytes aquatiques utilisés comme traceurs de la contamination en métaux lourds des eaux douces. *Mémoires de la Société Royale de Botanique de Belgique, 7,* 141–156.

Empain, A. (1977). Ecologie des populations de bryophytes aquatiques de la Meuse, de la Sambre et de la Somme. Relations avec la qualityé des eaux, écophysiologie comparée et étude de la contamination par métaux lourds. University of Liège, doctoral dissertation.

Empain, A. (1978). Relations quantitatives entre les populations de bryophytes aquatiques et la pollution des eaux courantes. Définition d'un indice de qualityé des eaux. *Hydrobiologia, 60,* 49–74.

Empain, A. (1985). Heavy metals in bryophytes from Shaba provinces. In R. R. Brooks & F. Malaisse (Eds.), *The heavy metal tolerant flora of southcentral Africa: A multidisciplinary approach.* (pp. 103–117). Boston: A. Balkema.

Empain, A., Lambinon, J., Mouvet, C., & Kirchmann, R. (1980). Utilisation des bryophytes aquatiques et subaquatiques comme indicateurs biologiques de la qualityé des eaux courantes. In Pesson P. (Ed.), *La pollution des eaux continentales. Incidence sur les biocénoses aquatiques* (pp. 195–223). Paris: Gauthier-Villars.

Faafeng, B. A., & Fjeld, E. (1998). Precision of seasonal average values of important eutrophication indicators as determined by the number of samples and by eutrophication level. 27th SIL Congress, Dublin, August 8–14. *Book of Abstracts, 36.*

Frahm, J.-P. (1974). Wassermoose als Indikatoren für die Gewässerverschmutzung am Beispiel des Niederrheins. *Gewässer und Abwässer, 53–54,* 91–106.

Frahm, J.-P. (1975). Toxitoleranzversuche an Wassermoosen. *Gewässer und Abwässer, 57–58,* 59–66.

Frahm, J.-P. (1976). Weitere Toxitoleranzversuche an Wassermoosen. *Gewässer und Abwässer, 60–61,* 113–123.

Frahm, J.-P. (1997). Zur Ausbreitung von Wassermoosen am Rhein (Deutschland) und an seinen Nebenflüssen seit dem letzten Jahrhundert. *Limnologica, 27,* 251–262.

Frahm, J.-P., & Abst, U. W. (1993). Veränderungen in der Wassermoosflora des Niederrheins 1972–1992. *Limnologica, 23,* 123–130.

Freiley, K. J. (1993). Allozyme diversity and population genetic structure in *Haplopappus gracilis* (Compositae). *Systematic Botany, 18,* 543–550

Fresco, L. F. M. (1982). An analysis of species response curves and of competition from field data sets: Some results from heath vegetation. *Vegetatio, 48,* 175–185.

Frisque, G. E., Galoux, M., & Bernes, A. (1983). Accumulation de deux micropolluants (les polychlorobiphenyles et les gamma-HCH) par des bryophytes aquatiques de la Meuse. *Mededelingen Faculteit Landbouwkundige und Toegepaste Biologische Wetenschappen Universiteit Gent, 48,* 971–983.

Glime, J. M. (1992). Effects of pollutants on aquatic species. In: J. W. Bates & A. Faramer (Eds.), *Bryophytes and lichens in a changing environment* (pp. 333–361). Oxford, UK: Clarendon Press.

Glime, J. M., & Vitt, D. H. (1987). A comparison of bryophyte species diversity and niche structure of montane streams and stream banks. *Canadian Journal of Botany, 65,* 1824–1837.

Goffinet, B., Bayer, R. J., & Vitt, D. (1998). Circumscription and phylogeny of the Orthotrichales (Bryopsida). inferred from RBCL sequence analyses. *American Journal of Botany, 85,* 1324–1337.

Gonçalves, E., Soaves, H., Boaventura, R., Machado, A., & Silva, J. (1994). Seasonal variations of heavy metals in sediments and aquatic mosses from the Cavado River Basin, Portugal. *Science of the Total Environment, 142,* 143–156.

Greven, H. C. (1992). Changes in the Dutch bryoflora and air pollution. *Dissertationes Botanicae, 194,* 1–237.

Haury, J., Peltre, M. C., Muller, S., Trémolières, M., Barbe, J., Dutartre, A., & Guerlesquin, M. (1996). Des indices macrophytiques pour estimer la qualité des cours d'eau français. Premières propositions. *Ecologie, 27,* 233–244.

Hedderson, T. A., Chapman, R. C., & Rootes, W. R. (1996). Phylogenetic relationships of bryophytes inferred from nuclear-encoded rRNA gene sequences. *Plant Systematic and Evolution, 200,* 213–224.

Hellmann, H. (1994). Load trends of selected chemical parameters of water quality and of trace substances in the River Rhine between 1955 and 1988. *Water Science Technology, 29,* 69–76.

Heywood, J. S., & Levin, D. (1985). Associations between allozyme frequencies and soil characteristics in *Gaillardia pulchella* (Compositae). *Evolution, 39,* 1076–1086.

Himmler, H., & Ness, A. (1992). Die Moosflora der Bäche des Nordschwarzwaldes. Landesanstalt für Umweltschutz Baden-Württemberg, Ökologisches Wirkungskataster Baden-Württemberg, Karlsruhe.

Holmes, R. W., Whiting, M. C., & Stoddard, J. L. (1989). Changes in diatom-inferred pH and acid neutralizing capacity in a dilute, high elevation, Sierra Nevada lake since A.D. 1825. *Freshwater Biology, 21*, 295–310.

Hussey, B. (1982). Moss growth on filter beds. *Water Research, 16*, 391–398.

Jackson, J. E. (1991). A user's guide to principal components. New York: Wiley.

Joliffe, I. T. (1982). A note on the use of principal components in regression. *Applied Statistics, 31*, 300–303.

Karttunen, K., & Toivonen, H. (1995). Ecology of aquatic bryophyte assemblages in 54 small Finnish lakes, and their change in 30 years. *Annales Botanici Fennici, 32*, 75–90.

Kelly, M. G. (1998). Use of the trophic diatom index to monitor eutrophication in rivers. *Water Research, 32*, 236–242.

Kelly, M. G., & Huntley, B. (1987). *Amblystegium riparium* in brewery channels. *Journal of Bryology, 14*, 792.

Kelly, M. G., & Whitton, B. A. (1998). Biological monitoring of eutrophication in rivers. *Hydrobiologia, 384*, 55–67.

Kooijman, A. M., & Bakker, C. (1993). Causes of the replacement of *Scorpidium scorpidioides* by *Calliergonella cuspidata* in eutrophicated rich fens: 2. Experimental studies. *Lindbergia, 18*, 123–130.

Kooijman, A. M., & Hedenäs, L. (1991). Differentiation in habitat requirements within the genus *Scorpidium*, especially between *S. revolvens* and *S. cossonii*. *Journal of Bryology, 16*, 619–627.

Kooijman, A. M., & Kanne, D. M. (1993). Effects of water chemistry, nutrient supply and interspecific interactions on the replacement of *Sphagnum subnitens* by *S. fallax* in fens. *Journal of Bryology, 17*, 431–438.

Kosiba, P., & Sarosiek, J. (1995). Disappearance of aquatic bryophytes resulting from water pollution by textile industry. *Cryptogamica Helvetica, 18*, 85–93.

Kovacs, M., & Podani, J. (1986). Bioindication: A short review on the use of plants as indicators of heavy metals. *Acta Biologia Hungarica, 37*, 19–29.

Landwehr, J., & Barkman, J. J. (1966). Atlas van de nederlandse bladmossen. Koninklijke Nederlandse Natuurhistorische Vereniging, Amsterdam.

López, J., & Carballeira, A. (1989). A comparative study of pigment contents and response to stress in five species of aquatic bryophytes. *Lindbergia, 15*, 188–194.

Malle, K.-G. (1996). Cleaning up the River Rhine. *Scientific American, 274*, 55–59.

Martens, H., & Naes, T. (1989). *Multivariate calibration*. New York: Wiley.

Martinez-Abaigar, J., Nunez-Olivera, E., & Sanchez-Diaz, M. (1993). Effects of organic pollutants on transplanted aquatic bryophytes. *Journal of Bryology, 17*, 553–566.

Morrison, B. R. S., & Wells, D. E. (1981). The fate of fenitrothion in a stream environments and its effect on the fauna, following aerial spraying of a Scottish forest. *Science of the Total Environment, 19*, 233.

Mouvet, C. (1985). The use of aquatic bryophytes to monitor heavy metals pollution of freshwaters as illustrated by case study. *Verhein des internationales Verein für Limnologie, 22*, 2420–2425.

Mouvet, C., Galoux, M., & Bernes, A. (1985). Monitoring of polychlorinated biphenyls (PCBs) and hexachlorocyclohexanes (HCH). in freshwater using the aquatic moss *Cinclidotus danubicus*. *Science of the Total Environment, 44*, 253–267.

Muotka, T., & Virtanen, R. (1995). Stream as habitat templet for bryophytes, distribution along gradients in disturbance and substratum heterogeneity. *Freshwater Biology, 33*, 141–149.

Okland, R. H. (1990). Vegetation ecology: Theory, methods and applications with reference to Fennoscandia. *Sommerfeltia*, Suppl. 1, 1–233.

Oksanen, J., Läära, E., Huttunen, P., & Meriläinen, J. (1988). Estimation of pH optima and tolerances of diatoms in lake sediments by the method of weighted averaging, least squares and maximum likelihood, and their use for the prediction of lake acidity. *Journal of Paleolimnology, 1*, 39–49.

Paffen, B. G. H., & Roelofs, J. G. M. (1991). Impact of carbon dioxide and ammonium on the growth of submerged *Sphagnum flexuosum*. *Aquatic Botany, 40*, 61–71.

Palm, R., & Iemma, A. F. (1995). Quelques alternatives à la régression classique dans le cas de la colinéarité. *Revue de Statistiques Appliquées, 43*, 5–33.

Penulas, J. (1984). Pigment of aquatic mosses of the river Muga, NE Spain, and their response to water pollution. *Lindbergia, 10*, 127–132.

Probst, A., Massabuau, J. C., Probst, J. L., & Fritz, B. (1990). Acidification des eaux de surface sous l'influence des précipitations acides: Rôle de la végétation et du substratum, conséquences pour les populations de truites. *Comptes-Rendus de l'Académie des Sciences de Paris, 3*, 405–411.

Reynolds, J. D. (1998). *Ireland's freshwaters*. Dublin: Marine Institute.

Robach, F., Thiébaut, G., Trémolières, M., & Muller, S. (1996). A reference system for continental running waters: Plant communities as bioindicators of increasing eutrophication in alkaline and acidic waters in north-east France. *Hydrobiologia, 340*, 67–76.

Roberts, D. A., Singer, R., & Boylen, C. W. (1985). The submersed macrophyte communities of Adirondack Lakes (New York, USA) of varying degrees of acidity. *Aquatic Botany, 21*, 219–235.

Roelofs, J. G. M. (1983). Impact of acidification and eutrophication on macrophyte communities in soft waters in The Netherlands: I, Field observations. *Aquatic Botany, 17*, 139–155.

Roelofs, J. G. M., Schuurkes, J. A. A. R., & Smith A. J. M. (1984). Impact of acidification and eutrophication on macrophyte communities in soft waters: II. Experimental studies. *Aquatic Botany, 18*, 389–411.

Roy, S., Pellinen, J., Sen, C. K., & Hänninen, O. (1995). Benzo(a) anthracene and benzo(a) pyrene exposure in the aquatic plant *Fontinalis antipyretica*: Uptake, elimination and the responses of biotransformation and antioxidant enzymes. *Chemosphere, 29*, 1301–1311.

Rühling, A., & Tyler, G. (1970). Sorption and retention of heavy metals in the woodland moss *Hylocomium splendens* (Hedw.) Br. et Sch. *Oikos, 21*, 92–97.

Ryan, T. P. (1997). Modern regression methods. Wiley, N.Y.

Sand-Jensen, K., & Rasmussen, L. (1978). Macrophytes and chemistry of acidic streams from lignite mining areas. *Botaniska Tidsskrift, 72*, 105–111.

Sandermann, H. (1994). Higher plants metabolism of xenobiotics: The "green liver" concept. *Pharmacogenetics, 4*, 225–241.

Sarosiek, J., & Samecka-Cymerman, A. (1987). The bioindication of ethylene glycol in water by the mosses *Fontinalis antipyretica* L. and *Platyhypnidium rusciforme* (Neck.) Fleisch. *Symposia Biologica Hungarica, 35*, 835–841.

Sarosiek, J., Wiewiorka, Z., & Mroz, L. (1987). Bioindication of heavy metal toxicity of water by the liverwort *Ricciocarpus natans* (L.) Corda. *Symposia Biologica Hungarica, 35*, 827–833.

SAS Institute, Inc. (1999). SAS/STAT software: Preliminary draft documentation. Release 7.01. Cary, NC: Author.

Satake, K., Nishikawa, M., & Shibata, K. (1989). Distribution of aquatic bryophytes in relation to water chemistry of the acid river Akagawa, Japan. *Archiv für Hydrobiologie, 116*, 299–311.

Say, P. J., & Whitton, B. A. (1983). Accumulation of heavy metals by aquatic mosses: 1. *Fontinalis antipyretica* Hedw. *Hydrobiologia, 100*, 245–260.

Schrenk, C., Pflugmacher, S., Brüggemann, R., Sandermann, H., Steinberg, C. E. W., & Kettrup, A. (1998). Glutathione s-transferase activity in aquatic macrophytes with emphasis on habitat dependence. *Ecotoxicology and Environmental Safety, 40*, 226–233.

Schuurkes, J. A. A. R., Kok, C. J., & Den Hartog, C. (1986). Ammonium and nitrate uptake by aquatic plants from poorly buffered and acidified waters. *Aquatic Botany, 24*, 131–146.

Schwoerbel, J., & Tillmans, G. C. (1974). Assimilation of nitrogen from the medium and nitrate reductase activity in submerged macrophytes: *Fontinalis antipyretica* L. *Archiv für Hydrobiologie, 2* (Suppl. 47), 282–294.

Shaw, A. J. (1985). The relevance of ecology to species concepts in bryophytes. *Bryologist, 88*, 199–206.

Shaw, A. J., & Allen, B. H. (2000). Phylogenetic relationships, morphological incongruence, and geographic speciation in the Fontinalaceae (Bryophyta). *Molecular Phylogenetics & Evolution, 16*, 225–237.

Shaw, A. J., Meagher, T. R., & Harley, P. (1987). Elecrophoretic evidence of reproductive isolation between two varieties of the moss, *Climacium americanum. Heredity, 59*, 337–343.

Sidhu, M., & Brown, D. H. (1996). A new laboratory technique for studying the effects of heavy metals on bryophyte growth. *Annals of Botany, 78*, 711–717.

Siebert, A., Bruns, I., Krauss, G. J., Miersch, J., & Markert, B. (1996). The use of the aquatic moss *Fontinalis antipyretica* L. ex Hedw. as a bioindicator for heavy metals: 1. Fundamental investigations into heavy metal accumulation in *Fontinalis antipyretica* L. ex Hedw. *Science of the Total Environment, 177*, 137–144.

Sjögren E. (1995). *Changes in epilithic and epiphytic moss cover in two deciduous forest areas on the Island of Öland (Sweden)—a comparison between 1958–1962 and 1988–1990.* Uppsala: Opulus Press.

Slack, N. G. (1990). Bryophytes and ecological niche theory. *Botanical Journal of the Linnaean Society, 104*, 187–213.

Steinman, A. D. (1994). The influence of phosphorus enrichment on lotic bryophytes. *Freshwater Biology, 31*, 53–63.

Stephenson, S. L., Studlar, S. M., McQuattie, C. J., & Edwards, P. J. (1995). Effects of acidification on bryophyte communities in West Virginia mountain streams. *Journal of Environment Quality, 4*, 116–124.

Tenenhaus, M. (1998). *La régression PLS. Théorie et pratique.* Paris: Technip.

Tenenhaus, M., Gauchi, J.-P., & Ménardo, C. (1995). Régression PLS et applications. *Revue de Statistiques Appliquées, 43*, 7–63.

ter Braak, C. J. F. (1985). Correspondence analysis of incidence and abundance data: Properties in terms of a unimodal response model. *Biometrics, 41,* 589–873.

ter Braak, C. J. F., & Looman, C. W. N. (1986). Weighted averaging, logistic regression and Gaussian response model. *Vegetatio, 65,* 3–11.

ter Braak, C. J. F., & Looman, C. W. N. (1987). Regression. In: R. H. G. Jongman, C. J. F. ter Braak, & O. F. R. Van Tongeren (Eds.), *Data analysis in community and lanscape ecology* (pp. 29–77). Wageningen: Pudoc.

ter Braak, C. J. F., & van Dam, H. (1989). Inferring pH from diatoms: A comparison of old and new calibration methods. *Hydrobiologia, 178,* 209–223.

Thiébaut, G., Vanderpoorten, A., Guerold, F., Boudot, J.-P., & Muller, S. (1998). Bryological patterns and streamwater acidification in the Vosges mountains (N.E. France): An analysis tool for the survey of acidification processes. *Chemosphere, 36,* 1275–1289.

Tittizer, T., & Krebs, F. (Ed.). (1996). Ökosystemforschung: Der Rhein und seine Auen. Eine Bilanz. Berlin–Heidelberg: Springer-Verlag.

Trémoliéres, M., Carbiener, R., Ortscheit, A., & Klein, J.-P. (1994). Changes in aquatic vegetation in Rhine floodplain streams in Alsace in relation to disturbance. *Journal of Vegetation Science, 5,* 169–178.

Tremp, H., & Kohler, A. (1993). Wassermoose als Versauerungindikatoren. Praxiesorientierte Bioindikationsvervahren mit Wassermoosen zur überwachung des Säurezustandes von pufferschwachen Fließgewässern. Landesanstalt für Umweltschutz Baden-Württemberg, Projekt "Angewandte Ökologie" 6, Karlsruhe.

Tremp, H., & Kohler, A. (1995). The usefulness of macrophyte monitoring-systems, exemplified on eutrophication and acidification of running waters. *Acta Botanica Gallica, 142,* 541–550.

Vanderpoorten, A. (1999a). Aquatic bryophytes for a spatio-temporal monitoring of the water pollution of the rivers Meuse and Sambre (Belgium). *Environmental Pollution, 104,* 401–410.

Vanderpoorten, A. (1999b). Correlative and experimental investigations on the segregation of aquatic bryophytes as a function of water chemistry in the Walloon hydrographic network (Belgium). *Lejeunia N.S., 159,* 1–17.

Vanderpoorten, A., & Durwael, L. (1999). Trophic response curves of aquatic bryophytes in lowland calcareous streams. *Bryologist, 102,* 720–728.

Vanderpoorten, A., Ector, L., & Hoffmann, L. (2000). Physico chemical profiles of *Amblystegium riparium, Fontinalis antipyretica* and *Rhynchostegium riparioides* in the Grand-Duchy of Luxembourg. *Nova Hedwigia, 72,* 209–221.

Vanderpoorten, A., & Empain, A. (1999). Morphologie, écologie et distribution comparées d'*Amblystegium tenax* et d'*A. fluviatile* en Belgique. *Belgian Journal of Botany, 132,* 3–12.

Vanderpoorten, A., & Klein, J.-P. (1999a). Variations of aquatic bryophyte assemblages in the Rhine rift related to water quality: 2. The waterfalls of the Vosges and the Black Forest. *Journal of Bryology, 21,* 109–115.

Vanderpoorten, A., & Klein, J. P. (1999b). A comparative study of the hydrophyte flora from the Alpine Rhine to the Middle Rhine: Application to the conservation of the Upper Rhine aquatic ecosystems. *Biological Conservation, 87,* 163–172.

Vanderpoorten, A., Klein, J.-P., Stieperaere, H., & Trémolières, M. (1999). Variations of aquatic bryophyte assemblages in the Rhine rift related to water quality: 1. The Alsatian Rhine floodplain. *Journal of Bryology, 21,* 17–23.

Vanderpoorten, A., & Palm, R. (1998). Canonical variables of aquatic bryophyte combinations for predicting water trophic level. *Hydrobiologia, 386,* 85–93.

Vanderpoorten, A., & Palm, R. (2001). A comparative study of calibration methods for inferring ammonium nitrogen concentrations in rivers from aquatic bryophyte assemblages. *Hydrobiologia, 452,* 181–190.

Vanderpoorten, A., Shaw, A. J., & Goffinet, B. (2001a). Testing controversial alignments in *Amblystegium* and related genera (Amblystegiaceae: Bryopsida). Evidence from rDNA ITS sequences. *Systematic Botany, 28,* 470–479.

Vanderpoorten, A., Thiébaut, G., Trémolières, M., & Muller, S. (2001b). A model for assessing water chemistry by using aquatic bryophyte assemblages in north-eastern France. *Verhandlungen des Internationales Verein für Limnologie, 27,* 807–810.

Vanderpoorten, A., & Tignon, M. (2000). Amplified fragments length polymorphism between populations of *Amblystegium tenax* exposed to contrasting pollution levels. *Journal of Bryology, 22,* 57–62.

Van Rossum, F., Vekemans, X., Meerts, P., Gratia, E., & Lefèbvre, C. (1996). Allozyme variation in relation to ecotypic differentiation and population size in marginal populations of *Silene nutans. Heredity, 78,* 552–560.

Vhrovsek, D., Martincic, A., & Kralj, M. (1984). The application of some numerical methods and the evaluation of bryophyte indicator species for the comparison of the degree of pollution between two rivers. *Archiv für Hydrobiologie, 100,* 431–444.

Vitt, D. H., & Glime, J. M. (1984). The structural adaptations of aquatic Musci. *Lindbergia, 10,* 95–110.

Vitt, D. H., Glime, J. M., & Lafarge-England, C. C. (1986). Bryophyte vegetation and habitat gradients of montane streams in Western Canada. *Hikobia*, *9*, 367–385.

Weisberg, S. (1985). *Applied linear regression*. New York: Wiley.

Wells, J. M., & Brown, D. H. (1995). Cadmium tolerance in a metal-contaminated population of the grassland moss *Rhytidiadelphus squarrosus*. *Annals of Botany*, *75*, 21–29.

Werner, J. (1996). Die Moosflora des Luxemburger Oeslings. *Travaux Scientifiques du Musée d'Histoire Naturelle de Luxembourg*, *24*, 1–88.

Westerbergh, A., & Saura, A. (1992). The effect of serpentine on the population structure of *Silene dioica* (Caryophyllaceae). *Evolution*, *46*, 1537–1548.

Whittaker, R. H. (1967). Gradient analysis of vegetation. *Biological Reviews*, *49*, 207–264.

Wilson, J. B., Steel, J. B., Newman, J. E., & Tangney, R. S. (1995). Are bryophyte communities different? *Journal of Bryology*, *18*, 689–705.

Yan, N. D., Miller, G. E., Wile, I., & Hitchin, G. G. (1985). Richness of macrophyte floras of soft water lakes differing pH and trace metal content in Ontario, Canada. *Aquatic Botany*, *23*, 27–40.

5

Plant Succession in Littoral Habitats
Observations, Explanations, and Empirical Evidence

Lisandro Benedetti-Cecchi

Introduction

The maturity of a scientific discipline is often evaluated in terms of its capability to make successful predictions. Ecological models can be very inaccurate in their predictions because they try to explain patterns in abundance of natural populations that are idiosyncratic in space and time. Defined as a deterministic, ordered sequence of replacement of species invading open space (e.g., Clements, 1916, 1928; Odum, 1969), ecological succession provides a framework for theoretical development, empirical testing, and, in some instances, successful prediction in ecology. As such, the concept of succession is central to the maturation of the discipline.

The relevance of succession to the practice of ecology is manifested by the large body of theoretical and empirical work summarized in various reviews (Connell & Slatyer, 1977; Sousa, 1984a; Anderson, 1986; Pickett et al., 1987a; McCook, 1994). Key topics include phenomenological aspects of succession, such as descriptions of the sequence of replacement of species at a site after disturbance, experimental tests of hypotheses about specific mechanisms driving the process of replacement, probabilistic models to attempt predictions about the direction and end point of succession, and the refinement of general theoretical frameworks.

Fervent discussions have originated around these topics as a result of the different and sometimes opposite views of leading ecologists. A historical analysis reveals several steps along which the concept and the analysis of succession have progressed (McIntosh, 1985; McCook, 1994). Earlier this century, Clements's (1916, 1928) view of succession as an ordered, deterministic process largely driven by facilitation was challenged by Gleason

---

Lisandro Benedetti-Cecchi • Dipartimento di Scienze dell'Uomo e dell'Ambiente, Università di Pisa, Via A Volta 6, I-56126 Pisa, Italy.

(1926, 1927), who emphasized the role of stochastic events, initial species composition, and contingency in determining patterns of succession. These points were further pursued by Egler (1954) and Drury and Nisbet (1973), who emphasized the importance of initial species composition and the role of life histories, respectively. These authors argued against the traditional view of succession as a saltatory change in favor of a view of gradual changes, also recognizing cycles and divergences as possible alternatives to the directional replacement of species envisioned by Clements. Only after the publication of the seminal paper by Connell and Slatyer (1977), in which the influence of early colonists on later ones was classified according to the well-known facilitation, tolerance, and inhibition models, that analyses of the causes of succession through rigorous manipulative experimentation became common. Intense experimentation allowed the refinement of early models (e.g., Farrell, 1991) and the formulation of new mechanistic theories that allowed for specific, quantitative predictions of patterns of replacement of species during succession (Noble & Slatyer, 1980; Tilman, 1985, 1990; Huston & Smith, 1987; Huston, 1994).

Debates on succession also had important ramifications for the analysis and understanding of other ecological issues. For example, Clements's (1928) view of succession as a process oriented toward a stable state—the climax—implied a view of the community as a highly structured, interdependent, and coevolved collection of species—the superorganism. In contrast, Gleason's (1926, 1927) emphasis on stochastic processes and contingency led to the individualistic concept of the community. Equilibrium versus nonequilibrium views of the structure and dynamics of assemblages have originated from these debates (Sousa, 1979a; Chesson, 1986; DeAngelis & Waterhouse, 1987). More recently, the recognition that ecological patterns and processes are temporally (and spatially) heterogeneous has provided a context for new theoretical and empirical work on natural variability (Wiens et al., 1986; Kotliar & Wiens, 1990; Landres et al., 1999).

As is the case for most, if not all, paradigmatic notions in ecology, there are problems with the concept of succession due to confusion in terminology and definition, overgeneralization, and lack of operationalization (Peters, 1991). Anderson (1986, p. 276), discussing the issue of replacement of species during succession wrote that

> it is not immediately obvious what the key diagnostic features of this fundamental ecological replacement might be. Must all individuals of a species have been replaced or supplanted before succession can be deemed to have occurred? Must the replacement process be totally and repeatedly predictable? Are there in all successional sequences simple features which serve to identify if not diagnose a stable, climax community where it allegedly occurs?

Making a concept operative requires explicit statements about how the concept can be evaluated. This is a critical step to avoid tautologies, to produce testable theories, and to increase the accuracy of predictions that these theories permit. Lack of consistency in terminology and vague usage of terms is an impediment to the testing and maturation of theories. Although these limitations are recognized in ecology (McIntosh, 1980; Lohele, 1988), they persist possibly as a consequence of the complexity of ecological patterns and processes, of constraints in the experimental analysis of theory, and, perhaps more importantly, of diverging philosophical views of ecology.

The purpose of this chapter is to review our current understanding of succession and to explore whether gaps of knowledge reflect inconsistencies in the logical structure of ecological research on this topic. This is done through the analysis of the logical components of a structured research program (e.g., Underwood, 1990, 1997), with the belief that ecological understanding and predictability can be better achieved by preserving the logical structure linking observations, explanations, hypothesis testing, and interpretation of results. I start with a review on the phenomenology of succession—the original observations that require explanation and motivate subsequent mechanistic research. Then, I consider some of the classical theories of succession and examine whether explanatory models accomplish to the logical requirement of being able to explain the original observations. I review the experiments used to test the proposed models and ask whether results have been interpreted coherently with respect to the proposed theory and the observations at hand. Finally, I discuss the importance of understanding ecological succession to address problems in environmental management and conservation. This chapter is not an exhaustive review of the literature. Rather, I illustrate these points using representative examples from the vast literature about plant succession in littoral habitats.

Phenomenology of Succession and Erratic Colonization

The Early Definition of Succession

Since the early studies on temporal changes in natural populations (Cowles, 1899; Cooper, 1913; Clements, 1916), it has become clear that once new space is provided in a given habitat, organisms start to invade and develop. A feature of the process of colonization that emerged more clearly than others is that of a sequence of replacement of species directed toward a "mature" state of the vegetation (Cowles, 1899; Clements, 1916, 1928). These observations were repeatedly obtained at different locations in different systems, appearing as a distinctive feature of many assemblages of terrestrial plants (Clements, 1928; Elton, 1958; Odum, 1969). This ordered, directional change in species composition was termed *succession* (after Thoreau, 1860). The definition circumscribed the range of possible observations about temporal changes in natural assemblages that could be appropriately described by the concept of succession.

Disturbance and the Provision of Open Space

According to the definition of *succession*, a phenomenological treatment of the subject should include a documentation of the processes contributing to the provision of open space. This description is, in fact, available for most littoral systems, at least in a narrative form. New space in sand dunes is originated by sand delivered to the shore by wind and waves, or by spring equinox tides killing resident organisms in a dune that is already colonized (Ranwell, 1972; Brown & McLachlan, 1990). In salt-marshes, frost (Hubbard & Stebbings, 1968), fire (Ranwell, 1972), and burial by wrack (Bertness & Ellison, 1987) are major disturbances causing the opening of new space in the established vegetation. On rocky shores, disturbance by storms, desiccation, sand scouring, and grazing are all causes of loss of sessile organisms from the substratum, resulting in new space available for colonization (Sousa, 1984a, 1985). Earthquakes and El Niño are episodic events that can initiate succession over large areas in rocky intertidal and subtidal habitats (Castilla, 1988; Dayton et al., 1984, 1992; Tegner & Dayton, 1987).

Natural disturbance is amenable to quantitative analyses. The frequency, timing, intensity, severity, and extent of natural disturbance are all measurable quantities that define the regime of disturbance. This is an important topic incorporated into the general theory of patch dynamics (Pickett & White, 1985). Here, I focus on the phenomenology of disturbance and its description (see Sousa 1984a, 1985, and references therein for a full account); the role of variable disturbance in explaining patterns of succession will be examined later. The frequency of disturbance is the number of events per unit of time. Timing refers to specific patterns in temporal distribution of disturbance. The intensity of disturbance is the magnitude of the disturbing force regardless of its effects on organisms. Severity is a measure of the sensitivity of natural populations to a disturbance, whereas extent is the area affected by a disturbance.

Despite the large numbers of descriptive accounts of natural disturbance, only a few studies have provided quantitative observations for littoral habitats. Most attempts focused on one or a few aspects of the regime of disturbance. A notable exception is the study by Bertness and Ellison (1987) on disturbance by wracks in a salt marsh in Rhode Island. They found that more wrack accumulated in the marsh in summer than winter, and that the extent of disturbance ranged from 10% in the highest part of the marsh up to 80% at the lower levels of the high marsh. The intensity of disturbance was expressed as the residence times of the wrack and the amount of biomass accumulated. These quantities changed in relation to height of the marsh, with resident times ranging from 2 days in the lower part up to 4 weeks in the highest part of the marsh, where most of the wrack accumulated. Although the intensity of disturbance was greater in the highest part of the marsh, its severity was evident in different habitats. Accumulation of wrack resulted in the loss of individuals from populations that dominated in different portions of the marsh, eventually creating patches of open space available for colonization. Bare patches ranged in size from fractions of a square meter to 50 m^2.

Quantitative estimates of natural disturbance have been obtained also in intertidal boulder fields. Disturbance was conveniently measured in these habitats by recording the frequency with which boulders of different size were overturned by waves, the force necessary to overturn them, the frequency of dislodgement from their spot, and the displaced distance (Osman, 1977; Sousa, 1979a; McGuinness, 1987). Overturning of boulders created new space available for colonization by exposing previously unoccupied portions of the substratum or through a combination of mechanical abrasion, shading, sand burial, and grazing by sea urchins on the organisms living on what was formerly the top surface. Sousa (1979a), working in a boulder field in California, found clear relationships between the size of boulders, susceptibility to disturbance, and structure of assemblages. Small boulders were overturned more frequently than larger ones and experienced a greater loss of algae. Similar results were obtained by McGuinness (1987) in an intertidal boulder field in New South Wales (Australia). In this study, however, neither size of rocks nor the provision of open space available for colonization were correlated with the frequency with which rocks were moved or with the mean distance moved.

Another common approach used to quantify disturbance in intertidal habitats is to measure the size distribution of patches of *open space* (generally defined as patches available for the colonization of erect organisms) in an otherwise continuous assemblage of algae and invertebrates. This procedure was used to quantify disturbance in mussel beds, where monopolization of the substratum by mussels prevented the colonization of algae (Paine & Levin, 1981; Sousa, 1984a), in mixed assemblages of algae and barnacles (Farrell, 1989), and in littoral rock pools (Sousa, 1980; Dethier, 1984; Benedetti-Cecchi & Cinelli, 1996). The frequency and intensity of disturbance can be estimated by recording the number of patches produced per unit of time and the size distribution of these patches, respectively. Turnover time, the time required for disturbance to remove all the sessile organisms in a given patch of habitat, can then be calculated from these measures (e.g.,

Farrell, 1989). This procedure is not free of problems, because it can confound disturbance with other causes of mortality of organisms, it confounds the severity with the intensity of disturbance, and it is biased toward the most intense events (only those that produce gaps in the assemblage are detected), possibly overestimating disturbance (McGuinness, 1987; Benedetti-Cecchi & Cinelli, 1996). Nevertheless, quantitative observations on the number, size, and shape of patches are relevant to the study of succession by definition, in that they refer to a necessary condition for succession to occur.

General Observations on Patterns of Succession

Succession is described for several littoral habitats, including sand dunes (Cowles, 1899; Olson, 1958), salt marshes (Ranwell, 1972), and marine rocky coasts (Hatton, 1932; Rees, 1940; Northcraft, 1948). The most common observations are about sequences of replacement of species, temporal trends in the life histories of species dominating at different stages of succession, and temporal trends in patterns of diversity (Margalef, 1963; Odum, 1969).

General descriptions of succession in sand dunes have documented the initial colonization by microorganisms and fungi followed by lichens and bryophytes (Webley et al., 1952). Vascular plants appearing early in succession include small grasses or forbs that are replaced by larger, dune-building species and eventually by woody plants that characterize the mature state of the assemblage (e.g., Olson, 1958). Replacement of species is paralleled by changes in the life-history traits of the dominant populations. Early species have physiological traits that allow them to tolerate high salinity and accumulation of sand, whereas later species have larger sizes and greater morphological complexity. Productivity and diversity of sand dunes tend to increase as succession proceeds (Carter, 1988).

General patterns of succession in salt marshes do not differ substantially from those illustrated for sand dunes, despite marked physical and biological differences between the two habitats (Chapman, 1974; Ranwell, 1972; Carter, 1988). Succession is initiated by salt-tolerant annual species that tend to dominate in the seaward part of the marsh and continues with the colonization of perennial, taller plants that dominate throughout the remaining portion of the marsh. Annual species occasionally occur in the high marsh but only in recently disturbed patches (Ellison, 1987). Although productivity increases with succession, there are no particular trends in patterns of species diversity with time, possibly as a consequence of the general low diversity of the salt marsh habitat (Carter, 1988).

There are, however, problems of spatial confounding with the analysis of succession in salt marshes and sand dunes. Many studies have used chronosequences to describe succession in these habitats. This choice is often dictated by the impossibility of observing succession over convenient and experimentally tractable time scales (Pickett, 1989). The basic assumption is that any pattern of variation from a younger portion of the habitat to an older one (e.g., a low marsh compared to a high marsh) reflects a sequence of replacement of species. This approach underlies a view of succession that is not consistent with its definition. It is an intrinsic property of salt marshes and sand dunes that colonization generates spatial variation along a tidal gradient, so that space and time are inherently confounded in these habitats, unless temporal changes are examined at fixed heights along the tidal gradient. Roozen and Westhoff (1985) documented changes in the vegetation of a salt marsh for over 30 years using fixed plots but did not find any evidence that assemblages converged toward those occurring at higher elevations of the marsh. Similarly,

De Leeuw et al. (1993) were able to reject the hypothesis that zonation reflects succession in one of two salt marshes in The Netherlands.

Descriptions of succession in rocky intertidal habitats documented the early colonization by bacteria and diatoms, then the appearance of filamentous green and brown algae, and ultimately the establishment of perennial algae (Rees, 1940; Northcraft, 1948; Dayton, 1971, 1975; Emerson & Zedler, 1978). Similarly to what was described for sand dunes and salt marshes, patterns of succession on rocky shores also involve the replacement of species with contrasting life-history traits. Many general accounts indicate that succession proceeds from the establishment of morphologically simple, fast-growing algae with large dispersal capabilities, to slow-growing species with more complex morphologies and limited dispersal (Littler & Littler, 1980; Santelices, 1990). Studies of succession in intertidal habitats have generally emphasized an increase in species diversity from early to midsuccessional stages and a decline thereafter (e.g., Paine, 1966).

In subtidal habitats, there are two main contexts within which observations on patterns of succession have been obtained. The first is provided by the release from grazing pressure due to episodic events of mass mortality of sea urchins; the second is accomplished through the submersion of artificial substrata for various purposes. Grazing by sea urchins is often considered a form of biological disturbance in subtidal habitats (e.g., Ayling, 1981), preventing the colonization of erect algae and maintaining wide patches of encrusting coralline algae, the so-called "barren habitat" (Lawrence, 1975). Many studies have reported rapid colonization of the barren habitat by erect algae following mass mortality of sea urchins, with no clear evidence of succession. In many cases, canopy species become established directly rather than as part of a sequence of replacements (e.g., Johnson & Mann, 1988, 1993). Succession on artificial structures generally proceeds from the colonization by filamentous green algae to a transient state dominated by erect, turf-forming species and eventually to the dominance of canopy algae. Similarly to what was described for intertidal habitats, diversity increases from early to midstages of succession due to the coexistence of ephemeral and perennial species, declining at later stages when early species disappear (Aleem, 1957; Fager, 1971; Foster, 1975). Some controversial evidence, however, originated also from studies on artificial structures (discussed in Foster, 1975), introducing us to the issue of variability in patterns of succession and to the problem of distinguishing successional from nonsuccessional temporal changes.

Variable Patterns of Succession

These simplistic generalizations reflect, at least in part, the intellectual need to identify regularities in nature and the difficulty to accommodate exceptions and natural variability in explanations of ecological patterns (Underwood & Denley, 1984). Variability in patterns of succession was suddenly recognized by plant ecologists (e.g., Cowles, 1899) but then neglected in favor of regularities that could be detected, and possibly predicted, more easily (Clements, 1916). Alternative views that acknowledged variation more explicitly have emerged several times in the history of succession (Gleason, 1926, 1927; Egler, 1954; Drury & Nisbet, 1973; Pickett & White, 1985; Connell, 1987; Huston & Smith, 1987). Variability in succession may take different forms. There can be spatial and temporal changes in the sequence of invasion and replacement of species, in the rate of succession, or in the specific combination of sequence and rate of colonization.

Variable patterns of succession were described for both sand dune and salt marsh habitats. Cowles (1899), in his early study of succession in the sand dunes of Lake Michigan, described variable rates of colonization of plants for different patches of the habitat. Observations included cases in which succession was "arrested" at an early stage of colonization, as well as spatial variability in the trajectory of succession. Other examples of variability come from descriptive studies in salt marshes indicating that early patterns of colonization can be dominated by one or a few species (e.g., *Spartina* spp.) or, alternatively, by a diversified assemblage of plants (reviewed in Ranwell, 1972). These patterns were sometimes related to the timing of reproduction of species and their capability to disperse, traits that can make succession contingent to the timing of formation of a disturbed patch and to its distance from the source of colonists. Ranwell (1972), commenting on Chapman's study (1959) on succession in Scolt Head Island, Norfolk, wrote

> But one has only to look at the slumped clods of main marsh level communities doomed to die in the bottom of a creek to realize that the probabilities of any particular square metre of salt marsh turf taking part in uninterrupted text-book succession may be very low indeed. Just what these probabilities are from site to site remains to be worked out.

There is indeed evidence that the probability of occurrence of species at any single point in time can be remarkably different from site to site during succession on sand dunes in the Netherlands (Olff et al., 1993).

Variable succession in salt marshes and sand dunes was also described in relation to changes in the distribution and abundance of herbivores. Early observations suggested that cattle, rabbits, and sheep were responsible for the maintenance of an early stage of succession by preventing the establishment of later species, either by trampling or foraging (reviewed in Ranwell, 1972; Carter, 1988). It was also suggested that grazers affected the trajectory of succession by selectively removing the most palatable species or by fertilizing the soil with their feces (Ranwell & Downing, 1959).

Observations relating grazing to succession have a long history on rocky shores as well. As described earlier, high densities of sea urchins can prevent the establishment of erect algae, allowing encrusting forms to dominate in subtidal habitats. Reductions in the abundance of these herbivores often result in the colonization of a diversified assemblage of filamentous, foliose, and canopy algae (see references above). There is also a plethora of studies reporting variable patterns of succession in relation to grazing in intertidal habitats. Most of these studies, however, involved the experimental exclusion of grazers (see review by Hawkins & Hartnoll, 1983, for a historical perspective), but what observations inspired such experiments is not always clear. Genuine correlative evidence suggesting that grazing can alter succession is limited for intertidal habitats. An example is provided by the succession of algae induced by mass mortality of limpets following the Torrey Canion oil spill along the western coasts of Britain (Southward & Southward, 1978). Whereas in the presence of limpets the rock was dominated by barnacles, the disappearance of grazers resulted in the early colonization of diatoms and unicellular algae, then in the establishment of dense mats of filamentous green algae (*Blidingia* and *Enteromorpha*), and finally in the dominance of large brown algae (*Fucus*).

Originally described for vascular plants in terrestrial systems, contingency of succession to the timing of reproduction and recruitment of organisms was also documented for rocky intertidal and subtidal habitats. Southward and Southward (1978) indicated that the early establishment of diatoms and unicellular algae was evident only in autumn and winter on the western coasts of Britain, succession proceeding with the establishment of filamentous green algae and *Fucus* in other seasons. In the Isle of Man (United Kingdom), Hawkins (1981) reported that *Fucus* could become established on barnacles in July, bypassing the stage of dominance by filamentous green algae characteristic of other times of the year. Studies on colonization of artificial substrata in subtidal habitats also documented considerable differences in patterns of establishment and succession of organisms in relation to the timing of submersion (e.g., Underwood & Anderson, 1994). Although some of these studies still involve some elements of succession in terms of sequential colonization and directionality, most of them documented variability in recruitment of populations to open space and idiosyncratic patterns of establishment and development of assemblages that did not conform to the definition of succession. These studies will be discussed in more detail in the next section.

Further variation in patterns of succession can occur when variable disturbances produce patches of different size. An important contribution to understand variability in succession in relation to the size of the disturbed patch is provided by a series of studies in mussel beds on intertidal rocky shores (Levin & Paine, 1974, 1975; Paine & Levin, 1981). In this system, strong waves remove mussels, opening gaps that are subsequently colonized by other invertebrates and algae. These studies described the demography of patches and provided a predictive model to relate patterns of succession to the size and age of the disturbed areas.

From the preceding analysis, it is clear that even the early studies on variability in succession were, to some extent, experimental in nature. Most of these investigations involved the artificial clearing of patches, the submersion of artificial structures, or the removal of grazers, regardless of the availability of quantitative observations about the potential relevance of these factors. Possibly, qualitative observations were clear enough to suggest likely processes responsible for variability in succession and to stimulate experimentation. The apparent scarcity of quantitative observations, however, may have limited the range of processes perceived as important causes of natural variability in succession.

Succession versus Erratic Colonization

A distinction seems appropriate at this stage between patterns of recovery that conform to the definition of succession and those that do not. Variability in recovery of disturbed patches may be in the form described earlier, in which a directional temporal replacement of species can still be recognized and eventually the same end point is achieved or, as discussed here, may be the expression of patterns that are neither ordered nor directional (referred to as colonization hereafter). A distinction between these two categories is implicit in many studies and reviews on ecological succession (e.g., Huston, 1994; McCook, 1994). There is, however, no general consensus on whether one should stick to the early definition of succession, or whether it should be continuously reformulated to incorporate the ever-increasing evidence of distinctive, erratic patterns in colonization of open space. For example, Van Andel et al. (1993) emphasized the need to distinguish between succession and fluctuations, and Huston and Smith (1987) categorized and modeled as succession several patterns of colonization of two species, including divergence, total suppression of one species by the other, convergence, and "pseudocyclic replacements," in addition to the traditional mode of sequential succession (see also the extensive review by McCook, 1994).

Several examples of nonsuccessional colonization are provided by studies of the development of assemblages on artificial substrata in marine littoral habitats (Sutherland, 1974; Sutherland & Karlson, 1977; Osman, 1977; Breitburg, 1985; Underwood & Anderson, 1994). Although most of these investigations involved some form of manipulation (see earlier discussion), they also had an important observational component and contributed more to a phenomenological than to a mechanistic understanding of colonization. These studies often revealed strong variability in colonization related to seasonal variation in recruitment of species, nonordered temporal changes, and multiple stable points at maturity. Therefore, species in these assemblages could not be consistently ranked in their capabilities to colonize, and neither a predictable sequence of replacement nor a particular structure of assemblages at maturity could be identified (Connell, 1987).

An example of these patterns is provided by Emerson and Zedler (1978) for algal colonization of artificial clearings in rocky intertidal habitats. These workers cleared patches of substratum at different times of the year and found that many species of algae could colonize patches only in a particular season. Colonization followed different trajectories depending on the timing of clearance. The dominant species in the assemblage, the coralline alga *Lithothrix aspergillum*, eventually colonized all the clearings, but there was no evidence for any particular sequence of replacement of species. Furthermore, there were also obvious seasonal changes in the structure and specific composition of the undisturbed assemblage, so that it was not clear what represented the "mature" state of that system.

A similar influence of seasonality leading to nonsuccessional colonization was documented in a study by Underwood and Anderson (1994) on intertidal estuarine fouling assemblages in oyster farms around Sydney (Australia). These authors tested hypotheses about temporal and spatial variation in colonization of species using replicate blocks of panels submersed in different seasons. They found that algae and barnacles dominated panels submersed from October to March, whereas panels submersed in January were colonized by the oyster *Saccostrea commercialis*. Therefore, oysters were able to colonize and grow rapidly, monopolizing the available space within 4–5 months on some panels but not others, depending on the time of submersion. Oysters did not establish on panels colonized by barnacles and algae that produced stable assemblages for the time scale of the investigation.

Studies conducted in subtidal habitats also provided examples of nonsuccessional patterns of colonization of open space. For example, Foster (1975) found that concrete blocks placed in a *Macrocystis pyrifera* forest were first colonized by ephemeral algae and then by perennial plants. However, the same ephemeral species colonizing the artificial substrata were also found as epiphytes in the surrounding mature assemblage, implying that no real replacement occurred when the dominant species, *Macrocystis*, developed. Another study documented a case in which a dominant alga, the articulated coralline *Calliarthon articulatum*, colonized first on substrata placed in subtidal areas where this alga was very abundant (Johansen & Austin, 1970). Again, there was no evidence that colonization proceeded through an ordered sequence of replacement of species.

The Detection of Succession

Given the large variability in colonization of open space documented by many studies of littoral habitats, it seems appropriate to ask which sort of data are needed to discriminate between successional and nonsuccessional changes. There are at least two reasons why this distinction is useful for the practice of ecology. The first is the need to maintain consistency in terminology, a necessary condition for a concept to be operationally useful (Peters, 1991). Redefining succession to include any possible pattern of invasion of open space makes the term superfluous in that it becomes synonymous with colonization. Second, succession implies a range of ecological processes (discussed later) that can be very different from those leading to nonsuccessional changes. Therefore, a distinction between succession and colonization is important in that it directs attention to the most likely processes influencing the establishment of species in open space.

A distinction was first proposed between succession and temporal fluctuations in populations and assemblages (Huisman et al., 1993; Prach et al., 1993; Van Andel et al., 1993). This distinction is unlikely to contribute much to this issue. Depending on how it is defined, temporal fluctuation does not necessarily exclude succession. For example, one may define temporal fluctuation as the temporal variance in abundance of a population around a certain mean within a period of time. This does not exclude that fluctuating populations of different species can succeed one another in originating an ordered sequence of replacements (Figure 5.1).

An alternative approach to distinguish between successional and nonsuccessional changes is to derive specific hypotheses from the definition of succession about temporal patterns in abundance of species. The approach requires us to be explicit about the sort of data that would unequivocally indicate succession *before* such data are collected. There are at least two classes of predictions that must be considered here. The first refers to the part of the definition of succession that requires an ordered sequence of replacement of species through time. The second refers to the deterministic part of the definition, which requires progression and convergence toward a mature state of the assemblage. Although restating these basic aspects of succession may appear superfluous, astonishingly few of the studies that claim to be about succession actually provide such evidence.

To test the hypothesis that a sequence of replacements occurs among species or higher taxonomical categories, it is necessary to have independent estimates of temporal changes in abundance for each of the response variables examined. This issue was addressed by Underwood and Anderson (1994), who noted that most studies on succession made use of fixed plots that yielded nonindependent estimates of temporal change. As discussed in the previous section, these authors investigated patterns of colonization of fouling assemblages on panels submersed in oyster farms. The experimental design included nine blocks of panels (each containing two replicate panels for each of four types of substrata) submersed in each season. Blocks were removed in groups of three after 1, 2, and 4–5 mo of submersion, and the abundance of the species present were recorded. Therefore, in this study, the length of time panels remained submersed was a factor in an analysis of variance, because the design allowed the collection of temporally independent estimates of abundance of populations.

The design used by Underwood and Anderson (1994) was an improvement over many other studies in which inferences about succession or colonization were made using temporally nonindependent data. However, such a design requires a further improvement in order to test whether a succession of species has occurred. The basic requirement is a test for an interaction between time and species (or any other relevant grouping), which can be done only if there are independent estimates of the abundance of each response variable at each time of sampling. Admittedly, such a test requires a large number of samples, but the analysis and interpretation of results is straightforward. For example, if preliminary observations are available, say, for an assemblage of algae in subtidal habitats, suggesting that invasion of open space proceeds from filamentous algae to fleshy erect algae and, ultimately, to large canopy species, one might use the analysis illustrated in

Succession in Littoral Habitats 107

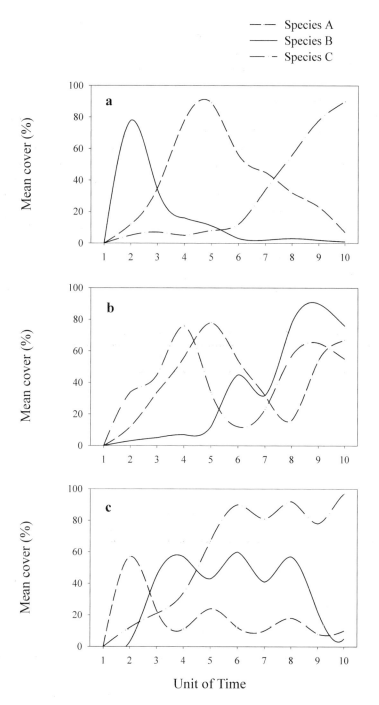

Figure 5.1. Examples of patterns of recovery of three species after disturbance: (a) Replacement of species occurs according to the definition of succession; (b) recovery is erratic; (c) there are temporal fluctuations in abundance of species but succession still occurs.

Table 5.1 to test whether the observation actually reflect succession. In order to do this, there must be independent estimates of abundance for each of the three morphological groups of algae at each time of sampling. Again, preliminary observations and knowledge of the natural history of the system may suggest the appropriate time scale to capture the major transitions in the developing assemblage. Suppose that preliminary information indicated major changes expected to occur after 6 mo, 1 year, and 2 years since the assemblage was disturbed. To test the hypothesis that these changes involve a directional sequence of replacements, one must design a sampling protocol providing independent sets of replicate observations for each of the three morphological groups at each time of sampling. Suppose that resources are available to sample five replicate units of each category at each of three times chosen to represent different stages of colonization. The total number of sampling units in this experiment would be 45, which seems logistically feasible even with limited resources.

To provide support for a model of succession, a significant interaction between species and time must occur in the data (Table 5.1a), and *post hoc* multiple comparisons must show dominance by different species at different times. In this example, SNK tests could be used to compare the abundance of morphological groups at each time of sampling, in addition to the temporal trajectory of each group separately. Replacement can

Table 5.1. Analysis of Variance of a Hypothetical Study on Recovery.

Source of variation	df	Expected MS	F	P
a (analysis of variance)				
Species	2	$\sigma_E^2 + 15K_S^2$	MS_S/MS_E	
Time	2	$\sigma_E^2 + 15K_T^2$	MS_T/MS_E	
Species × time	4	$\sigma_E^2 + 5K_{ST}^2$	MS_{ST}/MS_E	< 0.05
Residual	36	σ_E^2		
Total	44			

SNK tests within the Species × Time interaction (standard error for the comparison = $MS_E/\sqrt{5}$)
b (there is succession)

Within species		Within times	
Species	Time	Time	Species
Filamentous	6 > 12 > 24	6	Filamentous > fleshy > canopy
Fleshy	12 > 6 = 24	12	Fleshy > filamentous = canopy
Canopy	24 > 12 > 6	24	Canopy > fleshy > filamentous

c (there is erratic colonization)

Within species		Within times	
Species	Time	Time	Species
Filamentous	6 = 24 > 12	6	Filamentous > fleshy > canopy
Fleshy	12 = 24 > 6	12	Canopy = fleshy > filamentous
Canopy	12 > 24 > 6	24	Filamentous = fleshy > canopy

In the example, the abundance of three groups of algal species (filamentous, fleshy, and canopy algae) is sampled using five independent replicates at each of three times after disturbance (6, 12, and 24 mo). σ_E^2 is the variance among replicates, K_S^2 is the effect of species, K_T^2 is the effect of time, K_{ST}^2 is the species × time interaction. Species and time are fixed effects in the model. A necessary condition for succession to occur is that the interaction between species and time is significant (as indicated in *a*). Succession can then be distinguished from erratic colonization using *post hoc* multiple comparisons of the means (SNK tests in this case); *b*, succession is unequivocally identified; *c*, erratic colonization occurs (see text for further details).

be shown to have occurred if filamentous algae are significantly more abundant than other algae at early stages of colonization, whereas fleshy erect species dominate intermediate dates, and canopy algae become dominant at later stages (Table 5.1b). Similarly, comparison of the abundance of each group across time must show that filamentous algae are significantly more abundant during the first months of colonization than at any other time, that fleshy erect algae dominate only at midstages, and that canopy algae are dominant only at later stages (Table 5.1b). If, for example, the abundance of filamentous algae increases again at maturity, so that there is no difference between early and late stages, it is not possible to reject the null hypothesis of no replacement and to conclude that succession has occurred (Table 5.1c).

If there are enough resources to repeat the study in different places and/or different times (to examine whether or not a given pattern is general), it is also possible to apply a rank test to the results of the *post hoc* comparisons (e.g., Anderson test; Winer et al., 1991), to test the hypothesis that any particular species at any particular time ranks first in abundance significantly more often than expected by chance. This would provide an even more robust test of the hypothesis that colonization proceeds through a temporal sequence of replacements that is general and repeatable.

The approach advocated here relies on statistical significance to decide whether or not succession has occurred. Therefore, it is not necessary to show that all individuals of a given species have been killed or displaced to deem that succession has occurred, an unrealistic requirement for many assemblages. Rather, quantitatively relevant changes in the abundance of populations can provide evidence of succession. As usual in hypothesis testing, the main problem is to define a quantitative alternative to the null hypothesis; that is, what changes in abundance of populations can be considered relevant changes. Anticipating how large a change is to be expected may not be a simple task and requires a great deal of knowledge of the natural history of the system being investigated. Succession requires the replacement of species, thus temporal changes in patterns of dominance within an assemblage. Therefore, the focus is on those species that have the potential to monopolize space, and the expected effect size must be a large change in abundance.

As an example, Benedetti-Cecchi (2000a) found that grazing by limpets facilitated the replacement of filamentous algae by fleshy algae (*Rissoella verruculosa*) during recovery of disturbed patches on rocky shores in the northwest Mediterranean. Grazers reduced the cover of filamentous algae of about 50% compared to ungrazed patches, whereas the percentage cover of fleshy algae increased 70–80% in the presence of limpets. These data can be used to specify the expected effect size of temporal changes in abundance of algae (i.e., the quantitative alternative to the null hypothesis of no temporal change) in any future study of the influence of limpets on recovery in this assemblage.

Succession is deterministic; that is, the assemblage developing in a disturbed patch of habitat must converge toward the assemblage present in the surrounding environment. A useful measure to examine convergence is the percentage similarity between assemblages. The hypothesis being tested is that similarity increases through time until differences between assemblages are no longer significant. For example, Sousa (1980) used a measure of similarity (Whittaker, 1975) to examine changes in the structure of a disturbed assemblage of algae and invertebrates on intertidal boulders and compare them to the structure of the assemblage prior to disturbance. Foster (1992) reported a series of experiments conducted in the rocky intertidal habitats of central and northern California, where assemblages developing in artificially cleared plots were compared to those in control plots using the Bray–Curtis measure of similarity (Bray & Curtis, 1957). These measures were used to calculate the rate of recovery (percentage recovery/month) of algal assemblages in relation to density of grazers. These analyses indicated that recovery was delayed by herbivores.

An appropriate test of the hypothesis that recovery is deterministic requires a comparison between assemblages in disturbed patches and those of reference undisturbed sites *during* colonization. There is no logical basis for assuming that colonization of a disturbed patch produces the type of assemblage present prior to disturbance. Populations in reference sites may undergo drastic temporal changes as well, so that the original structure is not representative of what can occur at further points in time. It is necessary to establish a criterion to evaluate whether recovery has occurred when comparing the similarity between disturbed and reference assemblages. A meaningful approach is to compare this measure of similarity with the range of natural similarities within reference and disturbed sites. In particular, Chapman and Underwood (1998) used measures of dissimilarity (as opposed to similarity) to compare algal assemblages colonizing artificial clearings with those of control plots on intertidal rocky shores in New South Wales (Australia). They proposed a statistic based on the Bray–Curtis measure of dissimilarity among samples within the disturbed assemblage (Dis_{WA}), among samples from the controls (Dis_C), and on the average measure of dissimilarity between disturbed and control plots (Dis_B). The quantity $Dis_B - [(Dis_{WA} + Dis_C)/2]$ was used as a measure of the difference between the two assemblages and the average dissimilarity within these assemblages. When the statistic is near 0%, it indicates that assemblages have converged to natural within-assemblage variation (Chapman & Underwood, 1998). The same analysis can use measures of similarity rather than dissimilarity; in this case, convergence would be revealed by a value of the statistic near 100%. Although techniques are rapidly developing, few studies make use of new methodologies to provide formal tests of hypotheses about succession, particularly to test for temporal replacement of species and directionality.

Spatial and Temporal Scales of Observation

It is widely recognized that the scale of observation has a large influence on the detection of pattern in natural assemblages (Levin, 1992; Schneider, 1994). Many contrasting paradigms in ecology, such as the equilibrium versus nonequilibrium dichotomy or contingency versus generalization of patterns and processes can be reconciled by defining the appropriate scale to which models and concepts apply (McCook, 1994; Wu & Loucks, 1995). Therefore, the spatial and temporal scale(s) of observation must be explicitly stated in any analysis of succession or aimed at distinguishing between succession and variable colonization. Connell and Sousa (1983) discussed the importance of defining appropriate spatial and temporal scales to evaluate stability and persistence in assemblages. They suggested that an appropriate temporal scale to evaluate persistence and stability must include at least one complete turnover of the population with the longest life span in the assemblage. The relevant spatial scale is the one that embraces all the habitats where the populations constituting the assemblage can occur. Although these questions have stimulated intense research in several areas of ecology, the relationship between scale and succession has received less attention. Only a few studies have explicitly addressed the issue of scale in analyses of recovery of assemblages after disturbance (Dayton et al., 1992; McCook & Chapman, 1997; Chapman & Underwood, 1998). These studies provide an opportunity to attempt to answer the following questions: (1) What are the relevant scales over which succession should be investigated, and (2) are these scales uniquely determined for any type of assemblage and habitat?

Dayton et al. (1992) investigated disturbance and recovery at a number of sites 100–1000 m apart and over a period of 20 years in the Point Loma kelp forest (*Macrocystis pyrifera*) in southern California. This study, part of a long-term investigation on the ecology of this kelp forest (Dayton et al., 1984), included measures of the intensity and severity of disturbance, of the processes influencing recovery after disturbance, of the mechanisms allowing the regional persistence of patch structure, and an evaluation of spatial and temporal variability in these measures. The main results indicated considerable temporal variation in intensity of disturbance, remarkable between-site differences in rates of recruitment and recovery of algal populations (including several understory species), and the overwhelming competitive dominance of *Macrocystis pyrifera*. What is relevant to the present discussion is that patterns of recovery appeared variable at the local scale of the site and over short temporal scales (months to a few years). In contrast, at larger spatial and temporal scales, the system exhibited some degree of consistency in its internal patch structure, with several understory species dominating in recently disturbed areas but with *Macrocystis* dominating in the long term. However, the establishment of *Macrocystis* did not necessarily result in the disappearance of other species.

McCook and Chapman (1997) described patterns of algal colonization on intertidal rocky shores that were severely disturbed by sea ice in Nova Scotia (Canada). Disturbance generated patches of open space of various size along a 50-km stretch of the coast. Variability in recovery following disturbance was studied at two spatial scales: among locations 10's of km apart, and among sites within locations 1–100 m apart. Recovery was examined over a period of 3–5 years using five replicate quadrats (10×10 cm in size) at each site. The study documented the initial colonization of filamentous green algae and then the dominance of canopy algae (*Fucus*). Mussels and barnacles also became established in the disturbed patches but persisted at low densities. Variation in colonization was observed at both spatial scales and reflected different patterns in abundance of ephemeral algae and variation in timing and intensity of recruitment of *Fucus*. Therefore, the rate and trajectory of recovery differed among sites and locations. Interestingly, although small-scale patterns were apparently related to variation in recruitment of canopy algae, differences among locations in rate of recovery were correlated to size of the disturbed patches and severity of disturbance. Thus, different processes appeared to generate variation at different scales.

The study by Chapman and Underwood (1998) (see also the previous section) is another example of scales of variability in recovery. The study design included five shores distributed along about 800 km of coast. Replicate patches (14×16 cm in size) were cleared on each shore to simulate the effects of natural disturbance (mainly due to wave action), and the developing assemblages were compared with those in the surrounding habitat every 3 months over a period of 4 years. These data were used to test hypotheses about the relative importance of broad-scale biogeographic processes and small-scale historical processes in the development of algal assemblages. In addition, new clearings were produced on each shore every 3 mo for a period of 4 years, to test whether early patterns of colonization were consistent in time and space. Results of this study revealed considerable differences from shore to shore in recruitment and development of assemblages, with no clear broad-scale or biogeographic trend. Differences were also evident from time to time in the structure of assemblages at early stages of development. Much of this variability resulted from differences in the local abundance of a few species, including ephemeral green algae and *Corallina* spp. Despite initial differences, developing assemblages converged toward those in the surrounding habitat within 1–2 years, with ephemeral algae and sessile animals being replaced by perennial algae. These results suggested that local influences were more important than biogeographic processes in regulating recovery.

These few examples stimulate some considerations on the influence of scale in the analysis of succession or variable colonization of open space. Sousa (1984a) posed the question of whether small-scale patchiness averages to some form of equilibrium pattern when one examines large-scale or regional dynamics. The studies above suggest that consistency emerges over large spatial and/or temporal scales with respect to persistence of populations, convergence in patterns of recovery, and internal structure of assemblages (see also Paine, 1979; Connell, 1987). Therefore, the hypothesis can be proposed that ecological variance in recovery or succession within a given habitat is consistent across similar habitats over broad spatial and temporal scales. This seems to be an element of predictability in the dynamics of ecological assemblages that has not been sufficiently examined. A test of this hypothesis requires structured descriptive studies documenting invasion of open space at a hierarchy of spatial and temporal scales. This approach would also provide insights into the relevant ecological processes operating at different scales. As suggested by the study of McCook and Chapman (1997), the processes generating small-scale variability in succession are likely to be different from those responsible for large-scale variation. This provides a solid basis for proceeding with experimental analyses of models and hypotheses of succession.

With respect to the questions posed earlier, it seems clear that no single scale can be considered as the most appropriate for investigating colonization and succession in any type of habitat and for any type of assemblage. The issue is to identify the spatial and temporal resolution and extent of an investigation that is relevant to the ecological models and hypotheses being tested (Schneider, 1994). The spatial resolution of a study on succession is probably best determined by the size of the disturbed patches. This is a common choice in many experimental investigations of succession (see below). Admittedly, this criterion can be difficult to apply in cases in which disturbance produces very large patches, as in the study by McCook and Chapman (1997). Logistical constraints associated with the size of the sampling unit often dictate the resolution of such studies. As discussed earlier, the study should also be structured in such a way to incorporate a range of spatial extents, so that patterns of variation can be examined at a hierarchy of scales. When the resolution of a study is set, considerations on the geographical distribution of the focal assemblages, on the range of habitats where they occur, and on the life histories of the component organisms can all help in identifying the relevant scales to address. For studies of succession in littoral habitats, appropriate scales may range from a few meters up to hundreds of kilometers (see references above).

Considerations on temporal scales of succession follow along similar lines. As illustrated by these examples, temporal resolution of the order of months can capture relevant changes in algal assemblages on rocky shores. Less frequent sampling may be more appropriate for species that have slower growth rates and longer life spans, such as vascular plants. These variables also set the length of a study. Although complete recovery may occur in one or a few years in algal assemblages, much longer periods may be necessary for assemblages to reach maturity in salt marshes or sand dunes (Ranwell, 1972). If temporal variation in colonization or succession is of concern, then the same arguments developed above for the analysis of spatial patterns apply. Temporal variation can be examined at a range of scales, including seasonal, annual, and decadal variation.

The rationale for choosing one or more scales is that each of these addresses a different issue about temporal variation in recovery. For example, comparing patches produced in different seasons is appropriate to test hypotheses about the influence of small-scale temporal variability in reproduction and recruitment of species (e.g., Sousa, 1985; see also below). In contrast, comparisons over decades are appropriate to test hypotheses about the role of episodic events or climate shifts in influencing colonization and succession. Quantitative descriptions at multiple scales in space and time are therefore a prerequisite for investigating causal processes of colonization and succession.

Explanation of Succession and Colonization

Theoretical Background

Early explanations of colonization and succession were based mostly on interactions among species. Facilitation of later species by early colonists was central to Clements's view of succession (1916, 1928), whereas inhibition due to competition for resources was proposed by Gleason (1926, 1927) and reiterated later by Egler (1954). The importance of life-history traits in explaining succession was stressed by Egler (1954) and Drury and Nisbet (1973). These authors maintained that sequential replacement reflected interspecific differences in life-history traits such as growth rate, size, life span and dispersal, in addition to the capability of species to tolerate stress. Patterns of correlation among species' life-history traits were also proposed to explain erratic colonization (Drury & Nisbet, 1973).

Early propositions about causes of succession were later integrated into more coherent theoretical frameworks, including the well-known alternatives of facilitation, tolerance, and inhibition by Connell and Slatyer (1977), Tilman's model based on constraints and trade-offs (1985, 1990), and the analysis by Huston and Smith (1987) on the effects of physiology and life history (see McCook, 1994, for a detailed analysis of these and other models). Theory stimulated an intense experimental activity in the field, including tests of processes that were not yet part of a general theory of succession, such as the influence of size of patches, time of clearance, and consumers (reviewed in Sousa, 1984a, 1985; Farrell, 1991; Benedetti-Cecchi, 2000a); these processes are explicitly incorporated into the general framework of patch dynamics (Pickett & White, 1985).

Models were also proposed to explore the interactive effects of some combination of these processes. An example is the set of models proposed by Farrell (1991) to predict rates of succession as a function of the effects of consumers and the nature of the interaction between early and late colonists. I first review some of the experimental evidence related to the more general models of succession, then consider experiments that examined interactive effects among causal processes; finally, I discuss the relevance of experimental work to the issue of prediction of colonization and succession.

Experimental Analyses of General Models of Succession

Facilitation has been proposed as a prominent mechanism driving succession in salt marshes and sand dunes. Potentially important effects of early colonists on later species include stabilization of the substratum, accumulation of nutrients, and amelioration of physical stress through desalinization, aeration of the soil, and enhancement of pH (Olson,

1958; Olff et al., 1993; Van Andel et al., 1993). Some alternative explanations have also been postulated based on mechanisms other than facilitation, including suppression of early colonists by their own litter and accretion toward less saline and drier areas where other organisms can establish (e.g., Ranwell, 1972). The capability of discriminating among these models was limited by an excess of interest in classifying succession into distinct categories (e.g., primary vs. secondary, autogenic vs. allogenic). This has generated almost insoluble questions (see discussion in Sousa, 1984a, about the distinction between autogenic and allogenic succession), detracting from the important need to provide experimental tests of hypotheses that the proposed models originated.

As a result, most of the inferences on the causes of succession in sand dunes and salt marshes are based on correlative evidence. For example, the perceived importance of soil properties, nutrients, and light as causes of succession was examined by Olff et al. (1993) by comparing estimates of these variables across sites in sand dunes. Sites were placed close to vegetated areas of known age to represent different stages of succession. That several of the measured variables differed among sites was taken as evidence of their importance as determinants of succession (Olff et al., 1993, p. 705). Clearly, alternative interpretations of these results are possible. For example, differences among sites in physical variables might have been generated by, rather than being the cause of, age-specific patterns in species composition and structure of assemblages. The possibility of confounding effects due to spatial variation (pseudoreplication) was acknowledged by the authors (Olff et al., 1993, p. 696). This was not really a problem if the replicate plots representing each stage of succession were properly interspersed in space. If, in contrast, there was spatial segregation of areas representing different stages of succession, then any effect of age was confounded by possible intrinsic spatial variation in the variables being measured.

The problem of spatial confounding is present in all studies that use chronosequences. These studies are motivated by an operative definition of succession that encapsulates both the spatial and temporal component of heterogeneity. Thus, the causes of succession are often explored by comparing processes in different areas, under the assumption that these areas represent different stages of succession. This approach can affect the generality and validity of conclusions in at least two ways. First, differences among areas can reflect succession, but the pattern is specific for the particular salt marsh or sand dune investigated; that is, the degree to which space can be used as a surrogate of time changes from place to place. This problem could be obviated by replicating the study at several locations to test whether patterns and processes are consistent over large spatial scales. I am unaware of any study in salt marshes or sand dunes that incorporated the appropriate range of spatial scales to test this hypothesis properly. Second, different areas do not reflect succession, because the effect of age is confounded with intrinsic spatial variation in the response variables being measured. There is no simple solution to this problem. However, in analyses of the causes of succession, it should be possible to derive specific predictions from the models that have been proposed to explain the observed patterns and test these predictions experimentally. For example, several studies used Tilman's resource ratio model (1985, 1988) to explain progression of the vegetation in salt marshes and temporal replacement of species with contrasting life-history attributes (e.g., Olff et al., 1993; van Wijnen & Bakker, 1999). This is often done *a posteriori*, after spatial variation in composition of assemblages and species' life histories is detected. A more productive approach would be to derive predictions from the model and use experiments to test them. For example, according to Tilman's model (1985), one would predict that if species are transplanted from old to young areas of the habitat and are maintained without neighbors, survival is possible only if physical conditions are also ameliorated, for example, by adding nutrients and reducing abiotic stress. In contrast, if the transplant is done in the opposite direction, from young to aged areas, then survival is possible only if resident organisms are removed—this as a consequence of the negative patterns of correlation in life-history traits such as tolerance to stress and competitive capabilities assumed by the model. Some experiments involving the transplantation of seedlings or adult plants in salt marshes indicated that under stressful conditions, positive interactions among plants can lead to coexistence rather than replacement of species, and that competition generates spatial rather than temporal variation in assemblages (e.g., Bertness & Ellison, 1987; Bertness, 1991). Other studies showed that the experimental increase of nutrients in young marshes enhanced the biomass of resident species but did not lead to succession, contrary to what was expected from theory (Olff et al., 1993; van Wijnen et al., 1999).

Herbivory is another frequently invoked cause of succession in salt marshes and sand dunes. Experiments in which patches of substratum were maintained free of herbivores have a long history in these habitats (e.g., Farrow, 1917, cited in Paine, 2000; Ranwell, 1961, 1968). There are, however, considerable problems with inferences on herbivory and succession in terms of magnitude and direction of effects. Correlation is sometimes taken as evidence of causation, and spatial confounding is evident in some experiments (Olff et al., 1997; van Wijnen et al., 1999). Although the effects documented in such studies are often very large (see Paine, 2000), and the general conclusions probably correct, there is no logical basis for assuming that results are representative of the range of possible effects of herbivory on succession, unless properly replicated experiments are performed in these habitats. Repeated experimentation is therefore instrumental to test whether effects are consistent in space and time or, alternatively, to identify the scale(s) at which variation and interactive effects between grazing and other processes occur (e.g., Benedetti-Cecchi et al., 2000).

In contrast to data reported for salt marshes and sand dunes, cases of facilitation have been documented only occasionally in rocky intertidal and subtidal habitats. An example is the positive effect of turf-forming algae on the establishment of sea grasses (Turner, 1983; Williams, 1990). Conversely, several studies have indicated inhibition as a common trait of succession on rocky shores. For example, Sousa (1979b) examined the effects of competition on algal succession in intertidal boulder fields by comparing unmanipulated plots with those in which either early or intermediate species were selectively removed. These experiments showed that early colonists (*Ulva*) inhibited the establishment of mid- and late successional species (*Gigartina* spp.), whereas intermediate algae inhibited both early and late colonists. However, most of the evidence on inhibition is indirect and comes from experimental investigations on the influence of herbivores on macroalgae. In most of these experiments, recovery was investigated in artificially cleared patches where herbivores were either present at natural densities or excluded by fences, or in other ways (e.g., barriers of copper paint). Elimination of herbivores often resulted in the monopolization of space by ephemeral algae and the exclusion of perennial species (Lubchenco & Menge, 1978; Sousa, 1979b; Lubchenco, 1983). These results emphasized the importance of indirect effects of consumers in accelerating succession (see also Robles & Cubit, 1981; Underwood et al., 1983; Van Tamelen, 1987). According to Farrell (1991), consumers can accelerate succession only if they forage preferentially on early colonists and inhibit the establishment of later species. In contrast, consumers delay succession if there is facilitation regardless of their dietary preferences, if they prefer late successional species under any mode of interaction among colonists (inhibition, tolerance, or facilitation), or if there is no difference in palatability between early and late successional species that tolerate each other.

As discussed elsewhere (Sousa & Connell, 1992; Benedetti-Cecchi, 2000a), a model of indirect positive effects of consumers on succession can only be supported if (1) consumers forage preferentially on early colonists, (2) early colonists are superior competitors for space, and (3) there is no positive direct effect of consumers on later species. Independent tests of these alternative explanations have been performed only occasionally (see discussion in Sousa & Connell, 1992). Food preference of herbivores is often examined in the laboratory, whereas the possibility for consumers to exert positive effects on later colonists has received little attention in both the laboratory and the field. Apart from the detailed analysis by Sousa (1979b) discussed earlier, and a few other examples (e.g., Benedetti-Cecchi, 2000a), genuine evidence indicating that early colonists inhibit the establishment of later species and the way consumers impinge on the rate of succession is limited. These issues highlight the interactive nature of the processes that regulate succession and the need to use multifactorial experiments to unravel this complexity.

Experiments on Interactions among Processes Influencing Succession and Colonization

The need to explain spatial and temporal variability in succession and erratic colonization resulted in the proposition and analysis of complex models about the interaction among several causal processes, including the effects of size of patches, time of clearance, and consumers, among others (see reviews by Sousa, 1984a, 1985). There are, however, differences among studies and habitats in the specific combination of processes investigated, possibly reflecting the lack of a coherent theoretical framework within which multiple causality could be analyzed, the nature of the original observations that required explanation, or differences among researchers in interests and inspiration.

A series of factorial experiments were conducted in salt marshes of Rhode Island (Shumway & Bertness, 1992, 1994; Bertness & Shumway, 1993) to examine the proposition that positive interactions among colonists drive succession in stressful habitats while negative interactions are prevalent in benign environments. Physical stress was manipulated either by producing clearings of different size or by watering. Stress was higher in large patches, due to increased solar radiation and greater salinity, compared to small clearings, and lower in watered compared to unmanipulated patches. Patterns of recovery of several marsh plants maintained in isolation or in mixed assemblages with other species were compared under different regimes of physical stress. These experiments showed that the establishment of competitive dominant plants in large patches was facilitated by the early establishment of salt-tolerant species that reduced physical stress. In contrast, competition was prevalent in small patches, where physical conditions were not so stressful. Similarly, negative interactions among species were prevalent in watered patches, whereas positive interactions were prevalent in unmanipulated patches under stressful physical conditions. Collectively, these results indicated that the rate of recovery was governed by the interplay between biological interactions and physical stress, and that the actual mechanism of succession was dictated by the size of the disturbed patch or by any other process that reduced harshness within a patch.

Dormann and Bakker (2000) investigated the interactive effects of grazing and competition on flowering and survival of three species of vascular plants that characterized different stages of succession in a Dutch salt marsh. Grazing and competition were manipulated in orthogonal combinations, and replicate plots of each treatment were established for each species in each of three areas of the salt marsh, chosen to represent early, mid-, and late stages of succession. Although this experiment suffered problems due to spatial confounding and only a single marsh plant was used to represent species characteristic of each stage of succession, the results revealed interspecific differences in susceptibility to grazing and competition, and spatial variability in these effects. In general, however, all plants did better when both grazers and competitors were removed, suggesting additive rather than multiplicative effects of the two processes. Whether or not these patterns were able to explain succession in the salt marsh remained unclear.

Explanations of variability in recovery after disturbance in marine rocky intertidal and subtidal habitats were based on models of the effects of consumers, dispersal and recruitment of species, size of patches, time of clearance, and interactions among colonists. Sousa (1985) reviewed the circumstances under which these processes, in isolation or in combination, could affect succession in disturbed patches. For example, succession may differ between large and small patches, because the former sample a larger proportion of propagules from the water column. This may result in a larger proportion of species characterized by good dispersal capabilities in small clearings (Kay & Keough, 1981; Sousa, 1984b). Small patches also have a greater perimeter : surface ratio, so that vegetative encroachment of plants from the borders of the patch will be more important as a mechanism of recovery in small than large clearings. This implies that interactions between species that colonize through vegetative growth and those that recruit from propagules are more intense in small patches or toward the edge of large clearings (Farrell, 1989). The time at which a patch is produced may affect succession because of temporal variation in reproduction and growth of species (Foster, 1975; Sousa, 1979b; Hawkins, 1981; Paine, 1979; Dayton et al., 1984; Kim & DeWreede, 1996).

These processes can interact in several ways to generate variability in succession or erratic colonization. Only a few of the possible interactions have been investigated experimentally. A type of interaction for which there is substantial supporting evidence is the relationship between size of patches, density of herbivores, and rate of recovery of algae. It has been shown repeatedly that molluscan grazers such as limpets tend to be more abundant in small than large clearings, and this can have indirect consequences for recruitment and growth of algae (Suchanek, 1978; Sousa, 1984b; Farrell, 1989; Benedetti-Cecchi & Cinelli, 1993; Dye, 1993). This was clearly documented by Sousa (1984b) for patches cleared in intertidal mussel beds on rocky coasts in California. He used a factorial experiment in which the main grazers, limpets, were excluded with barriers of copper paint or left at natural densities in small (25×25 cm) and large (50×50 cm) patches. Results indicated that size of patches per se had no effect on rate of recovery of algae in the absence of grazers. In contrast, comparisons involving plots open to limpets indicated that recovery was significantly delayed in small patches due to the larger density of grazers compared to large clearings. Therefore, size of patches indirectly affected succession by influencing primarily the density of grazers.

The study by Sousa (1984b) is also one of the few examples in which hypotheses about the influence of recruitment and dispersal of species on succession were tested. Differences in recovery among patches were related to the composition of assemblages at increasing distances from the focus patches using multiple regression. Although the source of colonists was not manipulated experimentally, these analyses revealed a strong relationship between the identity of algae that invaded cleared plots and those present in the surrounding habitat within 1 m of the edge of the patches. Given the importance of dispersal and recruitment as causes of variation in these assemblages (e.g., Reed et al., 1988; Kendrick & Walker, 1991; Menge et al., 1993), it is surprising how few experimental studies have tested whether variable succession or erratic colonization reflects differences in the availability of colonists from nearby assemblages.

Some studies of rocky intertidal habitats investigated the interactive effects of grazing and physical stress on succession. Buschmann (1990), for example, manipulated the density of grazers (molluscan herbivores) and drainage of seawater in orthogonal combinations in a rocky intertidal habitat in southern Chile. Experimental plots were initially cleared of resident organisms and recovery examined under each treatment condition. Ulvoid algae became established in plots that had no drainage of seawater, with the largest cover attained in the absence of grazers. In contrast, the crustose brown alga *Ralfsia* sp. recovered in irrigated plots facilitated by the activity of grazers that reduced the cover of ulvoid algae. Therefore, grazers delayed recovery under stressful conditions, whereas they hastened succession when physical stress was ameliorated by irrigation. In a similar experiment on rocky shores in Hong Kong, Kaehler and Williams (1998) manipulated grazing and irrigation in mid- and low-shore habitats, both in the cold and the hot season. Macroalgae colonized high-shore plots only when these were irrigated, with encrusting species dominating in the presence of grazers and *Ulva* spp. becoming abundant where gastropods grazers and echinoids were excluded by fences. Irrigation had negligible effects in the low-shore habitat, where seasonal effects became important. Exclusion of herbivores in the cold season resulted in a mixed assemblage of erect coralline and foliose algae, the latter monopolizing space in the hot season. In contrast, erect coralline algae dominated plots exposed to herbivores in the cold season, whereas the encrusting brown algae *Ralfsia expansa* replaced encrusting corallines in the hot season. These results revealed complex interactions between physical and biological processes in regulating development of algal assemblages and the need for multifactorial experiments to unravel such complexity.

Few other studies have employed multifactorial experiments to investigate interactions among multiple causes of succession or colonization. Farrell (1989) examined the model in which recovery of patches in algal assemblages proceeded both through vegetative propagation from the borders of the patches and through recruitment of propagules from the plankton. He tested the logical prediction that succession was similar along the borders of large patches (16 × 16 cm) and in small clearings (4 × 4 and 8 × 8 cm), and that these patterns differed from succession in the center of large clearings. He further examined the interactive effects of grazing and size of patches as other possible causes of variation in recovery. Experimental plots were set up in a rocky intertidal habitat on the coast of Oregon (USA), and recovery followed for 2.5 years. Results supported these predictions, indicating that both vegetative propagation and the influence of grazers (limpets) contributed to the expected similarity in recovery between the borders of large patches and the small clearings. In fact, grazers were more abundant and effective in removing algae in these circumstances than in the middle of large clearings. A problem with the lack of independence of measures taken at the center and the edge of large patches, however, precluded a formal test of the interaction between grazing and position within a patch.

A similar experiment was conducted by Benedetti-Cecchi and Cinelli (1993) to test hypotheses about the interactive effects of grazing, size of patches, time of clearance, and spatial variation in these effects, in addition to position within large patches (22 × 22 cm), on rocky shores at Calafuria (northwest Mediterranean). Also this experiment had problems of nonindependence of measures taken at the edge and the center of large patches, and the effects of timing of clearance could not be related to seasonal effects, as initially proposed, because of lack of replication within seasons. Nevertheless, results after 1 year indicated considerable spatial and temporal variation in colonization of species, larger densities of limpets in small rather than large patches, and suggested positive indirect effects of grazers on the establishment of the perennial alga *Rissoella verruculosa* through removal of filamentous species. Most of the initial variability disappeared in the long run, but the beneficial effect of grazers on the establishment of *Rissoella* was supported by additional experiments appearing as a major process influencing the structure of the mature assemblage (Benedetti-Cecchi, 2000a). These patterns are in partial agreement with the results of other studies on rocky shores, indicating that variability can be large at early stages of recovery but decrease as succession progresses (reviewed by Sousa, 1985). This conclusion, however, may not apply to patterns observed at Calafuria, where indirect interactions can inject considerable variability in the structure of the "mature" assemblage. For example, dense patches of barnacles and *Rissoella* reduced the local abundance of limpets, facilitating the establishment of filamentous algae late in recovery (Benedetti-Cecchi, 2000a). Therefore, local variability in abundance of barnacles and *Rissoella* could result in different patterns of abundance of limpets and filamentous algae. Indirect effects of barnacles have been documented several times, but in most cases, these organisms provided a refuge from grazing for algae characteristic of late stages of succession (Van Tamelen, 1987; Farrell, 1991). This is different from the situation observed at Calafuria, where algae that benefited from the presence of barnacles were the same species that monopolized space early in recovery, when grazers were excluded. Such algae, then, could not be unequivocally qualified as early or late successional species, and there was no evidence that other colonists could limit their abundance (Benedetti-Cecchi, 2000a). Therefore, neither a true sequence of replacement of species nor directionality characterized recovery of patches in these assemblages; these patterns can be described more properly as variable colonization rather than true succession.

Hypotheses about the effects of size and location of patches, time of clearance, and intensity of disturbance on algal recovery were also tested in subtidal rocky reefs at Calafuria (Airoldi, 1998). It was proposed that trade-offs in species' capabilities to tolerate disturbance and to monopolize space influenced recovery. This model predicted that species with different life histories should dominate under different regimes of disturbance. Intensity of disturbance was manipulated by adding known amounts of sediment to experimental plots and leaving other plots untreated for comparison, for each combination of size, location, and time of clearance of patches. The results revealed interactive but transient effects of experimental factors on algae that recovered through sexual reproduction. With time, however, most of the experimental plots became monopolized by a filamentous algal turf that recovered quickly through vegetative propagation, regardless of treatment manipulation. These results reinforced the view that marine algae can integrate contrasting life strategies such as rapid growth and resistance to invasion, so that models based on trade-offs in life-history traits may not be the most appropriate to explain algal recovery in subtidal assemblages.

The preceding discussion suggests that the experimental analysis of interactive ecological processes is a powerful tool that has enhanced our understanding of succession and colonization considerably. Most of the studies reviewed here addressed explicit hypotheses, and results were discussed in relation to the questions posed. Although classical theories of succession were often used as a template for the discussion of results, even when no specific predictions from these theories were tested, it seems that most studies preserved the logical relationships among at least three components of the research: explanation, hypothesis testing, and conclusions (Underwood, 1997).

The Prediction of Succession and Colonization

What can we claim to have achieved with respect to predictability of recovery from disturbance after nearly a century of research since the first formal definition of succession as a predictable event (Clements, 1916)? Are there relevant traits of succession that can be consistently predicted in different habitats? What aspects of theoretical and empirical work require further development to improve accuracy and precision of predictions? The variable and interactive nature of the processes that affect recovery of disturbed patches makes these issues complex and their resolution challenging (Connell & Slatyer, 1977; Connell et al., 1987; Cattelino et al., 1979; Underwood et al., 1983; Pickett et al., 1987b). Berlow (1997) discussed the influence of historic effects and their interaction with actual processes as causes of indeterminacy in succession. He distinguished between contingent and canalized succession, drawing attention to the relative importance of deterministic versus stochastic processes. Deterministic factors leading to canalization include trade-offs between dispersal and competitive ability, strong biotic interactions, and functional redundancy within groups of species. Contingency may result from variable recruitment, indirect effects of consumers, prey escapes in size and density, any combination of these effects, and their interaction with abiotic processes. The difference between contingency and canalization is analogous to the distinction between succession and erratic colonization discussed so far. This is an important distinction, because it implies a different role of theory and different levels of accuracy and precision of predictions. Succession is deterministic and repeatable, by definition, so predictability is a necessary trait of succession, as documented in some systems (Clements, 1928; Odum, 1969; Paine, 1984). Understanding the causes of succession does not really make predictions more accurate in these circumstances: although theory retains its explanatory value, predictability is achieved anyway, because patterns are repeatable and deterministic. In the case of variable colonization, in contrast, explaining the causes of variation is the only way to attempt successful predictions. Because canalized versus contingent succession and the distinction between succession and colonization reflect the extremes of a continuum rather than true dichotomies, one general role of theory is that of anticipating the circumstances under which stochastic variation is reduced, so that successful predictions become more likely (Walker & Chapin, 1987; Berlow, 1997; Benedetti-Cecchi, 2000a). This requires explicit statements about the appropriate spatial and temporal scales of predictions, the basic processes that the theory incorporates, and how hypotheses can be tested.

Predictions can be attempted at several spatial and temporal scales. As already discussed (see "Phenomenology of Succession and Erratic Colonization: Spatial and Temporal Scales of Observation"), variability in colonization may be larger at small scales, so that predictions of local patterns over short periods of time can be difficult. Scaling-up from the small to the large may be problematic as well, because we know virtually nothing about the relative importance of different ecological processes in regulating patterns of colonization at different scales. Most of the experimental studies have investigated processes at single places and only one time (but see Berlow, 1997). There is an urgent need to understand mechanisms of recovery over a hierarchy of spatial and temporal scales, and to test whether processes are consistent or, alternatively, to detect the relevant scales of variation. Repeated experimentation is a necessary component of research to reduce uncertainty around predictions of succession or colonization at any particular scale.

Improving predictability requires a mechanistic theory that incorporates the relevant processes influencing colonization and some understanding of what makes these processes variable in space and time. Given these requirements, it is unlikely that a single theoretical framework will emerge that incorporates all the models necessary to explain and predict recovery in different systems and assemblages. Although some models may generate successful predictions in a given habitat (e.g., Tilman, 1985, 1990), they may not be readily applicable to other systems (e.g., Underwood & Anderson, 1994; Benedetti-Cecchi, 2000a). Considerations based on the same basic principles may, however, assist in developing models for distinct assemblages. Elsewhere, I have proposed a set of qualitative models applicable to different marine benthic habitats by considering how species' life-history traits affect the mode of exploitation of resources and the mechanism of interaction among species during recovery (Benedetti-Cecchi, 2000a). These propositions were derived from studies that emphasized the importance of life-history traits in colonization and as determinants of interspecific interactions (Drury & Nisbet, 1973; Connell, 1987; Huston & Smith, 1987; Tilman, 1990; McCook & Chapman, 1991; Wootton, 1993; Werner, 1994). These models generate testable predictions and can be improved by understanding the relevant scales at which different processes operate, and by testing quantitative alternatives to null hypotheses.

An example of this procedure is provided in a study of algal recovery in littoral rock pools in the northwest Mediterranean (Benedetti-Cecchi, 2000b). The main objective was to explain variable patterns of colonization and dominance at maturity of turf-forming and canopy algae documented in previous studies (e.g., Benedetti-Cecchi & Cinelli, 1996). Knowledge of the timing of reproduction and recruitment of the two groups of algae and their patterns of interaction provided the basis for a model of colonization in this assemblage. While turf-forming algae reproduced and recruited throughout the year, canopy-forming species reproduced mainly between April and July. It was also known that the interactions between the two groups of algae were nonhierarchical, depending upon the relative density and life stage of the interacting organisms rather than on their taxonomic identity, so that priority effects (*sensu* Paine, 1977) could be expected in this assemblage. A model based on the timing of disturbance with respect to the main period of recruitment of canopy algae was therefore proposed to explain variable colonization. From this, it was predicted that turf-forming species would colonize and dominate at different stages of recovery, if patches of substratum were cleared before or after the main period of recruitment of canopy algae. In contrast, if patches were cleared during the recruitment of canopy algae the following was expected: (1) both canopy and turf-forming species would appear early in recovery; and (2) canopy species would attain dominance, replacing turf-forming algae with time. These hypotheses were tested experimentally by clearing patches of substratum before, during, and after the main period of recruitment of canopy algae, at different times within periods to unconfound the effect of period on small-scale temporal fluctuations and, finally, by using replicate pools at each date to separate spatial from temporal effects. The results of this experiment supported the predictions of the model, so that both the trajectory of recovery and patterns of dominance at maturity could be predicted on the basis of the time when disturbance initiated colonization. This model, however, allowed for predictions only at a gross level of taxonomic resolution (turf-forming vs. canopy algae), whereas patterns of colonization of individual species within each category were unpredictable due to large variation in abundance at small spatial and temporal scales. Refinement of the model requires a better understanding of the ecology of single populations for this assemblage.

Although succession implies that accurate and precise predictions are possible about which species dominate at any given time and the structure of the assemblage at maturity, there are clear limits to the detail of such predictions in the case of variable colonization. From the large amount of experimental research performed in different habitats, however, it seems that ecologists have identified some of the relevant processes that affect recovery at any single time and place (e.g., the attributes of the disturbed patch, variable recruitment, and interactions among species). There is a prospect for increased capacity to predict variable patterns of colonization if theoretical and empirical studies will address the issue of what makes these processes variable in any given habitat. Although the fate of a single patch will probably remain unpredictable in variable assemblages, it could still be possible to anticipate the average values and variances that describe the frequency distributions of possible outcomes. This requires profuse experimentation to capture the spatial and temporal vagaries of the environment. The task is challenging, but this research is urgently needed given the implication that the capability of predicting recovery has in practical applications, such as the active restoration of degraded habitats.

Practical Implications of Succession and Colonization

There is increasing pressure on ecology to provide guidance for management of environmental problems. Anthropogenic disturbances cause the degradation and loss of many natural environments throughout the world. Restoration of altered environments and creation of new habitat are increasingly important managerial tools to mitigate human impacts (Zedler, 1996a, 1996b; Carr & Hixon, 1997). The primary goals of these interventions are to restore the biotic structure of natural assemblages as well as their functioning in terms of ecological processes and services. Understanding the mechanisms of succession and colonization of natural and managed habitats is of paramount importance to achieve these goals. Despite logical and methodological flaws in evaluating success of restoration efforts (Chapman, 1999), it is apparent that gaps in basic ecological understanding are the main problem hindering progress in this area of ecology. In fact, several experimental studies in managed littoral habitats have indicated that the structure and function of natural assemblages are difficult to restore. Most of these studies use basic ecological concepts of succession and variable colonization to explain the effects or, more often, the lack of intended effects of management and restoration.

Rehabilitation of salt marsh and sand dune habitats usually involves the reintroduction of dominant plant species and the management of physical attributes of the habitat (Race & Christie, 1982; Broome et al., 1988). Recovery of species, cycling of nutrients, and decomposition are evaluated to assess whether these interventions successfully restore the functioning of the managed environment. Although some studies reported successful recovery of some taxa compared to reference sites (e.g., van Aarde, 1996; Craft et al.,

1999), several investigations documented only partial recovery of plants and animals (e.g., Zedler & Langis, 1991; Levin et al., 1996).

Similar results are described for artificial structures in marine coastal habitats. Artificial reefs are constructed to mitigate for loss of hard substrata due to urban and industrial development (Carter et al., 1985), or to increase production of harvested species (Carr & Hixon, 1997). The extent to which these and related structures (e.g., breakwaters and sea walls) can be considered as surrogates of natural rocky shores is unclear. Comparative studies indicated that the assemblages developing on these structures differed to some extent from those found in the natural habitat (Connell & Glasby, 1999; Glasby, 1999). It is also unclear whether the artificial environment can restore the same ecological processes that operate in natural habitats (Bulleri et al., 2000). For example, Carter et al. (1985) reported that the giant kelp *Macrocystis pyrifera* failed to establish on a modular reef designed to reproduce kelp-dominated assemblages common of nearby rocky shores in southern California. Filamentous and turf-forming algae still dominated the reef after 3 years since deployment, suggesting that the assemblage was arrested at an early stage of colonization. Further studies in this habitat indicated that it may take 10–15 years for turf assemblages on artificial reefs to converge to those developing on natural reefs in the absence of canopy algae (Aseltine-Neilson et al., 1999).

Failure to reproduce the structure of natural assemblages and the underlying processes is usually explained with the timing of creation and size of the artificial habitat, continuity with natural environments, dispersal of organisms, and interaction among species; that is, the same processes influencing succession and colonization in natural habitats provide a mechanistic basis for an interpretation of experiments on restoration. Although the potential value of these experiments as tests of basic ecological theory has been emphasized (Bradshaw, 1987; Jordan et al., 1987), it seems that a better understanding of basic processes is urgently needed for restoration to become a useful procedure in environmental management. This requires more attention to the logical and methodological needs of the procedures used to assess restoration (Michener, 1997; Chapman, 1999), thorough experimental tests of hypotheses on processes influencing succession or colonization in natural and artificial habitats, and, ultimately, a better design of the restored environment to enhance recovery and to reproduce natural patterns of variation and diversity in assemblages (White & Walker, 1997).

Conclusions

The study of succession and colonization is an active area of research in ecology. Patterns of disturbance and recovery have been investigated in different habitats, under a wide range of environmental conditions, species' life histories and trophic structure of assemblages. Several theories have been proposed to explain succession, and a massive experimental evidence has accumulated on the importance of different physical and biological processes. There are, however, large gaps in our understanding of variable patterns of recovery that preclude accurate predictions of succession and colonization, limiting the applicability of basic ecological theory to management of environmental problems. The main issue here is to investigate whether these limitations reflect problems with the logical structure of research in terms of linkages among observations, explanations, hypothesis testing, and conclusions.

Despite the long history of the concept of succession, there are few quantitative descriptions of patterns of recovery following natural disturbance in littoral habitats. Most of the evidence is based on generalizations from the old literature, in which the spatial and temporal detail of replication provided by any single study was limited. The emphasis was on regular patterns that could be easily interpreted, whereas variability was considered noisy and of little ecological relevance (see also Chesson, 1986). There is a need for more structured research explicitly embracing a hierarchy of spatial and temporal scales of observations to stimulate better models and experiments on recovery. The observational basis underpinning experimental tests of theories of succession and colonization is not always clear. Many processes have become paradigmatic as explanations of recovery and are allegedly invoked even when there is no observational evidence for their support. Although experiments reveal strong effects of consumers, life history, dispersal and recruitment of colonists, interspecific interactions, size and time of formation of patches, and other physical attributes of the habitat, the relative importance of these processes as explanations of recovery in any specific context remains unclear. Progress in this direction requires more attention to the logical relationships between observation and theory (Underwood et al., 2000).

Conceptual and mechanistic models of succession play an important role in reducing uncertainty around explanations and predictions of recovery by stimulating experimental tests of causal processes (Tilman, 1985, 1990; Connell et al., 1987; McCook, 1994). Classical theories are, however, best suited to explain general patterns of replacement of species, but they cannot predict when and where specific outcomes occur. Simulation models can provide a mechanistic basis to explain variation, but there has been little experimental test of their predictions in littoral habitats. There is an urgent need to identify relevant criteria through which better theories for explaining and predicting variable colonization can be developed (Benedetti-Cecchi, 2000a, 2000c). This must be paralleled by profuse experimentation aimed at detecting the relevant scales of variation in ecological processes that affect recovery (Levin, 1992; Schneider, 1994; Underwood & Chapman, 1996). Repeated experimentation in space and time is necessary to understand the processes that more likely influence recovery at any particular scale and what attributes of assemblages can be consistently predicted under variable environmental conditions. Having identified the important mechanisms of succession and colonization, ecologists are now confronted with the major challenge of understanding the causes of variable colonization. Reductionism in hypothesis testing (Quinn & Dunham, 1983; Schoener, 1986; Inchausti, 1994) seems a more profitable approach than the unrealistic alternative of searching for grand, unifying theories of succession.

Acknowledgments

I wish to thank F. Micheli and I. Bertocci for comments and criticism on the manuscript. I. Bertocci, T. Fosella, and E. Maggi assisted with the editing of the manuscript. This review was conceived during a sabbatical at the Centre for Research on Ecological Impacts of Coastal Cities, at the University of Sydney. The Library of the University provided access to both the older and some of the recent literature. I wish to thank A. J. Underwood and M. G. Chapman for their kind hospitality during my staying in Sydney. This research was supported by a grant from the University of Pisa.

References

Airoldi, A. (1998). Roles of disturbance, sediment stress, and substratum retention on spatial dominance in algal turfs. *Ecology, 79*, 2759–2770.
Aleem, A. A. (1957). Succession of marine fouling organisms on test panels in deep-water at La Jolla, California. *Hydrobiologia, 11*, 40–58.
Anderson, D. J. (1986). Ecological succession. In: J. Kikkawa & D. J. Anderson (Eds.), *Community ecology: Patterns and processes* (pp. 269–285). Melbourne: Blackwell Scientific Publications.
Aseltine-Neilson, D. A., Bernstein, B. B., Palmer-Zwahlen, M. L., Riege, L. E., & Smith, R. W. (1999). Comparisons of turf communities from Pendleton Artificial Reef, Torrey Pines Artificial Reef, and a natural reef using multivariate techniques. *Bulletin of Marine Science, 65*, 37–57.
Ayling, A. M. (1981). The role of biological disturbance in temperate subtidal encrusting communities. *Ecology, 62*, 830–847.
Benedetti-Cecchi, L. (2000a). Predicting direct and indirect effects during succession in a midlittoral rocky shore assemblage. *Ecological Monographs, 70*, 45–72.
Benedetti-Cecchi, L. (2000b). Priority effects, taxonomic resolution, and the prediction of variable patterns of colonization of algae in littoral rock pools. *Oecologia, 123*, 265–274.
Benedetti-Cecchi, L. (2000c). Variance in ecological consumer-resource interactions. *Nature, 407*, 370–374.
Benedetti-Cecchi, L., Bulleri, F., & Cinelli, F. (2000). The interplay of physical and biological factors in maintaining mid-shore and low-shore assemblages on rocky coasts in the north-west Mediterranean. *Oecologia, 123*, 406–417.
Benedetti-Cecchi, L., & Cinelli, F. (1993). Early patterns of algal succession in a midlittoral community of the Mediterranean Sea: A multifactorial experiment. *Journal of Experimental Marine Biology and Ecology, 169*, 15–31.
Benedetti-Cecchi, L., & Cinelli, F. (1996). Patterns of disturbance and recovery in littoral rock pools: Nonhierarchical competition and spatial variability in secondary succession. *Marine Ecology Progress Series, 135*, 145–161.
Berlow, E. L. (1997). From canalization to contingency: Historical effects in a successional rocky intertidal community. *Ecological Monographs, 67*, 435–460.
Bertness, M. D. (1991). Interspecific interactions among high marsh perennials in a New England salt marsh. *Ecology, 72*, 125–137.
Bertness, M. D., & Ellison, A. M. (1987). Determinants of pattern in a New England salt marsh plant community. *Ecological Monographs, 57*, 129–147.
Bertness, M. D., & Shumway, S. W. (1993). Competition and facilitation in marsh plants. *American Naturalist, 142*, 718–724.
Bradshaw, A. D. (1987). Restoration: An acid test for ecology. In: W. J. Jordon, M. E. Gilpin, & J. D. Aber (Eds.), *Restoration ecology: A synthetic approach to ecological research* (pp. 23–29). Cambridge, UK: Cambridge University Press.
Bray, J. R., & Curtis, J. T. (1957). An ordination of the upland forest communities of Southern Wisconsin. *Ecological Monographs, 27*, 325–349.
Breitburg, D. L. (1985). Development of a subtidal epibenthic community: Factors affecting species composition and the mechanisms of succession. *Oecologia, 65*, 173–184.
Broome, S. W., Seneca, E. D., & Woodhouse, W. W., Jr. (1988). Tidal salt marsh restoration. *Aquatic Botany, 32*, 1–22.
Brown, A. C., & McLachlan, A. (1990). *Ecology of sandy shores*. Amsterdam: Elsevier Science.
Bulleri, F., Menconi, M., Cinelli, F., & Benedetti-Cecchi, L. (2000). Grazing by two species of limpets on artificial reefs in the northwest Mediterranean. *Journal of Experimental Marine Biology and Ecology, 255*, 1–19.
Buschmann, A. H. (1990). The role of herbivory and desiccation on early successional patterns of intertidal macroalgae in southern Chile. *Journal of Experimental Marine Biology and Ecology, 139*, 221–230.
Carr, M. H., & Hixon, M. A. (1997). Artificial reefs: the importance of comparisons with natural reefs. *Fisheries, 22*, 28–33.
Carter, J. W., Carpenter, A. L., Foster, M. S., & Jessee, W. N. (1985). Benthic succession on an artificial reef designed to support a kelp-reef community. *Bulletin of Marine Science, 37*, 86–113.
Carter, R. W. G. (1988). *Coastal environments: An introduction to the physical, ecological and cultural systems of coastlines*. London: Academic Press.
Castilla, J. C. (1988). Earthquake-caused coastal uplift and its effects on rocky intertidal kelp communities. *Science, 242*, 440–443.

Cattelino, P. J., Noble, I. R., Slatyer, R. O., & Kessell, S. R. (1979). Predicting multiple pathways of plant succession. *Environmental Management, 3*, 41–50.

Chapman, M. G. (1999). Improving sampling designs for measuring restoration in aquatic habitats. *Journal of Aquatic Ecosystem Stress and Recovery, 6*, 235–251.

Chapman, M. G., & Underwood, A. J. (1998). Inconsistency and variation in the development of rocky intertidal algal assemblages. *Journal of Experimental Marine Biology and Ecology, 224*, 265–289.

Chapman, V. J. (1959). Studies in salt marsh ecology: IX. Changes in salt marsh vegetation at Scolt Head Island, Norfolk. *Journal of Ecology, 47*, 619–639.

Chapman, V. J. (1974). *Salt marshes and salt deserts of the world*. New York: Wiley Interscience.

Chesson, P. L. (1986). Environmental variation and the coexistence of species. In: J. Diamond & T. J. Case (Eds.), *Community ecology* (pp. 240–256). New York: Harper & Row.

Clements, F. E. (1916). *Plant succession: An analysis of the development of vegetation*. Publication No. 242. Washington, DC: Carnegie Institute.

Clements, F. E. (1928). *Plant succession and indicators: A definitive edition of plant succession and plant indicators*. New York: Wilson.

Connell, J. H. (1987). Change and persistence in some marine communities. In: A. J. Grey, M. J. Crawley, & P. J. Edwards (Eds.), *Colonization, succession and stability* (pp. 339–352). Oxford, UK: Blackwell Science.

Connell, J. H., Noble, I. R., & Slatyer, R. O. (1987). On the mechanisms producing successional change. *Oikos, 50*, 136–137.

Connell, J. H., & Slayter, R. O. (1977). Mechanisms of succession in natural communities and their role in community stability and organization. *American Naturalist, 111*, 1119–1144.

Connell, J. H., & Sousa, W. P. (1983). On the evidence needed to judge ecological stability or persistence. *American Naturalist, 121*, 789–824.

Connell, S. D., & Glasby, T. M. (1999). Do urban structures influence local abundance and diversity of subtidal epibiota? A case study from Sydney Harbour, Australia. *Marine Environmental Research, 47*, 373–387.

Cooper, W. S. (1913). The climax forest of Isle Royale, Lake Superior, and its development. *Botanical Gazetin, 55*, 1–44.

Cowles, H. C. (1899). The ecological relations of the vegetation on the sand dunes of Lake Michigan. *Botanical Gazetin, 27*, 95–391.

Craft, C., Reader, J., Sacco, J. N., & Broome, S. W. (1999). Twenty-five years of ecosystem development of constructed *Spartina alterniflora* (Loisel) marshes. *Ecological Applications, 9*, 1405–1419.

Dayton, P. K. (1971). Competition, disturbance, and community organization: The provision and subsequent utilization of space in a rocky intertidal community. *Ecological Monographs, 41*, 351–389.

Dayton, P. K. (1975). Experimental evaluation of ecological dominance in a rocky intertidal algal community. *Ecological Monographs, 45*, 137–159.

Dayton, P. K., Currie, V., Gerrodette, T., Keller, B. D., Rosenthal, R., & Ven Tresca, D. (1984). Patch dynamics and stability of some California kelp communities. *Ecological Monographs, 54*, 253–289.

Dayton, P. K., Tegner, M. J., Parnell, P. E., & Edwards, P. B. (1992). Temporal and spatial patterns of disturbance and recovery in a kelp forest community. *Ecological Monographs, 62*, 421–445.

DeAngelis, D. L., & Waterhouse, J. C. (1987). Equilibrium and nonequilibrium concepts in ecological models. *Ecological Monographs, 57*, 1–21.

De Leeuw, J., De Munck, W., Olff, H., & Bakker, J. P. (1993). Does zonation reflect the succession of salt-marsh vegetation? A comparison of an estuarine and a coastal bar island marsh in The Netherlands. *Acta Botanica Neerlandica, 42*, 435–445.

Dethier, M. N. (1984). Disturbance and recovery in intertidal pools: Maintenance of mosaic patterns. *Ecological Monographs, 54*, 99–118.

Dormann, C. F., & Bakker, J. P. (2000). The impact of herbivory and competition on flowering and survival during saltmarsh succession. *Plant Biology, 2*, 68–76.

Drury, W. H., & Nisbet, I. C. T. (1973). Succession. *Journal of the Arnold Arboretum, 54*, 331–368.

Dye, A. H. (1993). Recolonization of intertidal macroalgae in relation to gap size and molluscan herbivory on a rocky shore on the east coast of southern Africa. *Marine Ecology Progress Series, 95*, 263–271.

Egler, F. E. (1954). Vegetation science concepts: I. Initial floristic composition—a factor in old-field vegetation development. *Vegetatio, 4*, 412–417.

Ellison, A. M. (1987). Effects of competition, disturbance, and herbivory on *Salicornia europaea*. *Ecology, 68*, 576–586.

Elton, C. S. (1958). *The ecology of invasions by animals and plants*. London: Methuen.

Emerson, S. E., & Zedler, J. B. (1978). Recolonization of intertidal algae: An experimental study. *Marine Biology, 44*, 315–324.

Fager, E. W. (1971). Pattern in the development of a marine community. *Limnology and Oceanography, 16*, 241–253.

Farrell, T. M. (1989). Succession in a rocky intertidal community: The importance of disturbance size and position within a disturbed patch. *Journal of Experimental Marine Biology and Ecology, 128*, 57–73.

Farrell, T. M. (1991). Models and mechanisms of succession: An example from a rocky intertidal community. *Ecological Monographs, 61*, 95–113.

Foster, M. S. (1975). Regulation of algal community development in a *Macrocystis pyrifera* forest. *Marine Biology, 32*, 331–342.

Foster, M. S. (1992). How important is grazing to seaweed evolution and assemblage structure in the north-east Pacific? In: D. M. John, S. J. Hawkins, & J. H. Price (Eds.), *Plant–animal Interactions in the marine benthos* (pp. 61–85). Oxford, UK: The Systematics Association, Clarendon Press.

Glasby, T. M. (1999). Interactive effects of shading and proximity to the seafloor on the development of subtidal epibiotic assemblages. *Marine Ecology Progress Series, 190*, 113–124.

Gleason, H. A. (1926). The individualistic concept of the plant association. *American Midland Naturalist, 21*, 92–110.

Gleason, H. A. (1927). Further views of the succession concept. *Ecology, 8*, 299–326.

Hatton, H. (1932). Quelques observations sur le repeuplement en *Fucus vesiculosus* des surfaces rocheuses denudes. *Bulletin du Laboratoire Maritime du Museum d'Histoire Naturelle, 9*, 1–6.

Hawkins, S. J. (1981). The influence of season and barnacles on the algal colonization of *Patella vulgata* exclusion areas. *Journal of the Marine Biological Association of the United Kingdom, 61*, 1–15.

Hawkins, S. J., & Hartnoll, R. G. (1983). Grazing of intertidal algae by marine invertebrates. *Oceanography and Marine Biology: An Annual Review, 21*, 195–282.

Hubbard, J. C. E., & Stebbings, R. E. (1968). *Spartina* marshes in Southern England: VII. Stratigraphy of the Keysworth marsh, Poole Harbour. *Journal of the British Grassland Society, 21*, 214–217.

Huisman, J., Olff, H., & Fresco, L. F. M. (1993). A hierarchical set of models for species response analysis. *Journal of Vegetation Science, 4*, 37–46.

Huston, M. A. (1994). *Biological diversity: The coexistence of species on changing landscapes*. Cambridge, UK: Cambridge University Press.

Huston, M. A., & Smith, T. (1987). Plant succession: life history and competition. *American Naturalist, 130*, 168–198.

Inchausti, P. (1994). Reductionist approaches in community ecology. *American Naturalist, 143*, 201–221.

Johansen, H. W., & Austin, L. F. (1970). Growth rates in the articulate coralline *Calliarthron* (Rhodophyta). *Canadian Journal of Botany, 48*, 125–132.

Johnson, C. R., & Mann, K. H. (1988). Diversity, patterns of adaptation, and stability of Nova Scotian kelp beds. *Ecological Monographs, 58*, 129–154.

Johnson, C. R., & Mann, K. H. (1993). Rapid succession in subtidal understorey seaweeds during recovery from overgrazing by sea urchins in eastern Canada. *Botanica Marina, 36*, 63–77.

Jordan, W. R., III, Gilpin, M. E., & Aber, J. D. (1987). Restoration ecology: Ecological restoration as a technique for basic research. In: W. R. Jordan, III, M. E. Gilpin, & J. D. Aber (Eds.), *Restoration ecology: A synthetic approach to ecological research* (pp. 3–21). Cambridge, UK: Cambridge University Press.

Kaehler, S., & Williams, G. A. (1998). Early development of algal assemblages under different regimes of physical and biotic factors on a seasonal tropical rocky shore. *Marine Ecology Progress Series, 172*, 61–71.

Kay, A. M., & Keough, M. J. (1981). Occupation of patches in the epifaunal communities on pier pilings and the bivalve *Pinna bicolor* at Edithburg, South Australia. *Oecologia, 48*, 123–130.

Kendrick, G. A., & Walker, D. I. (1991). Dispersal distances for propagules of *Sargassum spinuligerum* (Sargassaceae, Phaeophyta) measured directly by vital staining and venturi suction sampling. *Marine Ecology Progress Series, 79*, 133–138.

Kim, J. H., & DeWreede, R. E. (1996). Effects of size and season of disturbance on algal patch recovery in a rocky intertidal community. *Marine Ecology Progress Series, 133*, 217–228.

Kotliar, N. B., & Wiens, J. A. (1990). Multiple scales of patchiness and patch structure: A hierarchical framework for the study of heterogeneity. *Oikos, 59*, 253–260.

Landres, P. B., Morgan, P., & Swanson, F. J. (1999). Overview of the use of natural variability concepts in managing ecological systems. *Ecological Applications, 9*, 1179–1188.

Lawrence, J. M. (1975). On the relationships between marine plants and sea urchins. *Oceanography and Marine Biology: An Annual Review, 13*, 213–286.

Levin, L. A., Talley, D., & Thayer, G. (1996). Succession of macrobenthos in a created salt marsh. *Marine Ecology Progress Series, 141*, 67–82.

Levin, S. A. (1992). The problem of pattern and scale in ecology. *Ecology, 73*, 1943–1967.
Levin, S. A., & Paine, R. T. (1974). Disturbance, patch formation and community structure. *Proceedings of the National Academy of Sciences (USA), 71*, 2744–2747.
Levin, S. A., & Paine, R. T. (1975). The role of disturbance in models of community structure. In: S. A. Levin (Ed.), *Ecosystem analysis and prediction* (pp. 56–67). Philadelphia: Society for Industrial and Applied Mathematics.
Littler, M. M., & Littler, D. S. (1980). The evolution of thallus form and survival strategies in benthic marine macroalgae field and laboratory tests of a functional form model. *American Naturalist, 116*, 25–44.
Lohele, C. (1988). Philosophical tools: Potential contributions to ecology. *Oikos, 51*, 97–104.
Lubchenco, J. (1983). *Littorina* and *Fucus*: Effects of herbivores, substratum heterogeneity, and plant escapes during succession. *Ecology, 64*, 1116–1123.
Lubchenco, J., & Menge, B. A. (1978). Community development and persistence in a low rocky intertidal zone. *Ecological Monographs, 48*, 67–94.
Margalef, R. (1963). On certain unifying principles in ecology. *American Naturalist, 97*, 357–374.
McCook, L. J. (1994). Understanding ecological community succession: Causal models and theories, a review. *Vegetatio, 110*, 115–147.
McCook, L. J., & Chapman, A. R. O. (1991). Community succession following massive ice-scour on an exposed rocky shore: effects of *Fucus* canopy algae and of mussels during late succession. *Journal of Experimental Marine Biology and Ecology, 154*, 137–169.
McCook, L. J., & Chapman, A. R. O. (1997). Patterns and variations in natural succession following massive ice-scour of a rocky intertidal seashore. *Journal of Experimental Marine Biology and Ecology, 214*, 121–147.
McGuinness, K. A. (1987). Disturbance and organisms on boulders. *Oecologia, 71*, 409–419.
McIntosh, R. P. (1980). The relationship between succession and the recovery process in ecosystems. In: J. Cairns, Jr. (Ed.), *The recovery process in damaged ecosystems* (pp. 11–62). MI: Annual Arbor Science Publishers.
McIntosh, R. P. (1985). *The background of ecology: Concept and theory*. Cambridge, UK: Cambridge University Press.
Menge, B. A., Farrell, T. M., Olson, A. M., Van Tamelen, P. G., & Turner, T. (1993). Algal recruitment and the maintenance of a plant mosaic in the low intertidal region on the Oregon coast. *Journal of Experimental Marine Biology and Ecology, 170*, 91–116.
Michener, W. K. (1997). Quantitatively evaluating restoration experiments: Research design, statistical analysis, and data management considerations. *Restoration Ecology, 5*, 324–337.
Noble, I. R., & Slatyer, R. O. (1980). The use of vital attributes to predict successional changes in plant communities subject to recurrent disturbances. *Vegetatio, 43*, 5–21.
Northcraft, R. D. (1948). Marine algal colonization on the Monterey Peninsula, California. *American Journal of Botany, 35*, 396–404.
Odum, E. P. (1969). The strategy of ecosystem development. *Science, 164*, 262–270.
Olff, H., De Leeuw, J., Bakker, J. P., Platerink, R. J., Van Wijnen, H. J., & De Munck, W. (1997). Vegetation succession and herbivory in a salt marsh: Changes induced by sea level rise and silt deposition along an elevational gradient. *Journal of Ecology, 85*, 799–814.
Olff, H., Huisman, J., & Van Tooren, B. F. (1993). Species dynamics and nutrient accumulation during early primary succession in coastal sand dunes. *Journal of Ecology, 81*, 693–706.
Olson, J. S. (1958). Rates of succession and soil changes on Southern Lake Michigan sand dunes. *Botanical Gazetin, 119*, 125–170.
Osman, R. W. (1977). The establishment and development of a marine epifaunal community. *Ecological Monographs, 47*, 37–63.
Paine, R. T. (1966). Food web complexity and species diversity. *American Naturalist, 100*, 65–75.
Paine, R. T. (1977). Controlled manipulations in the intertidal zone, and their contribution to ecological theory. *Special Publications of the Academy of Natural Sciences, Philadelphia, 12*, 245–270.
Paine, R. T. (1979). Disaster, catastrophe, and local persistence of the sea palm *Postelsia palmaeformis*. *Science, 205*, 685–687.
Paine, R. T. (1984). Ecological determinism in the competition for space. *Ecology, 65*, 1339–1348.
Paine, R. T. (2000). Phycology for the mammalogist: Marine rocky shores and mammal-dominated communities—How different are the structuring processes? *Journal of Mammalogy, 81*, 637–648.
Paine, R. T., & Levin, S. A. (1981). Intertidal landscapes: Disturbance and the dynamics of pattern. *Ecological Monographs, 51*, 145–178.
Peters, R. H. (1991). *A critique for ecology*. Cambridge, UK: Cambridge University Press.
Pickett, S. T. A. (1989). Space-for-time substitution an alternative to long-term studies. In: G. E. Likens (Ed.), *Long-term studies in ecology: Approaches and alternatives* (pp. 110–135). New York: Springer-Verlag.

Pickett, S. T. A., Collins, S. L., & Armesto, J. J. (1987a). Models, mechanisms and pathways of succession. *Botanical Review*, 335–371.
Pickett, S. T. A., Collins, S. L., & Armesto, J. J. (1987b). A hierarchical consideration of succession. *Vegetatio*, *69*, 109–114,
Pickett, S. T. A., & White, P. S. (1985). Patch dynamics: a synthesis. In: S. T. A. Pickett & P. S. White (Ed.), *The ecology of natural disturbance and patch dynamics* (pp. 371–384). London: Academic Press.
Prach, K., Pyšek, P., & Šmilauer, P. (1993). On the rate of succession. *Oikos*, *66*, 343–346.
Quinn, J. F., & Dunham, A. E. (1983). On hypothesis testing in ecology and evolution. *American Naturalist*, *122*, 602–617.
Race, M. S., & Christie, D. R. (1982). Coastal zone development: Mitigation, marsh creation and decision making. *Environmental Management*, *6*, 317–328.
Ranwell, D. S. (1961). *Spartina* salt marshes in Southern England: I. The effects of sheep grazing at the upper limits of *Spartina* marsh in Bridgwater Bay. *Journal of Ecology*, *49*, 325–340.
Ranwell, D. S. (1968). Coastal marshes in perspective. *Regional Studies Group Bulletin Strathclyde*, *9*: 1–26.
Ranwell, D. S. (1972). *Ecology of salt marshes and sand dunes*. London: Chapman & Hall.
Ranwell, D. S., & Downing, B. M. (1959). Brent goose (*Branta bernicla* L.) winter feeding pattern and *Zostera* resources at Scolt Head Island, Norfolk. *Animal Behaviour*, *7*, 42–56.
Reed, D. C., Laur, D. R., & Ebeling, A. W. (1988). Variation in algal dispersal and recruitment: The importance of episodic events. *Ecological Monographs*, *58*, 321–335.
Rees, T. K. (1940). Algal colonization at Mumbles head. *Journal of Ecology*, *28*, 403–437.
Robles, C. D., & Cubit, J. (1981). Influence of biotic factors in an upper intertidal community: Dipteran larvae grazing on algae. *Ecology*, *62*, 1536–1547.
Roozen, A. J. M., & Westhoff, V. (1985). A study on long-term salt-marsh succession using permanent plots. *Vegetatio*, *61*, 23–32.
Santelices, B. (1990). Patterns of reproduction, dispersal and recruitment in seaweeds. *Oceanography and Marine Biology: An Annual Review*, *28*, 177–276.
Schneider, D. C. (1994). *Quantitative ecology: Spatial and temporal scaling*. San Diego: Academic Press.
Schoener, T. W. (1986). Mechanistic approaches to community ecology: A new reductionism? *American Zoologist*, *26*, 81–106.
Shumway, S. W., & Bertness, M. D. (1992). Salt stress limitation of seedling recruitment in a salt marsh plant community. *Oecologia*, *92*, 490–497.
Shumway, S. W., & Bertness, M. D. (1994). Patch size effects on marsh plant secondary succession mechanisms. *Ecology*, *75*, 564–568.
Sousa, W. P. (1979a). Disturbance in marine intertidal boulder fields: The nonequilibrium maintenance of species diversity. *Ecology*, *60*, 1225–1239.
Sousa, W. P. (1979b). Experimental investigations of disturbance and ecological succession in a rocky intertidal algal community. *Ecological Monographs*, *49*, 227–254.
Sousa, W. P. (1980). The responses of a community to disturbance: The importance of successional age and species' life histories. *Oecologia*, *45*, 72–81.
Sousa, W. P. (1984a). The role of disturbance in natural communities. *Annual Reviews of Ecological Systems*, *15*, 353–391.
Sousa, W. P. (1984b). Intertidal mosaics: Patch size, propagule availability, and spatially variable patterns of succession. *Ecology*, *65*, 1918–1935.
Sousa, W. P. (1985). Disturbance and patch dynamics on rocky intertidal shores. In: S. T. A. Pickett & P. S. White (Eds.), *The ecology of natural disturbance and patch dynamics* (pp. 101–124). London: Academic Press.
Sousa, W. P., & Connell, J. H. (1992). Grazing and succession in marine algae. In: D. M. John, S. J. Hawkins & J. H. Price (Eds.), *Plant–animal interactions in the marine benthos* (pp. 425–441). Oxford, UK: Clarendon.
Southward, A. J., & Southward, E. C. (1978). Recolonization of rocky shores in Cornwall after use of toxic dispersants to clean up the Torrey Canyon spill. *Journal of the Fisheries Research Board of Canada*, *35*, 682–706.
Suchanek, T. H. (1978). The ecology of *Mytilus edulis* L. in exposed rocky intertidal communities. *Journal of Experimental Marine Biology and Ecology*, *31*, 105–120.
Sutherland, J. P. (1974). Multiple stable points in natural communities. *American Naturalist*, *108*, 859–873.
Sutherland, J. P., & Karlson, R. H. (1977). Development and stability of the fouling community at Beaufort, North Carolina. *Ecological Monographs*, *47*, 425–446.
Tegner, M. J., & Dayton, P. K. (1987). El Niño effects on southern California kelp forest communities. *Advances in Ecological Research*, *17*, 243–279.

Thoreau, H. D. (1860). The succession of forest trees. Reprinted 1983 in *Excursions* (pp. 225–250). Boston: Houghton Mifflin.

Tilman, D. (1985). The resource-ratio hypothesis of plant succession. *American Naturalist, 125*, 827–852.

Tilman, D. (1988). *Plant strategies and the dynamics and structure of plant communities.* Princeton, NJ: Princeton University Press.

Tilman, D. (1990). Constraints and tradeoffs: Toward a predictive theory of competition and succession. *Oikos, 58*, 3–15.

Turner, T. (1983). Facilitation as a successional mechanism in a rocky intertidal community. *American Naturalist, 121*, 729–738.

Underwood, A. J. (1990). Experiments in ecology and management: Their logics, functions and interpretations. *Australian Journal of Ecology, 15*, 365–389.

Underwood, A. J. (1997). *Experiments in ecology: Their logical design and interpretation using analysis of variance.* Cambridge, UK: Cambridge University Press.

Underwood, A. J., & Anderson, M. J. (1994). Seasonal and temporal aspects of recruitment and succession in an intertidal estuarine fouling assemblage. *Journal of the Marine Biological Association of the United Kingdom, 74*, 563–584.

Underwood, A. J., & Chapman, M. G. (1996). Scales of spatial patterns of distribution of intertidal invertebrates. *Oecologia, 107*, 212–224.

Underwood, A. J., Chapman, M. G., & Connell, S. D. (2000). Observations in ecology: You can't make progress on processes without understanding the patterns. *Journal of Marine Biology and Ecology, 250*, 97–115.

Underwood, A. J., & Denley, E. J. (1984). Paradigms, explanations, and generalizations in models for the structure of intertidal communities on rocky shores. In: D. R. Strong, D. Simberloff, L. G. Abele, & A. B. Thistle (Eds.), *Ecological communities: Conceptual issues and the evidence* (pp. 151–180). Princeton, NJ: Princeton University Press.

Underwood, A. J., Denley, E. J., & Moran, M. J. (1983). Experimental analyses of the structure and dynamics of mid-shore rocky intertidal communities in New South Wales. *Oecologia, 56*, 202–219.

van Aarde, R. J., Ferreira, S. M., & Kritzinger, J. J. (1996). Successional changes in rehabilitating coastal dune communities in northern KwaZulu/Natal, South Africa. *Landscape and Urban Planning, 34*, 277–286.

Van Andel, J., Bakker, J. P., & Grootjans, A. P. (1993). Mechanisms of vegetation succession: A review of concepts and perspectives. *Acta Botanica Neerlandica, 42*, 413–433.

Van Tamelen, P. G. (1987). Early successional mechanisms in the rocky intertidal: The role of direct and indirect interactions. *Journal of Experimental Marine Biology and Ecology, 112*, 39–48.

van Wijnen, H. J., & Bakker, J. P. (1999). Nitrogen and phosphorous limitation in a coastal barrier salt marsh: The implications for vegetation succession. *Journal of Ecology, 87*, 265–272.

van Wijnen, H. J., van der Wal, R., & Bakker, J. P. (1999). Impact of herbivores on nitrogen mineralization rate: Consequences for salt-marsh succession. *Oecologia, 118*, 225–231.

Walker, R., & Chapin, F. S., III (1987). Interactions among processes controlling successional change. *Oikos, 50*, 131–135.

Webley, D. M., Eastwood, D. J., & Gimingham, C. H. (1952). Development of a soil microflora in relation to plant succession on sand dunes, including the "rizhosphere" flora associated with colonising species. *Journal of Ecology, 40*, 168–178.

Werner, E. E. (1994). Ontogenic scaling of competitive relations: Size-dependent effects and responses in two anuran larvae. *Ecology, 75*, 197–213.

White, P. S., & Walker, J. L. (1997). Approximating nature's variation: Selecting and using reference information in restoration ecology. *Restoration Ecology, 5*, 338–349.

Whittaker, R. H. (1975). *Communities and ecosystem.* New York: Macmillan.

Wiens, J. A., Addicott, J. F., Case, T. J., & Diamond, J. (1986). Overview: The importance of spatial and temporal scale in ecological investigations. In: T. J. Case (Ed.), *Community ecology* (pp. 145–153). New York: Harper & Row.

Williams, S. L. (1990). Experimental studies of caribbean seagrass bed development. *Ecological Monographs, 60*, 449–469.

Winer, B. J., Brown, D. R., & Michels, K. M. (1991). *Statistical procedures in experimental design.* New York: McGraw-Hill.

Wootton, J. T. (1993). Size-dependent competition: Effects on the dynamics vs. the end point of mussel bed succession. *Ecology, 74*, 195–206.

Wu, J., & Loucks, O. L. (1995). From balance of nature to hierarchical patch dynamics: A paradigm shift in ecology. *Quarterly Review of Biology, 70*, 439–466.

Zedler, J. B. (1996a). Ecological issues in wetland mitigation: An introduction to the forum. *Ecological Applications*, 6, 33–37.

Zedler, J. B. (1996b). Coastal mitigation in southern California: The need for a regional restoration strategy. *Ecological Applications*, 6, 84–93.

Zedler, J. B., & Langis, R. (1991). Comparison of constructed and natural salt marshes of San Diego Bay. *Restoration Management Notes*, 9, 21–25.

6

Phycoremediation

Algae as Tools for Remediation of Mine-Void Wetlands

Jacob John

Introduction

Wetlands have been declining in numbers and quality throughout the world in recent years. Freshwater bodies—lakes and rivers—have been experiencing symptoms of eutrophication, siltation, secondary salinization, and contamination by toxic chemicals and heavy metals as a result of deforestation, intense agriculture, mining, and industrial developments in the catchments. These problems are more intense in "dryland" countries such as Australia due to the declining sources of freshwater (Smith, 1998). To offset this imbalance, artificial wetlands are on the increase. Open cut mining for sands, clay, minerals, and coal results in voids or "pits" that intercept the watertable and inadvertently become wetlands, referred to as *mine-voids* or *pit-wetlands* (Parker & Robertson, 1999). These artificial wetlands are on the increase. In Australia, Canada, the United States, and other leading mining countries, the development of these mine pits into sustainable wetlands for conservation of biodiversity and aquaculture is a challenge for aquatic biologists in the 21st century, demanding innovative research approaches.

Limiting Factors

The mine voids are a potential refuge for endemic and migratory birds, especially in summer, when water is sparse. However, there are several factors limiting their development into functional wetlands, predominantly due to their "genesis." Acid drainage (AD) as a result of leaching from tailings and waste dumps is a long-lasting problem. Furthermore, if the aquifer is saline, these pits become "saltish" or "brackish." Acidity, salinity, low levels of nutrients—particularly phosphorus, and high concentrations of

---

Jacob John • Department of Environmental Biology, Curtin University, GPO Box U1987, Perth W.A., Australia.

ammonia, iron, magnesium, manganese, aluminium, zinc, copper, and other heavy metals are the major problems encountered in mine-void wetlands. Biologically, these problems are manifested through low primary productivity, poor development of food chains and lack of macrophytes, emergent vegetation and waterbirds, leading to an overall paucity of biodiversity.

Remediation

Enhancement of water quality and habitat are the two essential methods required for the remediation of pit-wetlands.

If acidity is the problem, as is often the case in coal mine-voids, lime (calcium carbonate) treatment to neutralize the acidic water is recommended. If elements such as Al, Fe, Mn, and Cu are in high concentrations, above safe levels for drinking by cattle and for aquatic biota, then these metals have to be removed or reduced. Habitat modification includes landscaping the pits by altering the morphometry, introducing islands and peninsulas, converting deep, steep shorelines into gentle slopes, and creating undulating margins. Vegetating the wetlands with appropriate native fringing and emergent plants does attract waterbirds. However, pivotal to the development of healthy, functional wetlands is the presence of macrophytes (submerged hydrophytes and macroalgae).

Attributes of ideal macrophytes for the creation of functional wetlands include their ability to provide habitats and food for macroinvertebrates, fish, and waterfowl (Søndergaard et al., 1997). Because the water level of the mine-voids is often controlled by the aquifer, they may dry up in severe summers. Under such circumstances, ideal macrophytes are those that are capable of adapting their life cycle to the water-level fluctuations.

Phytoremediation

The concept of using plants for remediation of the environment historically had its "genesis" in the discovery by Florentine botanist Andrea Cesalpino, who, in the 16th century, described *Alyssum bertolonii* as a hyperaccumulator of minerals from Tuscany, Italy (Brooks, 1998). The exact mechanism of hyperaccumulation is not known, and the definition of the term is based purely on the concentration of elements observed in the dry weight of plants in relation to that of the surroundings. The term *hyperaccumulator* was introduced to describe plants containing $> 1000\,\mu g/g$ (0.1%) minerals in their dry material (Brooks, 1998). The term *phytoremediation* refers to the remediation of the environment—mostly contaminated soil—by hyperaccumulator plants.

Phytoremediation plants accumulate heavy metals from contaminated soil and water. These plants can be grown and harvested economically to clean up the environment, leaving only residual levels of pollutants (Ensley, 2000; Glass, 2000). The closely related term *bioremediation* is used in a different sense and encompasses the concept of using microbial communities for the degradation of organic compounds (Ensley, 2000). *Phytomining* refers to the actual technology involving extraction of metals (phytoextraction) from hyperaccumulator plants (Blaylock & Huang, 2000). Many seaweeds qualify as hyperaccumulators of metals and other elements. For example, arsenic is accumulated by the brown seaweeds *Sargassum* and *Macrocystis* (Dunn, 1998). High levels of iodine have been found in the brown alga *Laminaria*. There is considerable potential for applying seaweed chemistry to the environmental monitoring of coastal areas (Dunn, 1998).

Phycoremediation

The term phycoremediation was introduced by John (2000a) to refer to the remediation carried out by algae, and the definition was extended beyond the process as a result of the high concentrations of minerals found in some algae. Phycoremediation was defined as the remediation of aquatic environments by the use of algae as tools, resulting in the enhancement of water quality, including reduction of heavy metals, leading to the sustainable development of aquatic systems. Although considerable research is required to establish the mechanism involved in hyperaccumulation of minerals by algae, they are potentially a cost-efficient and energy-friendly ("green") tool for remediation of mine-voids or degraded wetlands with similar problems.

My objective in this chapter is to give a short account of the use of green algae as tools in the removal of heavy metals from acid mine-voids, illustrated by two case histories: (1) a filamentous green alga—*Mougeotia* sp.—as part of an enhancement of coal mine-void water, and (2) charophytes, in improving the water quality, as well as enhancing the biodiversity of sand mine pit-wetlands.

Green Algae as Hyperaccumulators of Metals: A Case History

Most of the literature on metal accumulation by algae deals with seaweeds. Reports of freshwater filamentous green algae, cyanobacteria, benthic microbial communities, and "biofilms" acting as hyperaccumulators are relatively sparse (Tyrrell, 1995).

Collie, located 200 km southeast of Perth, the capital of Western Australia, has been a coal-mining town since the 1890s. More than 100 years of coal mining has left a legacy of several deep pits with stagnant acidic water (John, 1999a). These mine-voids have been rehabilitated into productive wetlands and aquaculture ponds since 1996 (John, 2000b). In some of the shallower mine-voids and on their shores, where acid drainage (AD) occurs, a green filamentous alga (*Mougeotia* sp.) with a "reddish rusty" color was observed. It formed an extensive "reddish" mat, floating on some of the more shallow acidic mine-voids. It was discovered that up to 25% of the dry weight of this alga contains iron. The green color of the alga could be seen only after repeated washings. After laboratory culture and field investigations, it was found that the green alga was capable of growing prolifically at a varying pH of 3–6, sequestering iron and aluminium (John, 2000a). A combination of oxidation (due to photosynthetic production of oxygen), adsorption, bacterial activity, and active uptake might be implicated in the mechanism of hyperaccumulation of these metals. Irrespective of the mechanism of hyperaccumulation, applying the definition of hyperaccumulator mentioned earlier in this chapter, this alga can be considered a "high accumulator." As a result of this discovery, the hyperaccumulator alga (Figure 6.1) was incorporated as part of a passive treatment system for the remediation of acidic coal mine-voids at Collie (Figure 6.2).

Water from a coal mine-void (pH = 4), with a capacity of 75,000 m^3, was pumped by photovoltaic power into a small pond (800 L capacity), in which the hyperaccumulator alga (*Mougeotia* sp.) was cultured. The effluent water from the algal pond was fed into an ALD (anoxic limestone drain) containing powdered limestone (96% calcium carbonate) before being fed into an organic pond and a wetland pond, finally returning the effluent water to the acidic coal mine-void. Three years of trials have shown that the algal pond was effective in reducing the concentrations of iron and aluminium by half. The algae accumulate iron precipitates around their thalii.

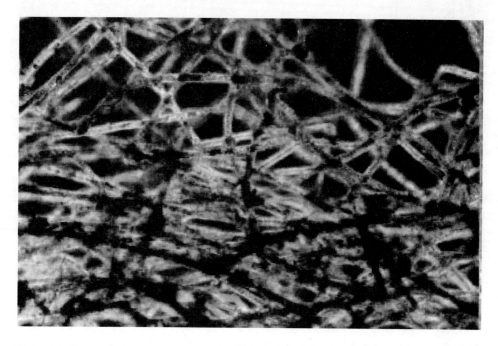

Figure 6.1. *Mougeotia* sp., a green alga that accumulates iron in shallow acidic mine-voids. (Photo by author.)

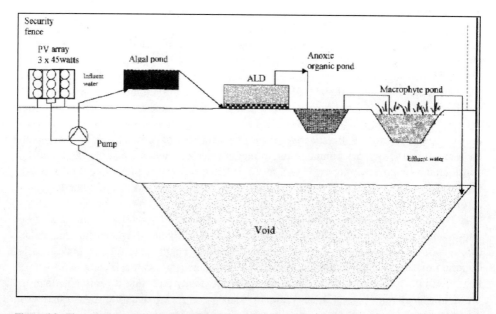

Figure 6.2. The passive treatment system driven by solar energy consisting of algal pond, anoxic limestone drain (ALD), organic (compost) pond, macrophyte pond.

Currently, a larger version of this treatment is being tested on a mine-void of 7 billion cubic meter capacity. An algal pond containing *Mougeotia* sp. is part of the treatment system. The goal of the project has been to provide water of good quality to an aquaculture pond, with the acidic mine-void as a source. The hyperaccumulator algae must be harvested periodically and disposed of to allow for a continuous, smooth flow of water. In this example of phycoremediation, the alga was used for the purpose of reducing Fe and Al—two elements whose concentrations of >1 mg/L are not recommended for a healthy wetland (Australian and New Zealand Environment and Conservation Council, 1992).

Charophytes as Tools for Phycoremediation: A Case History

Hutchinson (1975) was one of the first to report hyperaccumulation of metals by charophytes (stoneworts)—a group of algae that have a strong affinity to land plants. He reported that manganese was accumulated by charophytes. Additionally, some of the *Chara* species are able to sequester calcium carbonate, hence the common name "stoneworts."

Charophytes are submerged macrophytes providing ample shelter and food for macroinvertebrates, fish, and waterbirds; they are well-known pioneers in the colonization of artificial wetlands. In almost all newly established artificial wetlands in the Perth metropolitan area in Western Australia, within 2 to 3 years, charophytes began to appear (personal observations). However, as these lakes develop in the course of time, eutrophication sets in, leading to the elimination of charophytes and initiation of phytoplankton blooms. In The Netherlands, the elimination of species diversity of charophytes is directly linked to eutrophication (Simon, 1990). Lakes colonized by charophytes (specifically, *Nitella*) are often found to be crystal clear, with very little turbidity. However, there has been very little attempt to use them for remediation of degraded wetlands.

Discovery of *Nitella hyalina* as a Hyperaccumulator

Large meadows of the charophyte *Nitella hyalina* were observed in 1990 in a study of the ecology of Lake Leschenaultia in Western Australia. Lake Leschenaultia, a semiartificial lake constructed after damming a small stream in 1897, to provide water for railway locomotives, is located 40 km east of Perth. By 1900, increasing salinity made it unsuitable for this purpose. The lake has been subsequently developed for public recreation and is now part of a nature reserve. The lake, 40.5 ha in area, is less than 3 m deep for more than 50% of its area and has a maximum depth of 8 m. *N. hyalina* (DC) Ag. (Figure 6.3) covered almost 80% of the lake bed. Water clarity of the lake was excellent, with light penetration almost reaching the bottom for most of the year. Phytoplankton were sparse; the lake was circumneutral to slightly acidic and had low concentrations of nutrients in the standing water.

Presence of an ensheathment of mucilage is characteristic of the species *N. hyalina*. The mucilage envelope is easily stained violet metachromatically with Toluidine blue O (pH 1), due to its mucopolysaccharide structure (Kiernan, 1981) (Figure 6.4). P, N, Ca, Mg, Mn, Fe, Cu, and Zn in the water and sediments from the above lake were analyzed by standard methods. Chemical analysis of the intact shoots, with and without mucilage, was conducted for the nutrients and ions mentioned above (the mucilage sheath may be easily removed by squeezing). The results showed that mucilage and shoot tissue had very high levels of Ca, Mg, Mn, Fe, Cu, and Zn, in addition to N and P (Figures 6.5, 6.6, and 6.7).

Figure 6.3. *Nitella hyalina*, a charophyte that accumulates ions on the mucilage envelope in the growing meristem, useful in the remediation of shallow sand-mine voids. (Photo by author.)

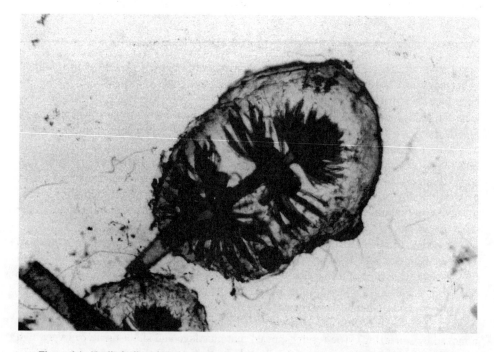

Figure 6.4. *Nitella hyalina* with the mucilage envelope stained with Toluidine B. (Photo by author.)

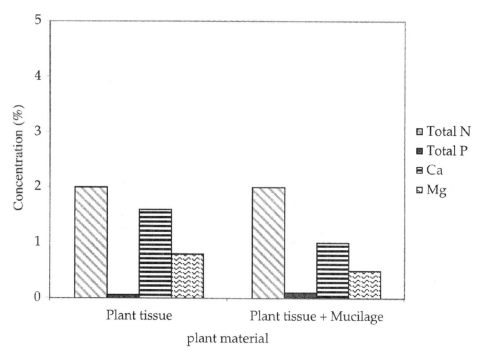

Figure 6.5. Concentrations of N, P, Ca, and Mg in *Nitella hyalina* in Lake Leschenaultia (Ca and Mg in standing water was 11 ppm and 35 ppm, respectively).

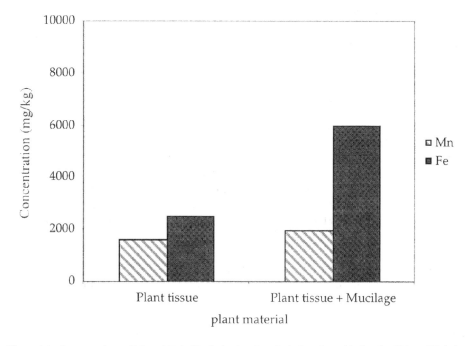

Figure 6.6. Concentrations of Mn and Fe in *Nitella hyalina* from Lake Leschenaultia (levels of Mn and Fe below detectable level in water).

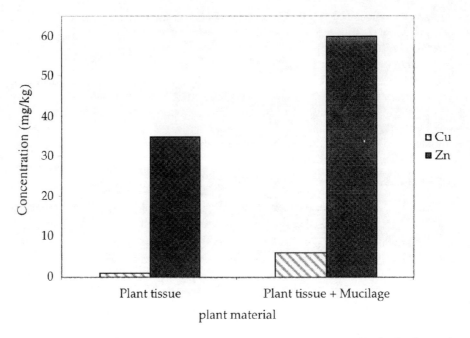

Figure 6.7. Concentrations of Cu and Zn in *Nitella hyalina* from Lake Leschenaultia (levels of Cu and Zn undetectable in water).

Additionally, it was discovered that the mucilage ensheathment harbored a thriving community of diatoms more indicative of the concentrations of minerals in the mucilage than those of the water outside (John & Gayton, 1994). The ability of *N. hyalina* to accumulate excessive amounts of several minerals was further investigated in a chain of sand-mine wetlands in Western Australia.

Nitella hyalina in Sand-Mine Wetlands

Western Australia is the leading producer of titaniferous minerals such as ilmenite and rutile. Sand minerals have been mined in Capel, 200 km south of Perth (33°S, 116°E) in Western Australia, since 1956. After mining, the mine-pits were converted into artificial wetlands between 1975 and 1979. These wetlands occupy an area of 50 ha, now known as the RGC Wetlands Centre (Rennison Goldfields Consolidated Mineral Sands, Ltd.) and comprise a chain of 15 lakes, mostly interconnected; the effluent water from the mine-processing plant was discharged into the lakes initially. The Wetlands Centre was established in 1985, with the objective of developing a self-sustaining ecosystem for conservation of waterbirds. In 1984 and 1985, initial studies showed that the system was not productive enough to attract birds, because the pH was mostly 2–4, ammonium levels were high, available phosphorus was too low, and the concentrations of iron, manganese, magnesium and sulphate were very high, unfavorable to the availability of phosphate (Doyle & Davies, 1998). Subsequently, the effluent water entering the chain of lakes from the mineral processing plant was buffered to a pH above 8. The lakes were landscaped, islands and peninsulas were created, and the steep shoreline was modified to make the margin convoluted and emergent; peripheral vegetation was established. These remediation measures considerably improved the water quality of the lakes. The pH of the lakes increased to above 7, with the concomitant increase in biodiversity.

The average annual rainfall of the region is 830 mm and the climate is Mediterranean, characterized by dry summers and wet winters, with a temperature range of 20–36°C in the summer and 8–15°C in the winter. The lakes range in depth from 1 m to more than 6 m. The water level drops drastically in severe summer periods. Within 6 months of increasing the pH of the effluent water, signs of improvement in productivity in all the lakes became obvious (John, 1993). The number of waterbirds visiting and nesting at the wetlands steadily increased. However, the general productivity still remained low compared to the nearby natural wetlands (Ward et al., 1997).

In 1992, while studying the ecology of the above wetlands, I discovered that two of the lakes were colonized by *Nitella hyalina*. The two lakes had the lowest electrical conductivity (EC) and the highest water clarity in the whole system. The mucilage covering the shoots of *Nitella* in these lakes had a thriving community of diatoms, different from that encountered in the mucilage of *Nitella* from Lake Leschenaultia. A chemical analysis of the charophyte shoot tissue and mucilage, the surrounding water, and the sediment from one of the lakes colonized by *Nitella* (Plover Lake) at Capel confirmed that *Nitella hyalina* was a hyperaccumulator of Ca, Mg, Mn, Fe, Cu, and Zn (Figures 6.8, 6.9, and 6.10). The tissue itself had accumulated Mn, Fe, Mg, and Ca (Table 6.1). Our records confirmed that the lake was colonized by *Nitella* in 1990. The other lake was probably colonized before that (John, 1999b).

In 1995, three years after the initial analysis, a chemical analysis of *N. hyalina*, its shoots, the standing water, and sediment of Plover Lake was conducted. Again, the result

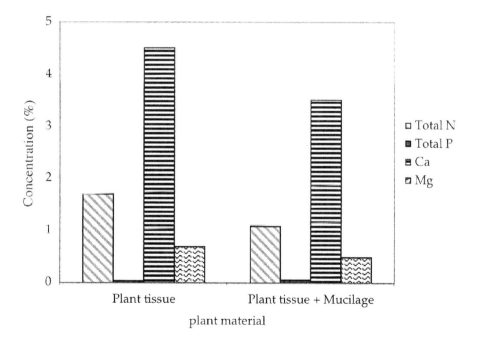

Figure 6.8. Concentrations of N, P, Ca, and Mg in *Nitella hyalina* from Plover Lake RGC Wetlands Centre 1992. Concentrations of Ca and Mg in standing water were 20 and 12 ppm respectively.

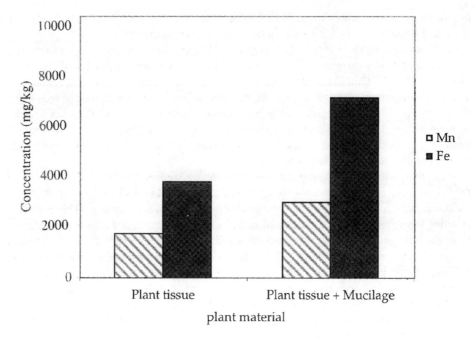

Figure 6.9. Concentrations of Mn and Fe in *Nitella hyalina* from Plover Lake. Mn and Fe concentrations in standing water were 0.07 and 0.05 ppm, respectively.

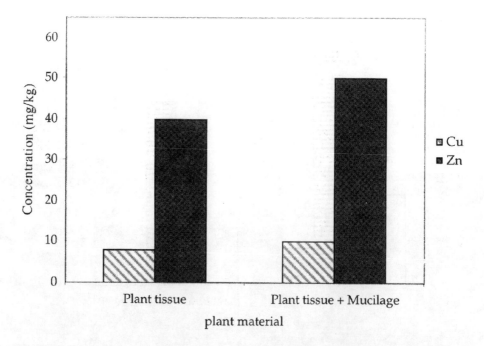

Figure 6.10. Concentrations of microelements of Cu and Zn in *Nitella hyalina* from Plover Lake (Cu and Zn < 0.01 ppm in standing water).

Table 6.1. Element Analysis of Charophytes from Plover Lake, 1 May 1995 (a Lake at Capel Wetlands Centre Colonized by *Nitella hyalina*)

Element	Site 1	Site 2
Total phosphorus (%)	0.06	0.05
Total sulfur (%)	0.28	0.22
Potassium (%)	0.4	0.91
Sodium (%)	0.18	0.11
Calcium (%)	5.6	8.3
Magnesium (%)	0.82	0.3
Copper (ppm)	3	4
Zinc (ppm)	8	80
Manganese (ppm)	4,000	5,000
Iron (ppm)	9,400	4,300

illustrated clearly the ability of *Nitella* to accumulate heavy metals (e.g., Al and Mn) in both the mucilage and in the shoot tissue (Figure 6.11). In 1996, yet another lake (Pobblebonk Lake) was colonized by *Nitella*. Chemical analysis once again confirmed that it was hyperaccumulating Al and Mn (Figure 6.12). Both plant tissue and mucilage had high concentrations of metals.

Invertebrates

Plover Lake was soon inhabited by two native fish species, whereas the nearby lakes, without *Nitella*, had no fish. A survey of the invertebrates in the two lakes colonized by *Nitella* and a neighboring lake, without *Nitella*, clearly showed that species richness and abundance of invertebrates were significantly higher in the former lakes (Figure 6.13).

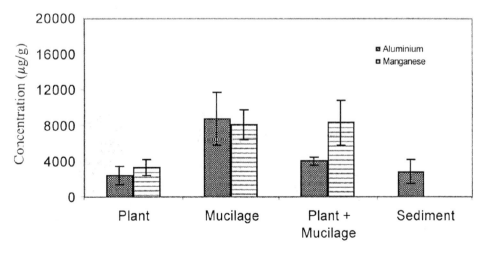

Figure 6.11. Concentrations of Al and Mn in the shoot, mucilage, and intact plant (shoot and mucilage) and sediment (Mn concentration in sediment is very low) in Plover Lake, colonized by *Nitella* over 6 years.

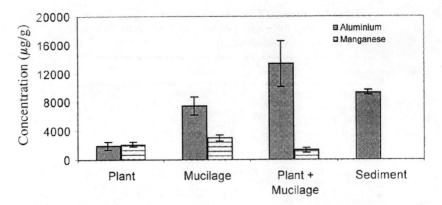

Figure 6.12. Concentrations of Al and Mn in thallus mucilage and intact plant (mucilage and thallus) and sediment in Pobblebonk Lake 2, colonized by *Nitella* for only 2 years. Mn in sediment was very low but Al was high.

During the summer months, the charophytes become an important source of food for waterfowl. Since the initial colonization by *N. hyalina* in the two lakes, two more species of charophytes have been observed in them (i.e., *Chara globularis* and *C. fibrosa*).

As the water quality improved considerably in Plover Lake (Figure 6.14), not only did the *Nitella* meadows become healthier, but they were also producing copious mucilage, inhabited predominantly by desmids, which are well known to prefer water of low EC. Because the lake recently colonized by *Nitella* (Pobblebonk Lake) had higher EC and mineral concentrations, not only were the *Nitella* shoots much smaller, but they also had a smaller volume of mucilage, predominantly inhabited by diatom species tolerant of the higher concentration of minerals in the water. An examination of the epiphytic diatom assemblages within the mucilage of the samples from the two lakes revealed that the

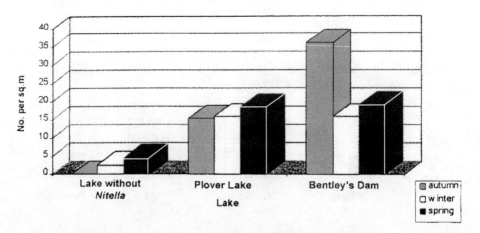

Figure 6.13. Invertebrates abundance per m^2 in Plover Lake and Bentley Dam (both colonized by *Nitella hyalina*, and a nearby lake with *Nitella*).

Figure 6.14. Plover Lake, a rehabilitated sand-mine pit showing high water clarity. (Photo by author.)

mucilage community in Plover Lake consisted of desmids, whereas that in Pobblebonk Lake had predominantly diatoms. This, again, can be interpreted as a reflection of changes in water quality in Plover Lake since its colonization by *Nitella* in 1990, as well as the difference in water chemistry between the two lakes; the EC in Plover Lake is much lower than that in Pobblebonk Lake.

Life Cycle of *Nitella hyalina*

The life-cycle pattern of the species has been investigated in both RGC Wetlands and in Lake Leschenaultia. The species is monoecious and behaves as an annual in the wetlands of RGC Wetlands Centre but continues as a perennial, with very few gametangia, in the deeper Lake Leschenaultia. The germination of oospores starts at the end of autumn and early winter, and growth proceeds throughout winter. The gametangia are produced by early spring. By summer, as the water level lowers, oospores mature and the charophytes complete their life cycle. In some years, the lakes dry up completely, leaving a rich "sporebank." With the first rains in autumn, the spores germinate. In deeper lakes such as Lake Leschenaultia, the *Nitella* meadows continue to grow by vegetative propagules. The pattern of the life cycle of *N. hyalina*, therefore, is adapted to both permanent water bodies and ephemeral pools. *N. hyalina* has been cultured in the laboratory for 5 years, and the stages in the life cycle have been verified.

Advertent Introduction of Charophytes

For the past 2 years, we have been focusing on the environmental conditions suitable for charophyte growth in the RGC Wetlands, with the intention of introducing *Nitella hyalina* to other lakes at the Centre. Principal component analysis (PCA) of the environmental factors and charophyte distribution in the RGC Wetlands has identified several factors associated with charophyte colonization: high concentration of magnesium, and low concentration of calcium and ammonium. Attempts to germinate in other lakes the oospores collected from Plover Lake met with considerable success.

Sediment containing the oospores was introduced to other lakes at the Wetlands Centre. Transplantation trials practiced over 2 years are currently yielding desirable results. Latest surveys indicate that charophytes—*N. hyalina* and three other species of charophytes—have now spread to six lakes. The water quality in these lakes has improved, with a reduction in EC and an increase in water clarity. The functional components of the aquatic systems have steadily expanded, leading to the development of productive wetlands from mine-voids.

Concluding Remarks

Algal mats consisting of cyanobacteria and diatoms are known to accumulate selected minerals. Although there are references in the literature on green algae such as *Microspora* and *Oedogonium* concentrating minerals, the case of *Mougeotia* sp. reported here is the first known example of its inclusion in a remediation project. Since the discovery of *Mougeotia* sp. as a remediation tool, I have observed this species in several other acidic lakes. Further research is essential not only to explore the mechanism involved in "hyperaccumulation" by these algae, but also to maximize its use in appropriate remediation procedures. At the moment, "harvesting" of the hyperaccumulator algae is practiced at regular intervals (3 to 6 months) in the passive treatment system for acidic mine-voids. The actual amount of algae required to remove a specific concentration of heavy metals has to be determined by developing predictive models, before bioaccumulators can be used most efficiently in the treatment of acidic mine-voids.

The concentration of minerals such as Fe, Al, Mn, and Ca in the mucilage and plant tissue of *Nitella hyalina* compares well (in some instances, even higher) with the concentrations of minerals reported in hyperaccumulators such as seaweeds and aquatic vascular plants. A thick boundary of mucilage has the advantage of minimizing the loss of nutrients into the surrounding water column. The presence of mucilage around the meristem of *N. hyalina* may be interpreted as an adaptive mechanism to maximize nutrient availability for the plant. This adaptation makes it an ideal tool for phycoremediation of mine-voids of relatively shallow depth.

Large areas of clear water associated with dense meadows of *Chara* have often been observed in several parts of the world. Reduced resuspension, nutrient uptake, and zooplankton using charophyte meadows as a refuge are some of the speculated reasons for the clarity of water in the lakes with dense charophyte beds (Van den Berg et al., 1997).

Outridge and Noller (1991) reported that Mn was most strongly absorbed by aquatic vascular plants. The other elements absorbed in descending order were Zn, Mo, Cu, and Pb. The water hyacinth (*Eichornia crassipes*) is one of the most commonly cited species of vascular plants for phytoremediation of polluted waters (Brooks & Robinson, 1998). Charophytes compare well to these plants in their role in remediation. Moreover, use of naturally occurring algae to clean up the aquatic environment is the most cost-effective treatment for degraded wetlands.

References

Australian and New Zealand Environment and Conservation Council. (1992). National water quality management strategy. Australian water quality guidelines for fresh and marine water. Australian and New Zealand Environment and Conservation Council.

Blaylock, M., & Huang, J. (2000). Phyto extraction of metals. In: I. Raskin & B. Ensley (Eds.), *Phytoremediation of toxic metals: Using plants to clean up the environment* (pp. 53–71) Wiley.

Brooks, R. (Ed.) (1998). Plants that hyperaccumulate heavy metals. CAB International, Oxon.

Brooks, R., & Robinson, B. (1998). Aquatic phytoremediation by accumulator plants. In: R. Brooks (Ed.), *Plants that hyperaccumulate heavy metals* (pp. 203–227). CAB International, Oxon.

Doyle, F., & Davies, S. (1998). Creation of a wetland ecosystem from a sand-mining site: A multidisciplinary approach. In: A. McComb & J. Davis (Eds.), *Wetlands for the future* (pp. 260–772). Adelaide: Gleneagles.

Dunn, C. (1998). Seaweeds as hyperaccumulators. In: R. Brooks (Ed.), *Plants that hyperaccumulate heavy metals* (pp. 119–133). CAB International, Oxon.

Ensley, B. (2000). Rationale for use of phytoremediation. In: I. Raskin & B. Ensley (Eds.), *Phytoremediation of toxic metals: Using plants to clean up the environment* (pp. 3–11). Wiley.

Glass, D. (2000). Economic potential of phytoremediation. In: I. Raskin & B. Ensley (Eds.), *Phytoremediation of toxic metals: Using plants to clean up the environment* (pp. 15–31). Wiley.

Hutchinson, G. (1975). A treatise on limnology. In *Limnological botany*, (pp. 31–77). New York: Wiley.

John, J. (1993). The use of diatoms in monitoring the development of created wetlands at a sand-mining site in Western Australia. *Hydrobiologia, 269/270*, 427–436.

John, J. (1999a). Acid mine drainage: Impacts and remediation—an Australian perspective. In *Proceedings of the 2nd International Conference on Contaminants in Soil Environment in the Australia–Pacific Region*, pp. 434–435. New Delhi.

John, J. (1999b). Mining pits into wetlands—a rehabilitation case-history from Western Australia. In *Proceedings of the 2nd International Conference on Contaminants in Soil Environment in the Australia–Pacific Region*, pp. 468–469. New Delhi.

John, J. (2000a). A self-sustainable remediation system for acidic mine voids. In: Proceedings of the 4th International Conference on Diffuse Pollution, pp. 506–511. Bangkok: International Association of Water Quality.

John, J. (2000b). Rehabilitation of acidic coal mine voids into productive wetlands: A case history from Western Australia. In: Proceedings of the International Conference on the Remediation and Management of Degraded Lands, pp. 60–62. Fremantle, Western Australia.

John, J. & Gayton, C. (1994). A comparative study of epiphytic diatom communities on aquatic macrophytes in two artificial wetlands. Technical Report No. 23, RGC Wetlands Centre, Capel, Western Australia.

Kiernan, J. (1981). *Histological and histochemical methods, theory and practice*. Oxford, UK: Pergamon Press.

Outridge, P., & Noller, B. (1991). Accumulation of toxic trace elements by freshwater vascular plants. *Reviews of Environmental Contamination and Toxicology, 121*, 1–63.

Parker, G., & Robertson, A. (1999). *Acid mine drainage*. Australian Minerals and Energy Environment Foundation, Occasional Paper II, 227 pp.

Simon, J. (1990). Decline of the Characeae community in the shallow peat Lake Botshol. In: J. Rozema, & J. A. C. Verkleij (Eds.), *Ecological responses to environmental stresses*. Kluwer Academic.

Smith, D. (1998). *Water in Australia: Resources and management*. Oxford, UK: Oxford University Press.

Søndergaard, M., Lauridsen, T., Jeppensen, E., & Brunn, L. (1997). Macrophyte waterfowl interactions: Tracking a variable resource and the impact of herbivory on plant growth. In: E. Jeppesen, M. Søndergaard, & K. Christoffersen (Eds.), *The structuring role of submerged macrophytes in lakes* (pp. 298–307). Ecological Studies No. 31. Springer.

Tyrrell, W. (1995). A review of wetlands for treating coal mine wastewater, particularly in low rainfall environments in Australia. AMEEF Occasional Paper. Melbourne: Australian Minerals and Energy Environment Foundation.

Van den Berg, M. S., Coops, H., Meijer, M., Scheffer, M., & Simons, J. (1997). Clear water associated with dense *Chara* vegetation in the shallow and turbid Lake Veluwemeer, The Netherlands. In: E. Jeppesen, M. Søndergaard, & K. Christoffersen (Eds.). *The structuring role of submerged macrophytes in lakes* (pp. 339–352). Ecological Studies No. 31. Springer.

Ward, M., Zilm, N., & John, J. (1997). *The role of charophytes, a group of submerged macrophytes in the rehabilitation of the RGC Wetlands Centre*. Technical Report No. 35, RGC Wetlands Centre, Capel, Western Australia.

7

UV-B Impact on the Life of Aquatic Plants

Donat-P. Häder

Introduction

Drastic stratospheric ozone depletion over both poles, as well as significant decreases in total ozone at high and midlatitudes, have been reported. The resulting increases in solar UV-B radiation (280–315 nm, Commission Internationale de l'Eclairage [CIE] definition) have been shown to affect the biota and cause increased incidences of skin cancer as well as other diseases in humans. This short-wavelength radiation has been found to impair terrestrial and aquatic ecosystems. It also affects the chemistry of the troposphere (e.g., photochemical smog formation) and influences biogeochemical cycles (Madronich et al., 1998). Solar UV radiation has been shown to penetrate to significant depths in many freshwater and marine ecosystems, where phytoplankton productivity is located (Smith et al., 1992; Häder, 1995; Booth et al., 1997; Coohill et al., 1996). Model predictions show that these trends will increase well into the current century. Furthermore, there is evidence that this trend alters the ratio of UV-B : UV-A : PAR (photosynthetic active radiation), which may affect the sensitive, light-dependent responses of aquatic plants, including photosynthesis, photoorientation, and photoprotection (Smith et al., 1992; Häder et al., 1995, 1997a, 1997b; Gerber et al., 1996; Jiménez et al., 1996) and pose a source of significant stress for the diverse aquatic ecosystems (International Arctic Science Committee [IASC], 1995). UV-B and UV-A impair growth and productivity by a number of mechanisms involving numerous molecular targets. However, most organisms possess effective protective and repair mechanisms to mitigate the damage inflicted by solar UV radiation.

Significant increases of solar UV radiation may result in decreased biomass productivity, which may affect all levels of the intricate food web. This might result in reduced food production for animal consumers and humans (Häder et al., 1995; Häder, 1997e; Häder & Worrest, 1997). Another consequence may be reduced sink capacity for

Donat-P. Häder • Friedrich-Alexander Universität Erlangen-Nürnberg, Institut für Botanik und Pharmazeutische Biologie, Staudtstr. 5, D-91058 Erlangen.

atmospheric carbon dioxide (Ducklow et al., 1995; Takahashi et al., 1995, 1997), with impacts on global warming (Sarmiento & Le Quéré, 1996; Thomson, 1997). Finally, changes in species composition and ecosystem integrity are being discussed. Important reviews on various aspects of UV effects on aquatic ecosystems include aquatic ecosystems in general (Nolan & Amanatidis, 1995; Häder, 1997c; Häder & Worrest, 1997), the role of mycosporine-like amino acids (MAAs) in marine organisms (Dunlap & Shick, 1998), phytoplankton (Cullen & Neale, 1997a, 1997b; Häder, 1997a), macroalgae (Franklin & Forster, 1997; Häder & Figueroa, 1997), primary and secondary producers (Shick et al., 1996; Lesser, 1996), lake acidification, and UV penetration (Williamson, 1995, 1996).

Ozone Layer and UV Radiation

Stratospheric ozone depletion results in an increased transmission of solar UV radiation to the Earth's surface. In addition to time of day and season, the level of solar UV radiation is affected by cloud pattern and atmospheric pollution, as well as latitude and elevation, and strongly depends on the total ozone column. Continuous ozone observations are performed from ground and satellite-based instruments (TOMS aboard the Nimbus 7 satellite or the recent Earth Probe).

Several ground-based networks have been installed, for example, by Biospherical Instruments, Inc. (San Diego, California). In order to guarantee quality control of the light measurements, considerable efforts are necessary (Seckmeyer et al., 1994). Another monitoring network is the European Light Dosimeter Network (ELDONET), with over 40 stations spread over Europe and additional stations in India, Japan, Egypt, New Zealand, and Argentina (Figure 7.1). In addition to high-altitude stations in the European Alps and the Spanish Sierra Nevada, seven instruments are located in the water column to monitor the transparency of major coastal waters. The instruments record solar radiation unattended in three channels (UV-B, UV-A, PAR). The data are transmitted to a central server in Pisa and are available on the Internet (http://www.eldonet.org).

The stratospheric ozone layer has been observed to decrease continuously during the last two decades over Antarctica and later the Arctic as well as Southern and Northern midlatitudes (Müller et al., 1997; Rex, 1997; Stolarski, 1997), which has resulted in significantly higher UV-B radiation levels on the ground (Kerr & McElroy, 1993). It is beyond any reasonable doubt that anthropogenic impact causes the depletion of stratospheric ozone. The main sources are the emission of halogen-containing compounds (World Meteorological Organization, 1994a; 1998). The ozone layer is measured in Dobson Units (DU), defined as the depth (in millimeters $\times$ 100) that the gaseous ozone layer would be if compressed to atmospheric pressure at the Earth's surface at 0°C (Madronich & Flocke, 1997).

Biologically Active UV Radiation

Extraterrestrial solar radiation < 280 nm (UV-C) is almost quantitatively filtered out by the oxygen in the atmosphere and therefore not ecologically relevant; also, the majority of the UV-B radiation (280–315 nm, CIE definition) is removed by atmospheric ozone. Most of the ozone is located in the stratosphere, with only a small fraction in the troposphere (Figure 7.2). UV-A (315–400 nm) and the visible or PAR (400–700 nm) are not absorbed by the ozone in the atmosphere.

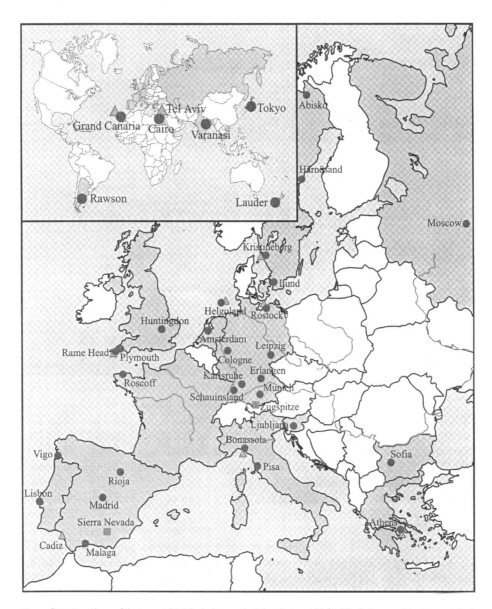

Figure 7.1. Locations of the terrestrial (circles), aquatic (triangles) and high altitude (squares) instruments in the ELDONET network of solar dosimeters.

All other factors being equal, the total ozone concentration is mainly controlled by the total ozone column (Kirchoff et al., 1997; Seckmeyer et al., 1997; Taalas et al., 1997). Shorter wavelengths depend more on ozone changes than longer wavelengths because of the decreasing absorption of ozone toward longer wavelengths.

Highest increases of UV-B radiation have been recorded under the Antarctic ozone hole (Seckmeyer et al., 1995); the UV radiation has even exceeded maximum summer values measured at San Diego (USA) (Booth et al., 1997). The size and depth of the Antarctic ozone hole have not changed much since the early 1990s; also, little change has

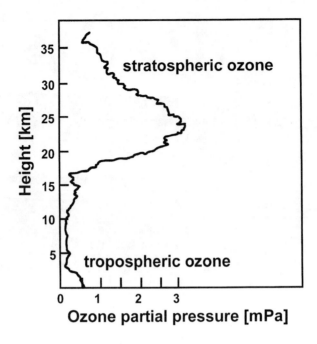

Figure 7.2. Ozone concentrations in the troposphere and stratosphere (in Dobson units).

been found at midlatitudes during the last 5 years. The decrease of stratospheric ozone since 1970 was calculated to be about 50% in the Antarctic during the Austral spring, 15% in the Arctic spring, 6% at Northern Hemisphere midlatitudes in winter and spring, 3% in summer and fall, and 5% at Southern Hemisphere midlatitudes throughout the year (Madronich et al., 1998). This caused increases in sunburning UV radiation by 130%, 22%, 7%, 4%, and 6%, respectively. No significant ozone changes have been measured at tropical and subtropical latitudes.

In order to determine the biological effects of increasing UV radiation, the spectral sensitivity of the biological effect must be determined. For this purpose, action spectra have been calculated for a number of physiological functions in several aquatic organisms (e.g., Boucher & Prezelin, 1996; Häder et al., 1994). Multiplication of the physically measured solar UV radiation with a specific action spectrum results in the biologically active UV irradiance or exposure for this organism or response (Madronich et al., 1998). The dependence of the biologically active irradiance is defined by the radiation amplification factor (RAF), expressed as the increase in biologically active UV irradiance that results from a 1% ozone reduction. RAFs are good indicators of the sensitivity of a biological process to ozone depletion, but this dependence is nonlinear for larger ozone changes. Figure 7.3 shows the solar spectral irradiance at the Earth's surface in the UV range and a biological action spectrum. The overlap of the two spectra covers the biologically active radiation.

Measurement of UV Radiation

UV measurements have been improved considerably during the last few years by the development of new instrumentation and better calibration procedures. Quality control has increased as a result of intercalibrations between instruments (Seckmeyer et al., 1994;

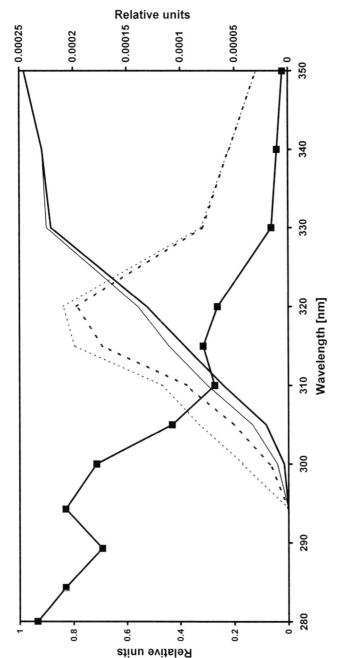

Figure 7.3. Solar emission spectra $E(\lambda)$ at 250 DU (fine line) and 350 DU (heavy line), and the biological action spectrum for the inhibition of photosynthesis $P(\lambda)$ in the cyanobacterium *Nodularia* (heavy line and square symbols, right ordinate) overlap in the UV-B range. The area under the product function $E(\lambda) \times P(\lambda)$ is defined as the biologically active fluence rate (broken lines).

Thompson et al., 1997; Kjeldstad et al., 1997; Webb, 1997; Leszczynski et al., 1998). Measurement errors of better than 5% in the UV-A range and 10% in the UV-B range are achieved. Recent measurements have confirmed the expected UV levels over Europe and North America predicted from ozone depletion values determined from satellites (Seckmeyer et al., 1995; World Meteorological Association, 1998). Aerosols may have a considerable effect on the level of solar UV radiation at the surface. Anthropogenic sulfate aerosols, resulting primarily from fossil fuel burning, have reduced ground UV-B irradiances by 5–18% in industrialized regions of the Northern Hemisphere (Varotsos and Kondratyev, 1995; Mims, 1997). There are significant differences between the Northern, polluted and Southern, less-polluted hemispheres (Seckmeyer & McKenzie, 1992; Seckmeyer et al., 1995). In addition, considerable vertical gradients can be measured on mountains (Madronich et al., 1995).

Predicted UV Radiation

Short-term forecasts of solar UV radiation are now available in several countries. The values are often made available to the public via the recently developed UV index (World Meteorological Association, 1994b) which is the physical irradiance weighted by the erythemal action spectrum (McKinlay & Diffey, 1987) and multiplied by a factor of 40. A UV index of 10 or more indicates very high UV irradiances.

Future ozone trends may be affected by natural events such as volcanic eruptions (e.g., the 1991 Mt. Pinatubo eruption). Predictions of the production and emission of pollutants are also difficult and need to include greenhouse gases and the emissions from intercontinental and supersonic aircraft. The recovery of the ozone layer may be delayed significantly due to interactions with increasing greenhouse gas concentrations (Schindell et al., 1998). The future scenarios are based on the 1997 Montreal Amendments. Stratospheric ozone levels will depend on whether the subscriber countries comply with the regulations, and whether illegal production and trafficking of ozone-depleting substances can be controlled. UV radiation is expected to return to pre-1980 levels by the middle of the current century. This is largely due to the long half-life times of the halocarbons currently present in the atmosphere.

UV Penetration into the Water

The penetration of solar UV and PAR into the water column needs to be determined (Montecino & Pizarro, 1995; Björn et al., 1996). Aquatic ecosystems show large temporal and spatial changes in the concentrations of absorbing substances. The ratio between the 0.1% penetration for UV-B and PAR can be utilized to evaluate the detrimental effects on aquatic organisms in the euphotic zone (Piazena & Häder, 1997). Recently developed instrumentation allows accurate determination of the underwater light distribution (Morrow & Booth, 1997). Gelbstoff (yellow substances), chlorophylls, and other photosynthetic pigments, as well as organic and inorganic particulate material, attenuate the UV-B radiation in the water column. Also, phytoplankton absorb radiation, and algal canopies modify the light quality by absorption and scattering of the incident light.

The underwater light climate can also be determined from satellite data (e.g., CZCS and SeaWiFS). Even though these instruments cover only the visible range, attempts are being made to extrapolate the data into the UV range. There have been great efforts to measure algal biomass by using remote sensors (Figure 7.4), quantifying phytoplankton

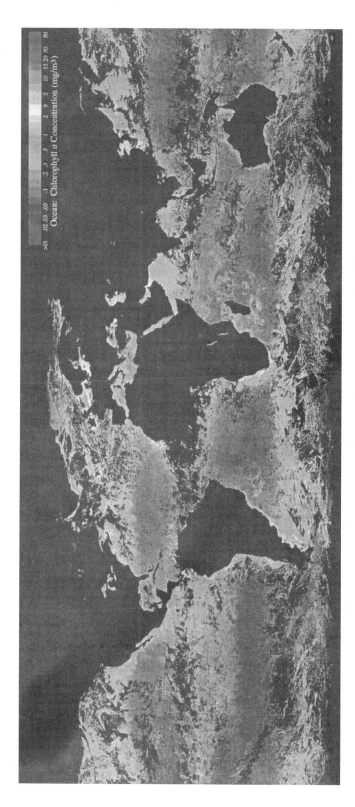

Figure 7.4. Chlorophyll concentration in the ocean, 29.3.–5.4.2000. Provided by the SeaWiFS Project, NASA/Goddard Space Flight Center.

chlorophyll (Brown et al., 1995). A major problem of remote monitoring is that overflying instruments mainly determine the surface signal, whereas signals from deeper layers contribute only little (Piazena & Häder, 1997). Therefore, profound knowledge of the vertical distribution of phytoplankton, as well as the composition of the algal communities, is necessary for a quantitative analysis of biomass productivity.

Dissolved Organic Matter and Solar UV Radiation

Dissolved organic carbon (DOC) is mostly of terrestrial origin and relatively resistant to chemical and bacterial degradation (Naganuma et al., 1996), but after photolytic splitting by UV radiation, the smaller products are readily consumed by bacterioplankton (Wetzel et al., 1995). DOC strongly absorbs UV radiation. Therefore, breakdown of DOC by UV and subsequent uptake by bacteria enhances the UV-B penetration into the water column. This process is partly offset by the damage of bacteria as a result of enhanced UV radiation at the surface (Herndl, 1997).

Behrenfeld and Falkowski (1997a) have evaluated models used to estimate photosynthetic rates derived from satellite-based chlorophyll concentration. In addition, they (1997b) have evaluated various primary productivity models, provided a classification scheme for these productivity models, and shown that many of these, apparently different, models fundamentally agree. If equivalent parameterizations are used for satellite-derived chlorophyll and the maximum chlorophyll-specific carbon fixation rate, then estimates of global annual primary production are found to be due primarily to these variables.

Bacterio- and Picoplankton

The productivity of bacterioplankton is comparable to or exceeds phytoplankton primary productivity (Herndl, 1997). Bacteria are no longer regarded solely as terminal decomposers of organic material. In the "microbial loop hypothesis," bacterioplankton has a central position in the food web (Pomeroy & Wiebe, 1988), playing a major role in the carbon flux in aquatic ecosystems by taking up DOC and remineralizing the carbon.

Bacterioplankton are affected by solar UV radiation, and the damage largely depends on the spectral attenuation coefficients in the water column and the time pattern of exposure and protection of the organisms given that they are passively moved in the mixing layer. Bacteria lack UV screening pigments such as mycosporines (Karentz et al., 1994; Garcia-Pichel, 1994). Therefore, they are more prone to UV-B stress than larger eukaryotic organisms, and exposure to solar radiation induces about double the amount of cyclobutane dimers (Jeffrey et al., 1996a, 1996b). The damage is at least partially offset by photoreactivation via the repair enzyme photolyase (Nicholson, 1995). UV/blue radiation (360–430 nm) is most effective in the induction of the activity. Bacterial ectoenzymes responsible for the cleavage of external organic matter are also affected by solar UV-B (Müller-Niklas et al., 1995).

The bacterioplankton provides food for heterotrophic picoplankton (<1 μm). Whereas the bacterial plankton is limited by UV damage, viruses and heterotrophic flagellates also show a high sensitivity to solar UV radiation (Aas et al., 1996; Sommaruga et al., 1995, 1996).

Cyanobacteria

Although the cosmopolitan prokaryotic cyanobacteria developed early in evolution, they possess a higher plant–type oxygenic photosynthesis and are major biomass producers in terrestrial and aquatic habitats as well as wetlands. In addition, many of these organisms are capable of fixing atmospheric nitrogen either as free-living organisms or as symbionts of many protists, animals, and plants (Sinha & Häder, 1997). They use the enzyme nitrogenase to reduce atmospheric nitrogen into ammonium ions. Excess production is made available for aquatic eukaryotic phytoplankton as well as higher plants (Kashyap et al., 1991; Sinha et al., 1996; Sinha & Häder, 1996; Kumar et al., 1996b). In agriculture (e.g., tropical rice paddy fields), cyanobacteria serve as biological fertilizers (Banerjee & Häder, 1996).

Despite the fact that cyanobacteria developed early in evolution, which is believed to be characterized by high levels of UV radiation, UV-B impairs processes such as growth, survival, motility, as well as the enzymes and key proteins of nitrogen metabolism and CO_2 fixation (Donkor & Häder, 1996, 1997). In many species, growth and survival decrease within a few hours of UV-B irradiation.

The photosynthetic pigments are sensitive targets of solar UV radiation; especially the phycobiliproteins are readily bleached and the proteins cleaved (Sinha et al., 1995, 1996; Aráoz & Häder, 1997). At lower UV doses, the energy transfer to the reaction center of photosystem II is hampered (Sinha et al., 1996). However, simultaneous with UV-induced breakdown, increased synthesis of phycobiliproteins has been observed under mild UV-B stress (Aráoz & Häder, 1997). These pigments strongly absorb in the UV-B range and might function as effective screening pigments, which may be an important feature, because they form a peripheral layer around the central part containing the sensitive DNA. The peripheral phycobilins have been found to intercept more than 99% of UV-B photons before they penetrate to the genetic material.

UV-B affects the photosynthetic activity in marine and freshwater cyanobacteria. In addition to the breakdown of pigments, Sinha et al. (1996) reported that Rubisco (ribulose-1,5-bis-phosphate carboxylase/oxygenase) activity is severely affected by UV-B treatment.

Ammonium uptake was reduced by 10% in cultures exposed to solar radiation. The nitrogen-fixing enzyme nitrogenase is affected by UV-B even after a few minutes of *in vivo* exposure, and a total loss of activity was found after less than 1 h of exposure to UV radiation (Kumar et al., 1996a). This inactivation has been interpreted to be due to impaired ATP synthesis. Also, the ammonia-assimilating enzyme glutamine synthetase (GS) is inhibited, whereas nitrate reductase was found to be stimulated by UV radiation in all studied strains (Sinha et al., 1995).

Cyanobacteria have developed a number of protective strategies to mitigate the negative effects of excessive radiation, including avoidance of brightly irradiated habitats, synthesis of UV screening pigments, and production of molecular scavengers that detoxify the highly reactive oxidants produced under the impact of UV radiation (Vincent & Roy, 1993).

Screening pigments in cyanobacteria include scytonemin and mycosporine-like amino acids (MAAs). In addition, a plethora of spectroscopically characterized but chemically unidentified water-soluble pigments have been found, such as a brown-colored pigment from *Scytonema hofmanii* and a pink extract from *Nostoc spongiaeforme* (Kumar et al., 1996a; Donkor & Häder, 1995). Many cyanobacteria (e.g., *Scytonema* and *Nostoc*) form filaments that are embedded in a mucilaginous sheath, into which the screening pigments are deposited during the late stationary phase of growth. The screening pigment from *Scytonema hofmannii* shows an absorption maximum at 314 nm. These organisms are more tolerant to UV-B exposure than those that do not contain such covering (Sinha et al., 1995); that is, other species of *Scytonema* that do not produce this pigment are unable to survive 2 h of UV-B (2.5 W m^{-2}). Other screening pigments such as scytonemins, carotenoids, and MAAs are incorporated into the cytoplasm to protect the central DNA from solar short-wavelength radiation (Karsten & Garcia-Pichel, 1996).

Phytoplankton

It is undisputed by most scientists in the field that environmental UV radiation is a considerable stress factor for phytoplankton, the primary producers in many freshwater and marine ecosystems, that influences growth, survival, and distributions of the organisms. Phytoplankton is the most important biomass producer in aquatic ecosystems on a global scale. The mostly unicellular organisms populate the surface layers of the oceans and freshwater habitats in order to receive sufficient photosynthetic active radiation (PAR) to satisfy their energy needs. This top layer in the water column is called the *euphotic zone*, the base of which is defined as the depth at which photosynthetic carbon fixation balances respiratory losses over a day (typically a depth to which 1.0 to 0.1% of PAR penetrates). Simultaneously, phytoplankton are exposed to solar UV radiation, in this layer.

In order to estimate the UV-induced damage, a number of questions need to be answered:

- What are the spectral characteristics of solar radiation penetrating to depth and its space–time variability?
- What is the biological weighting function (BWF) of the biological effect?
- How does vertical mixing influence variable irradiance exposures of phytoplankton and their physiological response?

We are far from a full quantitative understanding, but there have been advances in each of these areas during the past few years (Häder, 1997c; Cullen & Neale, 1997a, 1997b; Vernet & Smith, 1997).

The BWFs recently determined for a number of species (Cullen & Neale, 1997; Boucher & Prezelin, 1996; Neale et al., 1998a) indicate that although the largest effect is in the UV-B, the UV-A is also a significant component. BWFs vary by species, region, mixing characteristics of the water column, and other environmental factors. Therefore, a single, or even a few, BWFs may not be sufficient for a complete description of an ecosystem.

In order to estimate an effect–response curve (ERC), one must determine whether the measured damage is a function solely of cumulative exposure or a function of exposure rate. This difference is important and has a decisive impact on both the design and interpretation of experiments (Neale et al., 1998a). The results depend on the balance between damage and repair and, thus, on the time scale considered. Recent studies have indicated that both forms of ERC exist in phytoplankton from different environments, thus making reliable modeling of UV-related impacts more complex.

Several authors have developed models to estimate the impact of ozone depletion (Arrigo, 1994; Cullen & Neale, 1994, 1996; Boucher & Prezelin, 1996; Neale et al., 1998a). These models try to identify the most significant processes and unknowns and to evaluate the uncertainties. These modeling efforts also demonstrate the difficulty of using short-term observations to estimate longer term (days to years) ecological responses (Smith et al., 1992; Vincent & Roy, 1993; Bothwell et al., 1994; Cullen & Neale, 1994; Holm-Hanson, 1997; Neale et al., 1998a).

Solar UV radiation impairs growth and reproduction, photosynthetic energy, harvesting enzymes (Vassiliev et al., 1994; Herrmann et al., 1995b, 1996, 1997; Giacometti et al., 1996; Figueroa et al., 1997; Gieskes & Buma, 1997), and other cellular proteins, as well as photosynthetic pigments (Gerber & Häder, 1995a, 1995b; Buma et al., 1996b; Peletier et al., 1996; Häder, 1997a). As in phytoplankton and macroalgae, the uptake of ammonium and nitrate is affected by solar radiation in phytoplankton (Behrenfeld, 1995; Döhler, 1996, 1997; Döhler & Hagmeier, 1997; Döhler et al., 1995).

Under UV stress, phytoplankton produces heat-shock proteins and show changes in their cellular amino acid pools. DNA is one of the major targets in the cell, because it strongly absorbs in the short-wavelength range of solar radiation. Solar UV-B has been found to cause DNA damage, such as thymine dimers, and DNA synthesis is delayed (Scheuerlein et al., 1995; Buma et al., 1995, 1996a, 1997). UV-B effects have also been studied on the ecosystem level with the use of mesocosms (Santas et al., 1996; Wängberg & Selmer, 1997).

Macroalgae and Seagrasses

Most macroalgae and seagrasses are sessile and therefore restricted to their growth site (Lüning, 1990). They show a distinct and fixed pattern of vertical distribution in their habitat. Although some of these plants inhabit the supralittoral zone (coast above high water mark) exposed only to the spray from the surf, others populate the eulittoral zone (intertidal) characterized by the regular tidal changes (Häder, 1997d). Shade-adapted algae are never exposed to the intensive radiation at the surface and are restricted to the sublittoral zone. The range in irradiances can be substantial, from over $1,000 \text{ W m}^{-2}$ (total solar radiation) at the surface to less than 0.01% of the radiation that reaches the understory of a kelp habitat (Markager & Sand-Jensen, 1994).

Most macroalgae have developed mechanisms to regulate their photosynthetic activity to adapt to the changing light regimen and protect themselves from excessive radiation (Häder & Figueroa, 1997; Franklin & Forster, 1997) by the mechanism of photoinhibition, similar to higher plants, to decrease the photosynthetic electron transport at high light. Photoinhibition funnels excessive radiation into thermal dissipation and other relaxation mechanisms. Different algal species differ in their ability to cope with enhanced UV radiation (Häder & Figueroa, 1997). A large number of researchers studied the different adaptation strategies to solar radiation of ecologically important species of green, red, and brown algae from the North Sea, Baltic Sea, Mediterranean, Atlantic, and polar and tropical oceans (Markager & Sand-Jensen, 1994; Wiencke et al., 1994; Figueroa et al., 1996; Beach & Smith, 1996a, 1996b; Kirst & Wiencke, 1996; Häder & Figueroa, 1997; Porst et al., 1997).

Photoinhibition can be determined by oxygen exchange (Häder & Schäfer, 1994) or by PAM (pulse amplitude modulated) fluorescence measurements developed by Schreiber et al. (1986). PAM is based on transient changes of chlorophyll fluorescence. Surface-adapted macroalgae, such as several brown (*Cystoseira*, *Padina*, *Fucus*) and green algae (*Ulva*, *Enteromorpha*), display maximal oxygen production at or close to the surface (Herrmann et al., 1995a; Häder & Figueroa, 1997), whereas algae adapted to lower irradiances usually thrive best deeper in the water column (the green algae *Cladophora*, *Caulerpa*, most red algae) (Häder & Figueroa, 1997).

Photochemical and nonphotochemical quenching can be determined via the parameters derived during PAM measurements (Büchel & Wilhelm, 1993). A more recent development is an underwater PAM instrument designed for *in situ* measurement of the quantum yield of fluorescence by diving, which allows the study of algae at their growth site, under water. Nonphotochemical quenching operates in parallel to the violaxanthin cycle, which is believed to quench excess excitation energy in both algae and higher plants (Demmig-Adams & Adams, 1992; Häder & Figueroa, 1997). Deepwater algae and those adapted to shaded conditions are inhibited within less than 1 h when exposed to direct solar radiation, but even algae growing near the surface, where they are exposed to extreme irradiances, show signs of photoinhibition after extended periods of exposure (Figure 7.5). Large differences were also found in the recovery from excess irradiation between surface and deepwater species. Exclusion studies have shown that a large proportion of photoinhibition is due to PAR, but that solar UV-B and UV-A contribute significantly to this phenomenon (Herrmann et al., 1995b).

Chronic photoinhibition (also called photodamage) occurs when algae are exposed for extensive periods to excessive irradiance. It is characterized by damage of PS II reaction centers and subsequent proteolysis of the D1 protein (Critchley & Russell, 1994). Dynamic photoinhibition is readily reversible and follows a circadian pattern, with the lowest quantum yield around solar noon (Hanelt et al., 1994; Häder & Figueroa, 1997). Very low light compensation points for photosynthesis have been reported in Arctic and Antarctic algae (Gómez et al., 1995; Gómez & Wiencke, 1996; Wiencke, 1996).

In contrast to short-term effects, the long-term behavior under solar UV of the primary productivity of macroalgae still needs to be evaluated. Surface-water specimens in coral reefs undergo a 50% reduction in photosynthetic efficiency during the middle of the day but show a complete recovery by late afternoon.

The photoprotection by the xanthophyll cycle has been investigated mostly in microalgae (Schubert et al., 1994) and less in macroalgae, for example, the green alga *Ulva lactuca* (Grevby, 1996), and the brown algae *Dictyota dichotoma* (Uhrmacher et al., 1995) and *Lobophora variegata* (Franklin et al., 1996). The xanthophyll cycle has not been found in red algae.

The production of screening pigments such as carotenoids or UV-absorbing MAAs (Figure 7.6) represent another mechanism of protection from excessive UV radiation. MAAs have been found in green, red, and brown algae from tropical, temperate, and polar regions. These substances are chemically very stable and accumulate in the sediment of lakes. They have been used as a permanent record for past UV radiation levels (Leavitt et al., 1997). Enhanced levels of carotenoids and UV-absorbing compounds were detected in tissues from tropical algae. The concentrations were higher in the tissues from the canopy compared to tissues from understory locations in turf-forming rhodophytes (Beach & Smith, 1996a, 1996b). Recent results indicate that solar UV-B is a stress factor for macroalgae and seagrasses even at current levels; any further increases in UV-B may reduce biomass production and change species composition in macroalgae ecosystems.

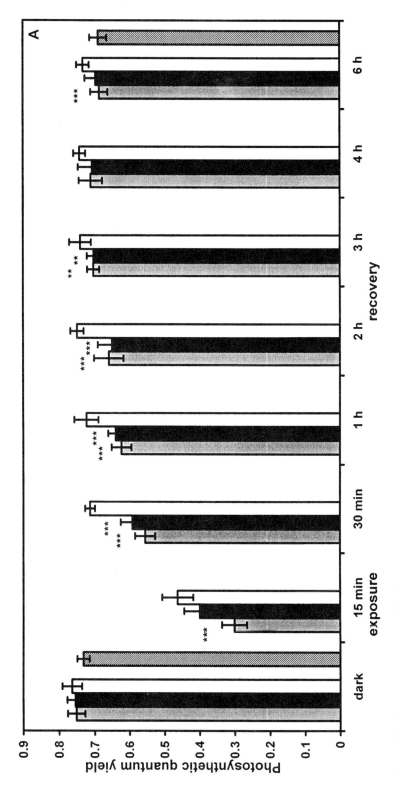

Figure 7.5. Photosynthetic quantum yield of the green alga *Cystoseira abies-marina* measured after 30 min dark adaptation, 15 min exposure, and after increasing recovery times in the shade calculated as $(F'_m - F_t)/F'_m$. The specimens were exposed either to unfiltered solar radiation (gray bars), radiation filtered through a 320 nm cutoff filter foil (black bars), or filtered through a 395 nm cutoff filter foil (white bars). Independent controls were subjected to the same treatment except for solar exposure and measured after the dark treatment and the recovery period, respectively (striped bars). For each data point, at least eight measurements were averaged and the standard deviation calculated. The values for unfiltered solar radiation and under the 320 nm cutoff filter treatments are statistically significantly different from the PAR only value (395 nm cutoff) in each set, with $p < .001$ (***) and $<.01$ (**), respectively, as indicated by the Student's *t* test.

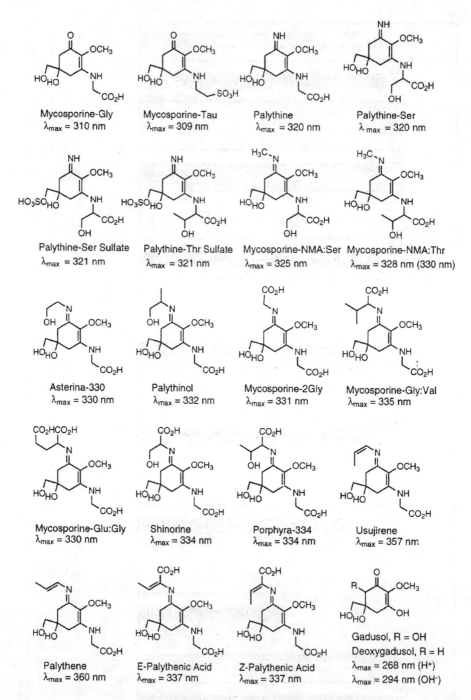

Figure 7.6. Molecular structures of MAAs found in marine organisms with their absorption maxima (after Dunlap & Shick, 1998).

Freshwater Ecosystems

The succession of periphytic and limnic algal communities is governed by complex external conditions, stress factors, and interspecies influences (Rai et al., 1996). Freshwater ecosystems are characterized by a high turnover, and the development of general patterns of community structure follows defined routes (Biggs, 1996). Even though UV-B penetrates considerably less deep into freshwater habitats than in oceanic waters, solar UV-B is a considerable stress factor affecting species composition and biomass productivity (Williamson, 1995, 1996; Häder & Häder, 1997; Piazena & Häder, 1997). The simultaneous exposure to UV-B and heavy-metal concentrations resulted in a synergistic inhibition of nutrient uptake, enzyme activity, carbon fixation, ATP synthesis, and oxygen evolution in several phytoplankton species (Rai et al., 1996; Rai & Rai, 1997).

A large number of freshwater algal species (Chlorophyta and Chromophyta) were screened to determine their UV-B sensitivity (Xiong et al., 1996). The species were chosen from different ecosystems, ranging from high-altitude lakes to thermal springs. In the most sensitive species, oxygen production was reduced by 30–50% during a 2-h UV-B exposure ($2 \, W \, m^{-2}$). UV-B–resistant species were found among those isolated from high mountain locations. They are characterized by solid cell walls encrusted with sporopollenin. In an exclusion experiment in a high-altitude mountain lake (Halac et al., 1997), no significant differences were found between the control (full sunlight) and the UV-B–depleted enclosure.

UV-A radiation has also been found to affect growth and photosynthesis (Kim & Watanabe, 1994). In other species, UV-A had been found to have a beneficial effect, partially counteracting UV-B inhibition (Quesada, 1995). Responses of ecosystems to elevated UV-B exposure cannot be predicted from single-species assessments. In an experiment by Bothwell et al. (1994), after some lag time, greater algal growth was observed in an artificial stream under UV-B exposure than in the control. The explanation of this surprising result was that the grazers, larval chironomids, were more sensitive to UV-B radiation than their food, the algae.

Antarctic

There is a large spatial and temporal variability in the productivity of the Southern Ocean (Sullivan et al., 1993; Arrigo, 1994; Smith et al., 1998). It is difficult to quantify the specific effects of solar UV radiation (Neale et al., 1998a; Karentz & Spero, 1995; Davidson et al., 1996). However, with increasingly dense observations, recent estimates of the effect of 50% ozone reduction on ecosystem productivity are relatively consistent between 0.7 and 8.5% depending on BWF, assumed mixing regimen, and cloudiness (Boucher & Prezelin, 1996; Neale et al., 1998b; Smith et al., 1992).

Recent investigations show convincing evidence that solar UV-B damages natural phytoplankton, but to determine long-term effects, acclimation and adaptation phenomena need to be assessed (Villafañe et al., 1995; Lesser, 1996; Helbling et al., 1996; Neale et al., 1998a). Models have been developed (Arrigo, 1994; Behrenfeld et al., 1994; Boucher & Prezelin, 1996; Neale et al., 1998a) to estimate reductions in ecosystem productivity. Vertical mixing is a major factor complicating the quantification of UV-B effects on phytoplankton. Only recently have the interactive effects of ozone depletion and vertical mixing been modeled (Neale et al., 1998a). Field results clearly demonstrate that photosynthesis of Antarctic phytoplankton is inhibited by ambient UV during incubation in fixed containers (Neale et al., 1998b; Smith et al., 1992; Helbling et al., 1994; Vernet et al., 1994). The problems arise in the generalization of these experimental results to Antarctic waters, where mixing significantly alters the exposure of phytoplankton to UV-B. Neale and coworkers (Neale et al., 1998a) found that near-surface UV strongly inhibits photosynthesis under all modeled conditions, and that inhibition of photosynthesis is enhanced or decreased by vertical mixing, dependent on the depth of the mixed layer. A sudden 50% reduction in stratospheric ozone could lower the water column photosynthesis by as much as 8.5%. This is consistent with the results of Smith and coworkers (1992), who studied the marginal ice zone (MIZ) with minimal vertical mixing. However, the estimated and measured results can be modified with vertical mixing by about ±37%, measured variable sensitivity of phytoplankton to UV by about ±46%, and cloudiness by about ±15%. We can conclude that independent of these natural interactions, UV radiation is a significant environmental stressor, and its effects are enhanced by ozone depletion.

Arctic

The Arctic aquatic ecosystem differs in many respects from its antipode (Weiler & Penhale, 1994; Wängberg et al., 1996). The Arctic Ocean is an almost closed water mass, characterized by little water exchange with the Atlantic and Pacific oceans. It comprises 25% of the global continental shelf. A considerable freshwater inflow causes pronounced stratification during most of the year and is responsible for high concentrations of particulate and dissolved organic carbon (POC and DOC); this influx strongly affects the penetration of solar UV into the water column. Other differences between the Arctic and the Antarctic include the greater importance of macroalgae in the Arctic.

The Arctic aquatic ecosystem is one of the most productive aquatic ecosystems; it is a source of fish and crustaceans for human consumption (Springer & McRoy, 1993). In the Bering Sea, the coastal communities contribute about 40–50% of the total productivity. Due to the shallow water and the prominent stratification of the water layer, phytoplankton are exposed to relatively high levels of solar UV-B.

The high concentrations of humic substances, which strongly absorb UV-B radiation, may alter the underwater light penetration significantly (Wängberg et al., 1996). UV-B has been found to be more detrimental for small phytoplankton organisms (Karentz et al., 1994) and even more so for the bacterioplankton (Herndl, 1997). In contrast, a recent study of size-fractionated phytoplankton in a lake indicated that cells larger than 2 μm were twice as sensitive to solar UV-B than smaller cells (Milot-Roy & Vincent, 1994).

The Arctic Ocean is often nutrient-limited, especially in inorganic nutrients such as nitrogen and phosphorus. Nitrogen and phosphorus uptake are UV-B sensitive (Döhler, 1992), which may augment the UV-B sensitivity of Arctic phytoplankton communities. Low doses of UV-B increase the uptake of phosphate, which is probably used for DNA repair, whereas UV-B decreases the uptake at higher doses. All these factors affect the biogeochemical cycles.

Conclusions and Consequences

Possible consequences of enhanced levels of exposure to UV-B radiation are reduction of biomass productivity, reducing aquatic food sources for humans. Furthermore, changes in species composition are being discussed. UV effects on cyanobacteria reduce the availability of nitrogen compounds. Reduced uptake capacity for atmospheric carbon dioxide may result in the potential augmentation of global warming. Although there is significant evidence that increased UV-B exposure is harmful to aquatic organisms, damage to ecosystems is still uncertain. Ecosystem responses will not be limited to simple decreases in primary production, and shifts in community structure may be more common and result in few detectable differences in ecosystem biomass.

Acknowledgments

Part of the work summarized in this chapter has been made possible by grants from the European Community (Grant Nos. ENV4-CT97-0580 and EV5V-CT94-0425; DG XII, Environmental Programme).

References

Aas, P., Lyons, M., Pledger, R., Mitchell, D. L., & Jeffrey, W. H. (1996). Inhibition of bacterial activities by solar radiation in nearshore waters and the Gulf of Mexico. *Aquatic Microbial Ecology, 11*, 229–238.

Aráoz, R., & Häder, D.-P. (1997). Ultraviolet radiation induces both degradation and synthesis of phycobilisomes in *Nostoc* sp.: A spectroscopic and biochemical approach. *FEMS Microbiology Ecology, 23*, 301–313.

Arrigo, K. R. (1994). Impact of ozone depletion on phytoplankton growth in the Southern Ocean, large-scale spatial and temporal variability. *Marine Ecology Progress Series, 114*, 1–12.

Banerjee, M., & Häder, D.-P. (1996). Effects of UV radiation on the rice field cyanobacterium, *Aulosira fertilissima*. *Environmental and Experimental Botany, 36*, 281–291.

Beach, K. S., & Smith, C. M. (1996a). Ecophysiology of tropical rhodophytes: I. Microscale acclimation in pigmentation. *Journal of Phycology, 32*, 701–710.

Beach, K. S., & Smith, C. M. (1996b). Ecophysiology of tropical rhodophytes: II. Microscale acclimation in photosynthesis. *Journal of Phycology, 32*, 710–718.

Behrenfeld, M. J. (1995). Ultraviolet-B radiation effects on inorganic nitrogen uptake by natural assemblages of oceanic plankton. *Journal of Phycology, 31*, 25–36.

Behrenfeld, M. J., & Falkowski, P. G. (1997a). Photosynthetic rates derived from satellite-based chlorophyll concentration. *Limnology and Oceanography, 42*, 1–20.

Behrenfeld, M. J., & Falkowski, P. G. (1997b). A consumer's guide to phytoplankton primary productivity models. *Limnology and Oceanography, 42*, 1479–1491.

Behrenfeld, M. J., Lee, H., II, & Small, L. F. (1994). Interactions between nutritional status and long-term responses to ultraviolet-B radiation stress in a marine diatom. *Marine Biology, 118*, 523–530.

Biggs, B. J. F. (1996). Patterns in benthic algae of streams. In: R. J. Stevenson (Ed.), *Algal ecology* (pp. 31–56).

Björn, L. O., Cunningham, A., Dubinsky, Z., Estrada, M., Figueroa, F. L., Garcia-Pichel, F., Häder, D.-P., Hanelt, D., Levavasseur, G., & Lüning, K. (1996). Technical discussion: I. Underwater light measurements and light absorption by algae. In: F. L. Figueroa, C. Jiménez, J. L. Pérez-Lloréns & F. X Niell (Eds.), *Underwater light and algal photobiology. Scientia Marina, 60*(Suppl. 1), 59–63.

Booth, C. R., Morrow, J. H., Coohill, T. P., Frederick, J. E., Häder, D.-P., Holm-Hansen, O., Jeffrey, W. H., Mitchell, D. L., Neale, P. J., Sobolev, I., van der Leun, J., & Worrest, R. C. (1997). Impacts of solar UVR on aquatic microorganisms. *Photochemistry and Photobiology, 65*, 252–269.

Bothwell, M. L., Sherbot, D. M. J., & Pollock, C. M. (1994). Ecosystem response to solar ultraviolet-B radiation: Influence of trophic level interactions. *Science, 256*, 97–100.

Boucher, N. P., & Prezelin, B. B. (1996). An *in situ* biological weighting function for UV inhibition of phytoplankton carbon fixation in the Southern Ocean. *Marine Ecology Progress Series, 114*, 223–236.

Brown, C. W., Esaias, W. E., & Thompson, A. M. (1995). Predicting phytoplankton composition from space—using the ratio of euphotic depth to mixed-layer depth: An evaluation. *Remote Sensing Environment, 53*, 172–176.

Büchel, C., & Wilhelm, C. (1993). *In vivo* analysis of slow chlorophyll fluorescence induction kinetics in algae: Progress, problems and perspectives. *Photochemistry and Photobiology, 58,* 137–148.

Buma, A. G. J., Engelen, A. H., & Gieskes, W. W. C. (1997). Wavelength-dependent induction of thymine dimers and growth rate reduction in the marine diatom *Cyclotella* sp. exposed to ultraviolet radiation. *Marine Ecology Progress Series, 153,* 91–97.

Buma, A. G. J., van Hannen, E. J., Roza, L., Veldhuis, M. J. W., & Gieskes, W. W. C. (1995). Monitoring ultraviolet-B-induced DNA damage in individual diatom cells by immunofluorescent thymine dimer detection. *Journal of Phycology, 51,* 314–321.

Buma, A. G. J., van Hannen, E. J., Veldhuis, M. J. W., & Gieskes, W. W. C. (1996a). UV-B induces DNA damage and DNA synthesis delay in the marine diatom *Cyclotella* sp. *Scientia Marina, 60*(Suppl. 1), 101–106.

Buma, A. G. J., Zemmelink, H. J., Sjollema, K., & Gieskes, W. W. C. (1996b). UVB radiation modifies protein and photosynthetic pigment content, volume and ultrastructure of marine diatoms. *Marine Ecology Progress Series, 147,* 47–54.

Coohill, T. P., Häder, D.-P., & Mitchell, D. L. (1996). Environmental ultraviolet photobiology: Introduction. *Photochemistry and Photobiology, 64,* 401–402.

Corn, P. S. (1998). Effects of ultraviolet radiation on boreal toads in Colorado. *Ecological Applications, 8,* 18–26.

Critchley, C., & Russell, A. W. (1994). Photoinhibition of photosynthesis *in vivo*: The role of protein turnover in photosystem II. *Physiologia Plantarum, 92,* 188–196.

Cullen, J. J., & Neale, P. J. (1994). Ultraviolet radiation, ozone depletion, and marine photosynthesis. *Photosynthesis Research, 39,* 303–320.

Cullen, J. J., & Neale, P. J. (1997). Effects of ultraviolet radiation on short-term photosynthesis of natural phytoplankton. *Photochemistry and Photobiology, 65,* 264–266.

Cullen, J. J., & Neale, P. J. (1997a). Biological weighting functions for describing the effects of ultraviolet radiation on aquatic systems. In: D.-P. Häder (Ed.), *The effects of ozone depletion on aquatic ecosystems* (pp. 97–118). Austin, TX: Environmental Intelligence Unit, Academic Press and R. G. Landes.

Cullen, J. J., & Neale, P. J. (1997b). Effects of ultraviolet radiation on short-term photosynthesis of natural phytoplankton. *Photochemistry and Photobiology, 65,* 264–266.

Davidson, A. T., Marchant, H. J., & de la Mare, W. K. (1996). Natural UV exposure changes the species composition of Antarctic phytoplankton in mixed culture. *Aquatic Microbial Ecology, 10,* 299–305.

Demmig-Adams, B., & Adams, W., III. (1992). Photoprotection and other responses of plants to high light stress. *Annual Reviews of Plant Physiology and Plant Molecular Biology, 43,* 599–626.

Döhler, G. (1992). Impact of UV-B radiation (290–320 nm) on uptake of ^{15}N-ammonia and ^{15}N-nitrate by phytoplankton of the Wadden Sea. *Marine Biology, 112,* 485–489.

Döhler, G. (1996). Effect of UV irradiance on utilization of inorganic nitrogen by the Antarctic diatom *Odontella weissflogii* (Janisch) Grunow. *Botanica Acta, 109,* 35–42.

Döhler, G. (1997). Impact of UV radiation of different wavebands on pigments and assimilation of ^{15}N-ammonium and ^{15}N-nitrate by natural phytoplankton and ice algae in Antarctica. *Journal of Plant Physiology, 151,* 550–555.

Döhler, G., & Hagmeier, E. (1997). UV-Effects on pigments and assimilation of ^{15}N-ammonium and ^{15}N-nitrate by natural marine phytoplankton of the North Sea. *Botanica Acta, 110,* 481–488.

Döhler, G., Hagmeier, E., & David, C. (1995). Effects of solar and artificial UV irradiation on pigments and assimilation of ^{15}N ammonium and ^{15}N nitrate by macroalgae. *Journal of Photochemistry and Photobiology B: Biology, 30,* 179–187.

Donkor, V. A., & Häder, D.-P. (1995). Protective strategies of several cyanobacteria against solar radiation. *Journal of Plant Physiology, 145,* 750–755.

Donkor, V. A., & Häder, D.-P. (1996). Effects of ultraviolet irradiation on photosynthetic pigments in some filamentous cyanobacteria. *Aquatic Microbial Ecology, 11,* 143–149.

Donkor, V. A., & Häder, D.-P. (1997). Ultraviolet radiation effects on pigmentation in the cyanobacterium *Phormidium uncinatum*. *Acta Protozoologica, 36,* 49–55.

Ducklow, H. W., Carlson, C. A., Bates, N. R., Knap, A. H., & Michaels, A. F. (1995). Dissolved organic carbon as a component of the biological pump in the North Atlantic Ocean. *Philosophical Transactions of the Royal Society of London B: Biological Sciences, 348,* 161–167.

Dunlap, W. C., & Shick, J. M. (1998). Ultraviolet radiation-absorbing mycosporine-like amino acids in coral reef organisms: A biochemical and environmental perspective. *Journal of Phycology, 34,* 418–430.

Figueroa, F. L., Jiménez, C., Lubián, L. M., Montero, O., Lebert, M., & Häder, D.-P. (1997). Effects of high irradiance and temperature on photosynthesis and photoinhibition in *Nannochloropsis gaditana* Lubián (Eustigmatophyceae). *Journal of Plant Physiology, 151,* 6–15.

Figueroa, F. L., Jiménez, C., Pérez-Lloréns, J. L., & Niell, F. X. (1996). Underwater light and algal photobiology. *Scientia Marina, 60*(Suppl. 1), 343 pp.

Franklin, L. A., & Forster, R. M. (1997). The changing irradiance environment: Consequences for marine macrophyte physiology, productivity and ecology. *European Journal of Phycology, 32*, 207–232.

Franklin, L. A., Seaton, G. G. R., Lovelock, C. E., & Larkum, A. W. D. (1996). Photoinhibition of photosynthesis on a tropical reef. *Plant, Cell and Environment, 19*, 825–836.

Garcia-Pichel, F. (1994). A model for the internal self-shading in planktonic organisms and its implications for the usefulness of ultraviolet sunscreens. *Limnology and Oceanography, 39*, 1704–1717.

Gerber, S., Biggs, A., & Häder, D.-P. (1996). A polychromatic action spectrum for the inhibition of motility in the flagellate *Euglena gracilis*. *Acta Protozoologica, 35*, 161–165.

Gerber, S., & Häder, D.-P. (1995a). Effects of enhanced solar irradiance on chlorophyll fluorescence and photosynthetic oxygen production of five species of phytoplankton. *FEMS Microbiology Ecology, 16*, 33–42.

Gerber, S., & Häder, D.-P. (1995b). Effects of artificial and simulated solar radiation on the flagellate *Euglena gracilis*: Physiological, spectroscopical and biochemical investigations. *Acta Protozoologica, 34*, 13–20.

Giacometti, G. M., Barbato, R., Chiaramonte, S., Friso, G., & Rigoni, F. (1996). Effects of ultraviolet-B radiation on photosystem II of the cyanobacterium *Synechocystis* sp. PCC 6083. *European Journal of Biochemistry, 242*, 799–806.

Gieskes, W. W. C., & Buma, A. G. J. (1997). UV damage to plant life in a photobiologically dynamic environment: The case of marine phytoplankton. *Plant Ecology, 128*, 16–25.

Gómez, I., & Wiencke, C. (1996). Photosynthesis, dark respiration and pigment contents of gametophytes and sporophytes of the Antarctic brown alga *Desmarestia menziessi*. *Botanica Marina, 39*, 149–157.

Gómez, I., Wiencke, C., & Weykman, G. (1995). Seasonal photosynthetic characteristics of the brown alga *Ascoseira mirabilis* from King George Island (Antarctica). *Marine Biology, 123*, 167–172.

Grevby, C. (1996). *Organisation of the light harvesting complex in fucoxanthin containing algae*. Doctoral thesis, Göteborg University, Sweden.

Häder, D.-P. (1995). Photo-ecology and environmental photobiology. In: W. M. Horspool & P.-S. Song (Eds.), *CRC handbook of organic photochemistry and photobiology* (pp. 1392–1401). Boca Raton, FL: CRC Press.

Häder, D.-P. (1997a). Effects of UV radiation on phytoplankton. In: J. G. Jones (Ed.), *Advances in microbial ecology* (Vol. 15, pp. 1–26). New York: Plenum Press.

Häder, D.-P. (1997b). Effects of solar UV-B radiation on aquatic ecosystems. In: P. J. Lumsden (Ed.), *Plants and UV-B: Responses to environmental change* (pp. 171–193). Cambridge, UK: Cambridge University Press.

Häder, D.-P. (1997c). UV-B and aquatic ecosystems. In: J. Rozema, W. W. C. Gieskes, S. C. van de Geijn, C. Nolan & H. de Boois (Eds.), *UV-B and Biosphere* (pp. 4–13). Dordrecht: Kluwer Academic.

Häder, D.-P. (1997d). Penetration and effects of solar UV-B on phytoplankton and macroalgae. *Plant Ecology, 128*, 4–13.

Häder, D.-P. (1997e). Stratospheric ozone depletion and increase in ultraviolet radiation. In: D.-P. Häder (Ed.), *The effects of ozone depletion on aquatic ecosystems* (pp. 1–4). Austin, TX: Environmental Intelligence Unit, Academic Press and R. G. Landes.

Häder, D.-P., & Figueroa, F. L. (1997). Photoecophysiology of marine macroalgae. *Photochemistry and Photobiology, 66*, 1–14.

Häder, D.-P., & Schäfer, J. (1994). *In-situ* measurement of photosynthetic oxygen production in the water column. *Environmental Monitoring and Assessment, 32*, 259–268.

Häder, D.-P., & Worrest, R. C. (1997). Consequences of the effects of increased solar ultraviolet radiation on aquatic ecosystems. In: D.-P. Häder (Ed.), *The effects of ozone depletion on aquatic ecosystems* (pp. 11–30). Austin, TX: Environmental Intelligence Unit, Academic Press and R. G. Landes.

Häder, D.-P., Worrest, R. C., Kumar, H. D., & Smith, R. C. (1994). *Effects of increased solar ultraviolet radiation on aquatic ecosystems*. UNEP Environmental Effects Panel Report, pp. 65–77.

Häder, D.-P., Worrest, R. C., Kumar, H. D., & Smith, R. C. (1995). Effects of increased solar ultraviolet radiation on aquatic ecosystems. *Ambio, 24*, 174–180.

Häder, M., & Häder, D.-P. (1997). Optical properties and phytoplankton composition in a freshwater ecosystem (Main-Donau-Canal). In: D.-P. Häder (Ed.), *The effects of ozone depletion on aquatic ecosystems* (pp. 155–174). Austin, TX: Environmental Intelligence Unit, Academic Press and R. G. Landes.

Halac, S., Felip, M., Camarero, L., Sommaruga-Wögrath, S., Psenner, R., Catalan, J., & Sommaragu, R. (1997). An *in situ* enclosure experiment to test the solar UVB impact on plankton in a high-altitude mountain lake: I. Lack of effect on phytoplankton species composition and growth. *Journal of Plankton Research, 19*, 1671–1686.

Hanelt, D., Jaramillo, M. J., Nultsch, W., Senger, S., & Westermeir, R. (1994). Photoinhibition as a regulatory mechanism of photosynthesis in marine algae of Antarctica. *Serie Cient. Inst. Antarct., Chil., 44*, 67–77.

Helbling, E. B., Chalker, B. E., Dunlap, W. C., Holm-Hansen, O., & Villafane, V. E. (1996). Photoacclimation of antarctic marine diatoms to solar ultraviolet radiation. *Journal of Experimental Marine Biology and Ecology, 204*, 85–101.

Helbling, E. W., Villafane, V., & Holm-Hansen, O. (1994). Effects of ultraviolet radiation on Antarctic marine phytoplankton photosynthesis with particular attention to the influence of mixing. In: C. S. Weiler & P. A. Penhale (Eds.), *Ultraviolet radiation in Antarctica: Measurements and biological effects.* (pp. 207–227). Antarctic Research Series No. 62, Washington, DC: American Geophysical Union.

Herndl, G. J. (1997). Role of ultraviolet radiation on bacterioplankton activity. In: D.-P. Häder (Ed.), *The effects of ozone depletion on aquatic ecosystems* (pp. 143–154). Austin, TX: Environmental Intelligence Unit, Academic Press and R. G. Landes.

Herrmann, H., Ghetti, F., Scheuerlein, R., & Häder, D.-P. (1995a). Photosynthetic oxygen and fluorescence measurements in *Ulva laetevirens* affected by solar irradiation. *Journal of Plant Physiology, 145*, 221–227.

Herrmann, H., Häder, D.-P., & Ghetti, F. (1997). Inhibition of photosynthesis by solar radiation in *Dunaliella salina*: Relative efficiencies of UV-B, UV-A and PAR. *Plant, Cell and Environment, 20*, 359–365.

Herrmann, H., Häder, D.-P., Köfferlein, M., Seidlitz, H. K. & Ghetti, F. (1995b). Study on the effects of UV radiation on phytoplankton photosynthetic efficiency by means of a sunlight simulator. *Med. Biol. Environ., 23*, 36–40.

Herrmann, H., Häder, D.-P., Köfferlein, M., Seidlitz, H. K., & Ghetti, F. (1996). Effects of UV radiation on photosynthesis of phytoplankton exposed to solar simulator light. *Journal of Photochemistry and Photobiology B: Biology, 34*, 21–28.

Holm-Hansen, O. (1997). Short- and long-term effects of UVA and UVB on marine phytoplankton productivity. *Photochemistry and Photobiology, 65*, 266–268.

IASC (1995). Effects of increased ultraviolet radiation in the Arctic. IASC Report No. 2, IASC Secretariat. *Journal of Plant Physiology, 148*, 42–48.

Jeffrey, W. H., Aas, P., Maille Lyons, M., Coffin, R. B., Pledger, R. J., & Mitchell, D. L. (1996a). Ambient solar radiation-induced photodamage in marine bacterioplankton. *Photochemistry and Photobiology, 64*, 419–427.

Jeffrey, W. H., Pledger, R. J., Aas, P., Hager, S., Coffin, R. B., Haven, R. V., & Mitchell, D. L. (1996b). Diel and depth profiles of DNA photodamage in bacterioplankton exposed to ambient solar ultraviolet radiation. *Marine Ecology Progress Series, 137*, 283–291.

Jiménez, C., Figueroa, F. L., Aguilera, J., Lebert, M., & Häder, D.-P. (1996). Phototaxis and gravitaxis in *Dunaliella bardawil*: Influence of UV radiation. *Acta Protozoologica, 35*, 287–295.

Karentz, D., Bothwell, M. L., Coffin, R. B., Hanson, A., Herndl, G. J., Kilham, S. S., Lesser, M. P., Lindell, M., Moeller, R. E., Morris, D. P., Neale, P. J., Sanders, R. W., Weiler, C. S., & Wetzel, R. G. (1994). Report of working group on bacteria and phytoplankton. In: C. E. Williamson & H. E. Zagarese (Eds.), *Impact of UV-B radiation on pelagic freshwater ecosystems Archiv für Hydrobiologie Beiheft, 43*[Special issue], 31–69.

Karentz, D., & Spero, H. J. (1995). Response of a natural *Phaeocystis* population to ambient fluctuations of UVB radiation caused by Antarctic ozone depletion. *Journal of Plankton Research, 17*, 1771–1789.

Karsten, U., & Garcia-Pichel, F. (1996). Carotenoids and mycosporine-like amino acid compounds in members of the Genus *Microcoleus* (Cyanobacteria): A chemosystematic study. *Systematic and Applied Microbiology, 19*, 285–294.

Kashyap, A. K., Pandey, K. D., & Gupta, R. K. (1991). Nitrogenase activity of the Antarctic cyanobacterium *Nostoc commune*: Influence of temperature. *Folia Microbiologica, 36*, 557–560.

Kerr, J. B., & McElroy, C. T. (1993). Evidence for large upward trends of ultraviolet-B radiation linked to ozone depletion. *Science, 262*, 1032–1034.

Kim, D. S., & Watanabe, Y. (1994). Inhibition of growth and photosynthesis of freshwater phytoplankton by ultraviolet A (UVA) radiation and subsequent recovery from stress. *Journal of Plankton Research, 16*, 1645–1654.

Kirchoff, V. W. J. H., Zamorano, F., & Casiccia, C. (1997). UV-B enhancements at Punta Arenas, Chile. *Journal of Photochemistry and Photobiology B: Biology, 38*, 174–177.

Kirst, G. O., & Wiencke, C. (1996). Ecophysiology of algae. *Journal of Phycology, 31*, 181–199.

Kjeldstad, B., Johnsen, B., & Koskela, T. (1997). The Nordic intercomparison of ultraviolet and total ozone instruments at Iana, October 1996, final report. Finnish Meteorological Institute Report No. 36, Helsinki.

Kumar, A., Sinha, R. P., & Häder, D.-P. (1996b). Effect of UV-B on enzymes of nitrogen metabolism in the cyanobacteria *Nostoc calcicola*. *Journal of Plant Physiology, 148*, 86–91.

Kumar, A., Tyagi, M. B., Srinivas, G., Singh, N., Kumar, H. D., Sinha, R. P., & Häder, D.-P. (1996b). UVB shielding role of $FeCl_3$ and certain cyanobacterial pigments. *Photochemistry and Photobiology, 64*, 321–325.

Leavitt, P. R., Vinebrooke, R. D., Donald, D. B., Smol, J. P., & Schindler, D. W. (1997). Past ultraviolet radiation environments in lakes derived from fossil pigments. *Nature, 388*, 457–459.

Lesser, M. (1996). Acclimation of phytoplankton to UV-B radiation: Oxidative stress and photoinhibition of photosynthesis are not prevented by UV-absorbing compounds in the dinoflagellate *Prorocentrum micans*. *Marine Ecology Progress Series, 132*, 287–297 (correction: *Marine Ecology Progress Series, 141*, 312).

Leszczynski, K., Jokela, K., Ylianttila, L., Visuri, R., & Blumthaler, M. (1998). Erythemally weighted radiometers in solar UV monitoring: Results from WMO/STUK intercomparison. *Photochemistry and Photobiology, 67*, 212–221.

Lüning, K. (1990). *Seaweeds: Their environment, biogeography and ecophysiology.* New York: Wiley.

Madronich, S., & Flocke, S. (1997). Theoretical estimation of biologically effective UV radiation at the Earth's surface. In: C. Zerefos (Ed.), *Solar ultraviolet radiation—modeling, measurements and effects* (NATO ASI Series, Vol. I52, Berlin: Springer-Verlag.

Madronich, S., McKenzie, R. L., Björn, L. O., & Caldwell, M. M. (1995). Changes in ultraviolet radiation reaching the Earth's surface. *Ambio, 24*, 143–152.

Madronich, S., Velders, G., Daniel, J., Lal, M., McCulloch, A., & Slaper, H. (1998). Halocarbon scenarios for the future ozone layer and related consequences. In: D. Albritton, P. Aucamp, G. Megie, & R. Watson (Eds.), *Scientific assessment of ozone depletion: 1998* Geneva: World Meteorological Organization.

Markager, S., & Sand-Jensen, K. (1994). The physiology and ecology of light-grown relationship in macroalgae. In: F. E. Round & D. J. Chapman (Eds.), *Progress in phycological research* (Vol. 10, pp. 209–298). Bristol, UK: Biopress.

McKinlay, A. F., & Diffey, B. L. (1987). A reference action spectrum for ultraviolet induced erythema in human skin. In: W. R. Passchier & B. F. M. Bosnjakovic (Eds.), *Human exposure to ultraviolet radiation: Risks and regulations.* Amsterdam: Elsevier.

Milot-Roy, V., & Vincent, W. F. (1994). UV radiation effects on photosynthesis: The importance of near-surface thermoclines in a subarctic lake. In: C. E. Williamson & H. E. Zagarese (Eds.), *Advances in limnology: Impact of UV-B radiation on pelagic ecosystems. Archiv für Hydrobiologie Beiheft, 43*, 171–184.

Mims, F. M., III. (1997). Biological effects of diminished UV and visible sunlight caused by severe air pollution. In: W. L. Smith & K. Stamnes (Eds.), *IRS '96: Current problems in atmospheric radiation. Proceedings of the International Radiation Symposium, Fairbanks, Alaska 19–24 August 1996*, pp. 905–908. Hampton, VA: Deepak.

Montecino, V., & Pizarro, G. (1995). Phytoplankton acclimation and spectral penetration of UV irradiance off the central Chilean coast. *Marine Ecology Progress Series, 121*, 261–269.

Morrow, J. H., & Booth, C. R. (1997). Instrumentation and methodology for ultraviolet radiation measurements in aquatic environments. In: D.-P. Häder (Ed.), *The effects of ozone depletion on aquatic ecosystems* (pp. 31–44). Austin, TX: Environmental Intelligence Unit, Academic Press and R. G. Landes.

Müller, R., Crutzen, P. J., Gross, J.-U., Brühl, C., Russel, J. M., III, Gernandt, H., McKenna, D. S., & Tuck, A. F. (1997). Severe chemical ozone loss in the Arctic during the winter of 1995–96. *Nature, 389*, 709–712.

Müller-Niklas, G., Heissenberger, A., Puskaric, S., & Herndl, G.J. (1995). Ultraviolet-B radiation and bacterial metabolism in coastal waters. *Aquatic Microbial Ecology, 9*, 111–116.

Naganuma, T., Inoue, T., & Uye, S. (1997). Photoreactivation of UV-induced damage to embryos of a planktonic copepod. *Journal of Plankton Research, 19*, 783–787.

Naganuma, T., Konishi, S., Inoue, T., Nakane, T., & Sukizaki, S. (1996). Photodegradation or photoalteration? Microbial assay of the effect of UV-B on dissolved organic matter. *Marine Ecology Progress Series, 135*, 309–310.

Neale, P. J., Cullen, J. J., & Davis, R. F. (1998a). Inhibition of marine photosynthesis by ultraviolet radiation: Variable sensitivity of phytoplankton in the Weddell-Scotia Sea during austral spring. *Limnology and Oceanography, 43*, 433–488.

Neale, P. J., Davis, R. F., & Cullen, J. J. (1998b). Interactive effects of ozone depletion and vertical mixing on photosynthesis of Antarctic phytoplankton. *Nature, 392*, 585–589.

Nicholson, W. L. (1995). Photoreactivation in the genus *Bacillus*. *Current Microbiology, 31*, 361–365.

Nolan, C. V., & Amanatidis, G. T. (1995). European Commission research on the fluxes and effects of environmental UVB radiation. *Journal of Photochemistry and Photobiology B: Biology, 31*, 3–7.

Peletier, H., Gieskes, W. W. C., & Buma, A. G. J. (1996). Ultraviolet-B radiation resistance of benthic diatoms isolated from tidal flats in the Dutch Wadden Sea. *Marine Ecology Progress Series, 135*, 163–168.

Piazena, H., & Häder, D.-P. (1997). Penetration of solar UV and PAR into different waters of the Baltic Sea and remote sensing of phytoplankton. In: D.-P. Häder (Ed.), *The effects of ozone depletion on aquatic ecosystems*, (pp. 45–96). Austin, TX: Environmental Intelligence Unit, Academic Press and R. G. Landes.

Pomeroy, L. R., & Wiebe, W. J. (1988). Energetics of microbial food webs. *Hydrobiologia, 159*, 7–18.
Porst, M., Herrmann, H., Schäfer, J., Santas, R., & Häder, D.-P. (1997). Photoinhibition in the Mediterranean green alga *Acetabularia mediterranea* measured in the field under solar irradiation. *Journal of Plant Physiology, 151*, 25–32.
Quesada, A. (1995). Growth of Antarctica cyanobacteria under ultraviolet radiation: UVA counteracts UVB inhibition. *Journal of Phycology, 31*, 242–248.
Rai, L. C., Tyagi, B., & Mallick, N. (1996). Alternation in photosynthetic characteristics of *Anabaena dolium* following exposure to UV-B and Pb. *Photochemistry and Photobiology, 64*, 658–663.
Rai, P. K., & Rai, L. C. (1997). Interactive effects of UV-B and Cu on photosynthesis, uptake and metabolism of nutrients in green alga *Chlorella vulgaris* and simulated ozone column. *Journal of General and Applied Microbiology, 43*, 281–288.
Rex, M. (1997). Prolonged stratospheric ozone loss in the 1995–96 Arctic winter. *Nature, 389*, 835–838.
Santas, R., Häder, D.-P., & Lianou, C. (1996). Effects of solar UV radiation on diatom assemblages of the Mediterranean. *Photochemistry and Photobiology, 64*, 435–439.
Sarmiento, J. L., & Le Quéré, C. (1996). Oceanic carbon dioxide uptake in a model of century-scale global warming. *Science, 274*, 1346–1350.
Scheuerlein, R., Treml, S., Thar, B., Tirlapur, U. K., & Häder, D.-P. (1995). Evidence for UV-B-induced DNA degradation in *Euglena gracilis* mediated by activation of metal-dependent nucleases. *Journal of Photochemistry and Photobiology B: Biology, 31*, 113–123.
Schindell, R. T., Rind, D., & Lonergan, P. (1998). Increased polar stratospheric ozone losses and delayed eventual recovery owing to increasing greenhouse-gas concentrations. *Nature, 392*, 589–592.
Schreiber, U., Schliwa, U., & Bilger, W. (1986). Continuous recording of photochemical and non-photochemical chlorophyll fluorescence quenching with a new type of modulation fluorometer. *Photosynthesis Research, 10*, 51–62.
Schubert, H., Kroon, B. M. A., & Matthijs, H. C. P. (1994). In vivo manipulation of the xanthophyll cycle and the role of the zeaxanthin cycle in the photoprotection against photodamage in the green alga *Chlorella pyrenoidosa*. *Journal of Biological Chemistry, 269*, 7267–7272.
Seckmeyer, G., Mayer, B., Bernhard, G., McKenzie, R. L., Johnston, P. V., Kotkamp, M., Booth, C. R., Lucas, T., & Mestechikina, T. (1995). Geographical differences in the UV measured by intercompared spectroradiometers. *Geophysical Research Letters, 22*, 1889–1892.
Seckmeyer, G., Mayer, B., Bernhard, G., Erb, R., Albold, A., Jäger, H., & Stockwell, W. R. (1997). New maximum UV irradiance levels observed in central Europe. *Atmospheric Environment, 31*, 2971–2976.
Seckmeyer, G., & McKenzie, R. L. (1992). Elevated ultraviolet radiation in New Zealand (45°S) contrasted with Germany (48°N). *Nature, 359*, 135–137.
Seckmeyer, G., Thiel, S., Blumthaler, M., Fabian, P., Gerber, S., Gugg-Helminger, A., Häder, D.-P., Huber, M., Kettner, C., Köhler, U., Köpke, P., Maier, H., Schäfer, J., Suppan, P., Tamm, E., & Thomalla, E. (1994). Intercomparison of spectral-UV-radiation measurement systems. *Applied Optics, 33*, 7805–7812.
Shick, J. M., Lesser, M. P., & Jokiel, P. L. (1996). Effects of ultraviolet radiation on corals and other coral reef organisms. *Global Change Biology, 2*, 527–545.
Sinha, R. P., & Häder, D.-P. (1996). Photobiology and ecophysiology of rice field cyanobacteria. *Photochemistry and Photobiology, 64*, 887–896.
Sinha, R. P., & Häder, D.-P. (1997). Impacts of UV-B irradiation on rice-field cyanobacteria. In: D.-P. Häder (Ed.), *The effects of ozone depletion on aquatic ecosystems* (pp. 189–198). Austin, TX: Environmental Intelligence Unit, Academic Press and R. G. Landes.
Sinha, R. P., Kumar, H. D., Kumar, A., & Häder, D.-P. (1995). Effects of UV-B irradiation on growth, survival, pigmentation and nitrogen metabolism enzymes in cyanobacteria. *Acta Protozoologica, 34*, 187–192.
Sinha, R. P., Singh, N., Kumar, A., Kumar, H. D., Häder, M., & Häder, D.-P. (1996). Effects of UV irradiation on certain physiological and biochemical processes in cyanobacteria. *Journal of Photochemistry and Photobiology, B: Biology, 32*, 107–113.
Smith, R. C., Baker, K. S., Byers, M. L., & Stammerjohn, S. E. (1998). Primary productivity of the Palmer long-term ecological research area and the Southern Ocean. *Journal of Marine Systems, 529*,
Smith & Cullen, 1996.
Smith, R. C., Prézelin, B. B., Baker, K. S., Bidigare, R. R., Boucher, N. P., Coley, T., Karentz, D., MacIntyre, S., Matlick, H. A., Menzies, D., Ondrusek, M., Wan, Z., & Waters, K. J. (1992). Ozone depletion: Ultraviolet radiation and phytoplankton biology in Antarctic waters. *Science, 255*, 952–959.
Sommaruga, R., Krössbacher, M., Salvenmoser, W., Catalan, J., & Psenner, R. (1995). Presence of large virus-like particles in a eutrophic reservoir. *Aquatic Microbial Ecology, 9*, 305–308.

Sommaruga, R., Oberleiter, A., & Psenner, R. (1996). Effect of UV radiation on the bacterivory of a heterotrophic nanoflagellate. *Applied and Environmental Microbiology, 62*, 4395–4400.

Springer, A. M., & McRoy, C. P. (1993). The paradox of pelagic food webs in the northern Bering Sea: III. Patterns of primary production. *Continental Shelf Research, 13*, 575–599.

Stolarski, R. (1997). A bad winter for Arctic ozone. *Nature, 389*, 788–789.

Sullivan, C. W., Arrigo, K. R., McClain, C. R., Comiso, J. C., & Firestone, J. (1993). Distributions of phytoplankton blooms in the Southern Ocean. *Science, 262*, 1832–1837.

Taalas, P., Damski, J., Kyro, E., Ginzberg, M., & Talamoni, G. (1997). Effect of stratospheric ozone variations on UV radiation and on tropospheric ozone at high latitudes. *Journal of Geophysical Research, 102*, 1533–1539.

Takahashi, T., Feely, R. A., Weiss, R. F., Wanninkhof, R. H., Chipman, D. W., Sutherland, S. C., & Takahashi, T. (1997). Global air-sea flux of CO_2: An estimate based on measurements of sea-air pCO_2 difference. *Proceedings of the National Academy of Sciences* (*USA*), *94*, 8282–8299.

Takahashi, T., Takahashi, T. T., & Sutherland S. (1995). An assessment of the role of the North Atlantic as a CO_2 sink. *Philosophical Transactions of the Royal Society of London B: Biological Sciences, 348*, 143–152.

Thompson, A., Early, E. A., DeLuisi, J., Disterhoff, P., Wardle, D., Kerr, J., Rives, J., Sun, Y., & Lucas, T. (1997). The 1994 North American interagency intercomparison of ultraviolet monitoring spectroradiometers. *Journal of the National Institute of Standards Technology, 102*, 279–322.

Thomson, D. J. (1997). Dependence of global temperatures on atmospheric CO_2 and solar irradiance. *Proceedings of the National Academy of Sciences* (*USA*), *94*, 8370–8377.

Uhrmacher, S., Hanelt, D., & Nultsch, W. (1995). Zeaxanthin content and the degree of photoinhibition are linearly correlated in the brown alga *Dictyota dichotoma*. *Marine Biology, 123*, 159–165.

Varotsos, C., & Kondratyev, K. Y. (1995). On the relationship between total ozone and solar ultraviolet radiation at St. Petersburg, Russia. *Geophysical Research Letters, 22*, 3481–3484.

Vassiliev, I. R., Prasil, O., Wyman, K. D., Kolber, Z. K., Hanson, A. K., Prentice, J. E., & Falkowski, P. G. (1994). Inhibition of PS II photochemistry by PAR and UV radiation in natural phytoplankton communities. *Photosynthesis Research, 42*, 51–64.

Vernet, M., Brody, E. A., Holm-Hansen, O., & Mitchell, B. G. (1994). The response of Antarctic phytoplankton to ultraviolet radiation: Absorption, photosynthesis, and taxonomic composition. In: C. S. Weiler & P. A. Penhale (Eds.), *Ultraviolet radiation in Antarctica: Measurements and biological effects* (pp. 207–227). Antarctic Research Series No. 62. Washington, DC: American Geophysical Union.

Vernet, M., & Smith, R. C. (1997). Effects of ultraviolet radiation on the pelagic Antarctic ecosystem. In: D.-P. Häder (Ed.), *The effects of ozone depletion on aquatic ecosystems* (pp. 247–265). Austin, TX: Environmental Intelligence Unit, Academic Press and R. G. Landes.

Villafañe, V. E., Helbling, W. W., Holm-Hansen, O., & Chalker, B. E. (1995). Acclimatization of Antarctic natural phytoplankton assemblages when exposed to solar ultraviolet radiation. *Journal of Plankton Research, 17*, 2295–2306.

Vincent, W. F., & Roy, S. (1993). Solar ultraviolet radiation and aquatic primary production: Damage, protection and recovery. *Environmental Reviews, 1*, 1–12.

Wängberg, S.-Å., & Selmer, J.-S. (1997). Studies of effects of UV-B radiation on aquatic model ecosystems. In: D.-P. Häder (Ed.), *The effects of ozone depletion on aquatic ecosystems* (pp. 199–214). Austin, TX: Environmental Intelligence Unit, Academic Press and R. G. Landes.

Wängberg, S.-Å., Selmer, J.-S., Ekelund, N. G. A., & Gustavson, K. (1996). *UV-B effects on Nordic marine ecosystems*. Denmark: TemaNord, Nordic Council of Ministers.

Webb, A. R. (1997). *Advances in solar ultraviolet spectroradiometry*. European Commission Air Pollution Research Report No. 63, Luxembourg.

Weiler, C. S., & Penhale, P. A. (1994). *Ultraviolet radiation in Antarctica: Measurements and biological effects*. Washington, DC: American Geophysical Union.

Wetzel, R. G., Hatcher, P. G., & Bianchi, T. S. (1995). Natural photolysis by ultraviolet irradiance of recalcitrant dissolved organic matter to simple substrates for rapid bacterial metabolism. *Limnology and Oceanography, 40*, 1369–1380.

Wiencke, C. (1996). Recent advances in the investigation of Antarctic macroalgae. *Polar Biology, 16*, 231–240.

Wiencke, C., Bartsch, I., Bischoff, B., Peters, A. F., & Breeman, A. M. (1994). Temperature requirements and biogeography of Antarctic, Arctic and amphiequatorial seaweeds. *Botanica Marina, 37*, 247–259.

Williamson, C. E. (1995). What role does UV-B radiation play in freshwater ecosystems? *Limnology and Oceanography, 40*, 386–392.

Williamson, C. E. (1996). Effects of UV radiation on freshwater ecosystems. *International Journal of Environmental Studies, 51*, 245–256.

World Meteorological Organization. (1994a). Scientific assessment of ozone depletion, 1994. In: D. L. Albritton, R. T. Watson, & P. J. Aucamp (Eds.), *Global Ozone Research and Monitoring Project Report No. 37*. Geneva: Author.

World Meteorological Organization. (1994b). *Report of the WMO meeting of experts on UV-B measurements, data quality and standardization of UV indices*. World Meteorological Organization Global Atmosphere Watch Report No. 95, Geneva.

World Meteorological Organization. (1998). Scientific assessment of ozone depletion, 1998. In: D. L. Albritton, P. J. Aucamp, G. Megie, & R. T. Watson (Eds.), *Global ozone research and monitoring project* Geneva: Author.

Worrest, R. C., & Kimeldorf, D. J. (1976). Distortions in amphibian development induced by ultraviolet-B enhancement (290–315 nm) of a simulated solar spectrum. *Photochemistry and Photobiology, 24*, 377–382.

Xiong, F., Lederer, F., Lukavsky, J., & Nedbal, L. (1996). Screening of freshwater algae (*Chlorophyta, Chromophyta*) for ultraviolet-B sensitivity of the photosynthetic apparatus. *Journal of Plant Physiology, 148*, 42–48.

8

The Significance of Ultraviolet Radiation for Aquatic Animals

Horacio E. Zagarese, Barbara Tartarotti, and Diego A. Añón Suárez

Introduction

The total amount of energy reaching the outer limit of the Earth's atmosphere is defined as the *solar constant*. This is an inappropriate name for a magnitude that is known to vary over time, but it very much reflects the quotidian perception that the sun's energy is virtually unchangeable. Such perceptions have started to change. To find the reasons for this change, we do not have to look at sun itself, but to alterations occurring within our own atmosphere. First, the release of greenhouse gases has increased the capacity of the atmosphere to capture heat. Second, ozone depletion has decreased the atmosphere's ability to filter out some of the most damaging wavelengths within the ultraviolet region.

The recognition that the balance and spectral composition of the electromagnetic energy reaching the Earth's surface are changing has provided the impetus for much of the research on the biological effects of ultraviolet radiation (UVR). The number of publications has increased roughly exponentially since the middle 1970s (Figure 8.1), accompanied by an extraordinary development of measuring instruments and incubation devices. Here, we have not attempted an exhaustive review of the available literature. Our aim is to provide a synthetic overview of the major advances in our knowledge of ultraviolet effects on aquatic animals.

An Overview of UVR Effects on Aquatic Animals

UVR is known to induce mortality in early life-history stages (embryos and larvae) of aquatic vertebrates, including several fish (Bell & Hoar, 1950; Eisler, 1961; Hunter et al.,

---

H. E. Zagarese • Centro Regional Universitario Bariloche, Universidad Nacional del Comahue, U.P Universidad, 8400 Bariloche, Argentina and Consejo Nacional de Investigaciones Científicas y Técnicas **Barbara Tartarotti** • Institute of Zoology and Limnology, University of Innsbruck, Technikerstrasse 25, 6020 Innsbruck, Austria D. A. Añón Suárez • Centro Regional Universitario Bariloche, Universidad Nacional del Comahue, 8400 San Carlos de Bariloche, Argentina.

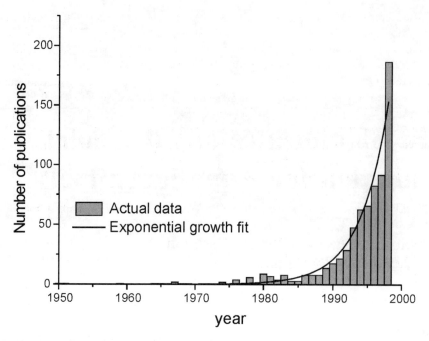

Figure 8.1. The evolution of the number of publications related to environmental effects of UVR. Source: http ://homepage.uibk. ac.at/homepage/c719/c71986/

1979; Kouwenberg et al., 1999a; Novales Flamarique & Harrower, 1999) and amphibian species (Blaustein et al., 1994, 1995; Ovaska et al., 1997; Anzalone et al., 1998; Hatch & Burton, 1998; Lizana & Pedraza, 1998). Based on differential sensitivity to UVR, Blaustein and coworkers proposed a link between increased levels of UVR and the decline of certain amphibian species (UV sensitivity hypothesis). Similarly, Walters and Ward (1998) suggested that exposure to UVR during the early developmental stages could be responsible for the declines in survival and recruitment rates of steelhead trout (*Oncorhynchus mykis*), coho salmon (*O. kisutch*), chinook salmon (*O. tschwytscha*), and Atlantic salmon (*Salmo salar*) in the Northern Hemisphere. They suggested that increasing exposure to radiation during freshwater rearing causes metabolic damage that results in increased mortality during the stressful period of smoltification, ocean migration, and adaptation to saltwater (metabolic impairment hypothesis). *In situ* incubations of fish eggs typically show increased survival with incubation depth and concentration of dissolved organic matter (DOM) in lake water (Williamson et al., 1997; Gutierrez Rodriguez & Williamson, 1999; Battini et al., 2000). Williamson and coworkers (1997) showed that the depth distribution of yellow perch (*Perca flavescens*) nests is deeper in low DOM lakes compared to high DOM lakes, and Battini et al. (2000) suggested that present levels of UVR may be sufficient cause to explain the absence of fish populations from highly transparent mountain lakes in southern Argentina. However, other researchers have found no evidence of UV-induced mortality of amphibian eggs when exposed to natural or enhanced levels of UVR (Grant & Licht, 1995; Ovaska et al., 1997; Corn, 1998; Crump et al., 1999; Cummins et al., 1999; Starnes et al., 2000). Such a lack of UV effects is thought to be due either to the high tolerance of the assayed species or to the efficient removal of damaging radiation by DOM rich waters (Crump et al., 1999).

The whole series of events leading to embryo mortality is not completely understood. At the subcellular level, one of the immediate consequences of UV irradiation is the formation of photoproducts, such as cyclobutane pyrimidine dimers (CPDs), which impair DNA replication and RNA transcription. Vetter and coworkers (1999) were able to demonstrate experimentally that the concentration of CDPs in early developmental stages of the Northern anchovy (*Engraulis mordax*) tracked the diel cycle of solar radiation. In experiments with amphibian embryos, Blaustein et al. (1994, 1995) showed that tolerance to UV exposure was related to the levels of photolyase, a DNA repair enzyme responsible for CPD removal. Embryonic malformations (teratogenesis) have been shown to correlate positively with the concentration of CPDs in embryos of *Xenopus laevis* (Bruggeman et al., 1998), thus supporting the interpretation that malformations, such as lesions to the eye, the olfactory bulb, and the brain, in anchovy and mackerel larvae reported in the earlier literature (Hunter et al., 1979), resulted from initial DNA damage.

Although mortality as a direct consequence of UV exposure is unusual in adult vertebrates, exposure to high levels of UVR may result in severe lesions of the most exposed organs and tissues, such as eye and skin. Cases of sunburn in fishes have been reported since as early as 1930 (Crowell & McCay, 1930). Sunburn is usually characterized by the development of gray necrotic areas on the dorsal surface and extensive erosion of the epidermis. Figure 8.2 shows the histological effects of irradiated epidermis in brown trout (*Salmo truta*). It must be pointed out that some of the observed effects, including the increased sloughing in the epidermis and variations in mucifications, are nonspecific responses that can be confounded with other pathologies, such as ectoparasite infestations or saprolegnasis (Noceda et al., 1997). Evidence of sunburn caused by natural radiation has been reported for juvenile rainbow trout (*Onchorynchus mykiss*) (Dunbar, 1959; Bullock & Coutts, 1985), juvenile plaice (*Pleuronectes platessa*) (Bergham et al., 1993), and juvenile paddlefish (*Polyodon spathula*) (Ramos et al., 1994), among others. Interspecific differences in susceptibility to sunburn can be significant. For example, lahontan cutthroat trout (*Oncorhynchus clarki henshawi*) proved to be sensitive to artificial ultraviolet-B radiation (UV-B) and exhibited grossly visible signs of sunburn upon exposure, whereas razorback suckers (*Xyrauchen texanus*) showed no gross signs after UV exposure (Blazer et al., 1997).

The eye lens opacity (cataracts) sometimes observed in UV-exposed fish (Cullen & Monteith-McMaster, 1993) appears to be mediated by the presence of peroxides. *In vitro* experiments have shown that UVR reduces the ability of the dogfish (*Mustelus canis*) lens to protect itself against H_2O_2 insult via catalase (Zigman & Rafferty, 1994). Retinal damage has been observed in *Rana cascadae* (Fite et al., 1998). In addition to anatomic damage, exposure to UV radiation has also been shown to affect important physiological parameters. For example, UVR has been reported to impair the respiratory control in plaice (*Pleuronectes platessa*) larvae (Freitag et al., 1998) and to compromise the immune system in fish (Salo et al., 1998, 2000).

Many studies have shown reduced survival in different zooplankton species after UV exposure (Siebeck & Böhm, 1991; Williamson et al., 1994; Zellmer, 1995, and references therein). However, the response to UVR shows enormous variability among species (Siebeck, 1978; Siebeck & Böhm, 1994; Williamson et al., 1994; Cabrera et al., 1997; Zagarese et al., 1997a, 1997b; Hurtubise et al., 1998; Tartarotti et al., 1999), life stages (Karanas et al., 1979; Zellmer, 1995), pigmented versus unpigmented morphs (Ringelberg et al., 1984) and even within single species (Karanas et al., 1981; Cabrera et al., 1997;

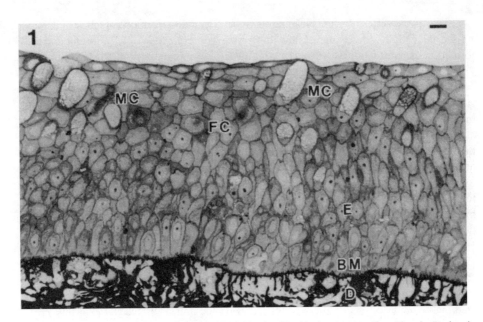

Figure 8.2. *Salmo trutta* epidermis. Plate 1, normal epidemis of healthy brown trout. E: epidermis; D: dermis; BM: basement membrane; FC: filament-containing cell; MC: mucous cell. Scale bar = 10 μm. Plates 2 to 7, epidermis of irradiated *Salmo trutta*. All scale bars = 10 μm. Plate 2, irradiated epidermis showing an increased sloughing involving several upper cell layers. Between the basal and the middle zones, another early alteration surface can be also seen (at level of arrows). E: epidermis; D: dermis; BM: basement membrane. Plate 3, irradiated epidermis with upper cell layers lost by sloughing. Intercellular spaces forming a cleavage surface appear between the basal and middle layers (asterisks). Usually, lymphocytes (arrowheads) can be seen in the cleavage surface. Plate 4, irradiated epidermis with a significant decrease in the number of cell layers as a consequence of the surface alterations. Some cells show large digestive vacuoles (arrows). Plate 5, epidermis regenerating. The restoration of the tissue integrity involved intensive mitotic proliferation (asterisk) and rapid sealing of the surface. Plate 6, epidermis regenerating. Mitotic proliferation (asterisk); few cell layers and some intracellular spaces are still visible, but the tissue surface already shows the features of normal epidermis. Plate 7, epidermis with abundant sunburn cells. Note the fragmentation of their nuclear material into several dense granules (arrows). Reprinted from Noceda et al. (1997) with permission of Inter-Research.

Zagarese et al., 1997a; Stutzman, 1999). For example, the cladocerans *Daphnia* and *Diaphanosoma* were sensitive to both UVB and UVA radiation, whereas the rotifer *Keratella* was not impaired by either waveband after short-term (3 d) exposure to solar UVR (Williamson et al., 1994). Differences in UV tolerance between the same genus (*Boeckella*) have also been observed (Zagarese et al., 1997a). Even among populations of a single species, but living in different habitats, differences have been found. For example, *Boeckella gracilipes*, a common calanoid copepod of southern South America, has been reported to be highly vulnerable to UV radiation in Patagonian lakes (41°S) (Zagarese et al., 1997a), but much more tolerant in the Andean lake Laguna Negra (33°S) (Cabrera et al., 1997; Tartarotti et al., 1999).

Because gene expression is more active in embryos and larvae, UV-induced damage seems to be more severe in those sensitive life stages (Naganuma et al., 1997). Karanas et al. (1979) observed an inverse relationship between the age of marine copepods and their UV sensitivity, the nauplii being the most vulnerable. Reproduction can also be affected by solar UV-B radiation. UV-B radiation eliminated reproduction and reduced ingestion by up

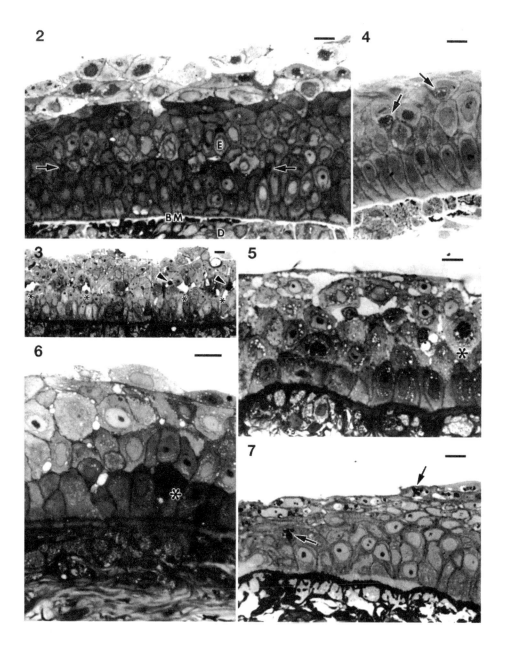

Figure 8.2. Continued

to 90% in the rotifer *Brachionus caliciflorus* (Preston et al., 1999). Williamson et al. (1994) observed a suppression of the reproduction of the freshwater copepod *Diaptomus* down to 6 m in an oligotrophic lake after short-term exposure to ambient levels of UVR. Furthermore, UV exposure can induce changes in the sex ratios of intertidal copepods (Chalker-Scott, 1995). On the other hand, long-term exposure (>3 weeks) of different calanoid copepods to *in situ* levels of solar UVR showed no measurable consequences in the reproduction of some species (Cabrera et al, 1997; Zagarese et al., 1997b). These results suggest that certain species may already be adapted to high levels of UVR.

Exposure to UVR can potentially affect the survival of bottom-dwelling organisms. For example, McNamara and Hill (1999) found that the survival of mayfly and chironomid larvae was inversely related to the dose of UV-B radiation. Moreover, they found that the lethal dose 50% (LD_{50}) was lower at high irradiances, an indication that the reciprocity principle[1] does not hold for these organisms. Marine littoral organisms are also affected by increased UVR (Dunlap et al., 1986; Masuda et al., 1993; Dunne & Brown, 1996; Johnsen & Kier, 1998; Santas et al., 1998). Arm loss was observed in ophiuroid echinoderms exposed to UVB radiation (Johnsen & Kier, 1998). Moreover, UVR is thought to be partly responsible for the bleaching of reef corals (Fitt & Warner, 1995; Lesser, 1997). UV-B irradiation has also been implicated in photosynthesis inhibition of the symbiontic zoonxanthellae in solitary coral *Fungia* spp. (Masuda et al., 1993) and noncoral anthozoans (Shick, 1991). In addition, Jokiel and York (1980, as cited by Shick et al., 1995), showed that exposure to solar UV reduces calcification in a Pacific reef coral. Given the strong association between corals and their algae (Ruppert & Barnes, 1996), it is presently unclear whether the bleaching of coral is exclusively due to the death of the animal tissue, or whether it is triggered or exacerbated by the lack or dysfunction of the symbiotic zooxanthellae (Fitt & Warner, 1995). The damaging effects of UVR, especially on photoautotrophic-symbiotic organisms, can be a direct consequence of UV exposure or can be mediated by toxic forms of oxygen (Shick, 1993; see below).

The Long Way from Extraterrestrial UVR to Biological Effects

A large number of factors affect the extent to which organisms are exposed to UVR in their natural habitats. Among these factors, we may list the levels of UVR at the surface, the attenuation of UVR by the water, and the distribution of the organisms with respect to the most exposed areas (depth distribution, shaded areas). Once a certain amount of UVR has reached the organism's outer tissue, there are two possible alternatives. UVR may be absorbed by specialized compounds that dissipate the energy without causing any harm to the organism, or it may hit critical structures (i.e., cell membranes) or molecules (i.e., DNA, proteins), resulting in damage to the organism. Finally, the last line of defense is the repair of the damage, which is undertaken by specialized mechanisms.

UV Radiation at Ground Level

The amount of solar radiation reaching the ground level is affected by a number of astronomical (geometric) and atmospheric (density) variables. The first ones are related to the angle at which the solar rays hit the Earth's surface: Latitude, day of the year, and time are the most important variables. The second set of variables determines the quantity and quality of the atmospheric filter between the extraterrestrial radiation and the Earth's

---

[1]The reciprocity principle states that the effect is a function of the dose (i.e., the product of irradiance and exposure time). For example, under the assumption of reciprocity, the same effect should be observed if the irradiance is doubled, as long as exposure time is reduced by half.

surface: elevation, clouds and moisture, ozone concentration, aerosols, dust. Geometric and density variables are also interrelated. For example, at low sun angles, the photons must traverse a thicker atmosphere than at high sun angles. Daily doses of UVR are also dependent on the duration of the day. The highest daily doses are observed not in the tropics, where maximal irradiance occurs, but at midlatitudes, where the time-integrated product of irradiance and day duration are maximal.

Penetration of UV Radiation in the Water Column

Several factors affect the penetration of UVR in the water. In the first place, there may be reflection at the water surface. This reflection is referred to as *albedo* and varies according to the sun angle, the roughness of the water surface, and so on (Whitehead et al., 2000). Attenuation within the water column typically obeys the Lambert–Beer law; that is, a constant proportion of photons is absorbed by each meter of water:

$$E_d(z) = E_d(0) \times e^{-k_d \times z},$$

where $E_d(z)$ is the downward irradiance at depth z meters, $E_d(0)$ is the downward irradiance just below the surface, and k_d is the attenuation coefficient for downward irradiance (Kirk, 1994). The value of the attenuation coefficient varies with wavelength. Within the UV range, there is an inverse relationship between k_d and wavelength.

The main factor controlling the attenuation of UVR in natural waters (i.e., by affecting the k_d value) has been found to be the concentration of dissolved organic matter (DOM) (Scully & Lean, 1994; Morris et al., 1995). The dramatic influence of DOM on the attenuation of UVR has led Williamson et al. (1996) to hypothesize that changes in DOC will probably be more important in controlling the UV environment in lakes than changes in stratospheric ozone. Other factors such as particulates, pH (Schindler et al., 1996b; Yan et al., 1996), salinity (Arts et al., 2000), and phytoplankton (Laurion et al., 2000) may contribute to explain the differences in UV attenuation within (Smith et al., 1999) as well as between water bodies (Arts et al., 2000).

Spatial Distribution of Organisms

One strategy to minimize UV-induced damage is to avoid exposure to high levels of UVR by staying in deep water layers or by undergoing diel vertical migration (DVM). Several authors have suggested that DVM is a natural response to UVR, because migration is also observed in habitats lacking predators (Damkaer, 1982; Hessen, 1993, 1994). Storz and Paul (1998) suggested that in addition to factors such as kairomones released by predators and changes in relative light intensity, the effect of UVR may be more substantial than currently assumed. However, Zagarese et al., (1997b) suggested that UV-B radiation is unlikely to be responsible for the deep vertical distribution during daytime, because the main daytime distribution of most zooplanklon is close or below the 10% attenuation depth of UV-B radiation.

The vertical distribution of planktonic organisms is greatly influenced by the turbulent vertical mixing of the water. In a well-mixed water column, the position of organisms is continuously changing due to the movements of the media, as well as their own passive (sinking) or active displacements. Several models can be used to incorporate the effects of mixing on the vertical distribution of individuals or populations (Neale et al., 1998). But the implications of mixing on photobiological effects go beyond the difficulties of predicting the instantaneous vertical position of individuals. Mixing causes planktonic organisms to be exposed to fluctuating radiation levels that may vary over a range of several orders of magnitude (Smith, 1989). If we consider an animal rotating vertically around a second individual that remains at a fixed depth, the radiation that they experience has been shown to differ in three important ways (Zagarese et al., 1998a): (1) Because radiation attenuates exponentially with depth, the vertically moving organism will be exposed to higher doses of any given wavelength; (2) because of the inverse relationship between attenuation and wavelength (Morris et al., 1995), the spectral composition experienced by the moving organisms will be shifted toward shorter wavelengths; (3) whereas a static organism will experience a relatively constant irradiance, the organism that moves up and down will be exposed to pulses of radiation many times higher than the mean, followed by periods of very low intensity. Zagarese et al. (1998b) have shown that knowing the spectral dose experienced by an organisms may be insufficient to predict the performance of vertically moving individuals. Thus, static incubations may offer a distorted picture of the effects of UVR in nature.

Solar UVR may also influence the distribution and abundance of bottom-dwelling organisms (Bothwell et al., 1993, 1994; Williamson & Zagarese, 1994; Vinebrooke & Leavitt, 1996). The effect of UVR is stronger in shallow streams and the littoral areas of clear lakes (De Nicola & Hoagland, 1996; Hill et al., 1997; Vinebrooke & Leavitt, 1999). For example, ambient UVB inhibits the colonization of substrata by grazers, such as chironomid larvae (Bothwell et al., 1994) and the mayfly *Baetis* sp. (De Nicola & Hoagland, 1996). Kiffney et al. (1997a) found that mayfly and caddisfly abundance was lower in substrates exposed to UVR and photosynthetic active radiation (PAR) than on substrates exposed only to PAR. Depth of habitation is regarded as the most common response to potential light exposure for Antarctic benthic organisms such as sea urchins (Karentz et al., 1997). It is not always clear which mechanisms allow the animals to avoid exposure by moving away from high UVR sites (Kiffney et al., 1997b). In lotic environments, larval drift is a common response of stream insects when exposed to anthropogenic stress, such as acidification (Hall et al., 1980). It has been suggested that some taxa of stream invertebrates such as mayflies and caddisflies respond to artificial UVB radiation by entering the drift (Kiffney et al., 1997b).

UV Protective Compounds

A common strategy to reduce the damage caused by solar radiation is the use of UV-screening compounds (Zagarese & Williamson, 1994). Protective compounds in zooplankton have long been known from the literature (Brehm, 1938). UV-screening compounds provide a passive method for the reduction of UV-induced damage and are widely distributed among aquatic animals. A common chemical characteristic of UV-protecting compounds is the presence of the π-electron system, which is found in linear chain molecules with alternating single and double bonds, and in many aromatic and cyclic compounds containing electron resonance (Cockell, 1998). However, many compounds that happen to have aromatic or conjugated groups may fortuitously absorb UVR without having a specific UV-protecting function (Cockell & Knowland, 1999).

A comprehensive review of UV-screening compounds has been recently published by Cockell and Knowland (1999). The three major types of photoprotective compounds in aquatic animals are carotenoids, mycosporine-like amino acids (MAAs), and melanins.

The role of carotenoids as passive UV screening compounds is doubtful. However, they act as oxygen free-radical quenchers and thus provide indirect photoprotection from both UV and high levels of visible light. In many algae, they are known to be photoinducible, and a role in photoreactivation has also been suggested. Approximately 19 MAAs have been found in aquatic organisms. These compounds have strong absorption maxima, between 310 and 360 nm, that correspond to biologically harmful wavelengths of UVR. The basic chromophore responsible for UV absorbance is derived from the Shikimate pathway. MAAs are hypothesized to act as sunscreens by absorbing the energy of UVR before it reaches cellular targets and harmlessly dissipating this energy as heat. Animals are unable to synthesize carotenoids and MAAs and must therefore acquire these compounds through their symbiotic algae (i.e., corals) or the diet. The acquisition of MAAs from the diet was experimentally confirmed in the sea urchin *Strongyiocentrotus droebachiensis* (Carroll & Shick, 1996) and the medaka fish (*Oryzias latipes*) (Mason et al., 1998). Melanins, on the other hand, are produced by a wide variety of animal groups, including vertebrates and invertebrates. For many melanins, the plot of log (absorbance) against wavelength gives an inverse linear relationship, which is consistent with their passive UV screening role, helping to selectively remove the most energetic short wavelengths of UVR. However, melanins also absorb in the visible wavelengths, which results in their dark color. Direct solar UV-B radiation has been shown to induce melanosome dispersion in the dermis of Atlantic salmon (*Salmo salar*) (MacArdle & Bullock, 1987). Examples of additional protective pigments include the silver, guanine-based reflective covering of the fish scales (Ahmed & Setlow, 1993) and ether unidentified colorless photoprotective substances, such as those found in medaka *Oryzias latipes* (Fabacher et al., 1999).

Carotenoids are commonly found in copepods and cladocerans. A large range of carotenoid concentrations can be found among different copepod species (Tartarotti et al., 1999), life stages (Hairston, 1978), vertical distribution (Hairston, 1980), and habitat (Hairston, 1979). Several studies have shown the photoprotective role of these compounds in freshwater copepods (Hairston, 1976; Byron, 1982). For example, unpigmented species are more vulnerable to solar radiation than pigmented ones (Hairston, 1976; Byron, 1982; Ringelberg et al., 1984; Siebec & Böhm, 1994). However, far higher ($\sim$10 times) total carotenoid concentrations are observed in freshwater calanoid copepods than in cladocerans (Hebert & Emery, 1990; Hessen & Sörensen, 1990; Hessen, 1993, 1994).

Melanin can be found in *Daphnia* and other cladocerans (Hessen & Sörensen, 1990). Melanic clones are less UV-sensitive to both natural and artificial UVR (Hebert & Emery, 1990), and cuticular melanization is commonly observed in cladocerans of high-elevation lakes (Löffler, 1969; Beaton & Hebert, 1988; Hebert & Emery, 1990; Hobæk & Wolf, 1991).

The presence of MAAs has been reported in marine *Calanus propinquus* and the freshwater *Cyclops abyssorum tatricus* (Karentz et al., 1991; Sommaruga & Garcia Pichel, 1999) copepods. However, information about the occurrence of MAAs in zooplankton is still very scarce. A survey of 15 lakes located in the Central Alps ($\sim$45°N) showed that 3 to 8 different MAAs were present in all copepod species and life stages examined. High variability in the concentration of MAAs was found within a single species, related to the different life stages and the optical properties of the collection site. An exponential positive relationship ($r^2 = 0.74$) was found between MAA concentration of the copepods and lake transparency. UV-absorbing compounds were also detected in rotifers, whereas no MAAs were found in the cladocerans *Daphnia*, *Bosmina*, and *Chydorus* studied (Tartarotti et al., 2001).

The presence of MAAs in bottom-dwelling animals has been reported for reef-building corals (Dunlap et al., 1986; Shick, 1991; Shick et al., 1995; Teai et al., 1998), solitary corals (Masuda et al., 1993; Drollet et al., 1997), anemones, medusas (Banaszak & Trench, 1995b), and colonial ascidians (Dionisio-Sese et al., 1997). MAAs are also present in a number of nonsymbiotic invertebrates species such as holoturoids and sea urchins (Shick et al., 1992; Carroll & Shick, 1996; Karentz et al., 1997). McClintock & Karentz (1997) also found MAAs in a number of Antarctic species, including Porifera, Cnidaria, Nemertina, Mollusca, Arthropoda, Bryozoa and Echinodermata. Drollet et al. (1997) observed that the amount of UV-absorbing compounds in the mucus of the solitary coral *Fungia* sp. was significantly correlated with the flux of incident solar UVR. The concentrations of UV-absorbing compounds (S-320) in *Acropora formosa* at Davies Reef (Australia) decline considerably with depth from 1 to 15 m (Dunlap et al., 1986). The relationship between spectral irradiance and concentration of MAAs in coral species suggests a physiological acclimatization to variation in UV exposure (Dunlap et al., 1986; Banaszak et al., 1998). In the Antarctic Peninsula, the concentration of MAAs in sea urchins is related to depth, with the highest tissue concentrations occurring in shallow waters (Karentz et al., 1997). Banaszak and Trench (1995a) suggested that for symbiotic marine invertebrates, the production or acquisition of UV photoprotectants is likely to be a condition for inhabiting in shallow waters.

Spectral Sensitivity and Biological Weighting Functions

Dose-dependent effects of artificial UVR have been shown for several marine copepod species (Karanas et al., 1979, 1981; Thomson, 1986; Dey et al., 1988; Naganuma et al., 1997). The highest spectral resolution of UV-induced damage in copepods is given by Kouwenberg et al. (1999b) and Tartarotti et al. (2000). In their studies, long-pass filters (Schott) were used to determine the mortality in *Calanus finmarchicus* eggs under sunlamps (Xenon-arc-lamp), and in late life stages of *Boeckella gracilipes* under solar radiation, respectively. In the former species, the strongest effects were observed under exposure to wavelengths below 312 nm. At the shortest wavelengths (<305 nm), UV-induced mortality was strongly dose-dependent. However, survival was higher when exposed to 312 nm, and exposure to wavelengths of ~335 nm showed no effect (Kouwenberg et al., 1999b). Dose-dependent mortality was also shown in *B. gracilipes*. The strongest effects occurred under exposure to wavelengths below 335 nm; survival increased when wavelengths shorter than 360 nm were filtered out. Exposure to wavelengths longer than ~360 nm showed no effects (Tartarotti et al., 2000).

Because UV-induced effects are strongly wavelength-dependent (Cullen et al., 1992), it is necessary to determine biological weighting functions (BWFs) to relate responses (e.g., mortality) quantitatively to UV exposure. BWFs are determined by exposure of, for example, DNA, membranes, cells, or whole organisms to polychromatic radiation at different wavelengths (Rundel, 1983; Caldwell et al., 1986). Weighted irradiance spectra show which wavelengths are most biologically effective, regardless of the shape of the irradiance spectrum (Smith et al., 1980). To our knowledge, there are only two studies giving BWFs for zooplankton. One is obtained for UV-induced (Xenon-arc-lamp) mortality in the eggs of the marine copepod *Calanus finmarchicus* (Kouwenberg et al., 1999b). The BWF exhibited a slope similar to Setlow's (1974) DNA action spectrum. Kouwenberg et al., (1999b) suggested that UVB-induced mortality in *C. finmarchicus* resulted from DNA damage, because no significant effect of UV-A radiation was observed.

The other study gives a BWF for the mortality of the freshwater copepod *Boeckella gracilipes*, derived from experiments with natural solar radiation (Tartarotti et al., 2000). The slope of the curve shows that the BWF for this species more closely resembles the action spectrum for UV-induced erythema in human skin (McKinlay & Diffey, 1987) than Setlow's (1974) action spectrum for naked DNA (Figure 8.3). In contrast to the marine copepod, *B. gracilipes* was highly vulnerable to both UV-B and UV-A radiation (<360 nm).

Hunter et al. (1981) used several action spectra to relate weighted UV-B exposure to the survival of northern anchovy eggs and larvae. They found that survival was best predicted when the UVB exposure was weighted by Setlow's (1974) action spectrum. Kouwenberg et al. (1999a) presented the first BWF generated for fish (*Gadus morthua*) eggs. Due to the similarity of their BWF and Setlow's (1974) action spectrum, Kouwenberg et al. (1999a) concluded that mortality was most likely a direct result of DNA damage.

Repair Mechanisms

Because DNA is one of the main targets of UVR, organisms have developed several repair mechanisms to correct lesions. The two primary DNA repair systems are photoenzymatic repair ["light repair," which requires the presence of UV-A radiation and blue light (Sutherland, 1981)] and nucleotide excision repair ("dark repair"). The presence or

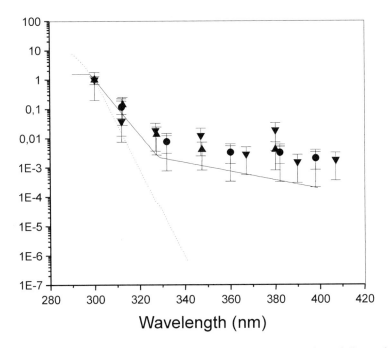

Figure 8.3. Biological weighting function for the mortality of the calanoid copepod *Boeckella gracilipes* under solar radiation conditions. The action spectra for wavelength-dependent damage to naked DNA (data redrawn from Setlow (1974) (dotted line), and for erythema in human skin (31) (continuous line) are plotted for comparison. The action spectra were normalized to 1 at 300 nm. Vertical error bars represent ± 1 standard deviation. Reprinted from Tartarotti et al. (2000), with permission of Elsevier Science B.V.

absence of photorepair mechanisms, such as repair of CPDs via the enzyme photolyase, can influence UV-induced impairment. UV-induced DNA damage (measured as CPDs) was observed in Antarctic zooplankton during periods of high UVB irradiance (Malloy et al., 1997). However, photorecovery is observed in several marine and freshwater zooplankton taxa (Siebeck, 1978; Siebeck & Böhm, 1991, 1994; Malloy et al., 1997; Naganuma et al., 1997; Zagarese et al., 1997a), whereas in others it is not (Ringelberg et al., 1984; Zagarese et al., 1997a). In the laboratory, substantial photoreactivation is observed even at relatively low light intensities, such as that from a single 40-W cool white fluorescent lamp (Damkaer & Dey, 1983; Dey et al., 1988; Malloy et al., 1997). Photoreactivation has been reported in several marine (Kaupp & Hunter, 1981; Regan et al., 1982; Malloy et al., 1997) as well as freshwater (Applegate & Ley, 1988; Funayama et al., 1993; Kouwenberg et al., 1999a) fish species. Cell cultures derived from goldfish (*Carassius auratus*) have been used extensively to examine photoreactivation (Shima & Setlow, 1984; Yasuhira et al., 1991, 1992; Ahmed & Setlow, 1993; Ahmed et al., 1993; Funayama et al., 1993; Uchida et al., 1995). In amphibians, the differential sensitivity to UVR exposure has been linked to interspecific differences in photoreactivation activity (Blaustein et al., 1994, 1999).

Constitutive versus Photoinduced Tolerance

Protection strategies, such as the accumulation of sun-screening compounds, can result in UV-induced physiological acclimation. In transparent aquatic ecosystems where solar UVR has been a significant environmental variable throughout the evolution, organisms have evolved adaptation strategies to cope with high UV fluxes. Acclimation to high ambient levels of UVR is observed in several zooplankton populations. For example, the calanoid copepods *Boeckella gibbosa* and *B. gracilipes* from clearwater lakes are tolerant to high levels of ambient UVR when exposed for 3 days at the lake surface.

Even after long-term (>3 weeks) *in situ* incubations, no negative effects on survival, fecundity, and development were found in these populations (Cabrera et al., 1997; Zagarese et al., 1997b).

Stutzman (1999) conducted a series of experiments to test the hypothesis that populations of *Diaptomus minutus* that routinely experience high levels of UVR were more tolerant of UVR than individuals routinely exposed to low-level of UVR. Tolerance was measured by exposing the animals to either lamp or solar UVR. Stutzman found that *Diaptomus minutus* from a high-UVR habitat was consistently more tolerant of lamp and solar UVR than those from a low UVR environment. Although such differences in UVR tolerance were apparent throughout the year in freshly collected animals, cultured animals from both habitats showed no differences in UVR tolerance.

The Real World: Interactions with the Environment

Under natural conditions, organisms are exposed to a multiplicity of environmental factors. The simultaneous exposure to UVR and a second environmental factor may result in one of several alternatives:

1. *Lack of interaction.* In this case, the effects of UVR and the other factor are simply added.

2. *Amelioration.* The presence of one factor enhances the performance of an organism when exposed to UVR. One of the most impressive examples of amelioration is photorecovery, a mechanism that greatly enhances the survival of UVR-exposed organisms when simultaneously exposed to recovery radiation.
3. *Potentiation or synergism.* In this case, the combined effect of UVR and a second environmental factor is greater than the sum of each factor acting alone.
4. *Photoactivation.* This is an indirect effect, in which UVR promotes the alteration of a chemical compound or changes in the biochemistry of food organisms. For example, the toxicity of polycyclic aromatic hydrocarbons (PAHs) is increased after irradiation with UVR, and the nutritional value of irradiated algae has been shown to be lower than that of algae growing without UVR.
5. *Photodeactivation.* This is essentially the opposite of photoactivation. Examples are the decreased virulence of pathogens after UV exposure, the reduction of grazer activity (Bothwell 1993, 1994), and so on. The concept is analogous to radiotherapy.

Combinations of different types of interactions have also been reported. For example, the phototoxic effect of PAHs is aggravated when the organisms are simultaneously exposed to photoactivated PAHs and UVR (Nikkila et al., 1999). Thus, the net effect is a combination of photoactivation and synergism. Figure 8.4 summarizes the potential interactions of UVR with a second environmental variable. The following paragraphs provide some examples of variables known or suspected to affect the outcome of UVR exposure.

Temperature

There are several reasons to suspect that the effects of UVR may be temperature-dependent. Temperature is known to affect the kinetics of enzyme reactions, with higher temperatures typically increasing the velocity of the reactions. The photorepair mechanism, which is catalyzed by the enzyme photolyase, may be expected to be more efficient at higher temperatures. Therefore, the net amount of damage may be expected to decrease with increasing temperature. Borgeraas and Hessen (2000) studied the survival of adult *Daphnia magna* after acute (<96 h) exposure to UVB radiation under various temperatures (6, 12, and 18°C). But contrary to expectations, the study showed that reduced temperature increased survival. Winckler and Fidhiany (1999) reported metabolic inhibition in the convict cichlid, *Cichlasoma nigrofasciatum*, exposed to artificial UVA radiation, but they found no evidence of interaction with temperature within the range of assayed temperatures (23 to 33°C).

Temperature may affect the results of UVR exposure through its effect on development duration. The hatching time of fish and amphibian eggs is inversely related to temperature. Thus, low temperatures could tend to increase the time during which the organisms are exposed to damaging radiation (Battini et al., 2000), Temperature is also known to affect the body size of aquatic animals, with lower temperature resulting in larger final sizes (Moore & Folt, 1993). Indirectly, this mechanism may affect the outcome of UVR exposure, because it seems unlikely that animals differing in size would display identical susceptibility to UVR. Finally, Drollet et al. (1995) suggested an association between elevated UV-B fluxes and seawater temperature, and Lesser (1997) postulates a synergic relationship between UVR and seawater temperature.

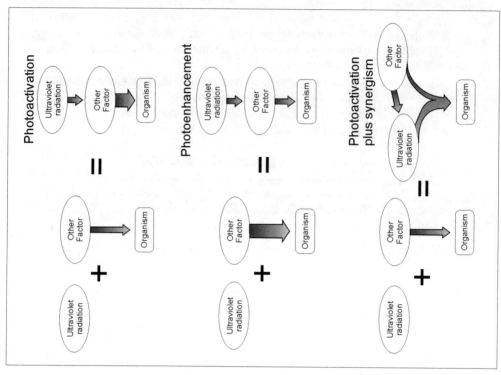
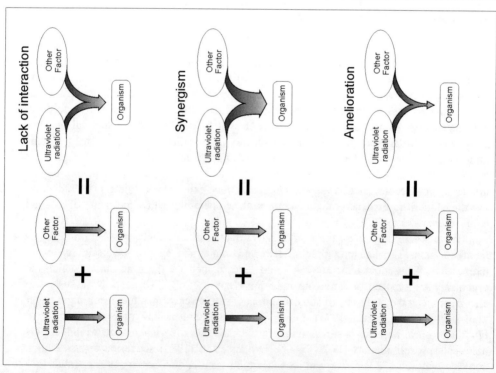

Acidity

A major effect of low water pH is the decrease in solubility of organic matter, which results in higher water transparency and deeper penetration of UVR (Schindler et al., 1996a; Yan et al., 1996). In addition, Long et al. (1995) found that exposure to enhanced UV-B radiation resulted in higher mortality in *Rana pipiens* in acidified water compared to circumneutral water, where they found no significant effects of UVR exposure.

PAHs and Other Toxicants

Aquatic contamination by PAHs is caused by petroleum spills, discharges and seepages, industrial and municipal wastewater, urban runoff, and atmospheric deposition (Nikkila et al., 1999). Most PAHs are not toxic below their water solubility concentration, but even at such low concentrations they may become highly toxic through photoactivation following UVR exposure. Thus, low levels of PAHs could act synergistically with environmental factors such as UVR to place young amphibians at risk (Hatch & Burton, 1998).

The photodynamic nature of these compounds results from their ability to absorb energy in the UV spectrum of sunlight, thereby resulting in excited singlet- and triplet-state molecules. The energy associated with these excited-state PAH molecules is released by nonradiative pathways, generating singlet oxygen and associated reactive oxygen species (ROS, such as superoxide anion, hydroxyl radical) (Weinstein et al., 1997, and references therein).

Monson et al. (1999) exposed larvae of the northern leopard frog (*Rana pipiens*) to different concentrations of flouranthene for a period of 48 h, and after that period to three different UVR (high UV, low UV, and no UV) intensities. After 48 h, the survival of the larvae that were not exposed to UV radiation was 100% at all assayed flouranthene concentrations. UVR exposure following flouranthene exposure significantly decreased survival. Survival was lower at higher UV irradiance (Figure 8.5).

Laboratory experiments revealed differences in the sensitivity to photoinduced toxicity of flouranthene among amphibian species. *Rana pipiens* and *Ambysoma maculatum* were not significantly affected at concentrations below the aqueous solubility of flouranthene in the presence of UV radiation. However, embryos of *Xenopus laevis* appeared to be particularly sensitive to the exposure of both UV radiation alone and to photoinduced toxicity of PAHs (Hatch & Burton, 1998).

In the absence of UVR, flouranthene is not acutely toxic to juvenile fathead minnows (*Pimephales pomelas*) below their water solubility but becomes toxic in the presence of artificial UVR (Weinstein et al., 1997). Histological examination of gill lamellae revealed that the mode of action of photo-induced flouranthene toxicity is a disruption of the mucosal cell membrane function and integrity, which eventually results in respiratory stress and death (Weinstein et al., 1997).

UV-B increased the toxicity of pentachlorophenol (PCP) and mercury to the rotifer *Brachionus caliciflorus* as much as fivefold depending on duration of UV-B exposure and toxicant concentration, and decreased the LC_{50} to PCP and mercury by 60% and 20%, respectively, independent of photochemical reactions with the toxicants (Preston et al., 1999). Pyrene is photoactivated when exposed to artificial UV-B radiation (Philips TL 40W/12RS tubes) and was more toxic to *Daphnia magna* in the presence of UV-B

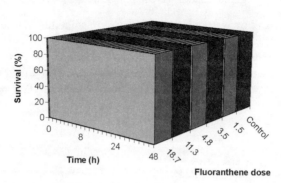

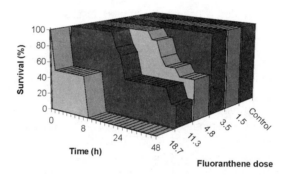

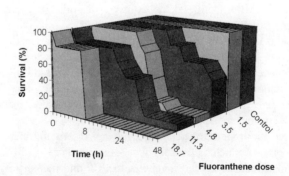

Figure 8.5. Time-dependent survival of *Rana pipiens* tadpoles exposed to various concentrations of flouranthene. Upper panel: control not exposed to UVR; middle panel: exposed to low (UVA: 4.17 to 4.75 µW cm^{-2}, UVB: 61 to 72 µW cm^{-2}) UV irradiance; and lower panel: exposed to high (UVA: 7.82 to 8.29 µW cm^{-2}, UVB: 130 to 153 µW cm^{-2}) UV irradiance. Modified from Monson et al. (1999), with permission of SETAC Press.

radiation. Dissolved organic matter (DOM) significantly reduced the photoenhanced toxicity of pyrene (Nikkila et al., 1999).

DOM/ROS

The exposure of natural waters to UVR in the presence of DOM results in the production of hydrogen peroxide and other reactive oxygen species (ROS) (Scully et al., 1995). It is highly probable that some of the effects of UVR are mediated by the formation of ROS. For example, UV-induced oxidative stress to the lens of fish results from multiple oxidative mechanisms, and it appears that near-UVR decreases the ability of the lens to detoxify H_2O_2 by reducing the activity of lens epithelial cell catalase (Zigman & Rafferty, 1994). In addition, UV irradiation can cause the production of ROS in host animals and their endosymbiotic algae. The main reason for this is the presence of photosensitizing molecules such as chlorophyll, flavins, and aromatic amines in the algal cells (Shick et al., 1995). Lesser (1997) shows that bleaching in corals due to elevated temperatures can be prevented by antioxidants, supporting the idea of the role of oxidative stress as an underlying mechanism for temperature-induced bleaching.

Food and Predators

UVR effects can potentially propagate through the food web (Hessen et al., 1997). In laboratory experiments, reduced digestibility of algae stressed by UV-B radiation was found in *Daphnia magna* (Van Donk & Hessen, 1995). Furthermore, UV-induced changes in nutrient deficiency, cell size, cell-wall properties, and fatty acids in phytoplankton affected life-history traits of *D. pulex*. The daphnids tended to be smaller in size, to produce less and smaller offspring, and to have a reduced population growth rate (De Lange & Van Donk, 1997).

Therefore, the energy transfer between different trophic levels might be negatively influenced by UV-B radiation as a consequence of changes in food quality combined with reduced digestibility (De Lange & Van Donk, 1997). Moreover, UVB-irradiated algae may become toxic and induce mortality to unexposed *Daphnia* (Scott et al., 1999). In addition to the previous examples of bottom-up propagation of UVR effects, Bothwell et al. (1994) reported an interesting case of top-down control. They found that grazers (chironomid larvae) tended to be less abundant in UV-B-exposed treatments. In turn, the exclusion of grazers resulted in an increase in the amount of accumulated biomass of primary producers (diatoms).

Pathogens

Kiesecker and Blaustein (1995) reported 40–50% increases in hatching success of toad embryos shielded from UV-B radiation. However, when the water mold *Saprolegnia ferax* was removed from the water, they obtain slightly greater survival in the UV-B-exposed than in the UV-B-shielded treatment. In addition, sunburn may have a significant role in the outbreaks of infectious diseases (Nowak, 1999).

Behavioral Responses to UVR

Behavior is undoubtedly one of the most salient animal characteristics. Thus, when comparing the effect of UVR on animals and plants, one of the first questions that comes to mind is whether animals show behavioral responses to elevated UVR levels. In order to react in response to UVR, an animal must be able either to detect UV radiation directly or perceive environmental clues that give indications of the levels of UVR. We refer to signals and proxy signals, respectively.

UV Vision in Aquatic Animals

Several conditions must be met to permit UV vision (Losey et al., 1999). First, to sustain vision, there must be a sufficient amount of photons in the environment; second, the retina must contain visual pigments sensitive to UVR; and third, the ocular media, and particularly the lens, must be transparent to UVR. The function of UV-blocking compounds in the cornea and lens of vertebrates and invertebrates (Shashar et al., 1998) is to protect the retina from UVR damage. The presence of such pigments in the light-passing components of the eye is considered a proof for the lack of UV vision (Losey et al., 1999). There is now ample evidence that many aquatic vertebrates and invertebrates can directly perceive UV radiation. An excellent review of the UV vision in fishes has recently been published by Losey et al. (1999). Among the several functions proposed for UV vision, the most relevant to the subject of this chapter are as follows:

1. *Avoidance of excessive UV photoexposure.* From an environmental perspective, this is perhaps the most appealing function for UV vision: If animals were able to detect damaging levels of radiation, they could avoid excessive exposure by moving to a more protected area of the habitat (deep water, shadowed places, etc.). UV-A vision has been demonstrated in rainbow trout (*Onchorhynchus mykiss*) (Beaudet et al., 1993), yellow perch (*Perca flavescens*) (Loew et al., 1993), Red Sea bream (*Pagrus major*) (Kawamura et al., 1997), and roach (*Rutilus rutilus*) (Downing et al., 1986), among other species. Several groups of invertebrates, including *Daphnia* (Smith & Macagno, 1990; Hessen, 1994), and even deep-sea decapod crustaceans (Frank & Widder, 1994a, 1994b), have been found to react to UV-A. *D. magna* has a photoreceptor with peak sensitivity in the UV-A range (Smith & Macagno, 1990; Storz & Paul, 1998). UV sensitivity of the compound eye of *Daphnia* was already suggested in the early works of Koehler (1924) and Merker (1930). So far, however, UV vision appears to be restricted to the longer and least energetic wavelengths within the UV-A portion of the spectrum (H. Browman, personal communication, 2000); therefore, its value in preventing UV damage seems quite limited. At first sight, the lack of receptors in the UV-B range may appear an unnecessary limitation, but it is important to consider that allowing UVR to strike the retina probably imposes some costs on the individual that must be offset by the benefits of UV vision. UVR, even the longer-wavelength UV-A, damages the retina and will increase the degree of chromatic aberration (Losey et al., 1999, and references therein). Another aspect is that perception of UVR does not have to be restricted to the eyes. Even if an animal is blind to UV-B and shorter UV-A, it still perceives their effects through less specialized receptors in the skin (see below).

2. *Improving contrast against the bright UV underwater space light.* At least some planktivorous fishes are able to feed under UV illumination alone (Browman et al., 1993; Loew et al., 1993). UV vision could be effective at larger ranges for detecting prey, predators, and competitors (Losey et al., 1999). Larval and juvenile fish have been reported to lose the ability to detect UVR during ontogenetic development. Interestingly, in the species studied so far, the loss of UV-A perception coincides with a diet shift from planktonic to benthic feeders (Beaudet et al., 1991; Loew et al., 1993; Wahl et al., 1993).
3. *Navigation and orientation using polarization patterns.* These have been proposed in salmonids and other fishes (Hawryshyn, 1992).

Regardless of the function that it might serve, what seems certain is that organisms capable of UV vision routinely experience exposure to UVR that may be damaging.

Responses to UV-B and Shorter UV-A

Although true UV vision appears to be restricted to the longer wavelengths within the UV-A portion of the spectrum, a number of researchers have reported cases in which the animals react in response to short UV-A and even UV-B radiation. At present, it is unclear whether the animals perceive these shorter wavelengths through nonvisual sensors or react in response to proxy signals that may indicate UVR levels in the surrounding media. However, experimental artifacts are difficult to sort out from true responses to UVR in nature. Potential sources of experimental artifacts include the following:

1. The artificial radiation used in lab experiments usually presents a tail extending toward the longer UV-A and even the visible region. Thus, it is often difficult to decide whether the animals react to UV-B or to the parasite longer wavelengths. In addition, even if illuminated with "pure" UV-B radiation, chances are that the animals could react to the visible light produced by fluoresce of algal cells or their own skin and teguments. (As a matter of fact, a routine method to test if a UV-B lamp is on is to look at the blue light produced by fluorescence in a white piece of paper.)
2. Many animals that possess UV-A vision are also able to detect polarized light patterns. Field experiments often make use of filters, such as Mylar, to block out the UV-B portion of the natural spectrum, but as far as we know, nobody has tested whether the optical filters affect the polarized light patterns. If so, the animals may be responding to changes in the polarized light patterns of long-wavelength UV-A rather than to changes in UV-B radiation.

With these caveats in mind, we briefly review cases in which animals have been reported to react to UV-B and short UV-A radiation.

UV-induced negative phototaxis has been observed in *Daphnia* (Becher, 1923; Merker, 1930; Storz & Paul, 1998) and several copepod species (Barcelo & Calkins, 1979; Aarseth & Schram, 1999). Moreover, active UV avoidance (312 nm) can be induced in *Daphnia* clones in laboratory experiments (Hessen, 1994). A rapid downward swimming was also observed in *D. magna* exposed to monochromatic UVR (Storz & Paul, 1998). The harpacticoid copepod *Tigriopus californicus* that lives in small, shallow tidepools has been reported to react in response to UVR. Field experiments using ambient light showed that individuals of *T. californicus* aggregate in areas of lower radiation at midday, yet have no preference at dawn and dusk. In lab experiments, *T. californicus* showed no preference between areas exposed only to visible radiation or shade, but aggregated in the shaded portion of a tank when exposed to UVB (Martin et al., 2000). Frank and Widder (1994a) demonstrated that the Hawaiian deep-sea shrimp *Syselapsis debilis* is sensitive to near-UVR. Laboratory and field experiments demonstrated that the eyes of the sandhopper *Talitrus saltator* (Amphipoda) are capable of perceiving not only the shorter wavelengths of the visible spectrum but also those in the UV region of the spectrum, which has been assumed to help in the orientation of *T. saltator* (Ugolini et al., 1993). The intertidal anemone *Anthopleura elegantissima* appears to possess a combination of behavioral and biochemical attributes that provide protection from the deleterious effects of high irradiance, UV, and PAR (Banaszak & Trench, 1995a). For example, during exposure at low tide, polyps contract, concealing the oral disc and tentacles, thereby shading the algal symbionts.

Concluding Remarks

Ultraviolet radiation is a strong environmental force whose effects have been shown to be significant even below several meters in clear environments (Regan et al., 1992). Following the discovery of the ozone hole, it is has become broadly acknowledged that UVR can affect most aquatic communities and ecological processes either directly or indirectly. But the ecological implications of UVR are by no means restricted to ozone phenomena. Since the beginning of the century, it has been recognized that UVR is a fundamental factor controlling the distribution of species across altitudinal gradients (Brehm, 1938; Thomasson, 1956). More recently, some studies have shown that several effects that had previously been ascribed to other factors, such as acidity, may in fact have resulted from changes in the UV environment (Schindler et al., 1996b; Yan et al., 1996). Acid rain, drought periods (Schindler et al., 1996b; Yan et al., 1996), land-use changes, eutrophication/restoration interventions, and climate-driven changes in tree line (Sommaruga et al., 1999; Zagarese et al., 2000) all have the potential to alter the UV environment of aquatic systems. Both within- and between-year variations in UV intensity and water transparency (Morris & Hargreaves, 1997) may influence seasonal successions in aquatic environments. The relationship between UVR and other environmental variables has only recently become acknowledged. But given the ubiquitous nature of UVR, it seems likely that new and unanticipated interactions will be reported in the future.

Acknowledgments

This work was supported by Universidad Nacional del Comahue (Proyecto B922), Agencia Nacional de Promoción Científica y Tecnológica (PICT No. 0 1-00002-00066), International Foundation for Science (Grant No. A/2325-3F) and Austrian Science Foundation (FWF 14153-BIO). Figure 8.2 was generously provided by Drs. C. Noceda, S. G. Sierra, and J. L. Martinez; and Figure 8.5 by Drs. P. D. Monson, D. J. Call, D. A. Cox, K. Liber, G. T. Ankley, and Ruben Sommaruga for the data of Figure 1.

References

Aarseth, K. A., & Schram, T. A. (1999). Wavelength-specific behavior in *Lepeophtheirus salmonis* and *Calanus finmarchicus* to ultraviolet and visible light in laboratory experiments (Crustacea: Copepoda). *Marine Ecology Progress Series, 186,* 211–217.

Ahmed, F. E., & Setlow, R. B. (1993). Ultraviolet radiation-induced DNA damage and its photorepair in the skin of the platyfish *Xiphophorus*. *Cancer Research, 53,* 2249–2255.

Ahmed, F. E., Setlow, R. B., Grist, E., & Setlow, N. (1993). DNA damage, photorepair, and survival in fish and human cells exposed to UV radiation. *Environmental and Molecular Mutagenesis, 22,* 18–25.

Anzalone, C. R., Kats, L. B., & Gordon, M. S. (1998). Effects of solar UV-B radiation on embryonic development in *Hyla cadaverina*, *Hyla regilla*, and *Taricha torosa*. *Conservation Biology, 12,* 646–653.

Applegate, L., & Ley, R. (1988). Ultraviolet radiation-induced lethality and repair of pyrimidine dimers in fish embryos. *Mutation Research, 198,* 85–92.

Arts, M. T., Robarts, R. D., Kasai, F., Waiser, M. J., Tumber, V. P., Plante, A. J., Rai, H., & deLange, H. J. (2000). The attenuation of ultraviolet radiation in high dissolved organic carbon waters of wetlands and lakes on the northern Great Plains. *Limnology and Oceanography, 45,* 292–299.

Banaszak, A. T., Lesser, M. P., Kuffner, I. B., & Ondrusek, M. (1998). Relationship between ultraviolet (UV) radiation and mycosporine-like amino acids (MAAs) in marine organisms. *Bulletin of Marine Science, 63,* 617–628.

Banaszak, A. T., & Trench, R. K. (1995a). Effects of ultraviolet (UV) radiation on marine microalgal-invertebrate symbioses: 1. Response of the algal symbionts in culture and in hospite. *Journal of Experimental Marine Biology and Ecology, 194,* 213–232.

Banaszak, A. T., & Trench, R. K. (1995b). Effects of ultraviolet (UV) radiation on marine microalgal-invertebrate symbioses: 2. The synthesis of mycosporine-like amino acids in response to exposure to UV in *Anthopleura elegantissima* and *Cassiopeia xamachana*. *Journal of Experimental Marine Biology and Ecology, 194,* 233–250.

Barcelo, J. A., & Calkins, J. (1979). Positioning of aquatic microorganisms in response to visible light and simulated solar UV-B irradiation. *Photochem. Photobiol., 29,* 75–83.

Battini, M., Rocco, V., Lozada, M., Tartarotti, B., & Zagarese, H. E. (2000). Effects of ultraviolet radiation on the eggs of landlocked *Galaxias maculatus* (Galaxidae, Pisces) in Northwestern Patagonia. *Freshwater Biology, 44,* 547–552.

Beaton, M. J., & Hebert, P. D. N. (1988). Geographical parthenogenesis and polyploidy in *Daphnia pulex*. *American Naturalist, 132,* 837–845.

Beaudet, L., Browman, H. I., & Hawryshyn, C. W. (1991). Ontogenetic loss of U.V. photosensitivity in rainbow trout, determined using optic nerve compound action potential recording. *Soc. Neurosci. Abstr, 17,* 299.

Beaudet, L., Browman, H. I., & Hawryshyn, C. W. (1993). Optic nerve response and retinal structure in rainbow trout of different sizes. *Vision Research, 33,* 1739–1746.

Becher, S. (1923). Die Sinnesempfindlichkeit für extremes Ultraviolett bei Daphnien. *Verh. Dtsch. Zool. Ges, 28,* 52–55.

Bell, G. M., & Hoar, W. S. (1950). Some effects of ultraviolet radiation on sockeye salmon eggs and alevins. *Canadian Journal of Research, 28,* 35–43.

Bergham, R., Bullock, A. M., & Karakiri, M. (1993). Effects of solar radiation on the population dynamics of juvenile flatfish in the shallows of Wadden Sea. *J. Fish Biol, 42,* 329–345.

Blaustein, A. R., Edmond, B., Kiesecker, J. M., Beatty, J. J., & Hokit, D. G. (1995). Ambient ultraviolet radiation causes mortality in salamander eggs. *Ecological Applications, 5,* 740–743.

Blaustein, A. R., Hays, J. B., Hoffman, P. D., Chivers, D. P., Kiesecker, J. M., Leonard, W. P., Marco, A., Olson, D. H., Reaser, J. K., & Anthony, R. G. (1999). DNA repair and resistance to UV-B radiation in western spotted frogs. *Ecological Applications, 9,* 1100–1105.

Blaustein, A. R., Hoffman, P. D., Kiesecker, J. M., & Hays, J. B. (1996). DNA repair activity and resistance to solar UV-B radiation in eggs of the red-legged frog. *Conserv. Biol., 10,* 1398–1402.

Blaustein, A. R., Hoffman, P. D., Hokit, D. G., Kiesecker, J. M. W., S. C., & Hays, J. B. (1994). UV repair and resistance to solar UV-B radiation in amphibian eggs: A link to population declines? *Proceedings of the National Academy of Science USA, 91,* 1791–1795.

Blazer, V. S., Fabacher, D. L., Little, E. E., Ewing, M. S., & Kocan, K. M. (1997). Effects of ultraviolet-B radiation on fish: Histological comparison of a UVB-sensitive and a UVB-tolerant species. *Journal of Aquatic Animal Health, 9,* 132–143.

Borgerass, J., & Hessen, D. O. (2000). UV-B induced mortality and antioxidant enzyme activities in *Daphnia magna* at different oxygen concentrations and temperatures. *Journal of Plankton Research, 22,* 1167–1183.

Bothwell, M. L., Sherbot, D., Roberge, A. C., & Daley, R. J. (1993). Influence of natural Ultraviolet radiation on lotic periphytic diatom community growth, biomass accrual, and species composition: Short-term versus long-term effects. *Journal of Phycology, 29,* 24–35.

Bothwell, M. L., Sherbot, D. M. J., & Pollock, C. M. (1994). Ecosystem responses to solar ultraviolet-B radiation: Influence of trophic-level interactions. *Science, 265,* 97–100.

Brehm, V. (1938). Die Rotfärbung von Hochgebirgsorganismen. *Biological Review, 13,* 307–318.

Browman, H. I., Novales-Flamerique, I., & Hawryshyn, C. W. (1994). Ultraviolet photoreception contributes to prey search behaviour in two species of zooplanktivorous fishes. *Journal of Experimental Biology, 186,* 187–198

Bruggeman, D. J., Bantle, J. A., & Goad, C. (1998). Linking teratogenesis, growth, and DNA photodamage to artificial ultraviolet B radiation in *Xenopus laevis* larvae. *Environmental Toxicology and Chemistry, 17,* 2114–2121.

Bullock, A. M., & Coutts, R. (1985). The impact of solar ultraviolet radiation upon the skin of rainbow trout, *Salmo gairdneri* Richardson, farmed at high altitude in Bolivia. *J. Fish Dis., 8,* 263–272.

Bryon, E. R. (1982). The adaptive significance of calanoid copepod pigmentation: A comparative and experimental analysis. *Ecology, 63,* 1871–1886.

Cabrera, S., Lopez, M., & Tartarotti, B. (1997). Phytoplankton and zooplankton response to ultraviolet radiation in a high-altitude Andean lake: Short- versus long-term effects. *Journal of Plankton Research, 19,* 1565–1582.

Caldwell, M. M., Camp, L. B., Warner, C. W., & Flint, S. D. (1986). Action spectra and their key role in assessing biological consequences of solar UV-B radiation change. In: *Stratospheric ozone reduction, solar ultraviolet radiation and plant life.* R. C. Worrest & J. M. Caldwell (Eds.), New York: Springer-Verlag.

Carroll, A. K., & Shick, J. M. (1996). Dietary accumulation of UV-absorbing mycosporine-like amino acids (MAAs) by the green sea urchin (*Stronglyocentrotus droebachiensis*). *Marine Biology, 124,* 561–569.

Chalker-Scott, L. (1995). Survival and sex ratios of the intertidal copepod, *Tigriopus californicus* following ultraviolet-B (290-320) radiation exposure. *Marine Biology, 123,* 799–804.

Cockell, C. S. (1998). Ultraviolet radiation, evolution and the pi-electron system. *Biological Journal of the Linnean Society, 63,* 449–457.

Cockell, C. S., & Knowland, J. (1999). Ultraviolet radiation screening compounds. *Biological Reviews of the Cambridge Philosophical Society, 74,* 311–345.

Corn, P. S. (1998). Effects of ultraviolet radiation on boreal toads in Colorado. *Ecological Applications, 8,* 18–26.

Crowell, M. F., & McCay, C. M. (1930). The lethal dose of ultraviolet light for brook trout (*Salvelinus fontinalis*). *Science, 72,* 582–583.

Crump, D., Berrill, M., Coulson, D., Lean, D., McGillivray, L., & Smith, A. (1999). Sensitivity of amphibian embryos, tadpoles, and larvae to enhanced UV-B radiation in natural pond conditions. *Canadian Journal of Zoology, 77,* 1956–1966.

Cullen, A. P., & Monteith-McMaster, C. A. (1993). Damage to the rainbow trout (*Onchorhynchus mykiss*) lens following acute dose of UV-B. *Current Eye Research, 12,* 97–106.

Cullen, J. J., Neale, P. J., & Lesser, M. P. (1992). Biological weighting function for the inhibition of phytoplankton photosynthesis by ultraviolet radiation. *Science, 258,* 646–650.

Cummins, C. P., Greenslade, P. D., & McLeod, A. R. (1999). A test of the effect of supplemental UV-B radiation on the common frog, *Rana temporaria* L., during embryonic development. *Global Change Biology, 5,* 471–479.

Damkaer, D. M. (1982). Possible influences of solar UV radiation in the evolution of marine zooplankton. In: J. Calkins (Ed.), *The role of solar ultraviolet radiation in marine ecosystems.* New York: Plenum Press.

Damkaer, D. M., & Dey, D. B. (1983). UV damage and photoreactivation potentials of larval shrimp, *Pandalus platyceros*, and adult euphausiids, *Thysanoesa raschii*. *Oecologia (Berlin), 60,* 169–175.

De Lange, H. J., & Van Donk, E. (1997). Effects of UVB-irradiated algae on life history traits of *Daphnia pulex*. *Freshwater Biology, 38,* 711–720.

De Nicola, D. M., & Hoagland, K. D. (1996). Effects of solar spectral irradiance (visible to UV) on a prairie stream epilithic community. *Journal of the North American Benthological Society, 15,* 155–169.

Dey, D. B., Damkaer, D. M., & Heron, G. A. (1988). UV-B dose/dose-rate responses of seasonally abundant copepods of Puget Sound. *Oecologia, 76,* 321–329.

Dionisio-Sese, M. L., Ishikura, M., Maruyama, T., & Miyachi, S. (1997). UV-absorbing substances in the tunic of a colonial ascidian protect its symbiont, *Prochloron* sp., from damage by UV-B radiation. *Marine Biology, 128,* 455–461.

Downing, J. E. G., Djamgoz, M. B. A., & Bowmaker, J. K. (1986). Photoreceptors of a cyprinid fish, the roach: Morphological and spectral characteristics. *J. Com. Physiol.* A, *159*, 859–868.

Drollett, J. H., Faucon, M., & Martin, P. M. V. (1995). Elevated sea-water temperature and solar UV-B flux associated with two successive coral mass bleaching events in Tahiti. *Marine and Freshwater Research*, *46*, 1153–1157.

Drollet, J. H., Teai, Faucon, M., & Martin, P. M. V. (1997). Field study of compensatory changes in UV-absorbing compounds in the mucus of the solitary coral *Fungia repanda* (Scleractinia: Fungiidae) in relation to solar UV radiation, sea-water temperature, and other coincident physico-chemical parameters. *Marine and Freshwater Research*, *48*, 329–333.

Dunbar, C. (1959). Sunburn in fingerling rainbow trout. *Progr. Fish-Cult*, 21, 74.

Dunlap, W. C., Chalker, B. E., & Oliver, J. K. (1986). Bathymetric adaptations of reef-building corals at Davies Reef, Great Barrier Reef, Australia: III. IV-B absorbing compounds. *Journal of Exp. Mar. Biol. Ecol*, *1014*, 239–248.

Dunne, R. P. & Brown, B. E. (1996). Penetration of solar UVB radiation in shallow tropical waters and its potential biological effects on coral reefs: Results from the central Indian Ocean and Andaman Sea. *Marine Ecology—Progress Series*, 144, 109–118.

Eisler, R. (1961). Effects of visible radiation on salmonoid embryos and larvae. *Growth*, 25, 281–346.

Fabacher, D. L., Little, E. E., & Ostrander, G. K. (1999). Tolerance of an albino fish to ultraviolet-B radiation. *Environmental Science and Pollution Research*, 6, 69–71.

Fite, K. V., Blaustein, A., Bengston, L., & Hewitt, P. E. (1998). Evidence of retinal light damage in *Rana cascadae*: A declining amphibian species. *Copeia*, 906–914.

Fitt, W. K., & Warner, M. E. (1995). Bleaching patterns of four species of Caribbean reef corals. *Biological Bulletin*, *189*, 298–307.

Frank, T. M., & Widder, E. A. (1994a). Comparative study of behavioural sensitivity thresholds to near-UV and blue-green light in deep-sea crustaceans. *Marine Biology*, *121*, 229–235.

Frank, T. M., & Widder, E. A. (1994b). Evidence for behavioural sensitivity to near-UV light in the deep-sea crustacean *Systellaspis debilis*. *Marine Biology*, *118*, 279–284.

Freitag, J. F., Steeger, H.-U., Storz, U. C., & Paul, R. J. (1998) Sublethal impairment of respiratory control in plaice (*Pleuronectes platessa*) larvae induced by UV-B radiation using a novel biocybernetic approach. *Marine Biology*, *132*, 1–8.

Funayama, T., Mitani, H., & Shima, A. (1993). Ultraviolet-induced DNA damage and its photorepair in tail fin cells of the medaka, *Oryzias latipes*. *Photochemical Photobiology*, *58*, 380–385.

Grant, K. P., & Licht, L. E. (1995). Effects of ultraviolet radiation on life-history stages of anurans from Ontario, Canada. *Canadian Journal of Zoology*, *73*, 2292–2301.

Gutierrez Rodriguez, C. & Williamson, C. E. (1999). Influence of solar ultraviolet radiation on early life-history stages of the bluegill sunfish. *Lepomis macrochirus*. *Environmental Biology of Fishes*, *55*, 307–319.

Hairston, N. G. (1976). Photoprotection by carotenoid pigments in the copepod *Diaptomus nevadensis*. *Proceedings of the National Academy of Science USA*, *73*, 971–974.

Hairston, N. G. (1978). Carotenoid photoprotection in *Diaptomas kenai*. *Verh. Internat. Verein. Limnol.*, *20*, 2541–2545.

Hairston, N. G. (1979). The adaptive significance of color polymorphism in two species of *Diaptomas* (Copepoda). *Limnol. Oceanogr.*, *24*, 15–37.

Hairston, N. G. (1980). The vertical distribution of diaptomid copepods in relation to body pigmentation. In: W. C. Kerfoot (Ed.), *Evolution and ecology of zooplankton communities*. Hanover, NH: University Press of New England.

Hall, R. J., Likens, G. E., Fiancé, S. B., & Hendrey, G. R. (1980). Experimental acidification of stream in the Hubbard Brook Experimental Forest, New Hampshire. *Ecology*, *61*, 976–989.

Hatch, A. C., & Burton, G. A. (1998). Effects of photoinduced toxicity of fluoranthene on amphibian embryos and larvae. *Environmental Toxicology and Chemistry*, *17*, 1777–1785.

Hawryshyn, C. W. (1992). Polarization vision in fish. *Am. Sci.*, *80*, 164–175.

Herbert, P. D. N., & Emery, C. J. (1990). The adaptive signficance of cuticular pigementation. *Daphnia. Funct. Ecol.*, *4*, 703–710.

Hessen, D. O. (1993). DNA-damage and pigmentation in Alpine and Arctic zooplankton as bioindicators of UV-radiation. *Verh. Internat. Verein. Limnol.*, *25*, 482–486.

Hessen, D. O. (1994). Daphnia responses to UV-B light. *Arch. Hydrobiol. Beih. Ergebn. Limnol*, *43*, 185–195.

Hessen, D. O., Delange, H. J., & Vandonk, E. (1997). UV-induced changes in phytoplankton cells and its effects on grazers. *Freshwater Biology, 38*, 513–524.

Hessen, D. O., & Sørensen, K. (1990). Photoprotective pigmentation in alpine zooplankton populations. *Aqua Fennica, 20*, 165–170.

Hill, W. R., Dimick, S. M., McNamara, A. E., & Branson, C. A. (1997). No effects of ambient UV radiation detected in periphyton and grazers. *Limnology and Oceanography, 42*, 769–774.

Hobæk, A., & Wolf, H. G. (1991). Ecological genetics of Norwegian Daphnia: II. Distribution of Daphnia longispina genotypes in relation to short-wave radiation and water colour. *Hydrobiologia, 225*, 229–243.

Hunter, J. R., Kaupp, S. E., & Taylor, J. H. (1981). Effects of solar and artificial ultraviolet-b radiation on larval northern anchovy, Engraulis mordax. *Photochemistry and Photobiology, 34*, 477–486.

Hunter, J. R., Taylor, J. H., & Moser, H. G. (1979). Effect of ultraviolet irradiation on eggs and larvae of the northern anchovy, *Engraulis mordax*, and the pacific mackerel, *Scomber japonicus*, during the embryonic stage. *Photochemistry and Photobiology, 29*, 325–338.

Hurtubise, R. D., Havel, J. E., & Little, E. E. (1998). The effects of ultraviolet-B radiation on freshwater invertebrates: Experiments with a solar simulator. *Limnology and Oceanography, 43*, 1082–1088.

Johnsen, S., & Kier, W. M. (1998). Damage due to solar ultraviolet radiation in the brittlestar *Ophioderma brevispinum* (Echinodermata: Ophiuroidea). *Journal of the Marine Biological Association of the United Kingdom, 78*, 681–684.

Jokiel, P. L., & York, R. H. (1980). Solar ultraviolet photobiology of the reef coral Pocillopora damicornis and symbiotic zooxantellae. *Bulletin of Marine Science, 207*, 1069–1071.

Karanas, J. J., Van Dyke, H., & Worrest, R. C. (1979). Midultraviolet (UV-B) sensitivity of Acartia clausii Giesbrecht (Copepoda). *Limnology and Oceanography, 24*, 1104–1116.

Karanas, J. J., Worrest, R. C., & Van Dyke, H. (1981). Impact of UV radiation on the fecundity of the copepod *Acartia clausii. Marine Biology (Berlin), 65*, 125–133.

Karentz, D., Dunlap, W. C., & Bosch, I. (1997). Temporal and spatial occurrence of UV-absorbing mycosporine-like amino acids in tissues of the Antarctic sea urchin *Sterechinus neumayeri* during springtime ozone-depletion. *Marine Biology, 129*, 343–353.

Karentz, D., McEuen, F. S., & Dunlap, W. C. (1991). Survey of mycosporine-like amino acid compounds in Antarctic marine organisms: Potential protection from ultraviolet exposure. *Marine Biology (Berlin), 108*, 157–166.

Kaupp, S., & Hunter, J. (1981). Photorepair in larval anchovy, *Engraulis mordax. Photochemistry and Photobiology, 33*, 253–256.

Kawamura, G., Miyagi, M., & Anraku, K. (1997). Retinomotor movement of all spectral cone types of red sea bream *Pagrus major* in response to monochromatic stimuli and UV sensitivity. *Fisheries Science, 63*, 233–235.

Kiesecker, M. M., & Blaustein, A. R. (1995). Synergism between UV-B radiation and a pathogen magnifies amphibian embryo mortaility in nature. *Proceedings of the National Academy of Sciences USA, 92*, 11049–11052.

Kiffney, P. M., Clements, W. H., & Cady, T. A. (1997a). Influence of ultraviolet radiation on the colonization dynamics of a Rocky Mountain stream benthic community. *Journal of the North American Benthological Society, 16*, 520–530.

Kiffney, P. M., Little, E. E., & Clements, W. H. (1997b). Influence of ultraviolet-B radiation on the drift response of stream invertebrates. *Freshwater Biology, 37*, 485–492.

Kirk, J. T. O. (1994). Optics of UV-B radiation in natural waters. *Archiv für Hydrobiologie Beihefte Ergebnisse der Limnologie, 43*, 1–16.

Koehler, O. (1924). Über das Farbensehen von *Daphnia magna* Strauss. *Zeitschrift Vergleichender Physiologie 1*: 84–174.

Kouwenberg, J. H. M., Browman, H. I., Cullen, J. J., Davies, R. F., St.-Pierre, J.-F., & Runge, J. A. (1999a). Biological weighting of ultraviolet-B induced mortality in marine zooplankton and fish: 1. Atlantic cod (*Gadus morthua* L) eggs. *Marine Biology, 134*, 269–284.

Kouwenberg, J. H. M., Browman, H. I., Runge, J. A., Cullen, J. J., Davis, R. F., & St-Pierre, J.-F. (1999b). Biological weighting of ultraviolet-B induced mortalilty in marine zooplankton and fish: 2. *Calanus finmarchicus* G. (Copepoda) eggs. *Marine Biology, 134*, 285–293.

Laurion, I., Ventura, M., Catalan, J., Psenner, R., & Sommaruga R. (2000). Attenuation of ultraviolet radiation in mountain lakes: Factors controlling the among- and within-lake variability. *Limnology and Oceanography, 45*(6), 1274–1288.

Lesser, M. P. (1997). Oxidative stress causes coral bleaching during exposure to elevated temperatures. *Coral Reefs, 16*, 187–192.

Lizana, M., & Pedraza, E. M. (1998). The effects of UV-B radiation on toad mortality in mountainous areas of central Spain. *Conservation Biology, 12*, 703–707.

Loew, E. R., McFarland, W. N. Mills, E. L., & Hunter, D. (1993). A chromatic action spectrum for planktonic predation by juvenile yellow perch, *Perca flavescens. Canadian Journal of Zoology, 71*, 384–386.

Löffler, H. (1969). High altitude lakes in Mt. Everest region. *Ver. Int. Ver. Limnol, 17*, 373–385.

Long, L. E., Saylor, L. S., & Soule, M. E. (1995). A pH/UV-B synergism in amphibians. *Conservation Biology, 9*, 1301–1303.

Losey, G. S., Cronin, T. W., Goldsmith, T. H., Hyde, D., Marshall, N. J., & McFarland, W. N. (1999). The UV visual world of fishes: A review. *Journal of Fish Biology, 54*, 921–943.

MacArdle, J., & Bullock, C. (1987). Solar ultraviolet radiation as a causal factor of "summer syndrome" in cage-reared Atlantic salmon, *Salmo salar* L.: A clinical and histopathological study. *J. Fish Dis, 10*, 255–264.

Malloy, K. D., Holman, M. A., Mitchell, D., & Detrich, H. W., III. (1997). Solar UVB-induced DNA damage and photoenzymatic DNA repair in antarctic zooplankton. *Proceedings of the National Academy of Science USA, 94*, 1258–1263.

Martin, G. G., Speekmann, C., & Beidler, S. (2000). Photobehavior of the harpacticoid copepod *Tigriopus californicus* and the fine structure of its nauplius eye. *Invertebrate Biology, 119*, 110–124.

Mason, D. S., Schafer, F., Shick, J. M., & Dunlap, W. C. (1998). Ultraviolet radiation-absorbing mycosporine-like amino acids (MAAs) are acquired form their diet by medaka fish (*Oryzias latipes*) but not by SKH-1 hairless mice. *Comparative Biochemistry and Physiology A: Molecular and Integrative Physiology, 120*, 587–598.

Masuda, K., Goto, M., Maruyama, T., & Miyachi, S. (1993). Adaptation of solitary corals and their zooxanthellae to low light and UV radiation. *Marine Biology, 117*, 685–691.

McClintock, J. B., & Karentz, D. (1997). Mycosporine-like amino acids in 38 species of subtidal marine organisms from McMurdo Sound, Antarctica. *Antarctica Science, 9*, 392–398.

McKinlay, A. F., & Diffey, B. L. (1987). A reference action spectrum for ultraviolet induced erythema in human skin. In: W. R. Passchler & B. F. M. Bosnajokovic (Eds.), *Human exposure to ultraviolet radiation: risks and regulations* (pp. 83–87). Amsterdam: Elsevier.

McNamara, A. E., & Hill, W. R. (1999). Effects of UV-B dose and irradiance: Comparison among grazers. *Journal of the North American Benthological Society, 18*, 370–380.

Merker, E. (1930). Sehen die Daphnien ultraviolettes Licht? *Zoo. Jahrb. Abt. allg. Zool. Physiol. Tiere, 48*, 277–348.

Monson, P. D., Call, D. J., Cox, D. A. Liber, K., & Ankley, G. T. (1999). Photoinduced toxicity of fluoranthene to northern leopard frogs (*Rana pipiens*). *Environmental Toxicology and Chemistry, 18*, 308–312.

Moore, M., & Folt, C. (1993). Zooplankton body size and community structure: Effects of thermal and toxicant stress. *Trends in Ecology and Evolution, 8*, 178–183.

Morris, D. P., & Hargreaves, B. R. (1997). The role of photochemical degradation of dissolved organic carbon in regulating the UV transparency of three lakes on the Pocono Plateau. *Limnology and Oceanography, 42*, 239–249.

Morris, D. P., Zagarese, H. E., Williamson, C. E., Balseiro, E. G., Hargreaves, B. R., Modenutti, B., Moeller, R., & Queimaliños, C. (1995). The attenuation of solar UV radiation in lakes and the role of dissolved organic carbon. *Limnology and Oceanography, 40*, 1381–1391.

Nagamura, T., Inoue, T., & Uye, S. (1997). Photoreactivation of UV-induced damage to embryos of a planktonic copepod. *Journal of Plankton Research, 19*, 783–787.

Neale, P. J., Davis, R. F., & Cullen, J. J. (1998). Interactive effects of ozone depletion and vertical mixing on photosynthesis of Antarctic phytoplankton. *Nature, 392*, 585–589.

Nikkila, A., Penttinen, S., & Kullonen, J. V. K. (1999). UV-B-induced acute toxicity of pyrene to the waterflea *Daphnia magna* in natural freshwaters. *Ecotoxicology and Environmental Safety, 44*, 271–279.

Noceda, C., Sierra, S. G., & Martinez, J. L. (1997). Histopathology of UV-B irradiated brown trout *Salmo trutta* skin. *Diseases of Aquatic Organisms, 31*, 103–108.

Novales Flamirique, I., & Harrower, W. L. (1999). Mortality of sockeye salmon raised under light backgrounds of different spectral composition. *Environmental Biology of Fishes, 55*, 279–293.

Nowak, B. F. (1999). Significance of environmental factors in aetiology of skin diseases of teleost fish. *Bulletin of the European Association of Fish Pathologists, 19*, 290–292.

Ovaska, K., Davis, T. M., & Flamarique, I. N. (1997). Hatching success and larval survival of the frogs *Hyla regilla* and *Rana aurora* under ambient and artificially enhanced solar ultraviolet radiation. *Canadian Journal of Zoology*, 75, 1081–1088.

Preston, B. L., Snell, T. W., & Kneisel, R. (1999). UV-B exposure increases acute toxicity of pentachlorophenol and mercury to the rotifer *Brachionus calyciflorus. Environmental Pollution*, 106, 23–31.

Ramos, K. T., Fries, L. T., Berkhouse, C. S., & Fries, J. N. (1994). Apparent sunburn of juvenile paddlefish. *Progressive Fish—Culturist*, 56, 214–216.

Regan, J. D., Carrier, W. L., Gusinski, H., Olla, B. L., Yoshida, H., Fujimura, R. K., & Wicklund, R. I. (1992). DNA as a solar dosimeter in the ocean. *Photochemistry and Photobiology*, 56, 35–42.

Regan, J. D., Carrier, W. L., Samet, C., & Olla, B. L. (1982). Photoreactivation in two closely related marine fishes having different longevities. *Mech of Aging and Dev*, 18, 59–66.

Ringelberg, J., Keyser, A. L., & Flik, B. J. G. (1984). The mortality effect of ultraviolet radiation in a red morph of *Acanthodiaptomus denticornis* (Crustacea: Copepoda) and its possible ecological relevance. *Hydrobiolgia*, 112, 217–222.

Rundel, R. D. (1983). Action spectra and estimation of biologically effective UV radiation. *Physiol. Plant*, 58, 360–366.

Ruppert, E. E., & Barnes, R. D. (1996). *Zoologia de los invertebrados* (6th ed.). Mexico: McGraw-Hill Interamericana.

Salo, H. M. Aaltonen, T. M., Markkula, S. E., & Jokinen, E. I. (1998). Ultraviolet B irradiation modulates the immune system of fish (*Rutilus rutilus*, Cyprinidae): 1. Phagocites. *Photochemistry and Photobiology*, 67, 433–437.

Salo, H. M., Jokinen, E. I., Markkula, S. E., & Aaltonen, T. M. (2000) Ultraviolet B irradiation modulates the immune system of fish (*Rutilus rutilus*, Cyprinidae): 2. Blood. *Photochemistry and Photobiology*, 71, 65–71.

Santas, R., Santas, P., Lianou, C., & Korda, A. (1998). Community responses to UV radiation: II. Effects of solar UVB on field-grown diatom assemblages of the Caribbean. *Marine Biology*, 131, 163–171.

Schindler, D. W., Bayley, S. E., Parker, B. R., Beaty, K. G., Cruickshank, D. R., Fee, E. J., Shindler, E. U., & Stainton, M. P. (1996a). The effects of climatic warning on the properties of boreal lakes and streams at the Experimental Lake Area, northwestern Ontario. *Limnology and Oceanography*, 41, 1004–1017.

Schindler, D. W., Curtis, P. J., Parker, B. P., & Stainton, M. P. (1996b). Consequences of climate warming and lake acidification for UV-B penetration in North American boreal lakes. *Nature*, 379, 705–708.

Scott, J. D., Chalker-Scott, L., Foreman, A. E., & D'Angelo, M. (1999). Daphnia pulex fed UVB-irradiated *Chlamidomonas reinharadtii* show decreased survival and fecundity. *Photochemistry and Photobiology*, 70, 308–313.

Scully, N. M., & Lean, D. R. S. (1994). The attenuation of ultraviolet radiation in temperate lakes. *Arch. Hydrobiol. Beih. Ergebn. Limnol*, 43, 135–144.

Scully, N. M. Lean, D. R. S. Mcqueen, D. J., & Cooper, W. J. (1995). Photochemical formation of hydrogen peroxide in lakes: Effects of dissolved organic carbon and ultraviolet radiation. *Canadian Journal of Fisheries and Aquatic Sciences*, 52, 2675–2681.

Setlow, R. B. (1974). The wavelengths in sunlight effective in producing skin cancer: A theoretical analysis. *Proceedings of the Academy of Science USA*, 71, 3363–3366.

Shashar, N., Harosi, F. I., Banaszak, A. T., & Hanlon, R. T. (1998). UV radiation blocking compounds in the eye of the cuttlefish *Sepia officinalis. Biological Bulletin*, 195, 187–188.

Shick, J. M. (1991). Ultraviolet radiation and photooxidative stress in Zooxanthellate Anthozoa: The sea anemone *Phyllodiscus semeni* and the octocoral *Clavularia* sp. *Symbiosis*, 10, 145–173.

Shick, J. M. (1993). Solar UV and oxidative stress in algal-animal symbiosis. In: A. Sima, Y. Fujiwara & H. Takebe (Eds.), *Frontiers of photobiology* (pp. 561–564). Amsterdam: Excerpta Medica.

Shick, J. M., Dunlap, W. C., Chalker, B. E., Banaszak, A. T., & Rosenzweig, T. K. (1992). Survey of ultraviolet radiation-absorbing mycosporine-like amino acids in organs of coral reef holoturoids. *Marine Ecology Progress Series*, 90, 139–148.

Shick, J. M., Lesser, M. P., Dunlap, W. C., Stochaj, W. R., Chalker, B. E., & Won, J. W. (1995). Depth-dependent responses to solar ultraviolet radiation and oxidative stress in the zooxanthellate coral *Acropora microphthalma. Marine Biology*, 122, 41–51.

Shima, A., & Setlow, R. B. (1984). Survival and pyrimidine dimmers in cultured fish cells exposed to concurrent sun lamp ultraviolet and photoreactivating radiations. *Photochemistry and Photobiology*, 39, 49–56.

Siebeck, O. (1978). Ultraviolet tolerance of planktonic crustaceans. *Verh. Internat. Verein. Limnol.*, 20, 2469–2473.

Siebeck, O., & Böhm, U. (1991). UV-B effect on aquatic animals. *Verh. Internat. Verein. Limnol.*, 24, 2773–2777.

Siebeck, O., & Böhm, U. (1994). Challenges for an appraisal of UV-B effects upon planktonic crustaceans under natural conditions with a non-migrating (*Daphnia pulex obtusa*) and a migrating cladocran (*Daphnia galeata*). *Arch. Hydrobiol. Beih, Ergebn, Limnol.*, *43*, 197–206.

Smith, K. C., & Macagno, E. R. (1990). UV photoreceptors in the compound eye of *Daphnia magna* (Crustacea: Branchiopoda): A fourth spectral class in single omatidia. *J. Com. Physiol.* A, *166*, 597–606.

Smith, R. C. (1989). Ozone, middle ultraviolet radiation and the aquatic environment. *Photochemistry and Photobiology*, *50*, 459–468.

Smith, R. C., Baker, K. S., Holm-Hansen, O., & Olson, R. (1980). Photoinhibition of photosynthesis in natural waters. *Photochemistry and Photobiology*, *31*, 585–592.

Smith, R. E. H., Furgal, J. A., Charlton, M. N., Greenberg, B. M., Hiriart, V., & Marwood, C. (1999). Attenuation of ultraviolet radiation in a large lake with low dissolved organic matter concentrations. *Canadian Journal of Fisheries and Aquatic Sciences*, *56*, 1531–1361.

Sommaruga, R., & Garcia Pichel, F. (1999). UV-absorbing mycosporine-like compounds in planktonic and benthic organisms from a high-mountain lake. *Archiv für Hydrobiolgie*, *144*, 255–269.

Sommaruga, R., Psenner, R., Schafferer, E., Koinig, K. A., & Sommaruga-Wögrath, S. (1999). Dissolved organic carbon concentration and phytoplankton biomass in high-mountain lakes of the Austrian Alps: Potential effect of climatic warming an UV underwater attenuation. *Arctic Antarctic and Alpine Research*, *31*, 247–253.

Starnes, S. M., Kennedy, C. A., & Petranka, J. W. (2000). Sensitivity of embryos of southern Appalachian amphibians to ambient solar UV-B radiation. *Conservation Biology*, *14*, 277–282.

Storz, U. C., & Paul, R. J. (1998). Phototaxis in water fleas (*Daphnia magna*) is differently influenced by visible and UV light. *J. Comp. Physiol. A*, *183*, 709–717.

Stutzman, P. L. (1999). A comparative study of ultraviolet radiation tolerance in different populations of *Diaptomus minutus*. *Journal of Plankton Research*, *21*, 387–400.

Sutherland, B. M. (1981). Photoreactivation. *BioScience*, *31*, 439–444.

Tartarotti, B., Cabera, S., Psenner, R., & Sommaruga, R. (1999). Survivorship of *Cyclops abyssorum tatricus* (Cyclopoida, Copepoda) and *Boeckella gracilipes* (Calanoida, Copepoda) under ambient levels of solar UVB radiation in two high-mountain lakes. *Journal of Plankton Research*, *21*, 549–560.

Tartarotti, B., Cravero, W., & Zagarese, H. E. (2000). Biological weighting function for the mortality of *Boeckella gracilipes* (Copepoda, Crustacea) derived from experiments with natural solar radiation. *Photochemistry and Photobiology*, *72*(3), 314–319

Tartarotti, B., Laurion, I., & Sommaruga, R. (2001). Large variability in the concentration of mycosporine-like amino acids among zooplankton from lakes located across an altitude gradient. *Limnol. Oceanogr.*, *46*(6), 1546–1552.

Teai, T., Drollet, J. H., Bianchini, J. P., Cambon, A., & Martin, P. M. V. (1998). Occurrence of ultraviolet radiation-absorbing mycosporine-like amino acids in coral mucus and whole corals of French Polynesia. *Marine and Freshwater Research*, *49*, 127–132.

Thomasson, K. (1956). Reflections on Arctic and Alpine lakes. *Oikos*, *7*, 117–143.

Thomson, B. E. (1986). Is the impact of UV-B radiation on marine zooplankton of any significance? In: J. G. Titus (Ed.), *Effects of changes in stratospheric ozone and global climate*, pp. 203–209. UNEP-USEPA, Vol. 2.

Uchida, N., Mitani, H., & Shima, A. (1995). Multiple effects of fluorescent light on repair of ultraviolet-induced DNA lesions in cultured goldfish cells. *Photochemistry and Photobiology*, *61*, 79–83.

Ugolini, A., Laffort, B., Castellini, C., & Beugnon, G. (1993). Celestial orientation and ultraviolet perception in *Talitrus saltator. Ethology Ecology and Evolution*, *5*, 489–499.

Van Donk, E., & Hessen, D. O. (1995). Reduced digestibility of UV-B stressed and nutrient limited algae by *Daphnia magna. Hydrobiologia*, *307*, 147–151.

Vetter, R. D., Kurtzman, A., & Mori, T. (1999). Diel cycle of DNA damage and repair in eggs and larvae of Northern anchovy, *Engraulis mordax*, exposed to solar ultraviolet radiation. *Photochemistry and Photobiology*, *69*, 27–33.

Vinebrooke, R. D., & Leavitt, P. R. (1996). Effects of ultraviolet radiation on periphyton in an alpine lake. *Limnology and Oceanography*, *41*, 1035–1040.

Vinebrooke, R. D., & Leavitt, P. R. (1999). Differential responses of littoral communities to ultraviolet radiation in an alpine lake. *Ecology*, *80*, 223–237.

Wahl, C. M., Mills, E. L., Mcfarland, W. N., & Degisi, J. S. (1993).Ontogenetic changes in prey selection and visual acuity of the yellow perch, *Perca flavescens. Canadian Journal of Fisheries and Aquatic Sciences*, *50*, 743–749.

Walters, C., & Ward, B. (1998). Is solar radiation responsible for declines in marine survival rates of anadromous salmonids that rear in small streams? *Canadian Journal of Fisheries and Aquatic Sciences, 55*, 2533–2538.

Weinstein, J. E., Oris, T. T., & Taylor, D. H. (1997). An ultrastructural examination of the mode of UV-induced toxic action of fluoranthene in the fathead minnow, *Pimephales promelas*. *Aquatic Toxicology, 39*, 1–22.

Whitehead, R. F., de Mora, S. J., & Demers, S. (2000). Enhanced UV radiation—a new problem for the marine environment. In: S. de Mora, S. Demers & M. Vernet (Eds.), *The effect of UV radiation in the marine environment* (pp. 1–34) Cambridge, UK: Cambridge University Press.

Williamson, C. E., Megzgar, S. L., Lovera, P. A., & Moeller, R. E. (1997). Solar ultraviolet ratiation and the spawning habitat of yellow perch, *Perca flavescens*. *Ecological Applications, 7*, 1017–1023.

Williamson, C. E., Stemberger, R. S., Morris, D. P., Frost, T. M., & Paulsen, S. G. (1996). Ultraviolet radiation in North American lakes: Attenuation estimates from DOC measurements and implications for plankton communities. *Limnology and Oceanography, 41*, 1024–1034.

Williamson, C. E., & Zagarese, H. E. (Eds.) (1994). Impact of UV-B radiation on pelagic freshwater ecosystems. *Arch. Hydrobiol. Beih. Ergebn. Limnol, 43*, xvii + 226 pp.

Williamson , C. E., Zagarese, H. E., Schulze, P. C., Hargreaves, B. R., & Seva, J. (1994). The impact of short-term exposure to UV-B radiation on zooplankton communities in north temperate lakes. *Journal of Plankton Research, 16*, 205–218.

Winckler, K., & Fidhiany, L. (1999). Temperature tolerance and metabolic depression of a convict cichlid under the influence of enhanced ultraviolet-A (320-400 nm) irradiation. *Aquaculture International, 7*, 13–27.

Yan, N. D., Keller, W., Scully, N. M., Lean, D. R. S., & Dillon, P. J. (1996). Increased UV-B penetration in a lake owing to drought-induced acidification. *Nature, 381*, 141–143.

Yasuhira, S., Mitani, H., & Shima, A. (1991). Enhancement of photorepair of ultraviolet damage by preillumination with fluorescent light in cultured fish cells. *Photochemistry and Photobiology, 53*, 211–215.

Yasuhira, S., Mitani, H., & Shima, A. (1992). Enhancement of photorepair of ultraviolet-induced pyrimidine dimers by preillumination with fluorescent light in the goldfish cell line: The relationship between survival and yield of pyrimidine dimers. *Photochemistry and Photobiology, 55*, 97–101.

Zagarese, H. E., Cravero, W., Gonzalez, P., & Pedrozo, F. (1998a). Copepod mortality induced by fluctuating levels of natural solar radiation simulating vertical water mixing. *Limnology and Oceanography, 43*, 169–174.

Zagarese, H. E., Diaz, M., Pedrozo, F., & Úbeda, C. (2000). Mountain lakes in Northwestern Patagonia. *Verh. Internat. Verein. Limnol*, 533–538.

Zagarese, H. E., Feldman, M., & Williamson, C. E. (1997a) UV-B induced damage and photoreactivation in three species of *Boeckella* (Copepoda, Calanoida). *Journal of Plankton Research, 19*, 357–367.

Zagarese, H. E., Tartarotti, B., Cravero, W., & Gonzalez, P. (1998b). UV damage in shallow lakes: the implications of water mixing. *Journal of Plankton Research, 20*, 1423–1433.

Zagarese, H. E., & Williamson, C. E. (1994). Modeling the impacts of UV-B radiation on ecological interactions in freshwater and marine ecosystems. In: R. H. Biggs & M. E. B. Joyner (Eds.), *Stratospheric ozone depletion/UV-B radiation in the biosphere* (pp. 315–328). New York: Springer-Verlag.

Zagarese, H. E., Williamson, C. E., Vail, T. L., Olsen, O., & Queimaliños, C. (1997b). Long-term exposure of *Boeckella gibbosa* (Copepoda, Calaanoida) to *in situ* levels of solar UVB radiation. *Freshwater Biology, 37*, 99–106.

Zellmer, I. D. (1995). UV-B-tolerance of alpine and arctic *Daphnia*. *Hydrobiologia, 307*, 87–92.

Zigman, S., & Rafferty, N. S. (1994). Effects of near UV radiation and antioxidants on the response of dogfish (*Mustelus canis*) lens to elevated H_2O_2. *Comparative Biochemistry and Physiology A—Physiology, 109*, 463–467.

9

Stress Responses in Cyanobacteria

Rajeshwar P. Sinha

Introduction

Cyanobacteria (blue-green algae) are phylogenetically a primitive group of Gram-negative prokaryotes, being the only bacteria to possess higher plant–type oxygenic photosynthesis. Fossil evidence dates their appearance to the Precambrian era (between 2.8 and 3.5×10^9 years ago). At that time, they were probably the main primary producers of organic matter and the first organisms to release oxygen into the then- oxygen-free atmosphere. Thus cyanobacteria were responsible for a major global evolutionary transformation leading to the development of aerobic metabolism and the subsequent rise of higher plant and animal forms (Walter et al., 1992; Schopf, 1993; Pace, 1997). As a group, cyanobacteria are thought to have survived a wide spectrum of environmental stresses, such as heat, cold, drought, salinity, nitrogen starvation, photooxidation, anaerobiosis, and osmotic and ultraviolet radiation stress (Fay, 1992; Tandeau de Marsac & Houmard, 1993; Sinha and Häder 1996a). The cosmopolitan distribution of cyanobacteria, ranging from hot springs to Antarctic and Arctic regions; their colonization of oceans, lakes, rivers, and various soils; and their presence as symbiotic organisms in fungi and plants demand high variability in adapting to diverse environmental factors.

A number of cyanobacteria are capable of buoyancy regulation due to their interior gas vacuoles and the formation and degradation of granules of reserve material of relatively high specific gravity. This allows them to change their position in the water column as environmental conditions change and thus always to face a nearly constant external milieu (Walsby, 1975). The sessile benthic cyanobacteria remain attached to the water–sediment interface, an environment that shows rapid and drastic fluctuations of light, pH, H_2S, and oxygen concentrations. Therefore, these organisms have to change the patterns of their metabolism rather than their location in the water column. In addition, benthic cyanobacteria often remain viable under aphotic conditions for many years, buried under the layers laid down as sediment above them annually (Jorgensen et al., 1988). This,

---

Rajeshwar P. Sinha • Institute for Botany and Pharmaceutical Biology, Friedrich-Alexander University, Staudtstraße 5, D-91058 Erlangen, Fed. Rep. Germany.

again, requires unique adaptive metabolic properties and maintenance mechanisms. One of these flexible metabolic patterns is the shift found in cyanobacteria from oxygenic to anoxygenic photosynthesis, from a photosystem I (PS I) plus PS II mechanism to photosynthesis driven by PS I only, with H_2S and not water as the electron donor (Cohen et al., 1975; Padan, 1979). Another example of alternative metabolic pathways is the ability of certain cyanobacteria to shift in the dark from aerobic (CO_2-producing) to anaerobic respiration and/or fermentation of polyglucose to lactate (Oren & Shilo, 1979). An additional example of metabolic versatility is the ability to use the same photosynthetic electron pathway alternatively for photoassimilation of CO_2, for production of hydrogen under overreduced conditions, or for the fixation of atmospheric nitrogen (Belkin & Padan, 1978). After a shift from anaerobic to aerobic conditions, a rapid induction of superoxide dismutase, up to a level 8- to 10-fold higher than under O_2 depleted conditions, occurs (Friedberg et al., 1979). Another unique feature of the cyanobacteria is their ability to adapt to the quality and low intensities of light reaching the benthic interface. This involves phototactic gliding to the sediment surface to prevent the burying of the benthic mat by sedimentation, enhanced synthesis of photosynthetic pigments, and an increase in the number or size of reaction centers. In addition, a number of cyanobacteria have the ability to vary their phycobiliprotein (phycocyanin/phycoerythrin) ratio, which allows regulation of the balance of wavelengths of light absorbed by cyanobacteria, a phenomenon known as *chromatic adaptation* (Tandeau de Marsac, 1977).

Because it is clear that microorganisms evolved and microbial mats were well established early in the Precambrian era, some mechanism(s) must have been functioning to protect these organisms from the deleterious effects of UV flux. Several mechanisms that might have afforded protection include the avoidance of brightly lit habitats; mating habits; absorbance of UV radiation by nitrates, nitrites, and organic substances in the water; and production of UV-absorbing/screening substances and repair mechanisms (Rambler & Margulis, 1980). This chapter covers the responses of cyanobacteria to various stress factors such as UV-B, heat, salinity, and so on.

Effects of UV-B Radiation on Cyanobacteria

The stratospheric ozone layer shields the Earth from the biologically most hazardous short-wavelength solar radiation. There is mounting evidence that the solar flux of UV-B radiation has increased at the Earth's surface due to the depletion of the ozone layer by anthropogenically released atmospheric pollutants such as chlorofluorocarbons (CFCs), chlorocarbons (CCs), and organobromides (OBs) (Blumthaler & Ambach, 1990; Crutzen, 1992; Kerr & McElroy, 1993; Lubin & Jensen, 1995). Ozone depletion has been reported in the Antarctic, as well as in the north polar regions, but it is most pronounced over the Antarctic, where ozone levels commonly decline by more than 70% during late winter and early spring. Recent TOMS (total ozone mapping spectrometer) data indicate an Antarctic ozone "hole" three times larger than the entire land mass of the United States (NASA, 2000). The "hole" had expanded to a record size of approximately 28.3 million square kilometers. Moreover, ozone depletion has been predicted to continue throughout this century and widespread severe denitrification could enhance future Arctic ozone loss by up to 30% (Toon & Turco, 1991; Elkins et al., 1993; Tabazadeh et al., 2000).

Nitrogen-fixing cyanobacteria form a prominent component of microbial populations in wetland soils, especially in rice paddy fields, where they significantly contribute to fertility as a natural biofertilizer (Venkataraman, 1972). The agronomic potential of free-living as well as symbiotic cyanobacteria, particularly *Azolla–Anabaena* symbiosis, has been recognized as one of the most promising biofertilizer systems for wetland soils (Singh, 1961; Roger & Kulasooriya, 1980; Kannaiyan et al., 1982; Watanabe, 1984; Sinha & Häder, 1996a; Sinha et al., 1998a; Vaishampayan et al., 1998). In nature, the process of nitrogen fixation counterbalances losses of combined nitrogen from the environment by denitrification (Kuhlbusch et al., 1991). In several instances, the availability of nitrogen has been found to be a limiting factor for productivity in natural habitats (Singh, 1961; Parker et al., 1978).

Light is one of the most important factors determining cyanobacterial growth in its natural habitats, because cyanobacteria are predominantly photoautotrophic organisms. In addition, light is used as an environmental factor controlling orientation and habitat selection. Many cyanobacteria are motile, using a gliding mechanism and show three types of photoresponses: (1) phototaxis, (2) photokinesis, and (3) photophobic (Häder, 1987a, 1987b). Phototactic movement is dependent on the light direction and can be either positive (movement toward the light) or negative (movement away from the light). Among prokaryotes, only cyanobacteria are capable of phototactic orientation. Cyanobacteria mostly show positive phototaxis at low irradiances and negative phototaxis at high irradiances. Thus, phototactic orientation enables the organisms to find and select habitats of optimal light intensities (Häder, 1987b). The dependence of the speed of movement of an organism on irradiance is defined as *photokinesis*. The speed of movement is normally higher at high irradiances (positive photokinesis), resulting in an accumulation of the organisms in shaded areas. Photokinetic responses in cyanobacteria seem to be due to a higher supply of energy at high irradiances (Häder, 1987a). A photophobic response of an organism is a reversal of the direction of movement as a result of a sudden change in the light intensity. These responses occur when organisms enter dark (step-down photophobic reaction) or too bright (step-up) areas. As a result, the organisms accumulate in light fields or avoid dark areas (Häder, 1987a).

The cyanobacterial populations in rice paddy fields particularly in the tropics are often exposed to high white light and UV irradiances. Considering the vital role of cyanobacteria as a biofertilizer in rice and other crop production, the fluence rate of UV-B (280–315 nm) radiation impinging on the natural habitats seems to be one of the major concerns, because UV-B radiation has been reported not only to impair motility and photoorientation (Donkor & Häder, 1991; Donkor et al., 1993) but also growth, survival, pigmentation, nitrogen fixation, photosynthesis and total protein profiles (Sinha & Häder, 1996a; Sinha et al., 1999a). The growth and survival of several rice-field cyanobacteria have been reported to be severely affected following UV-B irradiation for different durations. Growth and survival were seized within 120–180 min of UV-B irradiation, depending on the species type. Strains such as *Scytonema* sp. and *Nostoc commune*, filaments of which are embedded in a mucilagenous sheath, have been reported to be more tolerant in comparison to strains whose filaments do not contain such covering (e.g., *Anabaena* sp. and *Nostoc* sp.) (Sinha et al., 1995a; Sinha & Häder, 1998, 2000). Growth and survival depend very much on the genetic machinery of the organisms. Complete killing of the organisms following UV irradiation could be due to damage in the basic genetic material, DNA, absorbing strongly at 280 nm. Many workers have suggested that the cellular constituents absorbing radiation between 280–315 nm are destroyed by UV radiation, which may further affect the cellular membrane permeability and cause protein damage, eventually resulting in the death of the cell (Vincent & Roy, 1993; Sinha et al., 1995a, 1997).

The effect of UV-B on pigmentation of various rice-field cyanobacteria has revealed that the accessory light-harvesting pigment, phycocyanin (λ_{max} 620 nm) was bleached more rapidly and drastically than any other pigment such as chlorophyll a (Chl a) (λ_{max} 437 and 672 nm) or the carotenoids (λ_{max} 485 nm) (Sinha et al., 1995a, 1995b). This shows that UV-B radiation can photo-oxidize and thereby bleach all types of photosynthetic pigments and may also cause a depression of Chl a and carotenoids via reduced rates of biosynthesis. A decrease in the phycobiliprotein contents and disassembly of phycobilisomal complex following UV-B irradiation has been reported in a number of cyanobacteria, indicating impaired energy transfer from the accessory pigments to the photosynthetic reaction centers (Sinha et al., 1995b, 1995c). Spectroscopic analyses have shown drastic decline in both absorption and fluorescence, as well as a shift of the fluorescence peak toward shorter wavelengths (Figure 9.1), which are indicative of the disassembly of phycobilisomes and hence the impaired energy transfer from the phycobilisomes to the photosynthetic reaction centers (Sinha et al., 1995b, 1995c, 1997). This notion is further supported by sodium dodecyl sulphate-polyacrylamide gel electrophoresis (SDS-PAGE) analyses of the phycobiliproteins, which show a loss in the low molecular mass proteins between 16 and 22 kDa (phycobiliprotein α and β subunits) and high molecular mass rod and rod-core linker polypeptides (molecular masses between 24 and 45 kDa) as well as core membrane linker polypeptides (molecular masses around 66 kDa) (Figure 9.2). It seems probable that the supramolecular organization of the phycobilisomes is disassembled largely due to the loss of linker polypeptides after UV irradiation, as has

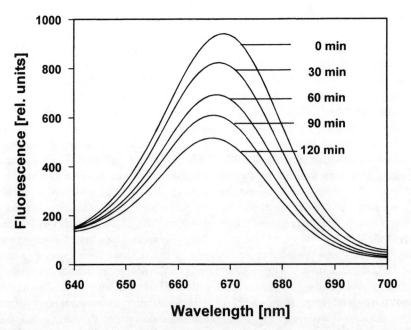

Figure 9.1. Fluorescence emission spectra of the cyanobacterium *Nostoc* sp. following increasing exposure time to UV-B when excited at 620 (phycocyanin) nm.

been shown by electrophoretic analyses (Sinha et al., 1995b, 1995c, 1997; Sinha & Häder, 2000). In fact, the assembly of the peripheral rods of the phycobilisomes is the consequence of a series of specific interactions between phycobiliproteins and linker polypeptides. The α and β subunits combine to form (αβ) monomers that assemble into trimers $(αβ)_3$ and hexamers $(αβ)_6$ and in turn stongly interact with linker polypeptides. The positions of the phycobiliproteins hexamers in peripheral rods are controlled by the linker polypeptides. The electrostatic interactions between the basic linker polypeptides and acidic biliproteins significantly stabilize the phycobilisome assembly and promote the unidirectional, highly efficient energy flow both within and from the phycobilisomes to the chlorophylls of the thylakoid membrane, approaching 100% efficiency (Grossman et al., 1993; Sinha et al., 1995b).

Differentiation of vegetative cells into heterocysts has also been reported to be severely affected by UV-B irradiation in a number of cyanobacteria (Blakefield & Harris, 1994; Sinha et al., 1996). Processes such as heterocyst and nitrogen fixation are metabolically very expensive, and during extreme stress conditions may require cellular resources for their maintenance. Because heterocysts are important primarily in supplying fixed nitrogen to the vegetative cells, apparently vegetative cells growing in areas without external nitrogen sources may be seriously affected. Most probably the C:N ratio is severely affected following UV-B irradiation, which in turn affects the spacing pattern of heterocysts in a filament (Sinha et al., 1996). In addition, major heterocysts polypeptides of around 26, 54, and 55 kDa have also been shown to decrease following UV-B irradiation, suggesting that the multilayered, thick wall of heterocysts may be disrupted, resulting in the inactivation of the nitrogen-fixing enzyme nitrogenase (Sinha et al., 1996). UV-B–induced membrane disruption leading to changes in membrane permeability and release of ^{14}C-labeled compounds have been observed in a number of rice-field cyanobacteria (Sinha et al., 1997).

UV-B–mediated inactivation of the nitrogen-fixing enzyme nitrogenase has been reported in many cyanobacteria (Kumar et al., 1996a; Sinha et al., 1996). Nitrogenase is a molybdoenzyme and requires ATP and reductant for its activity. There is a possibility that the inhibition of the nitrogenase activity might be due to reduced supply of reductant and ATP following UV-B irradiation. However, the loss of reductant and ATP does not occur immediately in cyanobacteria, because many nitrogen-fixing species are capable of nitrogenase activity at the cost of an endogenous pool. It has been suggested that UV irradiance causes complete inactivation/denaturation of the nitrogenase enzyme and thus appears to be a novel phenomenon (Kumar et al., 1996a; Sinha et al., 1996). Total protein profiles of several cyanobacteria show a decrease in protein content with increasing UV-B exposure time (Figure 9.2), indicating that cellular proteins are one of the main targets of UV-B, known to damage proteins and enzymes, especially those rich in aromatic amino acids such as tryptophan, tyrosine, phenylalanine, and histidine, all of which show strong absorption in the UV-B range from 270 to 290 nm (Sinha et al., 1995a, 1996; Kumar et al., 1996a).

The activity of the ribulose 1,5-bisphosphate carboxylase (Rubisco), the primary CO_2 fixing enzyme, has been reported to be inhibited by UV-B radiation in a number of cyanobacteria, which may be due to protein destruction or enzyme inactivation (Sinha et al., 1997). The control of Rubisco biosynthesis is strongly influenced by the prevailing light environment. During UV-B radiation, protein may undergo a number of modifications, including photodegradation, increased aqueous solubility of membrane proteins, and fragmentation of peptide chain, leading to inactivation of proteins (enzymes) and

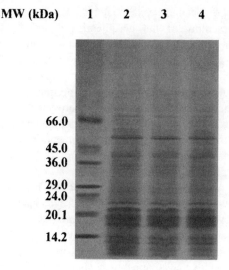

Figure 9.2. Vertical SDS-PAGE (gradient 5–20% T) protein profile of *Nostoc* sp. following increasing exposure time to UV-B. Lane 1: marker proteins; Lane 2: unirradiated control; Lanes 3 and 4: 30 min and 60 min of UV-B irradiation, respectively.

disruption of their structural entities (Sinha et al., 1997). UV-induced inhibition of $^{14}O_2$ uptake in various cyanobacteria has been reported (Figure 9.3), which could be due to the effect on the photosynthetic apparatus, leading to the reduction in the supply of ATP and $NADPH_2$ (Kumar et al., 1996a; Sinha et al., 1996). A disruption of the cell membrane and/or alteration in thylakoid integrity as a result of UV-B radiation may partly or wholly destroy the component required for photosynthesis and may thus affect the rate of CO_2 fixation (Sinha et al., 1996, 1997). In addition, UV-B-induced opening of the membrane-bound calcium channels has been demonstrated in the cyanobacterium *Anabaena* sp. (Richter et al., 1999). Thus, in natural habitats, avoidance of UV-B radiation seems to be of utmost importance for cyanobacterial growth and nitrogen fixation.

Photoprotection in Cyanobacteria

Cyanobacteria protect themselves from photodamage by adopting one or more of the following strategies:

1. Production of UV-absorbing/screening substances such as mycosporine-like amino acids (MAAs) and scytonemin (Sinha et al., 1998b; Cockell & Knowland, 1999; Gröniger et al., 2000).

2. Escape from UV radiation by migration into habitats having reduced bright light exposure. Such strategies include phototactic, photokinetic, and photophobic responses (Häder 1987a, 1987b), vertical migration into deeper strata of mat communities (Bebout & Garcia-Pichel, 1995), and sinking and floating behavior by a combination of gas vacuoles and ballast (Reynolds et al., 1987).

3. Production of quenching agents such as carotenoids (Burton & Ingold, 1984) or enzyme systems, such as those containing superoxide dismutase, that react with and

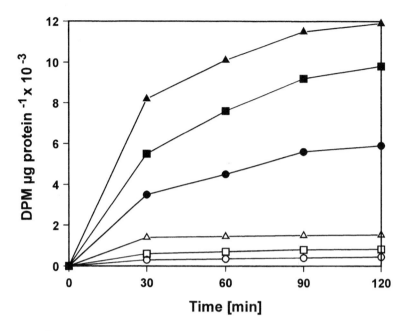

Figure 9.3. NaH^{14}CO$_3$ uptake by various cyanobacteria under fluorescent and UV-B light: triangle, *Scytonema* sp.; square, *Anabaena* sp.; and circle, *Nostoc* sp.; Solid symbols, fluorescent, and open symbols, UV-B light.

thereby neutralize the highly toxic reactive oxygen species produced by UV radiation (Vincent & Quesada, 1994).

4. Availability of a number of repair mechanisms such as photoreactivation and light-independent nucleotide excision repair of DNA (Britt, 1995; Kim & Sancar, 1995; Thoma 1999) and UV-A/-blue light-guided repair of the photosynthetic apparatus (Christopher & Mullet, 1994).

5. Chromatic adaptation (variation in phycocyanin/phycoerythrin ratio), which allows regulation of the balance of wavelengths of light absorbed (Tandeau de Marsac, 1977).

Mycosporine-Like Amino Acids

MAAs are water-soluble substances characterized by a cyclohexenone or cyclohex-enimine chromophore conjugated with the nitrogen substituent of an amino acid or its imino alcohol, having absorption maxima ranging from 310 to 360 nm and an average molecular weight of around 300 (Dunlap & Shick, 1998; Sinha et al., 1998b; Cockell & Knowland, 1999; Gröniger et al., 2000). A number of cyanobacteria isolated from freshwater, marine, or terrestrial habitats contain MAAs (Garcia-Pichel et al., 1993; Sinha et al., 1998b). Presence of MAAs has also been reported in the Antarctic (Quesada & Vincent, 1997), as well as in a community of halophilic cyanobacteria (Oren 1997). The occurrence of high concentrations of MAAs in the organisms exposed to high levels of solar radiation has been described as providing protection as a UV-absorbing sunscreen (Garcia-Pichel et al., 1993; Sinha et al., 1998b), but there is no conclusive evidence for the exclusive role of MAAs as sunscreen, and it is possible that they play more than one role in the cellular metabolism in all or some organisms (Vincent & Roy, 1993; Castenholz, 1997; Oren, 1997). It has been reported that MAAs may act as antioxidants to prevent cellular damage resulting from UV-induced production of active oxygen species (Dunlap & Yamamoto, 1995). Studies with cyanobacteria have shown that MAAs prevent 3 out of 10 photons from hitting cytoplasmic targets. Cells with high concentrations of MAAs are approximately 25% more resistant to UV radiation centered at 320 nm than those with no or low concentrations (Garcia-Pichel et al., 1993). This protection is unlikely to be effective for thin, solitary trichomes but may be especially important in some mat communities or large phytoplankton. The MAAs in *Nostoc commune* have been reported to be extracellular and linked to oligosaccharides in the sheath (Böhm et al., 1995). These glycosylated MAAs represent perhaps the only known example of MAAs that are actively excreted and accumulated extracellularly, and therefore act as a true screen (Ehling-Schulz et al., 1997). Experiments with rice-paddy cyanobacteria, *Anabaena* sp., *Scytonema* sp., and *Nostoc commune*, have revealed the existence and induction by UV-B radiation of a single MAA, shinorine (Figure 9.4), a bisubstituted MAA containing both a glycine and a serine group, with an absorption maximum at 334 nm (Sinha et al., 1999b, 2001). MAAs have been shown to be tolerant to UV-B and heat stress (Sinha et al., 2000). The single-cell sunscreen effect of intracellular MAAs in cyanobacteria is modest, and only 10–30% of incident photons were intercepted in a fairly large-celled cyanobacterium *Gloeocapsa* sp. (Garcia-Pichel et al., 1993), but the screening efficiency may be substantially increased in colony- and mat-forming cyanobacteria (Castenholz, 1997). There may be physiological limitations to the accumulation of osmotically active compounds such as MAAs within the cell, and it seems probable that the maximal specific content of MAAs in the cell is regulated by osmotic mechanisms, reflected by the fact that field populations of halotolerant cyanobacteria contain unusually high concentration of MAAs (Oren 1997). Action spectrum peaking at 310 nm for the induction of MAAs have been described in

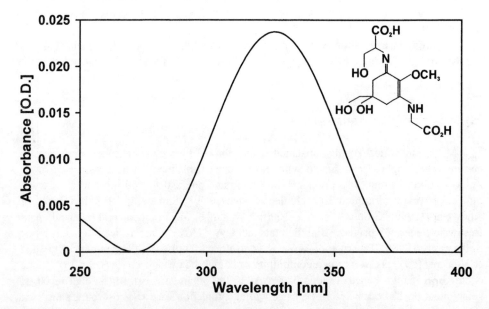

Figure 9.4. Absorption spectrum and molecular structure of mycosporine-like amino acid (MAA), shinorine (λ_{max} 334 nm), commonly found in cyanobacteria.

Chlorogloeopsis (Portwich & Garcia-Pichel, 2000) and a reduced pterin was proposed as a putative candidate for MAAs induction. Action spectra with a pronounced peak at 290 nm for the induction of MAAs has also been determined in rice-field cyanobacteria, *Anabaena* sp. (Sinha et al., 2002b) and *Nostoc commune* (Sinha et al., 2002a).

Scytonemin

Scytonemin is a yellow-brown, lipid-soluble, dimeric pigment located in the extracellular polysaccharide sheath of only some cyanobacteria. It has an *in vivo* absorption maximum at 370 nm (Garcia-Pichel et al., 1992), with a molecular mass of 544 Da and a structure based on indolic and phenolic subunits (Proteau et al., 1993). Purified scytonemin has an absorption maximum at 386 nm (Figure 9.5), but it also absorbs significantly at 252, 278, and 300 nm (Proteau et al., 1993; Sinha et al., 1998b, 1999c).

Strong evidence for the role of scytonemin as UV-shielding compound has been presented in several cyanobacterial isolates and collected materials from various harsh habitats, mostly exposed to high light intensities (Garcia-Pichel & Castenholz, 1991; Gröniger et al., 2000). Its role as a sunscreen was clearly demonstrated in the terrestrial cyanobacterium *Chlorogloeopsis* sp. (Garcia-Pichel et al., 1992). In cyanobacterial cultures, as much as 5% of the cellular dry weight may be accumulated as scytonemin. Naturally occurring cyanobacteria may have even higher specific content (Castenholz, 1997). The correlation between UV protection and scytonemin presence has been established under solar irradiance in a naturally occurring monospecific population of a cyanobacterium, *Calothrix* sp. It was shown that high scytonemin content is required for uninhibited photosynthesis under high UV fluxes (Brenowitz & Castenholz, 1997). Studies indicate that the incident UV-A radiation entering the cells may be reduced by around 90% due to the presence of scytonemin in the cyanobacterial sheaths (Garcia-Pichel & Castenholz, 1991; Proteau et al., 1993; Brenowitz & Castenholz, 1997). Once synthesized,

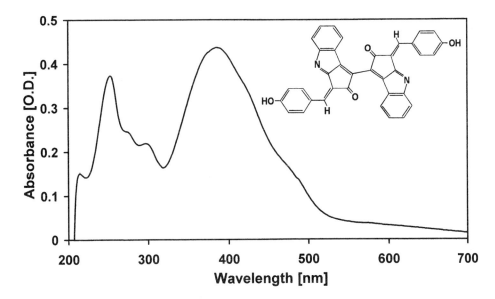

Figure 9.5. Absorption spectrum and molecular structure of cyanobacterial sheath pigment, scytonemin.

it remains highly stable and carries out its screening activity without further metabolic investment from the cell. Rapid photodegradation of scytonemin does not occur, evidenced by its long persistence in terrestrial cyanobacterial crusts or dried mats (Proteau et al., 1993; Brenowitz & Castenholz, 1997). This strategy may be invaluable to several scytonemin-containing cyanobacteria inhabiting harsh habitats, such as intertidal marine mats or terrestrial crusts, where they experience intermittent physiological inactivity (e.g., desiccation). During these metabolically inactive periods, other UV-protective mechanisms such as active repair or biosynthesis of damaged cellular components would be inaffective (Brenowitz & Castenholz, 1997; Castenholz, 1997; Sinha et al., 1998b, 1999c). In addition to these reports, another UV-protective agent with absorption maxima at 312 and 330 nm has been reported from the terrestrial cyanobacterium *Nostoc commune*, a species that also produces scytonemin (Scherer et al., 1988). A brown *Nostoc* species that produces three UV-absorbing compounds, with absorption maxima at 256, 314, and 400 nm, has been reported to be resistant to high light intensity and UV radiation (de Chazal & Smith, 1994). The shielding role against UV-B–induced damage of certain cyanobacterial pigments (a brown-colored pigment from *Scytonema hofmanii* and a pink extract from *Nostoc spongiaeforme*) has been demonstrated (Kumar et al., 1996b).

DNA Damage

The peak absorption of DNA lies in the UV-C range that is absorbed in the upper layers of the atmosphere and does not reach the Earth's surface. However, the absorption of UV-B radiation by DNA is sufficient to induce severe damage in cyanobacteria. Absorbed quanta of UV radiation can induce changes in the molecular structure of the DNA (Karentz 1994). The two major classes of mutagenic DNA lesions (Figure 9.6) induced by UV radiation are *cis-syn* cyclobutane pyrimidine dimers (CPDs) and pyrimidine (6-4) pyrimidone photoproducts (6-4PPs). 6-4PPs are formed at 20–30% of the yields of CPDs (Friedberg et al., 1995; Thoma, 1999). Both classes of lesions distort the DNA helix. CPDs and 6-4PPs induce a bend or kink of 7–9° and 44°, respectively (Wang & Taylor, 1991; Kim et al., 1995). The ability of UV radiation to damage a given base is determined by the flexibility of the DNA. Sequences that facilitate bending and unwinding are favorable sites for damage formation; for example, CPDs form at higher yields in single-stranded DNA at the flexible ends of poly(dA) · (dT) tracts, but not in their rigid center (Becker & Wang, 1989; Lyamichev, 1991). Bending of DNA toward the minor groove reduces CPD formation (Pehrson & Cohen, 1992). One of the transcription factors having a direct effect on DNA damage formation and repair is the TATA-box binding protein (TBP). TBP promotes the selective formation of 6-4PPs in the TATA-box, where the DNA is bent, but CPDs are formed at the edge of the TATA-box and outside, where DNA is not bent (Aboussekhra & Thoma, 1999). These DNA lesions interfere with DNA transcription and replication, and can lead to misreadings of the genetic code, causing mutations and death.

Repair of DNA

Photoreactivation

To remove DNA lesions generated by sunlight, many organisms have enzymes that specifically bind to CPDs (CPD photolyase) or 6-4PPs (6-4 photolyase) and reverse the

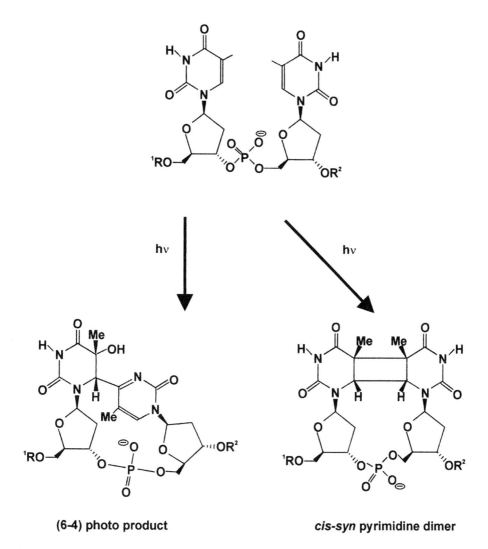

Figure 9.6. Formation of most abundant DNA lesions [cyclobutane pyrimidine dimers and (6-4) photoproducts] after UV exposure of DNA.

damage using the energy of light (photoreactivation). Photolyases contain flavin adenine dinucleotide (FAD) as a catalytic cofactor and a second chromophore as a light-harvesting antenna. The second chromophores are either 5,10-methenyl-tetrahydrofolate or 8-hydroxy-5-deazariboflavin, with absorption maxima of ~380 and ~440 nm, respectively. The crystal structure of CPD photolyase of *Escherichia coli* and *Anacystis nidulans* suggests that upon binding to DNA, the enzymes flip the pyrimidine dimer out of the duplex into a hole that contains the catalytic cofactor (Park et al., 1995; Tamada et al., 1997). The cyclobutane ring is then split by a light-initiated electron transfer reaction. CPD photolyases recognize CPDs with a selectivity similar to that of sequence-specific DNA-binding proteins (Sancar et al., 1987), which suggests that they could compete with histones for DNA accessibility in a manner similar to transcription factors.

Nucleotide Excision Repair (NER)

In contrast to photoreactivation, NER is a much more complex pathway that removes a wide variety of DNA-distorting lesions, including CPDs and 6-4PPs. It is present in most organisms and is not essential for viability, but defects in repair genes cause the sun-sensitive genetic disorders. NER is divided into two subpathways. Transcription-coupled repair (TC-NER) refers to the preferential repair of transcribed strands in active genes, whereas global genome repair (GG-NER) refers to repair in nontranscribed parts of the genome, including the nontranscribed strand of transcribed genes (Friedberg et al., 1995).

Heat and Salinity Stress

The physical environment of an organism is not constant; fluctuations occur in time and space. However, organisms have evolved to survive such conditions and show adaptation to the stressor(s). Each stress variable usually has a minimum and a maximum level beyond which an organism cannot survive (Levitt, 1980). There is also some interaction between the severity of a stressor and its duration. A long exposure at a moderately high temperature may be as injurious as a brief exposure to an extreme temperature. Temperature fluctuations in the environment can occur over a matter of minutes or seasons. Long-term environmental changes may also occur, such as the global warming that is suggested to occur because of increased greenhouse gas emissions (Parry et al., 1989; Hansen et al., 2000; Robertson et al., 2000; Spurgeon, 2000; Blunier & Brook, 2001; Monnin et al., 2001). It has been documented that most organisms including cyanobacteria respond to shock treatment by synthesizing a new set of proteins (Borbély & Surányi, 1988; Bhagwat & Apte, 1989; Thomas et al., 1990; Schubert et al., 1993; Sinha & Häder, 1996b). It has been postulated that particular proteins whose synthesis is induced by stress conditions are critical in the survival of the organism. Changes in gene activation, transcription, and translation often occur during the acclimation process and thus are thought to be involved in the induction of tolerance. It is a common phenomenon that above a threshold temperature, the normal pattern of protein synthesis is repressed and a new set of proteins is synthesized from newly transcribed mRNA, called heat-shock proteins (HSPs) (Schubert et al., 1993). There are a number of major families of HSPs, classified on the basis of their molecular weight; high molecular weight HSPs (80 kDa), the HSPs 70 family (molecular weight approximately 70 kDa), and low molecular weight HSPs (14–30 kDa). The heat-shock response is finely tuned to its thermal adaptation (Howarth & Ougham, 1993). Like other organisms, cyanobacteria have been reported to possess a heat-shock response (Borbély et al., 1985; Bhagwat & Apte, 1989; Nicholson et al., 1991; Sinha & Häder, 1996b).

Salinity has become an ever-increasing problem in irrigated agriculture. Certain bacterial strains such as *Rhizobium*, which nodulate a number of economically important crops, are highly sensitive to NaCl (Upchurch & Elkan, 1977). N_2 fixation in cyanobacteria is supposed to be the most sensitive process in response to enhanced salt concentrations, followed by photosynthesis; respiration is often increased in salt-adapted cyanobacteria (Tel-Or, 1980; Vonshak et al., 1988). Decreased photosynthesis and Chl *a* contents have been reported in *Microcystis firma* and *Synechocystis* sp. PCC 6803 (Erdmann et al., 1992) following salinity stress. Salinity-dependent limitation of photosynthesis and oxygen exchange have been reported in microbial mats consisting of cyanobacteria (Garcia-Pichel et al., 1999). Salt-induced sucrose accumulation mediated by sucrose-phosphate-synthase have been shown in cyanobacteria (Hagemann & Marin, 1999). Growth and protein synthesis have been reported to be significantly suppressed in a rice-paddy cyanobacterium, *Anabaena* sp., when the cells were subjected to elevated levels of salinity (Sinha & Häder, 1996b). The degree of suppression depends on the severity of the stressor. Suppression of normal protein synthesis during stress is commonly observed among other experimental systems (Ballinger & Pardue, 1982; Kimpel & Key, 1985).

Despite the widespread existence of shock proteins, very little is known about their precise physiological function(s). Several lines of evidence suggest that one of the major functions of the stress proteins is to enable the organisms to cope with the stress. Induction and accumulation of HSPs are closely correlated with the development of thermotolerance, that is, the ability of an organism to withstand an elevated, normally lethal temperature (Lin et al., 1984). During heat stress, there is not only a transcriptional activation of heat shock genes but also transcriptional repression of most previously active genes. This might be due to temperature induced alterations in the transcriptional machinery (Schöffl et al., 1987; Miller & Ziskin, 1989). An alternative/additional mechanism is suggested by Westwood et al. (1991), who found that heat-shock factor binds not only to puff sites on chromosomes associated with heat shock but also to a large number of other sites, suggesting that heat-shock factor binding may lead to the repression of normal gene activity. Repression of normal protein synthesis is also supposed to be under translational control exerted over the preexisting, normal cellular mRNAs. These are normally not degraded during heat stress but rather persist in a translatable form (Lindquist, 1981). On recovery from heat stress, these mRNAs are translated (Storti et al., 1980). Herman et al., (1995) reported that regulation of the heat-shock response depends on divalent metal ions in an *hflB* mutant of *E. coli*. It is hard to demonstrate unambiguously that a particular observed change in protein expression contributes to the survival process. Changes in the expression of large numbers of proteins occur due to stress, yet it is probable that only some of these proteins are directly involved in stress tolerance. It is possible that, in some cases, the synthesis of a protein indicates sensitivity to a stress rather than being part of a tolerance mechanism. In addition, of the previously reported stress factors, cyanobacteria often experience a long desiccation period in their natural habitats. The mechanisms of desiccation tolerance in cyanobacteria has been extensively reviewed by Potts (1994, 1999).

In conclusion, any substantial increase in these stressors in nature might be detrimental to the ecologically and economically important cyanobacterial communities, which in turn may affect the productivity of higher plants. There might be a setback in the agricultural economy of all countries in which cyanobacteria is being considered as an alternate natural source of nitrogenous fertilizers for rice paddies and other crops.

Acknowledgments

I thank Prof. D.-P. Häder for providing laboratory facilities. Thanks are also due to M. Schuster for excellent technical assistance.

References

Aboussekhra, A., & Thoma, F. (1999). TATA-binding protein promotes the selective formation of UV-induced (6-4)-photoproducts and modulates DNA repair in the TATA box. *The EMBO Journal, 18*, 433–443.

Ballinger, D. G., & Pardue, M. I. (1982). The subcellular compartmentalization of mRNA in heat shocked *Drosophila* cells. In: M. J. Schlesinger, M. Ashburner, & A. Tissieres (Eds.), *Heat shock from bacteria to man*. (pp. 183–190). Painview, NY: Cold Spring Harbor Laboratory.

Bebout, B. M., & Garcia-Pichel, F. (1995). UV-B induced vertical migrations of cyanobacteria in a microbial mat. *Applied and Environmental Microbiology, 61*, 4215–4222.

Becker, M. M., & Wang, Z. (1989). Origin of ultraviolet damage in DNA. *Journal of Molecular Biology, 210*, 429–438.

Belkin, S., & Padan, E. (1978). Hydrogen metabolism in the facultative anoxygenic cyanobacteria (blue-green algae). *Oscillatoria limnetica* and *Aphanothece halophytica*. *Archives of Microbiology, 116*, 109–111.

Bhagwat, A. A., & Apte, S. K. (1989). Comparative analysis of proteins induced by heat shock, salinity and osmotic stress in the nitrogen-fixing cyanobacterium *Anabaena* sp. strain L-31. *Journal of Bacteriology, 171*, 5187–5189.

Blakefield, M. K., & Harris, D. O. (1994). Delay of cell differentiation in *Anabaena aqualis* caused by UV-B radiation and the role of photoreactivation and excision repair. *Photochemistry and Photobiology, 59*, 204–208.

Blumthaler, M., & Ambach, W. (1990). Indication of increasing solar ultraviolet-B radiation flux in alpine regions. *Science, 248*, 206–208.

Blunier, T., & Brook, E. J. (2001). Timing of millennial-scale climate change in Antarctica and greenland during the last glacial period. *Science, 291*, 109–112.

Böhm, G. A., Pfleiderer, W., Böger, P., & Scherer, S. (1995). Structure of a novel oligosaccharide-mycosporine-amino acid ultraviolet A/B sunscreen pigment from the terrestrial cyanobacterium *Nostoc commune*. *Journal of Biological Chemistry, 270*, 8536–8539.

Borbély, G., & Surányi, G. (1988). Cyanobacterial heat-shock proteins and stress responses. *Methods in Enzymology, 167*, 622–627.

Borbély, G., Surányi, G., Korez, A., & Palfi, Z. (1985). Effect of heat shock on protein synthesis in cyanobacterium *Synechococcus* sp. strain PCC 6301. *Journal of Bacteriology, 161*, 1125–1130.

Brenowitz, S., & Castenholz, R. W. (1997). Long-term effects of UV and visible irradiance on natural populations of a scytonemin-containing cyanobacterium (*Calothrix* sp.). *FEMS Microbiology Ecology, 24*, 343–352.

Britt, A. B. (1995). Repair of DNA damage induced by ultraviolet radiation. *Plant Physiology, 108*, 891–896.

Burton, G. W., & Ingold, K. U. (1984). β-Carotene: An unusual type of lipid antioxidant. *Science, 224*, 569–573.

Castenholz, R. W. (1997). Multiple strategies for UV tolerance in cyanobacteria. *Spectrum, 10*, 10–16.

Christopher, D. A., & Mullet, J. E. (1994). Separate photosensory pathways co-regulate blue light/ultraviolet-A-activated psbD-psbC transcription and light induced D2 and CP43 degradation in barley (*Hordeum vulgare*). chloroplasts. *Plant Physiology, 104*, 1119–1129.

Cockell, C. S., & Knowland, J. (1999). Ultraviolet radiation screening compounds. *Biological Review 74*, 311–345.

Cohen, Y., Jorgensen, B. B., Padan, E. & Shilo, M. (1975). Sulphide-dependent anoxygenic photosynthesis in the cyanobacterium *Oscillatoria limnetica*. *Nature, 257*, 489–492.

Crutzen, P. J. (1992). Ultraviolet on the increase. *Nature, 356*, 104–105.

de Chazal, N. M., & Smith, G. D. (1994). Characterization of a brown *Nostoc* sp. from Java that is resistant to high light intensity and UV. *Microbiology, 140*, 3183–3189.

Donkor, V., Amewowor, D. H. A. K., & Häder, D.-P. (1993). Effects of tropical solar radiation on the motility of filamentous cyanobacteria. *FEMS Microbiology Ecology, 12*, 143–148.

Donkor, V., & Häder, D.-P. (1991). Effects of solar and ultraviolet radiation on motility, photomovement and pigmentation in filamentous, gliding cyanobacteria. *FEMS Microbiology Ecology, 86*, 159–168.

Dunlap, W. C., & Shick, J. M. (1998). Ultraviolet radiation-absorbing mycosporine-like amino acids in coral reef organisms: A biochemical and environmental perspective. *Journal of Phycology, 34*, 418–430.

Dunlap, W. C., & Yamamoto, Y. (1995). Small-molecule antioxidants in marine organisms, antioxidant activity of mycosporine-glycine. *Comparative Biochemistry and Physiology, 112*, 105–114.

Ehling-Schulz, M., Bilger, W., & Scherer, S. (1997). UV-B-induced synthesis of photoprotective pigments and extracellular polysaccharides in the terrestrial cyanobacterium *Nostoc commune*. *Journal of Bacteriology, 179*, 1940–1945.

Elkins, J. W., Thompson, T. M., Swanson, T. H., Butler, J. H., Hall, B. D., Cummings, S. O., Fisher, D. A., & Raffo, A. G. (1993). Decrease in the growth rates of atmospheric chlorofluorocarbons 11 and 12. *Nature, 364*, 780–783.

Erdmann, N., Fulda, S., & Hagemann, M. (1992). Glycosylglycerol accumulation during salt acclimation of two unicellular cyanobacteria. *Journal of General Microbiology, 138*, 363–368.

Fay, P. (1992). Oxygen relations of nitrogen fixation in cyanobacteria. *Microbiological Reviews, 56*, 340–373.
Friedberg, D., Fine, M., & Oren, A. (1979). Effect of oxygen on the cyanobacterium *Oscillatoria limnetica*. *Archives of Microbiology, 123*, 311–313.
Friedberg, E. C., Walker, G. C., & Siede, W. (1995). *DNA repair and mutagenesis.* Washington, DC: ASM Press.
Garcia-Pichel, F., & Castenholz, R. W. (1991). Characterization and biological implications of scytonemin, a cyanobacterial sheath pigment. *Journal of Phycology, 27*, 395–409.
Garcia-Pichel, F., Kühl, M., Nübel, U., & Muyzer, G. (1999). Salinity-dependent limitation of photosynthesis and oxygen exchange in microbial mats. *Journal of Phycology, 35*, 227–238.
Garcia-Pichel, F., Sherry, N. D., & Castenholz, R. W. (1992). Evidence for an ultraviolet sunscreen role of the extracellular pigment scytonemin in the terrestrial cyanobacterium *Chlorogloeopsis* sp. *Photochemistry and Photobiology, 56*, 17–23.
Garcia-Pichel, F., Wingard, C. E., & Castenholz, R. W. (1993). Evidence regarding the UV sunscreen role of a mycosporine-like compound in the cyanobacterium *Gloeocapsa* sp. *Applied and Environmental Microbiology, 59*, 170–176.
Gröniger, A., Sinha, R. P., Klisch, M., & Häder, D.-P. (2000). Photoprotective compounds in cyanobacteria, phytoplankton and macroalgae—a database. *Journal of Photochemistry and Photobiology B: Biology, 58*, 115–122.
Grossman, A. R., Schaefer, M. R., Chiang, G. G., & Collier, J. L. (1993). Phycobilisome, a light harvesting complex responsive to environmental conditions. *Microbiological Reviews, 57*, 725–749.
Häder, D.-P. (1987a). Photomovement. In: P. Fay & C. Van Baalen (Eds.), *The Cyanobacteria* (pp. 325–345). Amsterdam: Elsevier.
Häder, D.-P. (1987b). Photosensory behavior in prokaryotes. *Microbiological Reviews, 51*, 1–21.
Hagemann, M., & Marin, K. (1999). Salt-induced sucrose accumulation is mediated by sucrose-phosphate-synthase in cyanobacteria. *Journal of Plant Physiology, 155*, 424–430.
Hansen, J., Sato, M., Ruedy, R., Lacis, A. & Oinas, V. (2000). Global warming in the twenty-first century, An alternative scenario. *Proceedings of the National Academy of Sciences* (USA), *97*, 9875–9880.
Herman, C., Lecat, S., D'Ari, R., & Bouloc, P. (1995). Regulation of the heat-shock response depends on divalent metal ions in an *hflB* mutant of *Escherichia coli*. *Molecular Microbiology, 18*, 247–255.
Howarth, C. J., & Ougham, H. J. (1993). Gene expression under temperature stress. *New Phytologist, 125*, 1–26.
Jorgensen, B. B., Cohen, Y., & Revsbech, N. P. (1988). Photosynthetic potential and light-dependent O_2 consumption in a benthic cyanobacterial mat. *Applied and Environmental Microbiology, 54*, 176–182.
Kannaiyan, S., Govindarajan, K., Lewin, H. D., & Venkataraman, G. S. (1982). Influence of blue-green algal application on rice crop. *Madras Agriculture Journal, 69*, 1–5.
Karentz, D. (1994). Ultraviolet tolerance mechanisms in Antarctic marine organisms. In: C. S. Weiler & P. A. Penhale (Eds.), *Ultraviolet radiation in Antarctica: Measurements and biological effects. Antarctic Research Series, 62*, 93–110.
Kerr, J. B., & McElroy, C. T. (1993). Evidence for large upward trends of ultraviolet-B radiation linked to ozone depletion. *Science, 262*, 1032–1034.
Kim, J. K., Patel, D., & Choi, B. S. (1995). Contrasting structural impacts induced by *cis-syn* cyclobutane dimer and (6–4) adduct in DNA duplex decamers: Implication in mutagenesis and repair activity. *Photochemistry and Photobiology, 62*, 44–50.
Kim, S.-T., & Sancar, A. (1995). Photorepair of nonadjacent pyrimidine dimers by DNA photolyase. *Photochemistry and Photobiology, 61*, 171–174.
Kimpel, J. A., & Key, J. L. (1985). Heat shock in plants. *Trends in Biochemical Sciences, 10*, 353–357.
Kuhlbusch, T. A., Lobert, J. M., Crutzen, P. J., & Warneck, P. (1991). Molecular nitrogen emissions from denitrification during biomass burning. *Nature, 351*, 135–137.
Kumar, A., Sinha, R. P., & Häder, D.-P. (1996a). Effect of UV B on enzymes of nitrogen metabolism in the cyanobacterium *Nostoc calcicola*. *Journal of Plant Physiology, 148*, 86–91.
Kumar, A., Tyagi, M. B., Srinivas, G., Singh, N., Kumar, H. D., Sinha, R. P., & Häder, D.-P. (1996b). UVB shielding role of $FeCl_3$ and certain cyanobacterial pigments. *Photochemistry and Photobiology, 64*, 321–325.
Levitt, J. (1980). Responses of plants to environmental stress. *Chilling, freezing and high temperature stress*. Vol. 1, New York: Academic Press.
Lin, C. Y., Roberts, J. K., & Key, J. L. (1984). Acquisition of thermotolerance in soybean seedlings. *Plant Physiology, 74*, 152–160.
Lindquist, S. (1981). Regulation of protein synthesis during heat shock. *Nature, 293*, 311–314.
Lubin, D., & Jensen, E. H. (1995). Effects of clouds and stratospheric ozone depletion on ultraviolet radiation trends. *Nature, 377*, 710–713.

Lyamichev, V. (1991). Unusual conformation of (dA)n · (dT)n-tracts as revealed by cyclobutane thymine–thymine dimer formation. *Nucleic Acids Research, 19*, 4491–4496.

Miller, M. W., & Ziskin, M. C. (1989). Biological consequences of hyperthermia. *Ultrasound in Medicine and Biology, 15*, 707–722.

Monnin, E., Indermühle, A., Dällenbach, A., Flückiger, J., Stauffer, B., Stocker, T. F., Raynaud, D., & Barnola, J.-M. (2001). Atmospheric CO_2 concentrations over the last glacial termination. *Science, 291*, 112–114.

NASA encounters biggest-ever Antarctic ozone hole. (2000). *Nature, 407*, 122.

Nicholson, P., Varley, J. P. A., & Howe, C. J. (1991). A comparison of stress responses in the cyanobacterium *Phormidium laminosum*. *FEMS Microbiology Letters, 78*, 109–114.

Oren, A. (1997). Mycosporine-like amino acids as osmotic solutes in a community of halophilic cyanobacteria. *Geomicrobiology Journal, 14*, 231–240.

Oren, A., & Shilo, M. (1979). Anaerobic heterotrophic dark metabolism in the cyanobacterium *Oscillatoria limnetica*: Sulfur respiration and lactate fermentation. *Archives of Microbiology, 122*, 77–84.

Pace, N. R. (1997). A molecular view of microbial diversity and the biosphere. *Science, 276*, 734–740.

Padan, E. (1979). Facultative anoxygenic photosynthesis in cyanobacteria. *Annual Review of Plant Physiology, 30*, 27–40.

Park, H. W., Kim, S. T., Sancar, A., & Deisenhofer, J. (1995). Crystal structure of DNA photolyase from *Escherichia coli*. *Science, 268*, 1866–1872.

Parker, B. C., Heiskell, L. E., & Thompson, W. J. (1978). Nonbiogenic fixed nitrogen in Antarctica and some ecological implications. *Nature, 271*, 651–652.

Parry, M. L., Carter, T. R., & Porter, J. H. (1989). The greenhouse effect and the future of UK agriculture. *Journal of Royal Agricultural Society of England, 150*, 120–181.

Pehrson, J. R., & Cohen, L. H. (1992). Effects of DNA looping on pyrimidine dimer formation. *Nucleic Acid Research, 20*, 1321–1324.

Portwich, A., & Garcia-Pichel, F. (2000). A novel prokaryotic UVB photoreceptor in the cyanobacterium *Chlorogloeopsis* PCC 6912. *Photochemistry and Photobiology, 71*, 493–498.

Potts, M. (1994). Desiccation tolerance of prokaryotes. *Microbiological Reviews, 58*, 755–805.

Potts, M. (1999). Mechanisms of desiccation tolerance in cyanobacteria. *European Journal of Phycology, 34*, 319–328.

Proteau, P. J., Gerwick, W. H., Garcia-Pichel, F., & Castenholz, R. W. (1993). The structure of scytonemin, an ultraviolet sunscreen pigment from the sheaths of cyanobacteria. *Experientia, 49*, 825–829.

Quesada, A., & Vincent, W. F. (1997). Strategies of adaptation by Antarctic cyanobacteria to ultraviolet radiation. *European Journal of Phycology, 32*, 335–342.

Rambler, M. B., & Margulis, L. (1980). Bacterial resistance to ultraviolet radiation under anaerobiosis: Implications for pre-Phanerozoic evolution. *Science, 210*, 638–640.

Reynolds, C. S., Oliver, R. L., & Walsby, A. E. (1987). Cyanobacterial dominance: The role of buoyancy regulation in dynamic lake environments. *New Zealand Journal of Marine and Freshwater Research, 21*, 379–390.

Richter, P., Krywult, M., Sinha, R. P., & Häder, D.-P. (1999). Calcium signals from heterocysts of *Anabaena* sp. after UV irradiation. *Journal of Plant Physiology, 154*, 137–139.

Robertson, G. P., Paul, E. A., & Harwood, R. R. (2000). Greenhouse gases in intensive agriculture: Contributions of individual gases to the radiative forcing of the atmosphere. *Science, 289*, 1922–1925.

Roger, P. A., & Kulasooriya (1980). *Blue-green Algae and rice*. Los Banos, Laguna, Philippines: International Rice Research Institute.

Sancar, G. B., Smith, F. W., Reid, R., Payne, G., Levy, M., & Sancar, A. (1987). Action mechanism of *Escherichia coli* DNA photolyase: I. Formation of the enzyme-substrate complex. *Journal of Biological Chemistry, 262*, 478–485.

Scherer, S., Chen, T. W., & Böger, P. (1988). A new UV-A/B protecting pigment in the terrestrial cyanobacterium *Nostoc commune*. *Plant Physiology, 88*, 1055–1057.

Schöffl, F., Rossol, I., & Angermüller, S. (1987). Regulation of the transcription of heat shock genes in nuclei from soybean (*Glycine max*) seedlings. *Plant, Cell and Environment, 10*, 113–120.

Schopf, J. W (1993). Microfossils of the early archean apex chert: New evidence of the antiquity of life. *Science, 260*, 640–646.

Schubert, H., Fulda, S., & Hagemann, M. (1993). Effects of adaptation to different salt concentration on photosynthesis and pigmentation of the cyanobacterium *Synechocystis* sp. PCC 6803. *Journal of Plant Physiology, 142*, 291–295.

Singh, R. N. (1961). *Role of blue-green algae in nitrogen economy of Indian agriculture*. New Delhi: Indian Council of Agricultural Research.

Sinha, R. P., & Häder, D.-P. (1996a). Photobiology and ecophysiology of rice field cyanobacteria. *Photochemistry and Photobiology, 64,* 887–896.

Sinha, R. P., & Häder, D.-P. (1996b). Response of a rice field cyanobacterium *Anabaena* sp. to physiological stressors. *Environmental and Experimental Botany, 36,* 147–155.

Sinha, R. P., & Häder, D.-P. (1998). Effects of ultraviolet-B radiation in three rice field cyanobacteria. *Journal of Plant Physiology, 153,* 763–769.

Sinha, R. P., & Häder, D.-P. (2000). Effects of UV-B radiation on cyanobacteria. *Recent Research Developments in Photochemistry & Photobiology, 4,* 239–246.

Sinha, R. P., Ambasht, N. K., Sinha, J. P., & Hader, D.-P. (2002a). Wavelength dependent induction of a mycosporine-like amino acid in a rice-field cyanobacterium, *Nostoc commune*: role of inhibitors and salt stress. *Photochemical and Photobiological Sciences* (in press).

Sinha, R. P., Klisch, M., Gröniger, A., & Häder, D.-P. (1998b). Ultraviolet-absorbing/screening substances in cyanobacteria, phytoplankton and macroalgae. *Journal of Photochemistry and Photobiology B: Biology, 47,* 83–94.

Sinha, R. P., Klisch, M., Gröniger, A., & Häder, D.-P. (2000). Mycosporine-like amino acids in the marine red alga *Gracilaria cornea*—effects of UV and heat. *Environmental and Experimental Botany, 43,* 33–43.

Sinha, R. P., Klisch, M., & Häder, D.-P. (1999b). Induction of a mycosporine-like amino acid (MAA) in the rice-field cyanobacterium *Anabaena* sp. by UV irradiation. *Journal of Photochemistry and Photobiology B: Biology, 52,* 59–64.

Sinha, R. P., Klisch, M., Helbling, E. W., & Häder, D.-P. (2001). Induction of mycosporine-like amino acids (MAAs). in cyanobacteria by solar ultraviolet-B radiation. *Journal of Photochemistry and Photobiology B: Biology, 60* 129–135.

Sinha, R. P., Klisch, M., Vaishampayan, A., & Häder, D.-P. (1999c). Biochemical and spectroscopic characterization of the cyanobacterium *Lyngbya* sp. inhabiting Mango (*Mangifera indica*) trees: Presence of an ultraviolet-absorbing pigment, scytonemin. *Acta Protozoologica, 38,* 291–298.

Sinha, R. P., Kumar, H. D., Kumar, A., & Häder, D.-P. (1995a). Effects of UV-B irradiation on growth, survival, pigmentation and nitrogen metabolism enzymes in cyanobacteria. *Acta Protozoologica, 34,* 187–192.

Sinha, R. P., Lebert, M., Kumar, A., Kumar, H. D., & Häder, D.-P. (1995b). Disintegration of phycobilisomes in a rice field cyanobacterium *Nostoc* sp. following UV irradiation. *Biochemistry and Molecular Biology International, 37,* 697–706.

Sinha, R. P., Lebert, M., Kumar, A., Kumar, H. D., & Häder, D.-P. (1995c). Spectroscopic and biochemical analyses of UV effects on phycobiliproteins of *Anabaena* sp. and *Nostoc carmium*. *Botanica Acta, 180,* 87–92.

Sinha, R. P., Singh, S. C., & Häder, D.-P. (1999a). Photoecophysiology of cyanobacteria. *Recent Research Developments in Photochemistry & Photobiology, 3,* 91–101.

Sinha, R. P., Singh, N., Kumar, A., Kumar, H. D., Häder, M., & Häder, D.-P. (1996). Effects of UV irradiation on certain physiological and biochemical processes in cyanobacteria. *Journal of Photochemistry and Photobiology B: Biology, 32,* 107–113.

Sinha, R. P., Singh, N., Kumar, A., Kumar, H. D., & Häder, D.-P. (1997). Impacts of ultraviolet-B irradiation on nitrogen-fixing cyanobacteria of rice paddy fields. *Journal of Plant Physiology, 150,* 188–193.

Sinha, R. P., Sinha, J. P., Gröniger, A., & Häder, D.-P. (2002b). Polychromatic action spectrum for the induction of a mycosporine-like amino acid in a rice-field cyanobacterium, *Anabaena* sp. *Journal of Photochemistry and Photobiology B: Biology, 66,* 47–53.

Sinha, R. P., Vaishampayan, A., & Häder, D.-P. (1998a). Plant–cyanobacterial symbiotic somaclones as a potential bionitrogen-fertilizer for paddy agriculture: Biotechnological approaches. *Microbiological Research, 153,* 297–307.

Spurgeon, D. (2000). Global warming threatens extinction for many species. *Nature, 407,* 121.

Storti, R. V., Scott, M. P., Rich, A., & Pardue, M. L. (1980). Translational control of protein synthesis in response to heat shock in *D. melanogaster* cells. *Cell, 22,* 825–834.

Tabazadeh, A., Santee, M. L., Danilin, M. Y., Pumphrey, H. C., Newman, P. A., Hamill, P. J., & Mergenthaler, J. L. (2000). Quantifying denitrification and its effect on ozone recovery. *Science, 288,* 1407–1411.

Tamada, T., Kitadokoro, K., Higuchi, Y., Inaka, K., Yasui, A., de Ruiter, P. E., Eker, A. P., & Miki, K. (1997). Crystal structure of DNA photolyase from *Anacystis nidulans*. *Nature Structural Biology, 4,* 887–891.

Tandeau de Marsac, N. (1977). Occurrence and nature of chromatic adaptation in cyanobacteria. *Journal of Bacteriology, 130,* 82–91.

Tandeau de Marsac, N., & Houmard, J. (1993). Adaptation of cyanobacteria to environmental stimuli, new steps towards molecular mechanisms. *FEMS Microbiology Reviews, 104,* 119–190.

Tel-Or, E. (1980). Response of N_2-fixing cyanobacteria to salt. *Applied Environmental Microbiology, 40*, 689–693.

Thoma, F. (1999). Light and dark in chromatin repair, repair of UV-induced DNA lesions by photolyase and nucleotide excision repair. *The EMBO Journal, 18*, 6585–6598.

Thomas, S. P., Zaritsky, A., & Boussiba, S. (1990). Ammonia excretion by an L-methionine-DL-sulfoximine resistant mutant of the rice field cyanobacterium *Anabaena siamensis*. *Applied and Environmental Microbiology, 56*, 3499–3504.

Toon, O. B., & Turco, R. P. (1991). Polar stratospheric clouds and ozone depletion. *Scientific American, 264*, 68–74.

Upchurch, R. G., & Elkan, G. H. (1977). Comparision of colony morphology, salt tolerance and effectiveness in *Rhizobium japonicum*. *Canadian Journal of Microbiology, 23*, 1118–1122.

Vaishampayan, A., Sinha, R. P., & Häder, D.-P. (1998). Use of genetically improved nitrogen-fixing cyanobacteria in rice paddy fields, prospects as a source material for engineering herbicide sensitivity and resistance in plants. *Botanica Acta, 111*, 176–190.

Venkataraman, G. S. (1972). *Algal biofertilizers and rice cultivation*. New Delhi: Today and Tomorrows Printers and Publishers.

Vincent, W. F., & Quesada, A. (1994). Ultraviolet radiation effects on cyanobacteria, implication for Antarctic microbial ecosystems. In: C. S. Weiler & P. A. Penhale (Eds.), *Ultraviolet radiation in Antarctica*, Measurements and biological effects. *Antarctic Research Series, American Geophysical Union, 62*, 111–124.

Vincent, W. F., & Roy, S. (1993). Solar ultraviolet-B radiation and aquatic primary production: Damage, protection, and recovery. *Environmental Reviews, 1*, 1–12.

Vonshak, A., Guy, R., & Guy, M. (1988). The response of the filamentous cyanobacterium *Spirulina platensis* to salt stress. *Archives of Microbiology, 150*, 417–420.

Walsby, T. (1975). Gas vesicles. *Annual Review of Plant Physiology, 36*, 427–439.

Wang, C. I., & Taylor, J. S. (1991). Site-specific effect of thymine dimer formation on dAn · dTn tract bending and its biological implications. *Proceedings of the National Academy of Sciences* (USA), *88*, 9072–9076.

Walter, M. R., Bauld, J., Des Marais, D. J., & Schopf, J. W. (1992). A general comparison of microbial mats and microbial stromatolites bridging the gap between the modern and the fossil. In: J. W. Schopf & C. Klein (Eds.), *The Proterozoic biosphere* (pp. 335–338). Cambridge, UK: Cambridge University Press.

Watanabe, I. (1984). Use of symbiotic, free-living blue-green algae in rice culture. *Outlook on Agriculture, 13*, 166–172.

Westwood, J. T., Clos, J., & Wu, C. (1991). Stress-induced digomerization and chromosomal relocalization of heat-shock factor. *Nature, 353*, 822–827.

10

Biomonitoring and Bioindicators in Aquatic Ecosystems

Nándor Oertel and János Salánki

Introduction

In the past decades, an extremely large number of natural and synthetic chemical substances have been produced and used without knowledge of the possible environmental impact of the release of these chemicals, their impurities, and degradation products. Although most of these chemicals did not evoke an immediate effect in the aquatic environment, the total load of pollutants has certainly contributed to the observed changes in the structure and function of aquatic ecosystems.

For example, the enormous development of pesticides after World War II and the associated negative effects on consumers (birds and other organisms) of polluted prey, plus the emergence of diseases caused by toxic metals, such as Mina Mata and Itai-Itai, led to the shift in emphasis of environmental pollutants after 1962, the year in which Rachel Carson's *Silent Spring* was published (1962). Since then, there has been interest in the early detection of undesirable environmental effects of chemical pollutants.

The goal of aquatic biological monitoring is to give reliable and proper information on the possible effects of chemicals present in the water due to human activities, to enable the protection of the aquatic ecosystems, and particularly to provide scientific guidance for legislation and enforcement.

Both the new global problems (changes in global climate, greenhouse effect, ozone, etc.) and the awareness of the complexity of ecosystems in environmental policy demonstrate that additional information on biological systems is required. Most monitoring efforts started in the 1960s, concerned chemical and physical parameters. We have become aware that most monitoring programs can provide a sensitive indication of the concentration of substances that have been selected for study, but this information has

---

Nándor Oertel • Hungarian Danube Research Station of the Hungarian Academy of Science, Göd, Hungary
János Salánki • Balaton Limnological Research Institute of the Hungarian Academy of Science, Tihany, Hungary.

lacked essential biological information. Pollution implies deleterious effects, and these are usually assessed in relation to a biological system (Oertel, 1993a; de Zwart, 1995).

The biological objects used in monitoring the state of the environment, as well as the changes caused by the impact of environmental pollution, called *bioindicators* or *biomonitors*, cover a wide scale and level of biological systems and functions, from population to subcellular processes.

Emphasis is on monitoring biological variables, because biological monitoring systems can be cheap, require less sophisticated instruments and, most importantly, reflect the integrated expression of pollutional load. Nevertheless, parallel chemical monitoring must be done under all circumstances to reveal causative factors.

In this chapter, we make an effort to review the bioindicators and biomonitoring methods used for the analysis and evaluation of water quality. Because the topic is covered by lots of disciplines and the extent of this chapter is limited, only some selected conceptual and methodological aspects of the state-of-the-art branches of bioindication and biomonitoring in the aquatic ecosystems are expressed here.

Definition and Functions of Biomonitoring

In general, monitoring is a process of repetitive observation, for a defined purpose, of one or more elements of the environment, according to prearranged schedules in space and time, and using comparable methods for environmental sensing and data collection. Biomonitoring with the use of (aquatic) organisms as natural monitors provides additional and factual information concerning the present state and the past trends in environmental behavior. Generally, the monitoring systems may be descriptive in relation to emission (concentration or effects) or regulatory, comparable to control. According to this, these systems have some common functions:

- Signal function is used for the detection of adverse changes in environment, for early warning, and in searching for the causes of changes (localization of sources and emissions).
- Prediction function deals with the prediction of autonomous changes in the environment and contribution to the prediction of the effect of human activities to improve environmental (measures) or any other factor producing some load on the environment.
- Control function is frequently used for the observation of measurements or the achievement of goals concerning environmental quality.
- Scientific research for environmental management as activity–effect studies, or construction of compound balances and a response to the specific questions regarding the control of government concerning policy subjects, all correspond to the instrument function.

Biological monitoring, or biomonitoring, is the use of biological responses to assess changes in the environment, generally, changes due to anthropogenic causes. Biomonitoring programs can be qualitative, quantitative, or semiquantitative. According to the nature of the chemical compounds being monitored (fluctuating qualitatively, quantitatively, or both), the type of biosensors used to detect changes in water quality, and the treatment of response variables given by the organisms (chemical, electrical, computer analysis, etc.), one can differentiate biological monitoring also as "semicontinuous" or "continuous" methods. Biomonitoring, a valuable assessment tool, is receiving increased use in water quality monitoring programs of all types. According to the goals of a project, there are different types of biomonitoring. One type of biomonitoring is surveillance before and after a project is complete or before and after a toxic substance enters the water. The other type of biomonitoring ensures compliance with regulations and guidelines, or that water quality is maintained.

Biomonitoring involves the use of indicators, indicator species, or indicator communities. Generally, benthic macroinvertebrates, fish, and/or algae are used. Certain aquatic plants have also been used as indicator species for pollutants, including nutrient enrichment (Batiuk et al., 1992; Phillips & Rainbow, 1993). There are advantages and disadvantages to each method. Macroinvertebrates are the most frequently used biomonitors (Rosenberg & Resh, 1993). Biochemical, genetic, morphological, and physiological changes in certain organisms have been noted as being related to particular environmental stressors and can be used as indicators. The presence or absence of the indicator, or an indicator species or community, reflects environmental conditions. Absence of a species is not as meaningful as it might seem, because there may be reasons other than pollution that can result in its absence (e.g., predation, competition, or geographic barriers that prevented it from ever being at the site). Absence of multiple species of different orders with similar tolerance levels that were present previously at the same site is more indicative of pollution than absence of a single species. It is clearly necessary to know which species should be found at the site or in the system.

With the help of monitoring, human society tries to signal environmental changes that result from human activities. Its purpose is to maintain quality control in the biota of the ecosystem. It is important that this prospective signaling is done at an early stage in order to prevent undesirable ecosystem changes. This implies that immediate corrective action should be taken when deleterious effects are detected; otherwise, monitoring merely gives the illusion of protection. Retrospective monitoring is used to describe former environmental conditions (e.g., museum species); deep core sediment samples and deep core ice samples are used to register environmental conditions that preceded the industrial revolution. Many methods of biological and chemical monitoring have been described already; however, in future, there will be many more methods developed. They all share the necessity to predict possible hazards of chemicals for humans and the environment at an early stage.

Relation between Biomonitoring and Toxicology

Studies and description of the action of toxic substances on the ecosystem are the objective of ecotoxicology (Levin et al., 1989), a special combination of ecology and toxicology. Ecotoxicological approaches and methods are integral parts of biomonitoring of environmental pollution.

Ecology deals with the relationship and interaction of species and individuals present in a definite biological community, called an ecosystem, as well as the relationship of organisms with the surrounding physical world. Aquatic ecosystems composed of microbes, plants, and animals inhabiting a river, lake (reservoir), or sea, are characterized by a well-determined interrelationship and dynamic equilibrium. The quality of the outside world for the living is determined by the water itself and by its physical and chemical components. The physical and chemical characteristics of the water are under permanent influence and control of the wider environment and show significant changes and variations. Temperature, precipitation, water supply, and evaporation may result in significant alteration of these components. Aquatic organisms adapt to a certain degree to these changes by modifying the way of their life, such as movement, reproduction, feeding, metabolism, and others, in close relationship with the physical and chemical environment.

The adverse effects in living organisms caused by various chemical substances are considered toxic effects, studied in details by toxicology. Toxicity of chemicals manifests itself as a disturbing effect on life processes due to interference with metabolism at cellular and molecular levels. As a rule, toxic substances are present in the ecosystem in very low concentrations and are considered pollutants. Toxic chemicals that are concentrated, produced, or dissipated in the environment as a result of human activity are called anthropogenic pollutants. Some of them are natural elements of the geosphere (toxic metals), which in the course of civilization and industrial development appear in the environment in concentrations surpassing the geochemical background; others are products of biological systems (bacteria, plants, and animals) as well as inorganic or organic products and by-products of chemical industry and other technological processes.

Environmental pollutants may have a profound effect on marine and freshwater ecosystems, causing disturbances both in the organisms living in them, and in the interrelationship of different species. The effects can be observed at the level of individuals in some sensitive species but can be generalized to the ecosystem as a whole. Therefore, studies concerning the interaction between ecology and toxicology, namely, monitoring the presence and clarifying the biological effect of environmental pollutants, cover sub-individual, individual, population, and ecosystem levels equally.

Origin and Types of Toxic Chemicals in the Environment

The origin of anthropogenic environmental pollution, being the consequence of human activity, used to be divided into industrial, agricultural, communal, and airborne pollution not for didactical reasons only. This sort of grouping, besides the main activity types responsible for the pollution, refers also to some specificity of the release and to the methods of possible counteractions to eliminate adverse consequences.

For example, industrial activity is responsible for the production of a great number of toxic substances. Industrial technologies are involved in ore mining, separation, concentration, and procession of elements present in the geosphere (metals and others). Along with metal processing, in the production of various chemical substances, and use and treatment of by-products and wastes, a great number and large amounts of toxic substances become soluble or get into the air. Dissolved substances are released as components of raw or treated waste waters in higher or lower concentration into natural waters, representing a great part of environmental pollution. The best protection against such types of chemical pollution is application of closed, waste-free technologies, cleaning of polluted industrial waters at the place of production, and introduction of countermeasures for reducing the release of pollutants into the air.

Pollution as side effect of agricultural technologies is in fact also of industrial origin. Herbicides, fungicides, and pesticides are used in plant protection for their toxic biological effects against pests; however, when they reach rivers and lakes, they are considered toxic pollutants because of their harmful, unfavorable effect on the aquatic life and the ecosystem. Toxic chemicals used in plant protection or animal husbandry are applied diffusely in the field over large areas; therefore, collection and elimination of residues or toxic metabolites appearing as pollutants are practically impossible. In this case, reduction of the pollution is possible by limitation of the chemicals used and by application of drugs degrading rapidly after exerting their beneficial effects.

Toxic substances dispersing and reaching the environment from settlements are considered to be communal in origin. Chemicals used in households, such as detergents, bactericides, cosmetics, medicaments, stain removers, drugs of everyday life, wastes of heating and traffic (burning of coal and oil), corrosion of metal instruments, and construction waste belong to this category. A considerable amount of toxic substances is dispersed in the environment from spoils and waste dumps by wind and washing out, reaching also the surface waters. Reduction of settlement pollution in the environment may be achieved by canalization and sewage water purification, collection and proper treatment of wastes; also, limitation of the use of persistent toxicants can moderate pollution originating from municipalities. A significant pollution of rivers and lakes is caused by runoff waters from the settlements and roads.

Air pollution is basically of industrial and communal origin. Nevertheless, it is a special, independent category, because pollutants released into the air are transported for long distances and the fallout takes place far from their origin, either with precipitation or as dry sedimentation. Among toxic substances present in the air, both organic and inorganic chemicals may be present. At the area of the actual pollution, the quality and quantity of pollutants can be measured; however, reduction of the air pollution can be achieved only at the place of its release. Due to these special circumstances, it requires not only local or governmental but also often intergovernmental regulation and intervention.

Toxic chemicals occurring as pollutants can be grouped into two main categories:

- Substances of natural origin.
- Substances produced exclusively by human activity.

Heavy metals and their compounds, as well as some nonmetallic elements occurring in nature, belong to the first group. The most important toxic metals are mercury, cadmium, lead, chromium, aluminium, cobalt, nickel, and also some biologically important elements, such as copper, zinc, and sulphur. Radioactive elements represent a special category in this group. Low-background pollution is a natural phenomenon as a result of weathering of rocks and liberation of metals in the Earth's history. These elements were and are extracted in enormous quantity by mining and metallurgy since the beginning of human civilization, and together with their use, the possibility of pollution of the environment exists. While working with metals and using them in industry, agriculture, and for communal purposes, there are in each case wastes and byproducts produced that may in higher or lower concentration enter into the natural environment. In ionic form or in chemical or microbial processes, they may become soluble; in interaction, these substances may be toxic to living organisms.

Chemical industry produces many thousands of substances for use in industry, pharmacy, agriculture, and households, a great number of which are toxic in the environment.

The compounds themselves or their metabolic products may have adverse effects. Inorganic and organic chemicals with a metal component show close relationship with the former group, because their toxicity depends on the metal ion. The other substances, organic pollutants without metal ion, are the most important toxic compounds—the chlorinated hydrocarbons, polycyclic aromatic hydrocarbons, polychlorinated biphenyls, dioxins, nitrosamines, cyanides, organic esters.

Forms and Manner of Occurrence of Pollutants in the Aquatic Ecosystem

Most of the toxic pollutants reach surface waters in dissolved form at higher or lower concentration, depending on their origin. Others get into rivers and lakes, together with solid waste bound to particles and become deposited in the sediment. Solubility of substances in the water is often pH dependent, and with changes of pH, dissolved toxic substances may precipitate, or be released from their bound form and get into solution. Similar processes may occur as a result of complex-forming and microbial activity. As a consequence, the concentration of toxic substances dissolved in the water may change, resulting in modification of their accessibility for accumulation in living organisms and effectiveness for biological interference.

Due to water movements, diffusion, and mixing of the sediment, toxic substances usually are distributed uniformly in a large area and often also in a vertical direction. Nevertheless, the concentration is higher in the vicinity of point sources bearing diagnostic significance for localization of the origin of the pollution. In more compact sediments, stratification of toxic substances may be present, providing information about the history of the pollution. Also, in distribution and mixing of pollutants, living organisms may play a role, both disturbing physically the water and benthos, and by their metabolism, through turnover of chemicals.

Uptake of dissolved toxic substances into organisms can take place in various ways. They can reach the internal space through the membranes by absorption or with the participation of active transport mechanisms. Absorption requires specific membrane permeability, whereas for transport an energy-consuming carrier mechanism is needed. In accumulation of pollutants, both processes may be involved.

Toxic substances adsorbed to particles are taken up first of all by benthic organisms feeding in the sediment, and by planktonic animals filtering the suspended elements, but macroinvertebrates and fishes take them up also. As a result of digestive processes in the stomach and intestine, toxic substances may be released, solubilized, and passing the endothelium, reach the internal, extracellular fluid. Rooted aquatic plants may take up toxic substances directly from the sediment. Because the release of toxic substances is a slower process than uptake, accumulation of pollutants takes place. Many pollutants are taken up and accumulated in animals through the food chain, by consumption of polluted organisms (biomagnification). Accumulation of toxic substances in living organisms may result in an increase of the toxic effect, but they can go from decomposition to detoxification by binding, deposition in inactive form, or they can be excreted and therefore released back into the water or the sediment.

Effects of Toxic Pollutants on Aquatic Ecosystems

Pollution Effects on River Ecosystems

Due to strong directionality and interactions as a result of drainage basins, rivers are open systems. The special characters of the riverine ecosystems ultimately determine the effects of streamwater pollution on communities. The occurrence of organisms in running water strongly depends on spatial gradients and the temporal dimension.

- The longitudinal gradients—increasing depth and width of the riverbed, decreasing stream velocity and sediment grain size, increasing nutrient load—from headwaters to estuaries are a result of the aggregated effects of hydrological, morphological, geological, geographical, and climatic factors of the catchment areas. These abiotic factors of the longitudinal gradients fundamentally determine the ecological zonation of communities manifested in the structure and function of the river continuum.
- The aquatic zone of the riverine ecosystem is interlinked laterally with the riparian (littoral) and terrestrial (floodplains, wetlands) zones. A great variety of different habitats result from the transverse gradients of stream velocity, erosion, and sedimentation.
- The third spatial dimension appears in the vertical gradients of deepwater, in the sediment and interstitial waters of the river. The latter region is the indispensable life space of the special groundwater biota and the location of essential chemical processes or bank filtration of drinking water.
- Temporal variation—the fourth dimension—may be natural (floods, seasonal changes in water regimen) or human-induced. In the latter case, the time scale may be very short, and the effect of human impact (e.g., lack of natural events because of damming or fragmentation of a river) may be very drastic.

All these gradients result in great variety in habitats and in the species composition of communities along the rivers and watercourses. The actual aquatic community can be considered an integrated response to all existing biological relations and environmental conditions. Besides the hydromorphological dynamism, the temporary state of the riverine ecosystem is greatly affected by toxic pollution, eutrophication, canalization, and other river regulation training, land use on the catchment area, and the malfunction of natural processes as a consequence of human impacts. In order to evaluate the level of ecological integrity of river ecosystems, both man-induced and natural stress factors have to be addressed simultaneously during the monitoring activities of ecological water management.

On the one hand, our ecological knowledge of the structure and main functions of river ecosystems must be strengthened. Understanding the nature of underlying processes is a prerequisite to the correct selection of ecologically representative and sensitive-enough parameters for biological and ecological water quality assessment. On the other hand, biotypological classification of the rivers must be developed in order to define the natural variety in streamwaters. The scientifically valid and conceptual classification of running waters involving the reference sites and aquatic communities may be the only base of realistic biomonitoring in the practice of water management.

According to the classical concept, the river is divided into particular zones (Illies & Botosaneanu, 1963), whereas the continuum concept considers the watercourse as a continuum of communities (Vannote et al., 1980). Although sequential communities may differ from each other both in structure and function, they maintain the continuous flow of matter and energy along the river. The applicability of a continuum concept in large rivers—where discontinuum is prevailing opinion—is argued (Schönborn, 1992). The two concepts might be supplementary if one considers that a continuum approach is the conceptual base for understanding the function of the river ecosystems, whereas classification is the executable, practical way to assess water quality, zone by zone.

A realistic biotypological classification has to involve as many aspects as possible: It should have reference situations both in space and time; it should consider interstitial, riparian, and terrestrial zones (ecotops) besides the running water body; it should be valid for different scales (local–regional, national–international, transboundary, etc.), and last but not least, it should have a holistic approach regarding the whole watershed area.

In large rivers, the point-like and diffuse pollution combines with the geochemical background. Sometimes the individual effects can hardly be separated from each other (e.g., the merged load of point sources upstream appears as a diffuse effect at a downstream section of the studied river). The point sources, the so-called "hot spots," also need careful detection because of the "plume effect," which is directed by the changes in morphology, current, and convection according to the existing spatial and temporal gradients in the river.

Among the sophisticated and changeable conditions of the river ecosystem, those relative methods that seem to be reliable are based on comparison using controls. Translocation of aquatic organisms to monitor pollutants, one of the most frequently used active biomonitoring techniques, is the best example of this type of assessment. To eliminate entirely the uncertainties that accompany the study of local organisms is not possible, but they can be reduced significantly with the translocation techniques. Large rivers differ from the smaller streams not only in their dimension but also in their water-quality character, which can change within the length and width of the river itself, too. Therefore, the methods developed for smaller streams need careful investigation and confirmation during the process of validation in large rivers.

Pollution Effects on Lake Ecosystems

The main characteristics of lake ecosystems are similar to those of rivers; nevertheless, they bear some specific features of significance in monitoring toxic pollution. Lakes as recipients represent an open system for the catchment area, similar to rivers; however, due to limited or lack of outflow, they are closed systems, in which substances transported by inflow waters or precipitation remain for a long time. Although, depending on the size and depth of the lake, the mixing of substances caused by currents, wind effects, and diffusion can vary, dissolved or particulate chemicals disperse comparatively rapidly, and marked concentration gradients cannot remain even if there are differences in the level of pollution of inflow waters. Nevertheless, especially in large lakes, at the mouth of rivers or streams, or at the point sources (settlements, harbors, waste water treatment plants), local differences may be compared to the whole lake. This is less characteristic of the chemical composition of the water itself, but it can be detectable in the sediment as the slow process of dispersion of substances on the surface of the bottom.

Similar to rivers, in shallow or small lakes, the level and effect of pollution depends very much on the amount of precipitation; however, in large lakes, this is not a significant component except at the place of the imminent source of the pollution. In large and deep lakes, also, evaporation is not a significant factor in increasing the concentration of the pollutants. Therefore, striking, fast changes in the concentration of toxic substances cannot be observed in great lakes as a result of modification of water level. Contrary to that, due to mixing and dispersing of sediment, the concentration of toxic substances can rise significantly in the water.

In lakes, contrary to rivers, a great part of the water is eliminated by evaporation instead of outflow or water takeout, so toxic substances having arrived with inflow water will permanently be accumulated in the ecosystem. Reduction of toxic chemicals can take place only as a result of water outflow, sediment digging, and removal of biological products (fish, reed) containing accumulated toxicants. The elevation of the concentration of pollutants in the water is of low degree due to the main route of inactivation as accumulation, binding, and deposition in the sediment and the biomass. Because most living organisms are dying and decomposing in the lake, the accumulated toxic substances return again and again into the water, sediment, and organisms, resulting in a circulation of toxic chemicals in the ecosystem.

Accumulation and Synergism Resulting in Augmentation of the Effect

The ecological effect of the accumulation of toxic substances in surface waters is unambiguous: With increasing concentration, they endanger living organisms. Suffering of biological systems will occur, however, only when toxic substances are present in dissolved form and come in direct contact with regulatory mechanisms. When toxic chemicals are insoluble complexes, bound or adsorbed to particles, they sink down, increasing the pollution of the sediment, but do not cause any imminent adverse ecological effect. Due to chemical, enzymatic, and microbial processes taking place in the sediment, such an "optimal" situation is not endless; even complexes and bound forms mobilize and become more or less soluble. Besides that, in the mobilization of inactive toxicants, the whole ecosystem, especially benthic organisms, play a role through their metabolism. On the other hand, aquatic organisms are taking part also in accumulation, binding, inactivation, and sedimentation of toxicants dissolved in the water.

In terms of toxicity, the uptake and accumulation of harmful substances into living organisms are the key processes. To some degree, all dissolved substances may enter the organism through the membranes separating it from the surrounding water, while chemicals associated with particles are accessible by feeding, through the alimentary system, where they may become soluble, suitable for absorption. This is also the manner of uptake of substances incorporated previously into prey organisms. As a result, the concentration of the toxicants increases in the intercellular space and the blood; by binding to membranes and enzymes, they may adversely affect biochemical and physiological processes. Beside that, the binding of pollutants to specific locations may lead to inactivation, resulting in detoxification, deposition within the organism, or elimination by excretion. Uptake and accumulation in the organism depend on the concentration of the chemicals in the environment and the food, so the concentration present in the organism will signify the level of the pollution in the environment.

Besides signifying the level of the environmental pollution, accumulation of toxic substances also influences physiological effects. On the one hand, the deposited toxic substances may keep permanently a certain level of pollution within the organism; on the other, accumulated toxicants are not bound firmly forever, but due to internal processes, they can be mobilized and exert a toxic effect. This means that pollutants deposited in the organisms represent a permanent source of risk.

Various toxic substances occur in the waters not separately but, as usual, more substances are present together, although their concentration may be very different. On the other hand, the toxic concentration of various substances is rather dissimilar. Beside that, additive or antagonistic effects can be present between toxic chemical substances, especially if their targets are identical. Due to synergism of substances subthreshold concentrations may become effective, which can be revealed only in biological studies, not by chemical measurements of the concentrations.

Mechanisms and Consequences of the Action of Pollutants

Toxic substances exert their harmful effect by influencing the physiological processes in living organisms through interaction with internal regulatory mechanisms. The biological effect depends on whether the pollutant establishes an interaction with a regulatory system of the organism at molecular level, resulting in modification, disturbance, and inhibition of physiological mechanisms. Adverse effects occur if the toxic chemical agent comes in contact with superficial or internal receptors and binds to membranes and/or enzymes, resulting in alteration of their functioning and consequently the regulation of living processes. The effect can be very selective, acting on a specific system, such as photosynthetic activity, cell metabolism, movement regulation, food uptake and digestion, respiration, excretion, hormonal and neuronal regulation and others, whereas in other cases, a toxic substance may have parallel effects on more than one target. In extreme circumstances or at high concentrations, the biological effect of the toxic substances may be lethal, resulting in death of the organism. Clarification of the exact mechanisms of the effects require in each case an individual approach and detailed laboratory studies, which cannot be discussed here in detail.

Studying and clarifying ecological effects are somewhat different. Adverse effects caused by pollutants in the ecosystem seldom manifest a similar degree of degradation or decay of all organisms. Occasional, extreme pollution may result in simultaneous damage of a large number of species, but there is no total die out even in such cases, due to differences in reactivity, tolerance, and protection systems of various species. The sensitivity may be different even for individuals within the same species. The variation of reaction among different species can be partly explained by the fact that the physiological significance of the target of the pollutants is not the same in different organisms, and also that, in some species, the adverse effect causing another organism to perish is balanced by protective, compensatory mechanisms.

In case of chronic or sublethal pollution, the process is even more complex. In these cases, the accumulation and elimination or neutralization of the pollutants may play a role; as well, the varying sensitivity of species and the various developmental stages to oxygen deficit to temperature, food supply, and so on can have significance. In some species, resistance can be developed with long-lasting effects, either as a result of physiological adaptation of the individual or in the manner of selection, when the adaptation is manifested genetically and inherited in the whole population. The resistance of a species to pollutants can have also an effect on its predator organism. In this case, resistance means that the organism does not accumulate the pollutant or releases it in a very short time, so that a large amount of toxic substance does not enter into the predator. However, in case the mechanism of resistance is some inactivation process in the organism, namely, binding and/or deposition in the tissues, then, on the contrary, predators will be endangered to a higher degree by consuming food with concentrated toxicants. This happens with feeding on animals responding to heavy metal pollution by the protecting mechanism of metallothionein production and metal binding to it. Biological monitoring and laboratory testing can shed light on these specific properties and variations, important to the understanding of the effect of environmental pollution on the aquatic ecosystem.

The variable response and change of different species in reaction to pollutants result in deviation of the rate of abundance in the ecosystem; some species will be reduced, whereas others may increase and become dominant. The reduction or death of the population of sensitive species appears as a loss in the food web, with a negative effect on the feeding of predators. On the whole, this process adversely influences biodiversity, causing a reduction in it. Alterations occurring in an ecosystem—either in the proportion of species or in biodiversity—may occur for various reasons, and anthropogenic pollution may be one of those. This has been proven on several occasions, whereas in others, it is only suspected. The biological monitoring and detailed clarification of the effect of toxic substances can provide a basic information about the mechanism of action and in estimating their possible effects to the ecosystem.

Biological versus Chemical Monitoring in Water Quality Assessment (Advantages and Limitations)

Without understanding the effect of water quality changes on aquatic life and resources, the development of sound policies and their effective employment is virtually impossible. Because "biological versus chemical methods" is a central problem of water quality determination, therefore, it is necessary to emphasize some of the features of this question. In a compound-by-compound chemical analysis of an effluent or surface water, it is not realistic to analyze for all chemicals that may be present. Chemical analysis provides figures that require translation into possible biological effects on the basis of available toxicity data. The chemical approach cannot account any additive, synergistic, or antagonistic effect that might occur, and chemical data may be lacking for some constituents, particularly trace metabolites and reaction products. One always has to take into consideration that toxicity is not only an expression of compound interest qualities but also chemical and physical characteristics of the medium.

Chemical analysis of complex wastes was found to be inadequate for predicting their toxicity; therefore, living organisms were used as "reagents" to evaluate a biologically important property of water pollutants, their potency as toxicants. Such tests were termed and are now generally known as "toxicity bioassays." But the use of these bioassays requires very great extrapolation indeed, and we must learn which changes are adaptive (necessary to the continued success of organisms changed by pollution) and which are destructive (steps in disintegration of organisms).

Biological investigations of a polluted lake or stream have several advantages over chemical analysis. Animals and plants, acting as natural monitors, provide a record of the prevailing conditions (Warren, 1971; Cairns & Dickson, 1973; Gruber, 1989). Biological tests might overcome one part of the problems mentioned earlier in the case of chemical analysis. They integrate the effects of all effluent components and permit control of one limiting parameter, namely, effluent toxicity. They may indicate a likely biological response in the environment and may be more resource-oriented than a full, detailed chemical analysis. They also may determine additive, synergistic, or antagonistic interactions.

However, they also have limitations, because precision and reproducibility often cannot be expressed adequately. Biomonitoring can be a time-consuming process. The cause of toxic effect without additional data is not provided by the results. The extrapolation of results from laboratory to real environment is poorly developed at present.

Changes in biological indices are evidence of environmental changes. The idea of an index suggests that the biological indices themselves provide no direct evidence, whether the involved waters being used are endangered or not. Thus, not all changes in environmental conditions can be considered pollution, and not all changes in biological indices can be taken as evidence that pollutional changes occurred in the environment. But when water has been classified for particular uses, and when a known aquatic community persists under water quality conditions permitting those uses, changes in biological indices resulting from degradation of water quality may well be evidence of pollution (de Kruijf et al., 1988).

The hierarchical concept suggests that information on a certain time scale will not, or may only partly be useful at a higher scale. If, for instance, measurements are made on the flow and amount of water in a brook within a certain period, then that information cannot be used to estimate the temporal and spatial variation of hydraulics of a large river system. The type of information needed in monitoring is therefore also determined by the scale of the process involved. Temporal and spatial problems cannot easily be overcome by extrapolation techniques, because each hierarchical level has characteristics, which are more than the integrated sum of characteristics of the nearest lower level.

The future goal in examining biological monitoring systems is to determine their scientific justification; reliability; general acceptance of the methodology by the academic community, industry and regulatory agencies; and cost. If the methodology is found to be sound we must examine the reasons why it is not more widely accepted by regulatory agencies and industry.

An integrated approach of both chemical and biological monitoring must be incorporated into any water quality protection scheme. Biological monitoring does not replace chemical and physical monitoring. They all provide converging lines of information that supplement each other but are not mutually exclusive. The biomonitors would indicate whether a problem was present, and the chemical monitors would identify its nature.

Bioindicators in Monitoring Environmental Pollution

The development of water pollution biology has now spanned about 100 years. The growth of this field was uneven and slow until around the 1950s. This decade became a turning point. Public interest and resources became available to adequately support needed biological investigation. The past 20–30 years have seen the greatest growth in our knowledge of the biology of polluted waters.

The subject of pollution indicators cannot be treated exhaustively, because there are no limits to the methods of monitoring. We need imagination to develop the most appropriate way to reflect an early method of monitoring and meaningful change in the chemical and/or physical conditions of the environment. According to Cairns and van der Schalie (1980):

> Almost any species, community or parameter can be used for biological monitoring. Appropriate selection will depend on the conditions at each site. In addition to the scientific justification of the ecological importance of the species and/or parameters used, the sensor chosen should be reliable in terms of information produced and not particularly difficult to work with. If monitoring methods are to be widely employed, they must not be too difficult for a technician to use routinely.

From the early works (zones of pollution, saprobic systems), the idea of biological indication of pollution was developed. Analogous to the saprobic systems, there was a need for indicator organisms representative of the quality of the environment: pollution indicators. Effects following exposure to toxic pollutants can be detected via the changes in community and ecosystem structure, its dynamics and function. Effects will also be visible at a lower lever of biological organization: population, organism, and suborganism levels. Effects can be visible, for example, the disappearance of susceptible species, the increase of more tolerant species, changes in oxygen metabolism, nutrient cycling and energy flow in ecosystems, as well as feedback mechanisms.

The information received from the biomonitors might be organized according to the level of biological complexity, from the ecosystem level to the subspecies level (Warren, 1971; Cairns, 1982; Cairns & Dickson, 1973; Hart & Fuller, 1974; Herricks & Cairns, 1982; Salánki, 1989). In the following brief overview, pollution indication is given for different levels of biological organizations.

Pollution Indication on the Ecosystem Level

Any aquatic habitat is the result of interactions between physical, chemical, and biological factors. Man-made chemical factors such as toxic compounds can exert a more or less serious effect on aquatic ecosystems, mainly via a direct action on susceptible species, whereas via prey–predator–competitor interactions, the direct effect can result in a variety of indirect effects. Both direct and indirect effects can be visible from the structure and/or functioning of ecosystems. This means that one can use these parameters as a means of detecting effects, (e.g., some diversity index as a pollution indicator).

In evaluating the reliability of aquatic organisms as indicators of pollutional conditions and water quality, one must consider the different indicator organisms not separately, but as biological associations or communities. Because ecosystems are subject to evolutionary developments they are in a constant state of change. The short-term stability (homeostasis) in ecosystems is the result of cybernetic interaction between their many components. This homeostasis can be affected by many types of disturbances (e.g., by pollutants). Because of this homeostasis, it is very difficult to detect whether a change in ecosystem structure is the result of homeostasis or of a pollutant present. This means that the effects of pollutants can only be concluded when they exceed normal variation.

Biodiversity (changes in the number of taxonomic groups and the number of organisms) may be used for assessment of ecosystem disturbation, biomass, Chl a, plant pigments, RNA, DNA, and ATP, all of which may be useful biomonitor parameters as structural responses of ecosystems. Decomposition, primary production, community metabolism, flow of energy and matter, increased energetic costs, alteration of food webs, and so on, as functional parameters of the ecosystems should signal pollutant disruption of the nutrients cycle.

Evaluating both the advantages and shortcomings of bioindication methods on ecosystem level, Leeuwangh (1991) critically states:

> The state-of-the-art is that we have only limited understanding of community and ecosystem responses to chemical perturbations. Therefore it is impossible to recommend with certainly parameters that successfully predict adverse effects of chemical stress. Although the above parameters sometimes have been used successfully in describing ecosystem responses, their general utility for early hazard identification is not guaranteed.

Pollution Indication on the Population Level

Different indication parameters at the population level—for example, reproductive success, mortality, size distribution, population extinction, changes in composition, dominance switches, changes in diversity and similarity patterns, reduction in abundance and biomass, alteration of spatial structure, successional influences—can also be analyzed to monitor changes in population following pollution.

Pollution Indication on the Species Level

It was thought that susceptible ("sentinel") species play a role in identifying environmental toxic stress at an early moment. Unfortunately, no single species is more susceptible for any chemical. It is the best to monitor a set of species that are representative for all major taxa, routes of exposure, and physiology. Mortality in *in situ* biotests describes environmental toxicity most realistically. Bioaccumulation of toxicants is the result of uptake via water or food. The ultimate level of accumulation depends on the rate of uptake, metabolism, and elimination. Biochemical responses as sublethal effects occur after exposure to a toxicant: hormone levels, enzyme activity, induction of metallothioneins, macromolecules, enzyme induction. Morphological effects (cell and tissue damage, deformities), physiological processes, and behavior responses (avoidance–preference reactions, swimming behavior, feeding behavior, respiration activity, valve movements) are frequently used for bioindication and monitoring.

Biomonitoring Techniques at Different Levels of Biological Communities

Artificial substrate samplers provide a means of sampling in many locations where other samplers are not effective. They obtain an equally diverse fauna, with less variability between samples. This is an effective method of biological monitoring and evaluation of water pollution (Oertel, 1991). For example, in streams, the periphyton is often more useful than the plankton in evaluating the effects of pollution, because it reflects more accurately the conditions at a given sampling location. Development of artificial substrate samplers tightly paralleled the development of aquatic biology as a quantitative discipline. Artificial substrate samplers provide representative sample of native communities mainly in rivers and streams, using periphyton (algae, fungi, bacteria), macroinvertebrates (insects larvae, snails, free-living flatworms, amphipods, isopods, oligochaete, worms, bryozoans, hydroids, sponges), and rocky intertidal and subtidal communities. Mostly the following types of artificial substrates are used: rock-filled barbecue basket, multiplate substrates, or microscope slides.

Indices or scoring systems are useful in determining recovery from toxic stress, because they are expected to be to sensitive to indication of slow recovery processes. Environmental factors that are not related to pollution may influence the fauna and thus the index values or score. Because the lithoreophilic insects are considered to have the best (re)colonization possibilities, recovery from toxic stress may be determined by monitoring the rivers with artificial substrates at less frequent intervals than in the case of other methods.

Benthic macroinvertebrates, being relatively immobile animals, have often been preferred in indicating the water quality of aquatic ecosystems. As a biological measure of pollution stress (e.g., biodegradable organic wastes), generally, the type and number of species, as well as the number of specimens per species, are taken into account. The impact of toxic pollutants and the aquatic community response to this type of stress may be quite different (Armitage et al., 1983; Whaley et al., 1989). The results indicate that the tolerance of species is pollutant-specific, whereas the differences in their susceptibility to toxic conditions due to pollution by several toxicants are negligible. Therefore, the reliability of biological systems based on macrobenthos distribution to classify surface waters polluted with a variety of chemicals should be seriously doubted (Sloof, 1983).

Bivalves, as sedentary and sentinel organisms, have a known ability to concentrate sets of pollutants from ambient waters: the transuranic elements, halogenated hydrocarbons, petroleum hydrocarbons, and heavy metals (Goldberg et al., 1978). The advantages of the "mussel-watch" (Goldberg, 1986) in providing a measure of environmental pollutants can be summarized with the following arguments: Mussels are sedentary, filter-feeding, and widespread organisms (e.g., *Mytilus edulis* in marine environment; *Dreissena polymorpha* and *Unionidae* in freshwaters). They are usable to indicate ambient xenobiotic level because of their low metabolic activity and good accumulation capability (they could have a concentration factor of 10,000–100,000). Mussels are easy to sample (large populations), can be easily and readily transplanted, and because they are commercial products (oysters and mussels), they have direct connection to public health.

A surveillance program of U.S. coastal waters pointed out also some initial uncertainties of this method: the problem of temporal variability (seasonal concentration alterations, changes in biochemical activity, filtration rates, etc.) and the intercomparability of data from different species (*Mytilus edulis* and *M. californianus*). To overcome this problem, it is more reasonable to monitor a given site for an interval of few years to point out the highly industrialized areas, "hot spots," and to identify the "background" levels. The translocation technique, which was used in different monitoring programs in the marine environment along the Dutch coastal waters, has many advantages (de Kock, 1986). A translocation technique means that large samples of continuously submerged mussels of a selected size of class are translocated from a relatively unpolluted site to the target area and suspended from buoys in cages. The resolution power between geographical locations is increased by employing statistically similar groups of animals derived from common stock, with the chosen period of exposure known; then, it is possible to select exposure locations independent of the natural occurrence of the monitored species.

The Microtox test is developed for estimating acute, nonspecific toxicity in aqueous samples. The test organism involved is a strain of the bioluminescent bacterium *Photobacterium phosphoreum*. As a by-product of the Krebs cycle, this marine bacterium makes a compound named luciferase, which emits energy in form of light (Bulich, 1979). Whenever a toxic substance enters the bacteria and influences metabolism or membrane integrity, less luciferin will be produced, resulting in a reduction of luminescence. The reduction of light production is thought to be proportional to the toxicity of the sample tested. The main advantages of this Microtox test is that it requires small sample volume and has a short response time. The results obtained with the Microtox system correlated fairly well with the results from widely accepted standard tests, such as the acute Daphnia and fish tests. This technique was used with great success in monitoring toxicity of organic compounds dissolved in the river Rhine, as well as in monitoring toxicity of industrial wastewaters in Germany.

The nitrification rates as monitor for ecological recovery of the river Rhine were introduced by Botermans and Admiraal (1989). Nitrifying bacteria are sensitive to chemical pollution, and their ability to oxidize ammonium ions to nitrate may be impaired by concentrations of both organic and inorganic chemicals in ppm range. Because nitrification dominates other nitrogen processes in heavily industrialized and urbanized areas, chemical inhibition of nitrification in the river may create problems associated with the accumulation of ammonia. Otherwise, the increasing nitrification rates in the river, parallel with the decreasing concentration of toxic substances in the same period, would be suitable for monitoring the trend of ecological recovery of the river water.

Static and Dynamic Approaches in Biological Monitoring

In biological monitoring of the pollution and the toxic impact endangering the freshwater ecosystems, static and dynamic methods can he applied. A static approach considers and determines the level of the pollution (presence and concentration of pollutants) in organisms, with the supposition that this is indicative to the state of the environment in which they are living. With repeated measurements, the variation in pollution can be traced. In the most simple method of static monitoring, called passive monitoring, organisms inhabiting a biotop (native populations) are collected for measurement without any special manipulation. Static, but active, monitoring means that plants or animals are specially translocated from an unpolluted biotop to an area to be studied, and the accumulation of toxic substances, reflecting the level of the pollution in the experimental area, is measured. A dynamic approach in biological monitoring considers the adverse effects of pollutants in life and regulatory processes of individuals, of species communities, and functioning of the ecosystem as a whole.

Static Methods in Monitoring the State of the Environment

In evaluating the pollution of the environment, the concentrations of the toxic substances accumulated in the organisms are considered as a value characteristic of the surroundings (Crompton, 1997; Walker & Livingstone, 1992). This assumption is based on experimental findings that show a positive correlation between the concentration of toxic substances in the imminent environment and in the organisms living in it. It has been demonstrated both for toxic metals and for persistent organic compounds. This assumption is valid, however, only for sublethal concentrations. If the concentration of a pollutant is close to or above lethal level, it will damage severely physiological processes, including accumulation of substances, and the concentration measurable in the organisms will not give a correct value for the pollution in the biotop.

Static monitoring that reflects the pollution of the environment is most suitable for detecting the level of persistent substances and can be applied only with restriction for the control of easily degradable chemicals. Toxic metal ions and a large number of organic pollutants, such as organometals, chlorinated and polycyclic aromatic hydrocarbons, and polychlorinated biphenyls, belong to persistent pollutants having toxic effect and at the same time accumulating in the organisms. Static monitoring means measurement of the concentration of these substances in the plants and animal tissues.

It is important to know that different organisms may accumulate the same pollutant from their common environment to different degrees. For example, it was shown that animals living in Lake Balaton (Hungary) have varying concentrations of cadmium, copper, lead, and mercury in their bodies, with mercury accumulation prevalent in chironomids, copper in crustaceans, cadmium in molluscs (Salánki et al., 1982). The ability of microbes, plants, and animals to take up and store pollutants can be very different (Amiard-Triquet et al., 1993; Crompton, 1997), and even various parts and organs of an organism may have varying capacity for accumulation of a toxic substance (Scherer et al., 1997). In case of the fish in Lake Balaton (bream and pike-perch), among organs, the highest concentration of mercury was detected in the muscle, with lead in the gill and cadmium in the liver (Farkas et al., 2000). Also, the accumulation ability and storage of pollutants are different in various parts of aquatic plants (Kovács et al., 1985; Sawidis et al., 1995). All this refers to the fact that the concentration of a pollutant detected in an organism is characteristic not only of the level of pollution in the environment but also the specificity and accumulation ability of the organisms and tissues.

One can distinguish whether organisms are good or weak biological indicators. We consider good bioindicators the organisms that possess high accumulator capacity and the ability to store the substances taken up from the environment for long time.

Application and practical use of biological techniques in monitoring the state of the environment are based on selection of proper bioindicators and working out standardized measurement techniques and proper evaluation of results. When selecting suitable static biomonitors, one should take in consideration the following:

- The species used for monitoring should be widespread and available at any or most of the time in the studied aquatic ecosystem.
- It should be available for collection in necessary amounts.
- It should have a high capacity for accumulation and good storing ability of chemical substances.
- It should be tolerant of low level pollution.
- In case a special part or organ is used as biomonitor, its isolation should be simple and the quantity sufficient for chemical determination.

Any of the organisms matching these criteria can be used as a static bioindicator for aquatic and terrestrial biomonitoring alike.

The most popular freshwater organisms used for monitoring pollution of the aquatic environment are algae (Munawar et al., 1988), seaweed, and reed among higher plants (Kovács et al., 1984; Samecka-Cymerman & Kempers, 1996), the larvae of chironomides (Smock, 1983) and the mussels (Goldberg et al., 1978; de Kock, 1986) of the benthic animals, and the crustacea plankton (Muntau, 1981) and fishes living in the open water (Yasuno, 1995; Blanchard et al., 1997; Farkas et al., 1998). In the case of plants, besides the species identity, one should consider the stage of individual development during the vegetation period and the part of the plant used for measurements. Samples of algae collected from the water, as well as crustacea plankton consist, as a rule, not of a single species but a mixture of species. Alteration in the species composition may cause differences in the detected concentration of the pollutants even in the case of identical level of environmental pollution. Error can be reduced by quantitative analysis of species of the sample. Other individual specificities or circumstances should be taken into consideration when biological objects are used for monitoring environmental pollution. One of these is the relation of the pollutant concentration in the animal to its size and age (Cossa, 1989; Barak & Mason, 1990a, 199b), with a positive or negative regression for various substances (Farkas et al., 2000). In large-sized organisms (higher plants, molluscs, fish) not the total body but only a select, definite portion or organ should be used for measurement, partly because it is difficult and not necessary to prepare a homogenate from the whole animal, and partly because one should exploit the advantage that different organs accumulate substances to different degree, and one can select organs with the highest accumulating capacity.

At preparation of samples and in the measurement, the same procedure should be followed in each case. In comparing the results with earlier data or with values obtained in other laboratories, one should carefully consider the identity of the methodical circumstances, because differences in this respect may result in different values and incorrect conclusions. The use of authentic samples as reference materials can help to avoid such mistakes (Baudo et al., 1995); nevertheless, subsequent quantitative verification of earlier results obtained with differing methods is practically impossible.

As usual, there exists great variation between values obtained for the level of the pollution in individuals of the same species collected in the same biotop, even after careful preparation and measurement of samples. For obtaining acceptable results, one should perform a considerable number of parallel measurements (5 to 10 samples). Because the number of samples should remain, nevertheless, within reasonable limits, it seems practical to select a few of the best bioindicators for regular use, which can serve to monitor the level and timely changes of the pollution in the ecosystem under study.

Case Studies for Static Monitoring: Detection of Metal Pollution in the River Danube with Passive and Active Monitoring Techniques

The bioaccumulative capability of aquatic organisms during bioaccumulation monitoring was used to detect heavy metal levels in the ecosystem of the River Danube (Hungary) in the 1980s and 1990s.

In the early 1980s, *Cladophora glomerata*, a filamentous green alga, was used for control of the heavy metal levels in the Danube River. This attached, periphytic species was well known as a good indicator of heavy metals (Whitton et al., 1989), but we had to check this ability and the conditions in which it is valid in a large eutrophicated and industrially polluted river. Exposing *Cladophora* on an artificial floating substrate (AFS), we determined that the heavy metal concentration in the alga depends on its developmental stages and vertical position in the littoral zone. These changes in time and space strongly depended on and correlated with the dry weight of *Cladophora*. Besides the "weight dilution" effect, the most important internal factor, the ambient concentration of heavy metals proved to be the most dominant external factor that influences directly the heavy metal level in the alga. Other physicochemical factors (flood, light, temperature, conductivity, pH, ORP) indirectly influence—in most cases, through the changes of dry weight—the process of uptake and accumulation. After clarifying all these fundamental questions and comparing the actual heavy metal concentration in *Cladophora* to the base level that is characteristic for the alga on the studied river section, we could follow the pollution events. This could serve as a signal function and as a trend in long-term monitoring, and, moreover, as control function. After evaluating our results and comparing them to the literature, we could determine that the studied reach of the Danube was industrially, moderately polluted at the beginning of the 1980s (Oertel, 1991, 1993b).

As an example for the source orientation, the attached biomonitor, the *Cladophora* proved to be a good indicator of the dissolved form of heavy metals (e.g., Ag, Cd, Co), which can be easily taken up and originate from close pollution sources. Although attached algae proved to be a very sensitive and source-oriented indicator of the dissolved phase of heavy metals, Trichopteras living in the *Cladophora* web are in some cases one or two magnitudes higher concentrators of some heavy metals (e.g., Ag, Cd, Cu, Pb), than the alga itself. This group of macroinvertebrates as collector organisms prefers suspended particulate matter (SPM), which acts as the preconcentrator of heavy metals; therefore, they generally have higher metal concentration than different types of feeding groups of macroinvertebrates. Otherwise, in the case of animals, their mobility could challenge their validity to indicate and delimit the pollution sources (Oertel, 1994, 1995).

In our field experiments, started at the beginning of the 1990s on the Danube, an active biomonitoring device with zebra mussel was gradually developed to monitor heavy metal pollution. Over the years, the applicability of the chosen method was checked in as many aspects as possible to reduce the problems caused by variability and, finally, to offer an adequate technique that can be integrated into the water-quality monitoring systems of great rivers (Oertel, 1996, 1997). Equipment, named "Dreissena basket," was developed, partly utilizing the elements of former experiences gained by other experts studying smaller streams. The final goal was to prepare an instrument that can be successfully used in large rivers, such as the Danube. Nets of 8-mm mesh size were filled with 100-100 specimens of zebra mussels. These units of mussels-in-net were placed into outer containers, which were perforated cylindrical plastic tubes (30 cm in length, 10 cm in diameter, and 6 mm in mesh size). The latter, like shields, protect mussels against mechanical effects but allow appropriate flow of water and food. Finally, these double units of "Dreissena-baskets" were anchored 30 cm below the water level by means of ropes and weights (Oertel, 2000).

All the containers, ropes, and anchors are easily fastened together with carabiners. Because of the flexible, module-like structure of the "Dreissena basket," its parts are easy to transport and put together in the field, and, finally, it ensures the gentle handling of the mussels. The inner units, mussels-in-nets, are also simple to transport in water-filled containers; thus, the animals are easily replaced in the field during the consecutive periods. This easy to make and cheap instrument proved to be suitable in many respects both in deep and low waters of the Danube.

Variability common in native populations could be decreased to some degree by evaluating and utilizing the results of lethal and sublethal condition indices measured during monitoring activities. Using the correlation equations of shell length and body weight to calculate the differences between the measured and calculated body weight, we could differentiate between seasonal changes and changes in condition. Recovering the animals *in toto* and replacing them with new stocks of mussels according to the 5–7-week-long consecutive periods, we recorded both the shorter and longer events of pollution.

Using "Dreissena baskets" we could discover both the longer polluted sections and the hot spots in the Danube. In the 1990s, the entering border section (1,842 rkm), the sections of Medve-Nagymaros (1,806–1,695 rkm), and Paks-Mohács (1,532–1,447 rkm) proved to be polluted by most of the studied heavy metals. These sections are adjacent to the most industrialized regions of the Slovak Republic and Hungary. Because of their high concentration levels in mussels, Co, Ni, and Pb could be considered regular pollutants during the monitored period. The time-to-time high deviations also indicated intermittent appearance of these metals. The relatively high Fe concentration in zebra mussels, in all certainty, did not come from pollution but was the result of the high Fe content of suspended matter, which is typical for the Danube.

Due to its flexible construction, the gradually developed "Dreissena basket" is a very easy-to-handle device that can be used for active biomonitoring in the permanently changing hydrological circumstances of a large river. With this device, the larger part of the methodological problems can be overcome and the variability that accompanies passive techniques can be reduced. As a final conclusion, extensive use of "Dreissena-basket" as a reliable, cheap, and easy-to-handle device of active biomonitoring could be proposed to be integrated into the water quality monitoring systems of large rivers on the scale of drainage area.

Heavy metal body-burden values in mussels have similar importance compared to concentration values, because they confirm the real process of cumulative accumulation along the longitudinal section of the river. Although some of the authors consider body burden an artificial concept, according to our experiences, similar to that of Mersch and Pihan (1993), it may be very useful. Simultaneously evaluating the dry weight, concentration, and body burden values, the high concentration values originating from dry weight reduction, and not from pollution, can be separated, and the pollution caused by cumulative accumulation both in space and time can be determined.

Dynamic Methods in Monitoring the State of the Environment

Comparing the concentration of toxic substances in indicator organisms obtained by repeated measurements, one can judge stability or change in the level of environmental pollution depending on external sources as well as those occurring as a result of timely variations in the process of the pollution bound to seasonal, yearly, or several years' changes. Besides monitoring the temporal changes by static methods, namely, measuring the concentration of the pollutants in organisms repeatedly, dynamic, functional tests can also be applied. Dynamic methods reflect and evaluate the presence and level of pollution by monitoring the alterations caused in life functions of individuals, as well as changes at the level of population. Physiological, biochemical and morphological indices, movement, feeding, gas exchange, growth, reproduction, reactivity, and other functions are suitable as functional, dynamic tests. The research and literature in this field is very rich; a number of books, reviews (Salánki, 1989, 2000, Salánki et al., 1994; Wells et al., 1997; Girling et al., 2000), and periodicals (e.g., *Aquatic Toxicology, Ecotoxicology and Environmental Safety, Journal of Aquatic Ecosystem Stress and Recovery*) are devoted to this topic. We mention here only a few examples.

Mussels are not only good accumulators of toxic substances, and so are very suitable organisms for static biomonitoring, but also, due to their physiological response to pollutants, they can be used in dynamic monitoring of water pollution (Salánki & Varanka, 1978; Kramer et al., 1989; Salánki & V.-Balogh, 1989). The changes in the valve movement and the duration of active and rest periods is a measurable, excellent indicator of water quality. Also, water emission through the outflow siphon signal the presence of pollutants in the water, and can be used in monitoring water quality (Salánki et al., 1991; Kontreczky et al., 1997). Swimming behavior of the freshwater rotifer (Janssen et al., 1994), the color changes of the fiddler crab (Reddy & Fingerman, 1995), or changes in oxygen consumption and biochemical composition in a bivalve (Uma Devi, 1996) under the effect of toxicants are also examples of functional biomonitors. Changes in enzyme activity (Hänninen et al., 1991; Nemcsók et al., 1991; Cossu et al., 2000) and morphological alterations (Schwaiger et al., 1997, Teh et al., 1997), as well as growth and mortality of larvae, can also signal exposure to pollutants (Stenalt et al., 1998).

At the level of the ecosystem, the size of the population, diversity of species, ratio between species, and the structure of species community can be used as dynamic biomonitors. Pollutants and alterations in the degree of pollution of the aquatic environment may affect any of these characteristics that can be investigated in natural field studies or in model, artificial circumstances. To use the population for monitoring purposes require detailed knowledge of the characteristics of the species, and determination of the "control" status. It is not an easy task, especially because neither the individual nor the ecosystem is stable and rigid, but undergoes permanent alterations. There are a number of factors evoking modifications, and the selective influence of anthropogenic pollutants should be determined by specific studies.

Responses of the population to environmental pollutants can be interpreted as differences in sensitivity of various species to toxicants causing selective deterioration in the structure of the population, or as a result of molecular alterations resulting in genotoxic effects. For example, it has been suggested that chromosomal damage caused by mercury, PCBs, and other compounds present in industrial effluents are responsible for the changes in community structure of redbreast sunfish populations (Theodorakis et al., 2000). In outdoor microcosm experiments, the mortality of a fathead minnow population correlated significantly with brain acetylcholinesterase activity, inhibited by organophosphorus insecticides used for pest control in agriculture (Sibley et al., 2000).

Biological Early Warning Systems

Continuous monitoring permits rapid detection of responses to pollutant at stress levels as low as those found by chronic tests to be biologically safe. Techniques that make possible prediction of the toxicity of mixtures of chemicals fluctuating both qualitatively and quantitatively are called biological early warning systems (BEWSs). In BEWSs, short-term changes in toxicity are detected automatically on the basis of a response of the organism exposed on a frequent or continuous-flow basis to the water monitored.

BEWSs naturally need parameters that produce quick responses, not necessarily very specifically or with great exactness. On the other hand, trends in ecosystem development demand great carefulness with respect to the selection of adequate parameters indicating trends one wishes to follow. BEWSs are powerful in-plant and streamside monitoring techniques of toxicity of industrial effluents and receiving waters, and may protect aquatic organisms in receiving waters from sudden exposure to adverse conditions. BEWSs can be useful supplements to traditional monitoring methods based on chemical analysis. In these analyses, the number of parameters is limited and the measured concentrations do not always provide the desired information (i.e., the possible effects of pollution on aquatic organisms).

Automated biomonitors are self-regulating or self-acting means of assessing water quality, employing living organisms as sensors (from bacteria and invertebrates to vertebrates).

The following criteria can be summarized for BEWSs as fundamental operational requirements (Cairns & Gruber, 1979, 1980; Cairns & Schalie, 1980; Cairns, 1988; Gruber & Diamond, 1988; Gruber, 1989; Botterweg et al., 1989; Koeman et al., 1989):

- Operation of the system should be continuous and automatic, so the variable biological response can be quantifiable through computer analysis.
- The system must detect a wide range of developing toxic conditions rapidly and reliably.
- The number of false alarms due to nontoxic variations in water quality should be minimal.

- Appropriate methods for the analysis of data must be developed (i.e., statistical determination of normal range of response variables and of differences between normal variation and individual and/or diurnal variations).
- The monitoring system should be relatively easy to operate and the results easy to interpret.
- Aquatic organisms as biosensors should be inexpensive, easily cultured (e.g., commercially available).
- The apparatus should require as little maintenance and skilled human-labor as possible.

Types of Most Frequently Used BEWSs

Fish Monitor

Fish respiration, an ideal biomonitoring system for water quality assessment, should reliably detect rapidly a wide range of acute as well as chronic toxicological hazards on a continuous and automatic basis at low cost (Evans & Walwork, 1988). It seems likely that the monitor is capable of detecting most toxicants at levels that are at least a 10-fold the no observed effect level (NOEL). When the detection limits are compared with the NOEL for the same compounds, it is recognized that the system is probably only valuable in the prevention of accidental spills that produce acutely toxic conditions.

In most cases, monitoring of respiratory frequency is more sensitive than other methods available (positive rheotaxis, oxygen consumption of bacteria, oxygen production by algae, mobility inhibition of *Daphnia*). Application of fish respiration is possible in "in-plant monitor" (industries and sewage treatment plants, strategic location of watershed) but questionable for natural waters. Most successful applications seem to be monitoring effluent quality combined with other biological methods and adequate multidetectional chemicophysical monitoring systems, after defining the "response" and the "detection" (false detection, lag time, etc.) very precisely (Morgan & Kuhn, 1974; Hayward et al., 1988).

Daphnia Monitor

The system is based on the detection of changes in the activity of water fleas as a consequence of a pollution effect (Knie, 1978). The detection is based on the number of infrared light beam blocking events per unit of time caused by Daphnids in the test water column. During the test period of the Dynamic Daphnia Test, some practical problems had to be solved first (sediment removal, food concentration, and water temperature) on site (Caspers, 1988).

Mussel Monitor

Bivalves of marine and freshwater environments are capable of detecting high pollutant levels and are able to avoid pollutant stress by valve closure. Although such effective avoidance behavior may interfere with the results of accumulation monitoring programs, the repeatability and reversibility of the valve closure suggests that this response may be used in detecting rising toxicity conditions of the water (Salánki, 1979; Jenner et al., 1989; Borcherding, 1992; Borcherding & Volpers, 1994; Kramer et al., 1989; Kramer & Foekema, 2000).

Dreissena polymorpha valve response is used as a BEWS because this species is widely spread. Generally, the reaction to pollutants can be divided into two separate behavioral patterns: The activity of the mussel increases strongly (incomplete closure), and the frequency of complete valve closure at regular intervals increases. In contrast to the case of *Dreissena polymorpha*, in *Unio pictorum*, the periods of activity are caused by contraction of the muscles in the foot i.e., the mussel crawls through the top layer of the sediment). The observation of the reaction to isolate itself from damaging effect of toxic concentration confirms that this organism is capable of acting as an efficient indicator of chemical pollution.

Comparing the results obtained with trout, which is considered to be the most sensitive system presently available, the mussel appears to be slightly less susceptible to most compounds. The average detection limits of BEWSs based on respiratory frequency of fish (*Salmo gairdneri*) and valve closure response of mussel (*Dreissena polymorpha*) was thoroughly investigated by Sloof (1983).

Further advantages of the mussel system are as follows:

- A more direct method of measuring a response and less susceptible to noise problems, false alarms.
- Changes in temperature and photoperiod do not interfere with the behavior of the mussels.
- Easier to handle (e.g., no feeding is required).
- Mussels are self-maintaining.
- More feasible for widespread industrial application; can be installed and operated at much lower cost (ca. 20% of fish monitoring).

Mussels can be used in effluent monitoring of sewage or industrial outlet water, in intake monitoring of drinking water stations, in site-specific applications at drilling platforms in the sea for possible leakage of chemicals and in detection of contaminated bodies of water or of contamination gradients (Kramer & Botterweg, 1991).

Multisensor Monitors

As a part of the Rhine Action Program (RAP), it was recommended that several systems be applied at the same time. Together with a fish monitor, a system with water fleas and one with a photosynthetic organism (algae or cyanobacteria) should be implemented. The systems were installed at the monitoring station of Lobith in the Rhine River (Botterweg et al., 1989).

BASF toxicometer, a kind of model treatment plant, measures (semi)continuously the respiration activity of the microorganisms and sounds an alarm if the respiration decreases significantly. The toxicometer was constructed to indicate immediately acute and significant intoxication, but no slow or chronical deteriorations of the purification. If the toxicometer is far enough (ca. 3.5 km) from the sewage treatment plant, there is enough time to direct the incoming toxic wastewater, so as to protect the activated sludge basins (Pagga & Gunther, 1981).

Evaluation of Advantages and Shortages of BEWSs

- Those species that usually are easy to culture and/or maintain under artificial conditions are often not the most sensitive to toxicants. No species is equally sensitive to all possible pollutants. One possible solution to the problem of species sensitivity in a single-species system is to incorporate a sensor that is known to be

sensitive to the toxicant likely to occur in the water body being monitored. This, of course, presumes that the toxicants of concern are largely known beforehand (e.g., industrial wastewater). Certain species, such as *Daphnia pulex*, could be an overprotective sensor for a particular water body depending on biota indigenous to the system and the physical habitat. Generally, it can be said that problems of species sensitivity of BEWS may need to be observed on a case-by-case basis.
- With frequent rotation and replacement of sensor organisms, the problem of sensor acclimation can be solved, too.
- Maximum distance needs to allow sufficient time for alarm (hydraulics and turnover time).
- The monitored response must be one that changes as a prelethal symptom of poisoning, and it must be conducive to automation allowing for reliable detection of duration from normal conditions.
- The more easily observed responses are generally the less sensitive indicators. Therefore, an important conceptual issue in the development of any BEWS is the exact purpose for which the system is designed, along with an appropriate compromise of the trade-offs in cost versus system sensitivity.
- In a particular system (e.g., utilizing frequency, amplitude, cough pattern, and general activity of fish), the more behavioral and biological information utilized from the sensor, the more sensitive and rapid the response of the system as a whole.
- Considering the validation of the BEWSs, there is a big question for the "in-plant" system and how this signal of deleterious biological effect relates to the response of natural systems. All deleterious biological effects require corrective action even though the sensor may be somewhat more sensitive than the biota in the natural receiving systems.

It is stressed that if a BEWS does not give an alarm, there is still no guarantee of acceptable water quality. Toxic compounds can still be present, most probably at low concentrations, or in concentrations causing effects only after prolonged exposure. Environmental toxicant levels are generally far from the detection limits of these systems. Further on, only limited information is available on compound-specific sensitivity of the monitor systems. Besides, calamities will be often discovered by fish kills upstream of the monitoring station before harmful conditions are detected by biomonitors. Available experience indicates that the usefulness of BEWSs to warn of toxic conditions in natural surface waters should be accepted with reservation. However, these biomonitors can be applied successfully as "in-plant monitors" placed at major industries and sewage treatment facilities for the surveillance of effluent quality. In this respect, they can play an important role in protecting the aquatic environment from disastrous spills.

References

Amiard-Triquet, C., Jeantet, A. Y., & Berthet, B. (1993). Metal transfer accumulation in marine food chains: Bioaccumulation and toxicity. *Acta Biol. Hung*, *44*, 387–409.

Armitage, P. D., Moss, D., Wright, J. F., & Furse M. T. (1983). The performance of a new biological water quality score system based on macroinvertebrates over a wide range of unpolluted running-water sites. *Water Research*, *17*, 333–347.

Barak, N. A. E., & Mason, C. F. (1990a). Mercury, cadmium and lead in eels and roach: The effects of size, season and locality on metal concentrations in flesh and liver. *Science of the Total Environment*, *92*, 249–256.

Barak, N. A. E., & Mason, C. F. (1990b). Mercury, cadmium and lead concentrations in five species of freshwater fish from eastern England. *Science of the Total Environment, 92*, 257–263.

Batiuk, R. A., Orth, R. J., Moore, K. A., Dennison, W. C., Stevenson, J. C., Staver, L. W., Carter, V., Rybicki, N. B., Hickman, R. E., Kollar, S., Bieber, S., & Heasly, P. (1992). Chesapeake Bay submerged aquatic vegetation habitat requirements and restoration targets: A technical synthesis. Annapolis, MD: Environmental Protection Agency.

Baudo, R., Rossi, D., & Quevauviller, P. (1995). Validation of the use of aquatic bioindicators by means of reference materials. In: M. Munawar, O. Hänninen, S. Roy, N. Munawar, L. Kärenlampi, & D. Brown (Eds.), *Bioindicators of environmental health* (pp. 211–225). Amsterdam: SPB Academic Publishing.

Blanchard, M., Teil, M. J., Carru, A. M., Chesterikoff, A., & Chevreuil, M. (1997). Organochlorine distribution and mono-orthosubstituted PCB pattern in the roach (*Rutilus rutilus*) from the River Seine. *Water Research, 31*, 1455–1461.

Borcherding, J. (1992). Another early warning system for the detection of toxic discharges in the aquatic environment based on valve movements of the freshwater mussel *Dreissena polymorpha*. In: D. Neumann, & H. A. Jenner (Eds.), *Zebra mussel Dreissena polymorpha* (pp. 127–146). Stuttgart: Gustav Fischer Verlag.

Borcherding, J., & Volpers, M. (1994). The "Dreissena-Monitor." First results on the application of the biological early warning system in the continuous monitoring of water quality. *Water Science Technology, 29*, 199–201.

Botermans, Y. J. H., & Admiraal, W. (1989). Nitrification rates in the lower river Rhine as a monitor for ecological recovery. *Hydrobiologia, 188/189*, 649–658.

Botterweg, J., van de Guchte, C., & van Breemen, L. C. W. A. (1989). Bio-alarm systems: A supplement to traditional monitoring of water quality. H_2O, *22*, 778–794.

Bulich, A. A. (1979). Use of luminescens bacteria for determining toxicity in aquatic environments. In: L. L. Marking, & R. A. Kimerle, (Eds.), *Aquatic toxicology* (pp. 98–106). ASTM STP 667.

Cairns, J., Jr. (1982). *Biological monitoring in water pollution*. Oxford, UK: Pergamon Press.

Cairns, J., Jr. (1988). Validating biological monitoring. In: D. S. Gruber, & J. M. Diamond (Eds.), *Automated biomonitoring: Living sensors as environmental monitors* (pp. 40–48). Chichester, UK: Ellis Horwood.

Cairns, J., Jr., & Dickson, K. L. (1973). *Biological methods for the assessment of water quality*. Baltimore: American Society for Testing and Materials.

Cairns, J., Jr., & Gruber, D. (1979). Coupling mini- and microcomputers to biological early warning systems. *BioScience, 29*, 665–669.

Cairns, J., Jr., & Gruber, D. (1980). A comparison of methods and instrumentation of biological early warning systems. *Water Resource Bulletin, 16*, 261–266.

Cairns, J., Jr., & van der Schalie, W. H. (1980). Biological monitoring: Part I. Early warning systems. *Water Research, 14*, 1179–1196.

Carson, R. (1962). *Silent spring*. New York: Crest Book.

Caspers, N. (1988). Kritische Betrachtung des "Dynamischen Daphniatest." *Zeitschrift für Wasser- und Abwasser-Forschung, 21*, 152.

Cossa, D. (1989). A review of the use of *Mytilus* spp. as quantitative indicators of cadmium and mercury contamination in coastal waters. *Oceanologica Acta, 12*, 417–432.

Cossu, C., Doyotte, A., Babut, M., Exinger, A., & Vasseur, P. (2000). Antioxidant biomarkers in freshwater bivalves, *Unio tumidus*, in response to different contamination profiles of aquatic sediments. *Ecotoxicology and Environment Safety, 45*, 106–121.

Crompton, T. R. (1997). *Toxicants in the aqueous ecosystem*. Chichester, UK: Wiley.

de Kock, W. C. (1986). Monitoring bio-available marine contaminants with mussel (*Mytilus edulis* L.) in the Netherlands. *Environmental Monitoring and Assessment, 7*, 209–220.

de Kruijf, H. A. M., de Zwart, D., Viswanathan, P. N., & Ray, P. K. (1988). Manual on aquatic ecotoxicology. In *Proceedings of the Indo-Dutch Training Course on Aquatic Ecotoxicology*. Kluwer Academic Publishers.

de Zwart, D. (1995). Monitoring water quality in the future: Vol. 3. Biomonitoring. Bilthoven, The Netherlands: RIVM.

Evans, G. P., & Walwork, J. F. (1988). The WRc fish monitor and other biomonitoring methods. In: D. S. Gruber, & J. M. Diamond (Eds.), *Automated biomonitoring: Living sensors as environmental monitors* (pp. 75–90). Chichester, UK: Ellis Horwood.

Farkas, A., Salánki, J., & Varanka, I. (1998). Assessment of heavy metal concentrations in organs of two fish species of Lake Balaton. *Proceedings of the Latvian Academy of Science, 52*, 93–99.

Farkas, A., Salánki, J., & Varanka, I. (2000). Heavy metal concentrations in fish of Lake Balaton. *Lakes and Reservoirs: Research and Management, 5*, 271–279.

Girling, A. E., Pascoe, D., Janssen, C. R., Peither, A., Wenzel, A., Schafer, H., Neumeier, B., Mitchell, G. C., Taylor, E. J., Maund, S. J., Lay, J. P., Jüttner, I., Crossland, N. O., Stephenson, R. R., & Persoone, G. (2000). Development of methods for evaluating toxicity to freshwater ecosystems. *Ecotoxicology and Environmental Safety*, 45, 148–176.

Goldberg, E. D. (1986). The mussel watch concept. *Environmental Monitoring and Assessment*, 7, 91–103.

Goldberg, E. D., Bowen, V. T., Farrington, J., Harvey, G., Martin, J. H., Parker, P. L., Risebruogh, R. W., Robertson, W., Schneider, E., & Gamble, E. (1978). The mussel watch. *Environmental Conservation*, 5, 101–125.

Gruber, D. (1989). Biological monitoring and our water resources. *Endeavour* (New Series 13), 135–140.

Gruber, D. S., & Diamond, J. M. (1988). *Automated biomonitoring—living sensors as environmental monitors*. Chichester, UK: Wiley.

Hänninen, O., Lindström-Seppa, P., Personen, M., Huuskonen, S., & Muona, P. (1991). Use of biotransformation activity in fish and fish hepatocytes in the monitoring of aquatic pollution caused by pulp industry. In: D. W. Jeffrey, & B. Madden (Eds.), *Bioindicators and environmental management* (pp. 13–20). London: Academic Press.

Hart, C. W., Jr., & Fuller, S. L. H. (1974). *Pollution ecology of freshwater invertebrates*. New York, London: Academic Press.

Hayward, R. S., Reichenbach, N. G., Dickson, L. A., & Wildoner, T., Jr. (1988). Variability among bluegill ventilatory rates for effluent toxicity biomonitoring. *Water Research*, 22, 1311–1315.

Herricks, E. E., & Cairns, J., Jr. (1982). Biological monitoring: Part III. Receiving system methodology based on community structure. *Water Research*, 16, 141–153.

Illies, J., & Botosaneanu, L. (1963). Problemes et methodes de la classification et de la zonation ecologique des eaux courantes, consideres surtout du point de vue faunistique. *Mitt. Int. Verein. Limnol*, 12, 1–57.

Janssen, C. R., Ferrando, M. D., & Persoone, G. (1994). Ecotoxicological studies with the freshwater rotifer *Brachionus calyciflorus*: IV. Rotifer behavior as a sensitive and rapid sublethal test criterion. *Ecotoxicology and Environment Safety*, 28, 244–255.

Jenner, H. A., Noppert, F., & Sikking, T. (1989). A new system for the detection of valve-movement response of bivalves. *Kema Scientific and Technical Reports*, 7, 91–98.

Knie, J. (1978). Der dynaamischeen Daphnientest—ein automatischer Biomonitor zur Überwachung von Gewässer und Abwässer. *Wasser und Boden*, 12, 310–312.

Koeman, J. H., Poels, C. L. M., & Slooff, W. (1989). Continuous biomonitoring system for detection of toxic levels of water pollutants. In: Outzinger, O. (Ed.), *Aquatic pollutants, transformation and biological effects* (pp. 339–347). Oxford, UK: Pergamon Press.

Kontreczky, C., Farkas, A., Nemcsók, J., & Salánki, J. (1997). Short- and long-term effects of deltamethrin on filtering activity of freshwater mussel (*Anodonta cyngea* L.). *Ecotoxicology and Environment Safety*, 38, 195–199.

Kovács, M., Nyári, I., & Tóth, L. (1984). The microelement content of some submerged and floating aquatic plants. *Acta Biol. Hung*, 30, 173–185.

Kovács, M., Nyári, I., & Tóth, L. (1985). The concentration of microelements in the aquatic weeds of Lake Balaton. In: J. Salánki, (Ed.), *Heavy metals in water organisms* (pp. 67–81). 29th Symp. Biol. Hung, Akadémiai Kiadó, Budapest.

Kramer, K. J. M., & Botterweg, J. (1991). Aquatic biological early warning systems: An overview. *Bioindicators and Environmental Management*, 95–126.

Kramer, K. J. M., & Foekema, E. M. (2000). The "Musselmonitor" as biological early warning system: The first decade. In: Butterworth, F. M., Gonsebatt-Bonaparte, M. E., and Gunatilaka, A. (eds.), *Biomonitors and biomarkers as indicators of environmental change* (Vol. II pp. 59–87). New York: Kluwer Academic/Plenum.

Kramer, K. J. M., Jenner, H. A., & de Zwart, D. (1989). The valve movement response of mussels: a tool in biological monitoring. *Hydrobiologia*, 188/189, 433–443.

Leeuwangh, P. (1991). *Pollution and pollution indicators in the aquatic and terrestrial environment*. Delft, The Netherlands: International Institute for Hydraulic and Environmental Engineering.

Levin, S. A., Harwell, M. A., Kelly, J. R., & Kimball, K. D. (1989). *Ecotoxicology: Problems and approaches*. New York: Springer-Verlag.

Mersch, J., & Pihan, J.-C. (1993). Simultaneous assessment of environmental impact on condition and trace metal availability in zebra mussels *Dreissena polymorpha* transplanted into the Wiltz River, Luxemburg: Comparison with the aquatic moss Fontinalis aantipyretica. *Arch. Environ. Contam. Toxicol.*, 25, 353–364.

Morgan, W. S. G., & Kuhn, P. C. (1974). A method to monitor the effects of toxicant upon breathing rate of largemouth bass (*Micropterus salmoides* Lacepede). *Water Research*, 8, 67–77.

Munawar, M., Wong, P. T. S., & Rhee, G.-Y. (1988). The effects of contaminants on algae: An overview. In: N. W. Schmidtke (Ed.), *Toxic contamination in large lakes* (pp. 113–160). Chelsa, MI: Lewis.

Muntau, H. (1981). Heavy metal distribution in the aquatic ecosystem "Southern Lake Maggiore": II. Evaluation and trend analysis. *Mem. Ist. Ital. Idrobiol.*, *38*, 505–503.

Nemcsók, J., Albers, C., Benedeczky, I., Götz, K. H., Schricker, K., Kufcsák, O., & Juhász, M. (1991). Effect of ecological factors on the toxicity of CuSO4 in fishes. In: D. W. Jeffrey, & B. Madden (Eds.), *Bioindicators and environmental management* (pp. 365–377). London: Academic.

Oertel, N. (1991). Heavy-metal accumulation in *Cladophora glomerata* (L.) Kütz in the River Danbue. *AMBIO*, *20*, 264–268.

Oertel, N. (1993a). *Application of biomonitoring techniques in pollution control*. Final Report, Community's Action for Cooperation in Sciences and Technology with Central and Eastern European Countries. Ref. No. ERB3511PL922924, Prop. No. 12924.

Oertel, N. (1993b). The applicability of *Cladophora glomerata* (L.) Kütz in an active bio-monitoring techniques to monitor heavy metals in the river Danube. *Science of the Total Environment*, *2*, 1293–1304.

Oertel, N. (1994). Bio-monitoring in water quality control, with particular reference to bio-monitoring techniques used in the river Danube for detection of heavy metals. *Acta. Biol. Debr. Oceol. Hung.*, *5*, 81–90.

Oertel, N. (1995). Plants and animals as biomonitors of heavy metal level in the aquatic ecosystem of the River Danube. *Supplement of Archives of Toxicology*, *18*, 404–416.

Oertel, N. (1996). Use of zebra mussel (*Dreissena polymorpha*) to assess heavy metal pollution in the River Danube (Hungary). 31st Konferenz der IAD, Baja-Ungarn, Wissenschaftliche Referate, pp. 405–410.

Oertel, N. (1997). Active biomonitoring with zebra mussel (*Dreissenea polymorpha*): A tool for the control of heavy metals in the River Danube. 32nd Konferenz der IAD, Wien/Österreich, Wissenschaftliche Referate, pp. 19–24.

Oertel, N. (2000). "Freissena-Basket"—a powerful technique to monitor and control heavy metals in the River Danube. *International Association of Danube Research*, *33*, 383–390.

Pagga, U., & Gunthner, W. (1981). The BASF toximeter—a helpful instrument to control and monitor biological waste water treatment plants. *Wat. Sci. Techn.*, *13*, 233–238.

Phillips, D. J. H., & Rainbow, P. S. (1993). *Biomonitoring of trace aquatic contaminants*. New York: Elsevier Applied Science.

Reddy, R. S., & Fingerman, M. (1995). Effect of cadmium chloride on physiological color changes of the fiddler crab, *Uca pugilator*. *Ecotoxicology and Environmental Safety*, *31*, 69–75.

Rosenberg, D. M., & Resh, V. H. (1993). *Freshwater biomonitoring and benthic macroinvertebrates*. New York: Chapman & Hall.

Salánki, J. (1979). Behavioural studies on mussels under changing environmental conditions. In: J. Salánki, & P. Bíró (Eds.), *Human impacts on life in fresh waters* (pp. 169–176). 19th Symp. Biol. Hung, Akadémiai Kiadó, Budapest.

Salánki, J. (1989). New avenues in the biological indication of environmental pollution. *Acta Biol. Sci. Hung*, *40*, 295–328.

Salánki, J. (2000). Invertebrates in neurotoxicology. *Acta Biol. Hung*, *51*, 287–307.

Salánki, J., & V.-Balogh, K. (1989). Physiological background for using freshwater mussel in monitoring copper and lead pollution. *Hydrobiologia*, *188/189*, 445–454.

Salánki, J., V.-Balogh, K., & Berta, E. (1982). Heavy metals in animals of Lake Balaton. *Water Research*, *16*, 1147–1152.

Salánki, J., Jeffrey, D., & Hughes, G. M. (1994). *Biological monitoring of the environment. Manual of methods*. Wallingford, UK: CAB International.

Salánki, J., Turpajev, T. M., & Nechaeva, M. (1991). Mussel as a test animal for assessing environmental pollution and the sub-lethals effect of pollutants. In: D. W. Jeffrey, & B. Madden, (Eds.), *Bioindicators and environmental management* (pp. 235–244). London: Academic Press.

Salánki, J., & Varanka, I. (1978). Effect of some insecticides on the periodic activity of the freshwater mussel (*Anodonta cygnea* L.). *Acta Biol. Sci. Hung*, *29*, 173–180.

Samecka-Cymerman, A., & Kempers, A. J. (1996). Bioaccumulation of heavy metals by aquatic macrophytes around Wroclaw, Poland. *Ecotoxicology and Environmental Safety*, *35*, 242–247.

Sawidis, T., Chettri, M. K., Zachariadis, G. A., & Stratis, J. A. (1995). Heavy metals in aquatic plants and sediments from water systems in Macedonia, Greece. *Ecotoxicology and Environmental Safety*, *32*, 73–80.

Scherer, E., McNicol, R. E., & Evans, R. E. (1997). Impairment of lake trout foraging by chronic exposure to cadmium: A black-box experiment. *Aquatic Toxicology*, *37*, 1–7.

Schönborn, W. (1992). *Fließgewässerbiologie*. Stuttgart: Fisher Verlag.

Schwaiger, J., Wanke, R., Adam, S., Pawert, M., Honnen, W., & Triebskorn, R. (1997). The use of histopathological indicators to evaluate contaminant-related stress in fish. *Journal of Aquatic Ecosystem Stress and Recovery, 6*, 75–86.

Sibley, P. K., Chappel, M. J., George, T. K., Solomon, K. R., & Liber, K. (2000). Integrating effects of stressors across levels of biological organization: Examples using organophosphorus insecticide mixtures in field-level exposures. *Journal of Aquatic Ecosystem Stress and Recovery, 7*, 117–130.

Sloof, W. (1983). Biological effects of chemical pollutants in the aquatic environment and their indicative value, Ph.D. Thesis, Utrecht University, Utrecht, The Netherlands.

Smock, L. A. (1983). Relationships between metal concentrations and organism size in aquatic insects. *Freshwater Biology, 13*, 313–321.

Stenalt, E., Johansen, B., Lillienskjold, S. V., & Hansen, B. W. (1998). Mesocosm study of *Mytilus edulis* larvae and postlarvae, including the settlement phase, exposed to a gradient of tributyltin. *Ecotoxicology and Environment Safety, 40*, 212–225.

Teh, S. J., Adams, S. M., & Hinton, D. E. (1997). Histopathologic biomarkers in feral freshwater fish populations exposed to different types of contaminant stress. *Aquatic Toxicology, 37*, 51–70.

Theodorakis, C. W., Swartz, C. D., Rogers, W. J., Bickham, J. W., Donnelly, K. C., & Adams, S. M. (2000). Relationship between genotoxicity, mutagenicity, and fish community structure in a contaminated stream. *Journal of Aquatic Ecosystem Stress and Recovery, 7*, 131–143.

Uma Devi, V. (1996). Changes in oxygen consumption and biochemical composition of the marine fouling dreissinid bivalve *Mytilopsis sallei* (Recluz) exposed to mercury. *Ecotoxicology and Environment Safety, 33*, 168–174.

Vannote, R. L., Minshall, G. W., Cummins, K. W., Sedell, J. R., & Cushing, C. E. (1980). The River continuum concept. *Can. J. Fish. Aquat. Sci., 37*, 130–137.

Walker, C. H., & Livingstone, D. R. (1992). *Persistent pollutants in marine ecosystems*. Oxford, UK: Pergamon Press.

Warren, C. E. (1971). *Biology and water pollution control*. Philadelphia/London/Toronto: Saunders.

Wells, P. G., Lee, K., & Blaise, Ch. (1997). *Microscale testing in aquatic toxicology: Advances, techniques, and practice*. Boca Raton, FL: CRC Press.

Whaley, M., Garcia, R., & Sy, J. (1989). Acute bioassay with benthic macroinvertebrates conducted *in situ. Bull. Environm. Contam. Toxicol, 43*, 570–575.

Whitton, B. A., Burrows, I. G., & Kelly, M. G. (1989). Use of *Cladophora glomerata* to monitor heavy metals in the rivers. *Journal of Applied Phycology, 1*, 293–299.

Yasuno, M. (1995). Long-term biomonitoring of organochlorine and organotin comopounds along the coast of Japan by the Japan Environment Agency. In: M. Munawar, O. Hänninen, S. Roy, N., Munawar, L. Kärenlampi, & D. Brown (Eds.), *Bioindicators of environmental health* (pp. 179–193). Amsterdam: SPB Academic Publishing.

11

The Ecology of Wetlands Created in Mining-Affected Landscapes

Arthur J. McComb and Jane M. Chambers

Introduction

This review deals with the establishment of wetlands on land affected by mining, but it is useful to recall that the mining industry may have an impact in a number of ways, including direct impacts on existing wetlands, for example, physical obliteration, altered groundwater contours, heavy metal contamination, and eutrophication (nutrient enrichment and its consequences).

Our emphasis is not so much on these impacts, or on techniques to rehabilitate wetlands after such disturbance, but on creating wetlands as a form of rehabilitation of mined lands rather than returning them to agricultural or forest production. However, some of these impacts leave a heritage that must be taken into account when attempting to create wetlands in a landscape affected by mining.

We do not attempt a comprehensive review of the large literature on this topic, but rather select examples to illustrate the points being made. In many instances, we give Australian examples to complement the extensive literature that comprises examples from the United States and Europe.

We begin by briefly reviewing the kinds of voids left after mining, examining why it may be desirable to create wetlands after mining, outlining steps in establishing wetland ecosystems, and discussing criteria that might be used to judge whether a newly created wetland ecosystem will function in perpetuity with minimal maintenance. To clarify the concepts, we detail a case study concerned with the establishment of wetlands in a landscape created by mineral sand mining.

Just as wetlands around the world vary in character from bogs to coastal marshes, the development of wetland creation technology is not limited to inland, depression-type wetlands. To extend this review to the limits of wetland type, we note the relevance of

---

Arthur J. McComb and Jane M. Chambers • School of Environmental Science, Murdoch University, Perth, Western Australia 6150.

research on nearshore marine systems. We give an example of establishing an ecosystem after dredging for carbonate sand in nearshore seagrass meadows, pointing out analogies between this example and wetlands created after mining in inland environments.

The importance of wetlands created after mining has grown strongly in recent years. The situation in the United Kingdom has been reviewed by Peberdy (1998). He notes that some 15,000 ha of wetland habitats have been created in the United Kingdom this century as a result of mineral extraction, compensating a little for the impact on biodiversity of the loss of natural wetlands, and that more than 250 postindustrial sites, including mineral gravel extraction sites, are managed as wetland nature reserves.

The relative ecological value of these new wetlands is increasing; for example, the proportion of wildfowl counted in Britain that were recorded on mineral extraction sites has risen dramatically over the last 200 years (Figure 11.1).

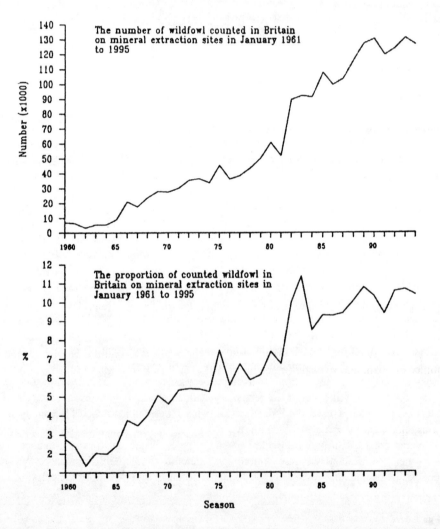

Figure 11.1. Waterbird usage of postindustrial wetland sites in Great Britain. Reproduced, with permission, from Peberdy (1998).

Postmining Environments with Potential for Wetland Creation

A diversity of voids may result from mining, many of which, containing water, have potential for wetland creation. These include open shafts, lakes resulting from opencut mining and quarrying, undulating landscapes intersecting the water table, and tailings ponds.

Open Shafts

These may be filled with water, usually because they intersect a deep or near-surface aquifer, but sometimes because of flooding by surface water. The water in the shaft may be static or continue to rise and even overflow from the shaft if the pressure head in the shaft is above that of the regional water table (in which case it is referred to as "artesian water"). Because shafts are typically steep-sided, there is little opportunity for the establishment of higher plants at the water margins, and at depth, there is insufficient light even for algal growth. Aquatic plants are usually limited to rock faces near the water surface, where there is sufficient light for photosynthesis to exceed respiration, and where there are sites for algal attachment. In our context, therefore, such shafts are mainly of interest as water sources rather than ecological wetland systems; the water they contain can be pumped out, or if artesian, be allowed to flood out, and used for human purposes or wetland establishment. Two Australian examples illustrate such systems.

Bibbawarra Bore, near Carnarvon in northwestern Australia, results from test drilling for coal in 1903. Although no significant coal was found, when the drill reached 914 m in the Carboniferous series, it hit an aquifer, which produced 2.4×10^6 L of hot water each day, and continues to flow, at a lower rate. Located in an arid part of the continent, the potential for watering stock was quickly realized, though the water was too hot to be used. In 1940, a sufficiently long, open trough was constructed to allow the water to cool for drinking by stock. The bore and trough are now managed for conservation (Figure 11.2).

The second example is near Mariborough in Victoria, where a now-disused gold mine became flooded with groundwater and continues to provide abundant water to irrigate summer crops in a profitable market garden (J. A. McComb, personal data).

It is possible in natural systems to find analogues for the limited ecological development typical of open shafts. In South Australia are the adjoinings Ewens and Piccaninne ponds, which communicate with a substantial aquifer. The system, favored for visits by cave divers, reaches a depth of 80 m. The water is extraordinarily clear, with a visibility of 30 m, and the sides slope at 45–50°. There is a zonation of algae attached to the walls, with some larger attached plants trailing downwards to 4 m. Below 4 m are calcareous sands carrying blue-green algae. Water can be seen flooding into the base of the ponds between limestone blocks and rubble. The discharge is high, and water residence time in the series of ponds is 1.5–6.0 h. The area is of considerable conservation significance (Hallam, 1985; McComb & Lake, 1990). Mention should also be made here of Mound Springs, which are well described for northern South Australia, at sites where water leaks to the surface along the edge of the very large Great Artesian Basin. Like the mining-created systems mentioned earlier, groundwater may pass to the surface at very high rates, and sometimes at high temperatures, and has been used for many years to support agriculture in an otherwise arid landscape. For a description and further references see Ponder (1986), and McComb and Lake (1990).

Figure 11.2. Bibbawarrah Trough, near Carnarvon, northwestern Australia. Constructed in 1940, it is fed by hot artesian water from an aquifer intercepted during exploratory drilling for coal. Constructed on a stock route to water flocks of sheep, it had to be sufficiently long to allow the water to cool to drinkable temperature. At 180 m, it is reputed to be "the longest trough in the southern hemisphere." Photography by J. McComb (2000).

Lakes Resulting from Quarrying

These typically share many of the properties of natural lakes and reservoirs, in that they often have a large surface area in relation to the volume of water they contain, and, like natural lakes, their management can include maintenance of the water level, avoidance of algal blooms, and reduction of turbidity brought about by suspended clay. They usually have steep sides, providing little area of substratum suitable for the establishment of peripheral vegetation, and so for the support of waterbird populations.

Waterbirds may use the open water for "loafing" secure from predators, but there is little food and few nesting sites. Such water bodies have the potential to contain significant fish populations, because they can support phytoplankton-based food webs, but, like reservoirs, they are generally not of great conservation significance; they may, however, be important as a water resource for human use.

In contrast to these generalizations about deep quarries, there is a diversity of water depths in various lands subjected to opencast mining, and a good example is among the case studies included by Pederby (1998). The complex of sites on a former opencast coal mining area at Druridge Bay near Newcastle, England, and including the area of Druridge Pools, was created during the "restoration" of former opencast coal mines the 1970s. It is landscaped and managed primarily for conservation and birdwatching, with some pools managed for fishing.

An example from Australia is the Blue Pool near Yamba in New South Wales. It results from quarrying granite in 1862–1889, and filled with water when an underground spring was disturbed. It is a tourist destination used for recreational boating and walking in the surrounds. And in areas of low and intermittent rainfall, as in northwestern Australia, depressions created by mining have created important refuge areas for waterbirds, and so are of ecological significance (Masini, 1988; Masini & Walker, 1989).

Wetlands in Gravel Pits

As noted in the introduction, gravel pits have been especially significant for the creation of wetlands in Britain, where the extractive industry produces the largest area of wetlands per annum. Some 37,000 ha of sand and gravel have been worked and reclaimed in the past 40 years, including ~110 sites now designated as nature reserves in England (Pederby, 1998).

A notable example is the Great Linford Wildfowl Reserve, managed in a cooperative arrangement between Amey Roadstone Corporation and the Game Conservancy Council, and leased to Milton Keynes Borough Council as an environmental education center. Sand and gravel were extracted from the area in the 1940s and the 1980s, and a research program into wetland creation began in 1970. A research facility and visitor center were established. The establishment, management, and conservation significance of the area are summarized in Peberdy (1998). Research funding ended in 1993, and the center remains primarily as a managed education center.

Wetlands on Mined Landscapes Intersecting the Water Table

These structures occur where mining has taken place on sandy soils, for example, as a result of gravel extraction or, as in the case study detailed below, sand mining. Sand mining involves removing sand rich in minerals by excavation or dredging, extracting the minerals, and returning the processed sand to the mined area. They may intersect the water table to form a landscape with shallow water bodies having extensive littoral areas suitable, in principle, for the establishment of macrophyte vegetation, and therefore the support of food webs.

In other relevant areas, peat has been harvested from peat bogs or fens, activities which might arguably be included as a form of "mining." However, extraction takes place in a wetland, and conservation practices, which consist mainly of allowing the fen or bog to regenerate, might better be thought of as rehabilitation of an existing land form rather than wetland creation. For further discussion, see Moore and Bellamy (1974).

Tailings Ponds

These are areas to which process water used in mining operations has been allowed to flow or has been pumped; they may function as settling ponds to allow particulates to settle out before water is used for other purposes, or discharged to drainage. The water in tailings ponds may contain undesirable concentrations of cyanide, used in gold mining, or metal ions (as in uranium mining; Finlayson et al., 1988). Dealing with a heritage of such chemicals provides particular problems if the ponds are to function as wetland ecosystems.

An instructive example is afforded by the tailings pond of the Ranger Uranium site in the Northern Territory of Australia. There, a major uranium mine has been operating in an area of significant cultural and conservation significance, so significant that the area was declared a World Heritage area (National Parks and Wildlife Service, 1980). Land ownership is under the National Park Service and the Aboriginal Land Trust. It includes spectacular landscape values, archaeological and cultural sites, unusual plants, animals and plants of conservation significance, and extensive wetlands.

Approval to mine uranium in such an area has aroused considerable controversy and has been conditional on the basis of extensive environmental impact studies and the implementation of monitoring programs. The Alligator Rivers Research Institute has been set up to design and implement the required environmental studies, under a Commonwealth Government statutory authority, the Office of the Supervising Scientist.

The tailings pond lies in the catchment of the East Alligator River, which flows through, and in part floods into, the significant wetland area, with excess water passing onto the ocean. The approach has been to design the tailings pond such that it will only overflow after periods of exceptionally heavy rainfall, and any such discharges are allowed only when river flow is high, ensuring rapid mixing and wide dispersion. For illustrations and further details see Anonymous (1987), McComb and Lake (1990), and Finlayson et al. (1988).

The tailings pond itself is not of ecological importance as a wetland habitat but is relevant in our context, primarily because of its role in protecting the ecological attributes of wetlands downstream.

Reasons for the Creation of Wetlands after Mining

There are several reasons why a decision might be made to establish wetlands in a void created by mining activity, where sufficient water is available for the purpose.

Obligations in Relation to Original Land Use

Sometimes a mining venture is required to return land to a use compatible with what was in the area before mining commenced, and the following possibilities spring to mind.

Agriculture

Where land is to be returned to agricultural use, creation of wetlands may well be appropriate in part of the area, being compatible with groundwater recharge and general water conservation for agricultural purposes.

Conservation

It sometimes happens that mining has taken place in an area of high conservation value, so much so that the area had been set aside as a nature reserve or even as part of a national park, or, as with the uranium mine referred to earlier, a World Heritage area. For such areas, it is very likely that approval to mine was conditional on meeting stringent requirements for restoration of the original ecosystem, or for substitution of similar environmental values.

The creation of wetlands may well be an appropriate use for an area of conservation significance, especially if there are otherwise few sites in the region for the feeding and breeding of wetland vertebrates, and especially waterbirds.

Education and Recreation

Wetlands are widely regarded as appropriate for education and passive recreation, often because of their high landscape value. Inclusion of a view over a wetland usually increases the value of land in a metropolitan area, and members of the public enjoy walking along paths beside or in a wetland, especially in the company of family, friends, and pets. And wetlands typically attract bird life, which becomes a focus for human interaction with wildlife. This interaction may range from simply feeding birds, through checking observations on a list, to serious study. Popular bird observatories and visitor centers are features of wetlands, some of them highly managed, for example, Herdsman Lake in Western Australia (Conservation Commission, 2001), Wicken Fen in Cambridge (National Trust, 1959).

The increasing popularity of this form of recreation and education is shown in the success of the newly created Wetland Centre at Barnes Elms in London.

Particular interest by the public attaches to observing waterbirds that have migrated from one continent to another, for example, the wave of waterbirds that moves into Australia from the Northern Hemisphere each year. The birds and their habitats may be protected by international agreements, such as the Japan–Australia Migratory Birds Agreement (1974).

There are other ways in which wetlands are of educational significance. Here, students are presented with a rich diversity of aquatic life in a compact ecosystem. They can readily learn to make systematic observations, test hypotheses, understand plant succession, frame objectives for management, and come to understand reasons for controlling water pollution. Small wonder that the establishment of wetlands after mining has significant educational value, especially in urban or other areas with high densities of students at all levels of the school and university systems.

Vacant Crown Land

This term is applied in Australia to land that has not been vested for conservation or management in a particular agency or authority, nor alienated as private land. Nevertheless, mining activities in such areas can only take place under stringent conditions, that will not be unduly prejudicial to future land uses. Not many years ago, there were few restrictions on mining activities in such areas, but now there are obligations to ameliorate at least the visual impacts of mines and spoil heaps, and to render such sites safe for future visitors. Again, the establishment of wetlands where the site intersects the water table is an appropriate direction for management.

Wetlands for Contaminant Removal

The ability of wetlands to remove a variety of contaminants from water passing through them has been exploited since the 1950s. Through chemical-, microbial-, and plant-mediated pathways, reductions can occur in biological oxygen demand (BOD), chemical oxygen demand (COD), nutrient concentrations (N and P particularly) and suspended solids. Early use of wetlands relied on natural systems, but now, constructed wetlands, specially designed to remove target contaminants, provide a more effective, controllable, and environmentally friendly alternative.

In their comprehensive treatise on treatment wetlands, Kadlec and Knight (1996) note that constructed wetlands are now used at more than 300 sites in the United States to increase pH and reduce concentrations of iron or manganese at coal mine sites, reducing the impacts of acid mine drainage.

Steps in Creating Wetlands

Wetland creation is a relatively new technology, and much of the early progress in this area was *ad hoc* and poorly documented. In recent years, more structured research and compilation of material has resulted in several books (e.g., Hammer, 1992; Kadlec & Knight, 1996) and an increase in our knowledge on how wetlands can be created. Although basic principles can be broadly applied, the nature of wetlands requires that much of the work needs to be site-specific, with due reference to local requirements. Whatever the type of the wetland being created, it is vital that an understanding of what is to be created, limitations to the development, a knowledge of appropriate technologies, and a clear plan of the steps involved is conceived before work proceeds.

Definition of Objectives

An objective for the created wetland must be clearly articulated. Is it to meet specific regulatory requirements, meet landscape aims, support waterbird populations, and so on? Are there primary and secondary objectives for the wetland? Are they mutually compatible? Such aims set the creation of the wetland firmly on track, define appropriate development techniques, and should ideally be reflected in a set of time lines and milestones, leading to the achievement of specified completion criteria. In most cases, a program of research would be needed to guide management toward the specified objective.

Modification of Landscape

This is usually a vital first step. It might seem trite to say that wetlands are defined by water, but if the hydrology is not appropriate to the type of wetland to be created, the project will fail. Modification of the landscape is usually required to create water bodies more suitable for the objectives. Water levels might be adjusted by controlling throughflow though the wetland, modifying inflow (perhaps mine process water), and/or controlling outflow with the use of weirs or active contouring of land surface to allow suitable conditions for wetland creation. It is clearly an advantage if such controls can be brought into play when required, as a tool for ongoing management.

Control of Water Quality

The water quality to be aimed for will relate to the objectives for the wetland. Desirable water qualities can be specified for different designated purposes, such as water supply, conservation, or irrigation with appropriate guidelines (e.g., Australia and New Zealand Environment and Conservation Council, 1992). Alternatively, there may be the more immediate aim of just making the water tolerable for supporting plant growth. As with hydrology, water quality is often a limiting factor in wetland development, because it determines the possible suite of plants and animals that might colonize the wetland.

Water quality problems encountered in mining-created wetlands include the following:

- Turbidity through suspended sediment.
- High concentrations of ammonia/ammonium (the more toxic form of ammonia occurring at high pH).
- High concentrations of iron or manganese, which can immobilize phosphate and reduce plant growth.
- High concentrations of sulphate, which can lead to problems of "acid mine drainage" (see previous discussion) with sulphuric acid and hydrogen sulphide formation.
- Heavy metals such as zinc, copper, cadmium, and lead, which can be toxic to animals and, at high concentrations, even to plants.

Levels of nitrogen and phosphorus can cause problems, because the availability of these nutrients can "limit" phytoplankton growth, especially phosphorus in freshwater ecosystems, and nitrogen in more marine situations. If N and/or P availability is increased, large phytoplankton blooms may be generated, leading to anoxia, and the death of invertebrates and fish. In extreme situations, blooms of toxic blue-green algae may be encouraged.

Establishment of Riparian and Aquatic Plants

Plants provide a number of functions in a wetland. They promote the physical structure of the wetland by binding the sediment and baffling wave energy, thus maintaining bank stability and reducing erosion. They modify hydrology through increasing water percolation through the soil and through evapotranspiration and shading of water.

They provide a fundamental role in the water quality and nutrient dynamics of the wetland, causing sedimentation of fine particles and absorbing nutrients from water, quite apart from their ecological role of providing habitat for fauna, and fueling the food chain either directly, through grazing (phytoplankton and macroalgae), or indirectly, through the detrital pathway (aquatic and riparian macrophytes).

The choice of plants in a wetland is then fundamental to the functioning of the ecosystem. Plants might be used as tools to develop the appropriate processes required to fulfil the wetland objective. Wise choice of plants can increase the speed and ease the management of wetland creation.

In considering emergent macrophytes for wetland creation, we have found it useful to categorize wetland plants as "seeders" or "sprouters" according to their response in germination and rhizome transplantation trials. In a survey of nine species of wetland plants from the Cyperaceae, Juncaceae, and Typhaceae, all occurring naturally in an area of coastal plain, species either showed high germination rates or could be readily transplanted as rhizome segments; rarely did they respond well to both propagation methods (Chambers & McComb, 1994). The different life histories require a very different approach to establishment, and often the more labor-intensive "sprouter" species are the ecologically valuable, native species.

The establishment of food webs requires primary producers (plants) to provide the necessary fixed carbon and energy to power food webs. The primary producers in wetlands include phytoplankton, benthic microalgae, periphyton, and submerged and emergent macrophytes. All require light for photosynthesis, and this factor is a major determinant in the growth and establishment of wetland plants. Light penetration determines the lower limit at which submerged plants will survive and is affected by water turbidity, which in turn is a reflection of suspended sediment and the occurrence of phytoplankton blooms.

Submerged Macrophytes

These are rooted in the sediments, which, as noted earlier, they bind, preventing erosion and reducing turbidity. Some (such as charophytes) clarify the water by trapping particulates in a mucus-like gel surrounding the leaves and stems. Submerged aquatics are readily decayed, supporting detrital pathways, and they provide habitat for invertebrates and small fish. In shallow water, they may be grazed by waterbirds.

Emergent Plants

Also rooted in the sediments, plants of this growth form have shoots or leaves that extend above the water surface (a few aquatics, such as water lilies, are intermediate in producing leaf blades at the water surface). Typically emergent macrophytes form very dense stands of vegetation, providing large amounts of organic material, and supporting detrital food webs. They also provide shelter and nesting material for birds, and are viewed as highly desirable components of created wetlands.

Finally, mention should be made of trees and shrubs, which can be considered emergent macrophytes. They produce leaf litter and shade the water, reducing temperature and light, therefore controlling algal growth. They may be used as roosting sites by waterbirds and for this reason alone can be of conservation significance, quite apart from their contribution to biodiversity.

Setting Up Food Webs

Broadly, we can recognize food chains in which primary producers, such as phytoplankton, are grazed by secondary producers, such as planktivorous fish or invertebrates, which are in turn consumed by organisms higher in the food chain and, ultimately, by birds and mammals. Rarely are things so simple. Instead, complex food webs are built up through interactions at several levels of food chains. And whereas simple food chains are easy to grasp, in fact, most energy and carbon transfer in wetlands takes place through detrital pathways, in which microorganisms break down organic material in decay processes and are consumed by other organisms (detritivores) that make energy available for food webs involving larger fish, birds, and mammals.

One has only to think of a complex stand of sedge vegetation with a high productivity, giving rise to large amounts of plant material each year, little direct grazing, yet a fairly constant biomass from one year to the next, to realize that most of the high production is being lost through decay, and driving detritus-based food webs.

Setting up appropriate food webs can largely be instigated by manipulating the flora of the wetland, but if specific species are required (e.g., fish, certain species of birds), then a knowledge of their diet, habitat, and breeding requirements must be taken into account.

Completion Criteria

The creation of wetlands as a process of rehabilitating mined lands is sufficiently new that practical techniques to assess its progress and development are not yet available. However such indicators are essentials to ensure that wetland development is carried out in an ecologically sustainable manner, to provide warnings about elements of the wetland that are not developing in an appropriate fashion and, most importantly, to inform managers when the created wetland no longer requires developmental support. At this stage, the created wetland might be said to be "complete," and ongoing management typical of any natural wetland may then be more appropriate.

Completion of a created wetland can be defined by regulatory provisions or, more problematically, by ecological sustainability of the desired wetland type. Regulatory provisions tend to focus on measurements such as the quantification of organisms added to the wetland during the creation process, for example, survivorship of planted vegetation in the first year (Garono & Kooser, 1994), but these are not particularly indicative of whether a functional ecosystem has been created. Their function is more to ensure that obligations of the people undertaking wetland mitigation are being followed through.

Several authors have promoted the use of succession theory as a model for restoration assessment (Van der Valk, 1998; Cole, 1998). Van der Valk (1998) states that restoration can be thought of as accelerated succession; although this may be valid in a general way, there are fundamental problems in adapting natural succession to artificial systems. For example, Poach and Faulkner (1998) point out that comparisons of created and natural wetlands are typically confounded by differences in wetland age. Created systems are generally younger than their natural counterparts. Succession in a natural wetland may already be supported by a significant seed bank, improved soil texture with high organic content, or established hydrological regimens. Rarely do natural systems have as disturbed an environment as created systems when they are being established. In fact, most successional work in wetlands does not describe primary succession but transitions from one "stable" state to another, due to an increase in water level or change in a water quality parameter such as, for example, salinity. Such succession is quite different from modifying a mine site to become a wetland, and there are different influences on the systems during succession. Observed differences between natural and created wetlands may be due to age or an inability to create wetland functions (Poach & Faulkner, 1998).

Successional theory becomes even harder to implement when the natural wetland state is based on perennial vegetation. In cold temperate climates, where recolonization after winter is a common attribute, created wetlands are more likely to approximate their natural counterparts rapidly (e.g., the Oleotangey wetlands of Ohio; Mitsch et al., 1998). However, in southwestern Australia, for example, where wetlands are typified by long-lived perennials, successional events are rare (e.g., after fire) and generally occur only in preconditioned environments.

This does not deny the value of employing successional theory to develop completion criteria, but it is wise to be cautionary in its application. Monitoring viability and robustness (i.e., the health and well-being of the system and its individual components) requires a comparative baseline for that system or comparative information from a similar system (Hammer, 1992). Once the type of wetland to be created has been determined, the best indicator of the parameters to which the created wetland should evolve might well be found in a nearby, natural, comparable system.

Ideally, development of completion criteria should begin by determining effective indicators of ecological development. The majority of indicators used to assess wetland health have been developed to assess disturbance of a natural system (via eutrophication, pollution, etc.). Such indicators focus on degradation of an established system (going backwards, if you like) rather than an increase in complexity, as is typical of positive succession. For example, the use of invertebrates as bioindicators can readily establish which species are common in pristine systems, and those which indicate eutrophic conditions. However, there is little indication of typical coloniser species and communities associated with young or undeveloped ecosystems.

Indicators of ecological development would probably best be focused on aspects that show development of the structural complexity of a system. These might be indicators of successful development of the food chain (e.g., species diversity, abundance, and breeding success of waterfowl), appropriate physical complexity (e.g., structural complexity of vegetation in both areal and vertical planes), or indicators of "wetland health" (e.g., macroinvertebrate indicators). Before any of these indicators could be successfully used, it would be imperative to ensure that the site morphology (water depth, slope, etc.) and hydrology is appropriate to the form of wetland to be created. This should obviously be carried out before the outset of wetland rehabilitation. But if diversions from the required succession are apparent, it is as necessary to ensure at least that basic physical requirements are being met.

At present, there is little work on which to base any of these criteria and a great need to develop ecologically based criteria in a simple, methodical assessment framework, so that created wetlands can be better and more effectively developed, monitored, and assessed.

Case Study: The Capel Wetlands, Western Australia

We explore these concepts by dealing in more detail with a specific example of a wetland created after mining. The example is from Western Australia, at a site subjected to the mining of sand rich in the metals ilmenite and rutile. Geologically, the deposit was laid down as a strandline along an ancient seashore. The metals in these strandlines have been harvested (1975–1979) by dredging out the sand, separating out the required minerals, and returning discarded sand and fine particulates to the mined void. The result was an undulating landscape in sandy soil, the lower levels intersecting the water table giving rise to a chain of shallow lakes, which came to form the Capel Wetlands Centre. The history of the Centre has been outlined on several occasions at different stages of its development (Brooks, 1991; Chambers & McComb, 1994; Doyle & Davies, 1998). The present account summarizes and extends these earlier publications.

Definition of Objectives

The land was in part owned by the mining company, and in part vested in a statutory Authority, and managed on its behalf of by a government instrumentality, the Department of Conservation and Land Management. The potential of the area was recognized for supporting waterbird populations, aiding in their conservation, and creating more wetlands in a state where their conservation and loss has become a matter of considerable public concern, as it is internationally (Lane & McComb, 1988). Such an approach was compatible with the requirements of the Department of Conservation and Land Management, as well as meeting obligations incumbent on the company under the terms of the Mining Act, as administered by another instrumentality, the Department of Minerals and Energy. The objectives were therefore to develop self-sustaining wetland ecosystems to facilitate research into wetland ecosystems and to develop facilities for public education and recreation, and to develop and demonstrate rehabilitation techniques for wetlands created by human activity. It was anticipated that, in the long term, it would be possible for the company to withdraw from the area and leave a functioning ecosystem requiring minimal ongoing maintenance.

In the absence of sufficient local information to guide management, experience was at first sought elsewhere. In particular, the development of the wetlands centre at Great Linford in the United Kingdom, mentioned earlier, provided inspiration.

A research program was set up involving staff from several universities in the state (the University of Western Australia, Curtin University of Technology and Murdoch University), as well as the Royal Australasian Ornithologists Union (RAOU; now Birds Australia) and a number of consultants, who addressed particular topics. Financial support was provided by the mining company, initially Renison Goldfields Consolidated, now Iluka Resources, with additional support for particular projects from other funding agencies. Considerable in-kind support was provided by the universities, research students, and volunteer organizations.

A steering committee was set up, chaired by the company, with representatives from the universities, government departments, and consultants.

The feasibility of developing the Centre was examined by the RAOU, and when that report was favorable, the research and management project was launched. The direction of the research and management approximated the steps outlined in the section on creating wetlands in this review. The research program is producing an ongoing series of technical reports, 52 to date, including 16 addressing birds, 10 for invertebrates, 8 for algae, 7 for water quality and groundwater, 6 for fish, 4 for other vertebrates, and 3 for macrophytes.

Mention should be made of the important role of a site warden, responsible for overseeing the landscaping program, collaborating in and expediting research and monitoring programs, looking after the activities of volunteers, and facilitating visits to the Centre.

Modification of Landscape

Considerable modification of topography was undertaken. The wetlands comprised a chain of shallow lakes running approximately north–south (Figure 11.3). Mine process water flowed into the chain from the north, took a convoluted path through the lakes, and excess water flowed over a weir to the Ludlow River; this weir provided the opportunity to adjust the water level in the system.

The topography of the lakes required modification. The banks were sloped to render the shores more suitable for plant establishment, to create a diversity of depths for different waterbirds, and to help ensure visitor safety. Landscaping was progressively carried out over several years, calling on the equipment and expertise of the mining company. Landscaping included construction of islands, access paths, bridges, and "frog hollows" (shallow depressions to encourage the breeding of frogs and other animals that favor shallow water).

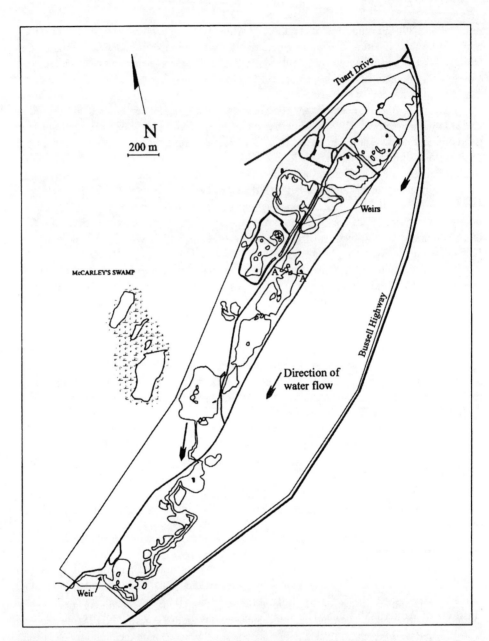

Figure 11.3. The Capel Wetlands Centre, near Capel, southwestern Australia. This is the concept plan for the Centre, and has been largely implemented. Reproduced, with permission, from Doyle et al. (1998).

Control of Water Quality

While the wetlands were in their infancy, several water quality problems were highlighted. These included low pH (3–4), high ammonium concentrations (22–83 mg L^{-1}), low phosphorus concentrations (0.000 to 0.015 mg L^{-1}) and high concentrations of iron and manganese (Gordon & Chambers, 1987).

The mine process water entering the chain of lakes from the north was the source of the ammonium. Fortunately, it was possible to alter the treatment of the outflow from the processing plant to markedly reduce the concentration of this ion, though the concentration remains high.

A series of studies that examined aspects of water quality ranged from laboratory studies of sediment phosphate adsorption and release to the use of microsystems (intact cores of water and sediment removed from the lake and studied in laboratory or glasshouse), mesocosm experiments (using plastic enclosures around sediment and water in a lake), and whole lake experiments (involving manipulation of the lake environment). These studies combined manipulating water quality with an aim to improve algal productivity of the lakes. The results are described in the section on setting up food webs.

Establishment of Riparian and Aquatic Plants

The landscape after mining had little plant life compared to other wetlands in the area, which have provided a useful yardstick against which to measure progress in establishing food webs in the newly created wetlands of the Centre. Establishment of aquatic plants quickly emerged a major priority because of their importance for the establishment of food webs, particularly those supporting waterbirds, the encouragement of which became a major aim of the Centre.

The lakes at first had little accumulation of organic-rich sediments, and had flocculent, mineral-rich "slimes" at depth. Coupled with low primary production, neither autotrophic nor heterotrophic pathways were supported, and there were few macroinvertebrates (Cale & Edward, 1990a).

Attention was directed to encouraging phytoplankton, which could be grazed by planktivorous fish and invertebrates, other algae such as periphyton, which could be consumed by invertebrates, and submerged aquatics, which would provide shelter for invertebrates. Particular attention was given to emergent plants such as *Typha* and sedges, which would provide shelter and nesting materials for birds, as well as an important source of detritus to support microbial and detrital-based food webs.

When choosing aquatic plants, preference was given to those that occurred in nearby, natural wetlands; as far as practicable, local plants were used, raised when necessary in a plant nursery set up on the site.

A major study of the environmental requirements and propagation of a suite of southwestern Australian wetland plants was undertaken. This study included sites other than Capel and was funded by a separate organization, the Minerals and Energy Research Institute of Western Australia. The study was summarized in a handbook by Chambers et al. (1995). Considerations similar to those used in the choice of wetland plants also applied to the selection and propagation of plants to be established on the "uplands" surrounding the lakes.

Setting Up Food Webs

The very low concentration of available phosphorus was particularly important. Low phosphorus is usually regarded as a very desirable attribute in freshwater systems in which plant growth is often phosphorus-limited. Addition of this element in available form can lead to severe eutrophication problems, with algal blooms (sometimes of toxic cyanobacteria) and deoxygenation of bottom waters leading to the death of invertebrates and fish. In contrast, at Capel, there was so little available phosphorus that phytoplankton populations were extremely low, providing little organic carbon generation to support food webs. It emerged from studies with microsystems that the sediments (containing high levels of iron and manganese) strongly bound any available phosphorus, which could only be released by the sediments under anoxic conditions. Phytoplankton blooms could only be generated through phosphorus addition, and the effect did not last long; a conclusion supported by experiments using enclosures set out in the wetlands and further trials in which phosphorus was added to one of the lakes.

An alternative was to add barley straw to the lakes, a method found in some overseas studies to suppress phytoplankton. In our studies, meadow hay supported the growth of phytoplankton and of microscopic algae on the sediment surface, and also acted as an excellent habitat for invertebrate populations. The immediate impact of an increased food source was reflected in waterbird numbers. In a sense, the straw was acting as a "slow-release fertilizer" and its effects lasted some months. Eventually, the benefit was reduced, however, as the straw disintegrated and phosphorus was in part flushed from the system, and in part bound to sediments. Much better, it was argued, to establish fringing plants, which would provide detritus to the system for the long term. Work on the invertebrates has clearly established that their numbers reflect the amount of particulate organic material present; a conclusion borne out by circumstantial observations of invertebrate populations within and away from clumps of macrophytes (Cale & Edward, 1990b).

Following the basic study of wetland plant establishment, plants were progressively and systematically established in suitable habitats throughout the Centre.

Waterbird populations have increased, though not to levels as high as in neighboring wetlands in adjoining farming country; these are quite eutrophic, supporting large phytoplankton and invertebrate populations, which in turn support large bird populations. Nevertheless, there is an increasing number of birds raising young at the Centre. A complicating factor is that we are dealing with a relatively small area of wetland close to other wetland habitats. Inevitably, the number of birds found at the Centre largely reflects regional shifts in waterbird populations, brought about, for example, by seasonally changing patterns in rainfall distribution.

Groundwater

It was critical to understand the dependence of the lakes on groundwater, especially because there was a long-term plan to cease mining operations, and so stop the flow of mine process water into the lakes. What effect might this have on lake water levels, and on the viability of the Centre in the long term?

It turns out that there are two components. First, there is a regional groundwater flow moving slowly toward the coast from higher ground inland. This is reflected in contours drawn through groundwater heights measured in a network of bores in and around the wetlands centre (Figure 11.4). Similar sets of data, recorded at different times and under different rainfall regimens, could be related to lake water levels fed into a computer program, and from that, predictions were made about what changes in water levels would occur with different inflows. There is a close interaction between lake water levels and groundwater. Under some weather patterns, the lakes of the Centre contribute water to the aquifer but at other times remove water. And if the water from the mine ceases, the lakes will not dry under normal rainfall patterns (Qiu et al., 2001).

Ecology of Wetlands Created in Mining-Affected Landscapes 263

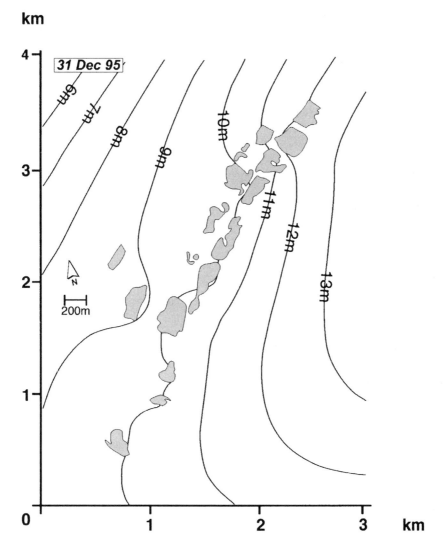

Figure 11.4. Groundwater levels in the region of the Capel Wetlands Centre, Western Australia. Contours are meters above sea level (Australian Height datum). The algorithm used to construct the contours considers regional groundwater flow, recharge from rainfall, evapotranspiration, and local effects, with lakes acting as "pumps." The data used were from 1996/97 (Qiu et al., 2000).

Public Participation

The potential for interesting school children in the Centre was recognized early in the research and management program, and this potential was realized by the involvement of science teachers, through the activities of the Science Teachers Association of Western Australia (STAWA). Teachers are represented on the management committee, arrange school excursions, and help design and distribute activity programs appropriate to schoolchildren and relevant school curricula.

Visitor numbers are assessed on a regular basis. The main source of visitors is schools, but the numbers from other sources are increasing, including visits arranged through mining companies, and through organisations concerned with adult education or birdwatching. The Centre is increasingly visited by private individuals and tour buses (especially on tours with a natural history flavor). The Centre is now included on a list of sites in Western Australia collectively referred to as Living Windows, and marketed together, with considerable cost savings.

For a time, the area was governed by mine-site regulations, some of which were restrictive to the general public (wearing hard hats and appropriate footwear, and being accompanied by staff). These have been relaxed, and casual visitors can be admitted more freely. The number will increase, and provision of facilities and guides is another call on resources.

Public participation also comes through the involvement of volunteers and trainee programs. This involvement has greatly improved the rate of wetland vegetation establishment and the assessment of bird populations.

Overview of Capel Wetlands

The Centre is at an interesting stage of development, with bird populations increasing, invertebrate communities rising and increasing in complexity, and macrophytes increasing. And there is a sound understanding of management issues. There has been some rationalization in land tenure, the company having purchased some of what was private land within Centre boundaries, and arrangements are being made to rationalize some of the remaining boundaries.

One of the expectations of the mining company was that it would be able to meet its obligations and ultimately withdraw from the situation, leaving a functional wetland ecosystem that would be essentially self-sustaining and require minimal ongoing maintenance. This situation has not been fully achieved given that acceptable completion criteria have not yet been defined and met.

Whatever the outcome, some ongoing maintenance will be required, and facilities kept up for visitors. Under discussion is the formation of a trust, set up with company funding but generating additional funding from visitors, land management agencies, research funding agencies, and tourism and industry, with in-kind support from tertiary institutions.

Supplementary Case Study: The Creation and Restoration of Nearshore Seagrass Meadows

It is instructive to draw attention to the problems faced by a company that dredges for shellsand in a nearshore area off the coast near Perth, Western Australia. The dredged shellsand is rich in carbonate and used in the manufacture of cement and lime. The resource is of great economic importance but underlies seagrass meadows of considerable ecological significance. The issues are very relevant to the theme of this review.

Like more conventional "terrestrial" wetlands such as fens, seagrass meadows may consist of large areas of monospecific stands, are dominated by rhizomatous monocotyledons with strap-like leaves, have high productivities (reflected in high leaf turnover rates), and are not extensively grazed but support detrital-based food webs. But they differ from conventional wetlands most obviously in the fact that seagrass meadows in Australian waters are submerged continuously beneath several meters of high-energy marine water. This leads to problems in resource recovery, accomplished with purpose-designed dredges, barges, and boats. It also leads to challenges for research, which must be carried out underwater using scuba gear.

Despite this obvious difference, the approaches and challenges follow almost exactly the steps we have outlined for the creation of more conventional wetlands after mining in the terrestrial environment.

The purpose of management is to minimize impacts of dredging on these important habitats. Dredging is carried out in specified lease areas under obligations specified by government instrumentalities. In contrast to the terrestrial environment, there is a strong sense of communal ownership and responsibility in marine situations; for example, many commercially or recreationally important fish swim over seagrasses, but it is difficult to establish the interdependence of the two.

Having defined the aims of management, a major research program was set up to achieve the following:

- Document (using geographical information systems, GIS) the distribution of seagrasses and the way this changes with time.
- The ecological significance of seagrasses.
- The dependence of fish and fisheries on seagrass meadows.
- The relationship between seagrass growth and sediment deposition.
- The possibility of locating alternative sources of high-quality carbonate close to refinery and market.
- And perhaps most challenging of all, the development of methods to re-create seagrass meadows by transplanting submerged seagrass from donor areas.

The work has resulted in new approaches to mapping attributes of ecological significance based on GIS and the development of technology to transplant functioning sods of seagrass meadow from one locality to another. An underwater seagrass harvesting and planting machine (Figure 11.5) was designed to extract and plant, with minimal disturbance, a large sod of seagrass ($0.25 \, m^2$ in area and 0.5-m deep). Good survival has been achieved, particularly for *Posidonia* sp., despite the high-energy conditions in these nearshore areas. The larger size of the transplants appears to be a key to survival, making them more robust and better able to withstand disturbance (Paling et al., 2001).

Participation of the public in discussions and briefings has been important, especially in view of strong interest from conservation groups and the fishing industry.

Drawing on results from the research program, an Environmental Review and Management Plan has been drawn up for another area proposed for leasing and dredging, a plan at present under consideration by the appropriate authorities (Lord, 2000).

General Discussion and Conclusions

The benefits accruing to a mining company through creating wetlands in mining voids include meeting statutory obligations under mining and environmental legislation, and achieving valuable public relations kudos.

Because of the potential importance of establishing wetlands after mining, it would clearly be an advantage if the ultimate use of mined voids for this (as indeed any other)

Figure 11.5. "Ecosub 1." Machine for transplanting intact sods of functioning seagrass meadow from areas to be dredged for shell sand in Owen anchorage, near Cockburn Sound, Western Australia. The machine is submerged to the sea floor, where it is operated by Scuba divers. Reproduced, with permission, from Lord (2000).

purpose were built into a "mine-closure plan" as part of the mine planning and approvals process. From the perspective of the company, this would provide a clear, long-term direction for planning and management, and should shorten the approvals process. From the perspective of government authorities concerned with giving approval to the mine, it would help in decision making. And from the perspective of the wider community, it would be possible to plan for relevant conservation activities and obviate landscape disfigurement. In the past, the general public has helped to create wetlands "after the event," coming into fortuitously help establish wetlands in mined areas (or postindustrial sites) that would otherwise be wasteland, but could become areas of conservation and landscape significance.

The availability of appropriate technology to assist wetland creation continues to increase, as does our understanding of wetland ecology. However, successful wetland creation, as shown by the case studies outlined here, requires clear objectives, focused planning, an understanding of local conditions, and rigorous attention to development, while being sufficiently flexible to allow the modification of techniques to suit the environment or changing circumstances. Our understanding of the successional changes that occur in newly created wetlands is still in its infancy, with the result that indicators of ecological development or completion criteria (in an ecologically meaningful sense) are difficult to establish. This is an area in which future research will assist in creating ecologically valuable wetlands.

References

Anonymous. (1987). *Ranger uranium environmental inquiry, second report.* Canberra: Australian Government Publishing Service.
Australian and New Zealand Environment and Conservation Council. (1992). *Australian water quality guidelines for fresh and marine waters.* National Water Quality Management Strategy.
Australian National Parks and Wildlife Service. (1980, May). *Nomination of Kakadu National Park for inclusion in the World Heritage List.*
Brooks, D. R. (1991). *Progressive developments of wetland habitats at the RGC Wetlands Centre, Capel, Western Australia.* RGC Technical Report Series No. 17.
Cale, D. J., & Edward, D. H. D. (1990a). *Macro-invertebrate fauna: A quantitative study of Boulder Lake, Capel, Western Australia from December 1987 to December 1988.* RGC Technical Report Series No. 9.
Cale, D. J., & Edward, D. H. D. (1990b). *The influence of aquatic macrophytes on macroinvertebrate communities in Swamphen Lake, Capel, Western Australia.* RGC Technical Report Series No. 10.
Chambers, J. M., & McComb, A. J. (1994). Establishment of wetland ecosystems in lakes created by sand mining in Western Australia. In W. J. Mitsch (Ed.), *Global wetlands: Old World and New* (pp. 431–441). Amsterdam: Elsevier.
Chambers, J. M., Fletcher, & McComb, A. J. (1995). *A guide to emergent wetland plants of south-western Australia.* Perth: Marine and Freshwater Research Laboratory, Murdoch University.
Cole, C. A. (1998). Theoretical function or functional theory? Issues in wetland creation. In A. J. McComb & J. A. Davis (Eds.), *Wetlands for the future* (pp. 720–736). Adelaide, South Australia: Gleneagles Press.
Conservation Commission. (2001). *Herdsman Lake Regional Park draft management plan 2001–2011.* Prepared by Conservation Commission of Western Australia and City of Stirling.
Doyle, F. W., & Davies, S. J. J. F. (1998). Creation of a wetland ecosystem from a sand mining site: A multidisciplinary approach. In A. J. McComb & J. Davis (Eds.), *Wetlands for the future* (pp. 761–772). Adelaide, South Australia: Gleneagles Press.
Eger, P., Melchert, D., Antonson, & Wagner, (1993). The use of wetland treatment to remove trace metals from mine drainage. In G. A. Morishiri (Ed.), *Constructed wetlands for water quality improvement* (pp. 283–292). Boca Raton, FL: Lewis.
Finlayson, C. M., Bailey, B. J., Freeland, W. J., & Fleming, M. R. (1988). Wetlands of the Northern Territory. In A. J. McComb & P. S. Lake (Eds.), *The conservation of Australian Wetlands* (pp. 103–126). Chipping Norton, New South Wales: Surrey Beatty.
Garono, R. J., & Kooser, J. G. (1994). Ordination of wetland insect populations: Evaluation of a potential mitigation monitoring tool. In W. J. Mitsch (Ed.), *Global wetlands: Old World and New* (pp. 509–516). Amsterdam: Elsevier.
Hallam, N. (1985). The biology of Ewens and Piccaninnie Ponds, South Australia. *Habitat, 13*(1), 18–22.
Hammer, D. A. (1992). *Creating freshwater wetlands.* Boca Raton, FL: Lewis.
Kadlec, R. H., & Knight, R. L. (1996). *Treatment wetlands.* Boca Raton, FL: CRC Press.
Lane, J. A. K., & McComb, A. J. (1988). Western Australian wetlands. In A. J. McComb, & P. S. Lake (Eds.), *The Conservation of Australian Wetlands* (pp. 126–146). Chipping Norton, New South Wales: Surrey Beatty.
Lord, D. A. (2000). *Long-term shellsand dredging, Owen Anchorage.* Environmental Review and Management Programme, prepared by D. A. Lord and Asssociates on behalf of Cockburn Cement, Ltd., 192 pp.
Lothian, J. A., & Williams, W. D. (1988). Wetland conservation in South Australia. In A. J. McComb & P. S. Lake (Eds.), *The conservation of Australian Wetlands* (pp. 147–166). Chipping Norton, New South Wales: Surrey Beatty.
Masini, R. J. (1988). *Inland waters of the Pilbara, Western Australia, Part 1,* Technical Series No. 10, Environmental Protection Authority, Western Australia, 40 pp.
Masini, R. J., & Walker, B. A. (1989). *Inland waters of the Pilbara, Western Australia, Part 2.* Technical Series No. 24, Environmental Protection Authority, Western Australia, 42 pp.
McComb, A. J., & Lake, P. S. (Eds.). (1988). *The conservation of Australian Wetlands.* Chipping Norton, New South Wales: Surrey Beatty.
McComb, A. J., & Lake, P. S. (1990). *Australian wetlands,* North Ryde, Australia: Collins/Angus and Robertson.
Mitsch, W. J., Wu, X., Nairn, R. W., Weine, P. E., Wang, N., Deal, R., & Boucher, C. E. (1998). Creating and restoring wetlands. *Bioscience, 48,* 1019.
Moore, P. D., & Bellamy (1974). *Peatlands.* New York: Springer-Verlag.
National Parks and Wildlife Service (1980). Nomination of Kakadu National Park for inclusion in the World Heritage list. Canberra, ANPWS, May 1980.

National Trust. (1959). *A guide to Wicken Fen* (3rd ed.), London: Author.

Paling, E. I., van Keulen, M., Wheeler, K., Phillips, J., & Dyhrberg, R. (2001). Mechanical seagrass transplantation on Success Bank, Western Australia. *Ecological Engineering, 16*, 331–339.

Peberdy, K. J. (1998). Wetland creation for nature conservation in post-industrial landscapes—examples from the UK. In A. J. McComb & J. A. Davis (Eds.), *Wetlands for the future* (pp. 720–736). Adelaide, South Australia: Gleneagles Press.

Poach, M. E. and Faulkner, S. P. (1998). SOIL Phosphorus characteristics of created and natural wetlands in the Atchafalaya Delta, L. A Estuarine Coastal and Shelf Science, 46:195–203, 1998.

Ponder, W. F. (1986). Mound Springs of the Great Artesian Basin. In P. De Deckker & W. D. Williams (Eds.), *Limnology in Australia* (pp. 403–420). Melbourne: CSIRO and Dordrecht: Junk.

Qiu, S., Scott, W., & McComb, A. J. (2001). Predicting the effects of inflow diversion on future water levels in the Capel Wetlands Centre. Technical Report Series, No. 53. Capel Wetlands Centre.

Van der Valk, A. G. (1998). Succession theory and restoration of wetland vegetation. In A. J. McComb & J. A. Davis (Eds.), *Wetlands for the future* (pp. 720–736). Adelaide, South Australia: Gleneagles Press.

12

Conservation of Soil and Nutrients through Plant Cover on Wetland Margins

R. S. Ambasht and N. K. Ambasht

Introduction

Wetlands roughly cover 6% of the land surface, occupying 8.6 million km^2 (Maltby & Turner, 1983). In addition, rice fields are also essentially wetlands, because during the growing season, they have enough free water. About half of the human population depends primarily on rice as staple food.

It is only during the last 30 years that the term *wetland* has acquired popularity and worldwide attention of not only ecologists but also naturalists, environmentalists, ornithologists, water resource managers, and pollution scientists. The term gained importance through the Ramsar Convention, when wetland scientists representing 18 nations gathered in Ramsar, a small city near the Caspian Sea in Iran, to reach intergovernmental agreements or treaties on conservation of wetlands of international importance, particularly as habitats for waterfowl. The convention was adopted on February 2, 1971 (signed the next day); therefore February 2 is celebrated as International Wetland Day.

Wetland conservation activities have been undertaken by various scientific bodies and notable advancements have been made. The International Association of Ecology (INTECOL) has created a section on wetlands that organizes at 4-year intervals the International Wetland Conferences, held so far at New Delhi (India, 1980), Rennes (France, 1988), Columbus, Ohio (United States, 1992), Perth (Western Australia, 1996) and Quebec (Canada, 2000). A number of other organizations connected with wetland studies also joined the Quebec 2000 conference, such as the Society of Wetland Scientists (21st annual meeting), the International Peat Society (11th Congress), the International Mire Conservation Group (9th symposium), and some other partner agencies such as

R. S. Ambasht and N. K. Ambasht • Department of Botany, Banaras Hindu University, Varanasi, India.

Wetland International and the World Conservation Union (IUCN). World Wildlife Fund for Nature (WWF) is also actively involved in wetland conservation. The Ramsar Convention has grown very strong and 118 countries now constitute the contracting parties, and as per stipulations, every contracting country has to designate one or more wetlands of international importance as Ramsar sites. Other terms of the contract are promoting wetland conservation; wise use; and encouragement of research and exchange data on flora and fauna. There are over 1,000 Ramsar sites the world over, because wetlands are of universal occurrence. Among the Indian Ramsar sites are Lake Chilika (116,500 ha), Keoladeo National Park (2,873 ha), Sambhar Lake (19,000 ha), Loktak Lake (28,890 ha), Harike Lake (4,100 ha), and Wular Lake (18,900 ha).

Definitions

With the growing importance and diversity of scientists and professionals involved in wetland studies, no single definition of wetland was suitable for all. Realizing this, the Ramsar Convention standardized one definition, as quoted by Dugan (1990): "Areas of marsh, fen, peatland or water, whether natural or artificial, permanent or temporary, with water that is static or flowing, fresh, brackish or salt including areas of marine water, the depth of which at low tides does not exceed six metres." Smith (1996) has quoted another definition given by Cowardin et al. (1979) as "land where water table is at, or above the land surface for long enough each year to promote formation of hydric soil and to support hydrophytes." Wetlands have been regarded as marshlands or transition lands between truly terrestrial and truly aquatic zones, or the drying, exposed phase of shallow water bodies, including seasonal rivulets. But later, the term grew in its scope to cover shallow water bodies and shallow zones of deeper bodies, including rivers and sea coasts. Mitsch and Gosselink (1993) have provided 26 common names of different wetlands. Bog, usually dominated by *Sphagnum*, is acidic and without noticeable inflow–outflow systems, but fen has inflow-outflow systems. Marsh is commonest kind. Bottomland, lagoon, mire, moor, peat, potholes, reedswamps, wet meadows, pools, puddles, flood plains (*Varzea* of Amazon, and *Diarrah* of Ganga) are other common kinds. Ambasht (1998a, 1988b) and Ambasht and Ambasht (1998) have provided details about Indian wetlands.

Ramsar (1990) has classified wetlands into (1) marine and coastal wetlands (11 types), and (2) inland wetlands (16 types) and man-made wetlands (8 types). Cowardin et al. (1979), referring to the U. S. Fish and Wildlife Service, have divided wetlands into (1) marine, (2) estuarine, (3) riverine, (4) lacustrine, and (5) palustrine.

Wetlands are the meeting ground of aquatic and terrestrial systems, and therefore have advantageous habitat features of both. In such ecotones, there is always high biodiversity. The ecotones also show a very high level of bioproductivity; therefore, despite only about 6% of space sharing of terrestrial systems, they contribute much more to total organic productivity and other human welfare needs.

Conservation Aspects

This chapter is mainly confined to the role of ecotonal vegetation in conserving soil against erosive forces, water runoff, and nutrient movement, along with soil and water draining down into the receiving water body. Conservation of both lacustrine and riverine ecotones is being considered. *Riparian* refers to stream or river banks. Babbit (1985), in his foreword of a report on riparian ecosystem management, has stated that until 15 years ago, the term *riparian* was unknown to the majority of people, layman and scientists alike, but there is sudden recognition of the existence of this valuable and controversial parcel of land, which critically needs study and conservation steps. He has advocated a comprehensive legislative mandate for protection, conservation, and rehabilitation of riparian ecosystems. The term *riparian* is derived from Latin word *rip*, meaning the banks of a water course. Lowrance et al. (1985) regard riparian ecosystems as complex assemblage of organisms and their environment existing adjacent to and near flowing water. According to Ewel (1978), riparian ecosystems have laterally flowing water, the level of which rises and falls at least once every year, and a high degree of connectedness with other ecosystems. The riparian habitats, being elongated in shape, has a high edge-to-area ratio and is very much open to interchanges of matter such as soil, water, and nutrients. They experience pulses of floods, exposure, erosion, silting, and deposition of live material such as diaspores, and so on. The habitat has strong linkages with other systems for both biotic and abiotic components.

Lowrance et al. (1985) have discussed the role of vegetal cover in the ability to filter excessive flow of nutrients into the water body. Lowrance et al. (1984) have provided information on the quantity of NO_3-N and SO_4-S input through rainfall and their runoff from a river watershed in Georgia, USA and Ambasht (1985) gave quantities of nitrogen and phosphorus input through rainfall and their retention by vegetal cover in the watershed of Chandraprabha River in India. Loss of vegetal cover increases severalfold the erosion of soil, water, nutrients, and runoff as quantified by Ambasht and his coworkers in a series of experiments carried out in field and culture conditions. Overgrazing, land clearing, dumping of solid wastes, and excessive application of fertilizers and biocides during the dry phase, and so on, are main factors of habitat degradation. People on holiday and picnickers visit wetland ecotones in ever-increasing numbers and frequency for recreational purposes and remove herbage cover to prepare clear sitting and cooking space. Devegetation and human congregation have chain effects on biotic and abiotic status, and the avifauna migrate to safer distances. Triquet et al. (1990) have recorded a sharp fall in richness of plant and animal species as the ecotone vegetation is cut. Lazarus (1990) has estimated that about 25,000 million tons of soil are being washed away each year from land surfaces on a global scale. Wetlands have become the main discharge point for heavy silt load and nutrient enrichment. Modern anthropogenic activities add pollutants from both point and nonpoint sources. In India alone, about 700 million rupees (Rs) worth of annual soil loss through erosion is estimated (Anonymous, 1982).

Verry and Timmons (1982), in their study, found that between 36 and 60% of annual nutrient input is held in streamside vegetation. Riparian forests are reported to take up 89% of nitrogen in Maryland (Peterjohn & Correll, 1984) and 86% in North Carolina (Cooper et al., 1986). High levels of phosphorus are taken up by vegetation in Florida (Boyt et al. 1977), and in this way, the nutrients are conserved against losses. Kadlec and Kadlec (1979) regard natural wetland and river margin lands as natural treatment systems for polluted waters. With increasing levels of habitat disturbances in 10 watersheds, Byron and Goldman (1989) found correlation of accelerated runoff of nitrates, soluble phosphorus, and total phosphorus. Schwer and Clausen (1989) have reported 89% phosphorus and 92% nitrogen conservation by a strip of vegetation on a 2% slope. Van der Valk et al. (1979), Blackburn and Wood (1990), and Young et al. (1987) have made similar studies on nutrient filtering by wetland margin vegetation. Lyngby and Brix (1982) have made observations on the role of wetland plants in regulating the level of metal ions such as copper, lead, and zinc in water. Srivastava and Ambasht (1990) and Ambasht and Srivastava (1994) found that in the Renukoot industrial belt, vegetation on the banks of G. B. Pant Sagar Wetland and Rihand River, dense patches of *Polygonum amphibium* accumulate in its vegetative body iron up to 1,875 µg/g, and as the river flows down 25–30 km to Obra, the iron content in plants was reduced to 1,340 µg/g. Other plant species in this region do not show such a property of harvesting out the heavy metal.

Agricultural practices with heavy input of chemicals and fertilizers are increasing on uplands circumscribing lentic wetlands and riparian uplands. This has exposed water bodies to severe stresses, and it is all the more important now to preserve the vegetation and integrity of concerned ecotonal ecosystems. It would be interesting to review briefly the experimental studies on quantification of conservation of soil, water, and nutrients by plants. The term *conservation* is derived from Latin words *con*, meaning "together," and *servare*, meaning "guard." In this root word, *conservation* means "to guard together." In ecology, conservation refers to wise use of resources in such a way that sustainability is ensured. Protection and preservation are part of conservation. The ability of plants in protecting the soil against erosion, in helping the rain water to infiltrate more and run down less, and in preventing soil nutrient erosion and runoff losses are therefore parts of conservation potential of herbage cover. Ambasht (1962, 1970) provided a concept of *conservation value percentage* to give a quantitative dimension. This refers to the percentage of resource retained by vegetation out of the losses that would occur in the event of total exposure of land without plant cover. Thus, measurements of soil, water, or nutrient erosional losses are measured in identical pairs of plots, one with intact vegetation and another of equal size without plants, from which runoff material is collected, weighed, and analyzed. He has provided the formula $Cv = 100(1 - SP/SO)$ to calculate the conservation value percentage (Cv) from the quantity of soil washed down from a plant-covered plot (SP), and SO is the quantity washed from the bare plot under identical erosive stress.

Some dominant and characteristic species on the riparian slopes of Ganga at Varanasi were transplanted on culture plots made of riparian topsoil at a gradient of 13°. Plants of seven different species were raised on separate plots, and one bare plot was maintained. Both plots were subjected to equal simulated rainfall, and the runoff soil was collected in separate tanks at the base of sloping plots. Ambasht (1962) found that *Cynodon dactylon*, a ubiquitous grass, in a natural rainfall event conserved soil up to 90%. Subsequently, as it achieved full luxuriance of herbage cover, the soil conservation percentage rose to 95–97%. The fibrous roots very effectively bound the soil, so much so that even after the aboveground herbage was scraped, the Cv due to roots alone was 77%. Another very ubiquitous plant *Cyperus rotundus*, a sedge common in fallow and cultivated fields, is reported to conserve soil between 82 and 93% under varying rainfall or water-showering treatments. A very dense, tussock-forming tall grass, common on riparian upland embankments, *Saccharum benghalensis*, is certainly the most efficient soil conserver, with 92–96.5% soil Cv. This hardy and rough grass is not liked by cattle and suffers least from biotic stresses. Its herbage is removed after the rainy season to prepare huts and thatch rooftops. Dicotyledonous species *Euphorbia hirta*, with its poorly branched taproot, has a soil Cv of only 10–12%, whereas *Alhagi camelorum*, with well branched rhizomatous parts, has 30–35% soil Cv. However, *Scoparia dulcis*, a dicot plant with the ability to produce adventitious roots under anaerobic situation on waterlogging, has 65–84% soil Cv. Thick, tuberous roots with a secondary network of branches enable

Ruellia tuberosa to bind soil up to 80–90% (i.e., comparable to grasses). Ambasht et al. (1984) found that *Phyla nodiflora*, a mat-forming shoot cover on the banks of the Ganga and Gomati rivers in India has almost 94% soil Cv. Reported water runoff from *Phyla* plot was 22% compared to 81% from a bare plot; thus, the water Cv for this species is 73%. In this work, Ambasht et al. (1984) have reported 95% and 74% soil and water Cv, respectively, for *Cynodon dactylon*. A common leguminous herb, *Crotalaria medicaginea*, has the respective soil and water Cv of 53.0% and 27.3%. In all these conservation experiments, Ambasht et al. have found that in the process of soil conservation, riparian plants conserve clay fragments better than coarser particles compared to soil erosion from bare plots. Not all grasses are high soil conservers. *Digitaria adscendens*, another densely growing grass, has 87% soil and 65% water Cv.

Kumar et al. (1992a, 1992b) have made similar studies on an extended scale of five common riparian weeds transplanted from the banks of G. B. Pant Sagar reservoir and Rihand River to sloping culture plots in the Banaras Hindu University. They have studied the movement of soil, water, nitrogen, and phosphorus across vegetated plots under the five selected species, and under bare conditions. The species are *Leonotis nepetaefolia*, *Cassia tora*, *Ageratum conyzoides*, *Parthenium hysterophorus*, and *Sida acuta*. Both these nutrient elements, nitrogen and phosphorus, are responsible for increasing the crop productivity and, therefore, are frequently applied in agricultural fields. Runoff water from these fields is nutrient-rich and causes eutrophication of receiving water bodies. The role of ecotonal vegetation acting as sinks is important for preventing eutrophication.

Kumar et al. (1992a, 1992b) carried out simulated rainfall treatments at the stage when sloping plots were fully covered with foilage. In order to generate high kinetic energy to cause erosion, they showered at the rate of 30 cm/hr^{-1} for 8.5 min on each plot a number of times, at intervals of 10 days. Water and soil removed from bare and vegetated plots were collected in separate receiving reservoirs. Soil and water were weighed and analyzed for phosphorus (total, inorganic and organic) and nitrogen (total, nitrate and ammonium). The quantities of soil, water, nitrogen, and phosphorus eroded from vegetated and bare plots were used to calculate the respective Cv percentage of each of the five species.

Leonotis, a stout, tall herb of quadrangular stem, of the family Lamiaceae, is reported to conserve 84% soil and 50% water. *Leonotis* plot had 98% canopy cover, 51 g/m^2 of litter mass, 12% antecedent soil moisture, and 188 g/m^{-2} standing crop dry biomass. Nitrogen Cv for this species was 71%, and an organic form of nitrogen contributed most of it. Total phosphorus Cv was 63%.

The soil and water Cv, respectively, of the other four species reported are *C. tora*, 69% and 34%; *A. conyzoides*, 57% and 44%, *P. hysterophorus*, 45% and 25%; and *S. acuta*, 33% and 19%. In all these plots at the showering time, the level of antecedent soil moisture ranged between 8 and 11%, and litter mass between 88 and 122 g/m^{-2}. Highest biomass was in *C. tora* (347 g/m^{-2}). From the *L. nepetaefolia* plot, the loss of total nitrogen was lowest at 1.3 kg/ha^{-1}, whereas from the bare plot it was 3.5 kg/ha^{-1}. Lowest total N Cv was for *S. acuta* (26%). But with respect to soluble nitrogen Cv efficiency, *A. conyzoides* was 82%, and this species had 97% nitrate/nitrogen conservation efficiency. Thus, different species show different efficiencies for various forms of nitrogen and phosphorus. White and Williams (1977) have reported that nutrient transport in runoff was significantly related to the quantity of soil and water losses. Kumar et al. (1992a) have found that a higher fraction of phosphorus runoff is through water (62% of total phosphorus, 53% of inorganic phosphorus, and 72% of organic phosphorus) than through soil.

Kumar et al. (1997) further elaborated findings on the reduction of nitrogen losses through erosion by *L. nepetaefolia* and *S. acuta* under different intensities of simulated rainfalls. Seedlings of these species were collected from the riparian belt of Rihand River and raised in the sloping experimental plots of the Botanical Garden of the Banaras Hindu University. The simulations are 20, 25, 31, 38, 46, and 51 mm water fall/h applied for 50 min on each vegetated and bare plot. It is interesting that as the rain intensity was increased, the soil conservation value of *Leonotis* increased from 63 to 88% and that of *Sida* from 27 to 57%. This should not suggest that with higher rain intensity the quantity of soil eroded is reduced. In fact, it increases, but because Cv is obtained from the quantities eroding down the vegetated and bare plots, the percentage increases due to much greater loss from the bare plot. Ecological features of the *Leonotis* plot were as follows: number of terminal rootlets, 4.67×10^5 m^2; canopy cover, 98%; litter mass, 46 g/m^{-2}; standing crop dry biomass, 196 g/m^{-2} and moisture level, 12%. The corresponding values for the *Sida* plot were 1.9×10^5 m^{-2} (rootlet numbers), 79% canopy cover, 82 g/m^{-2} litter mass, 150 g/m^{-2} biomass, and 11% soil moisture. Even though the Cv percentage increased with increase in rain intensities, the actual quantity of nitrogen losses increased. The results show that vegetation has greater control in ammonium nitrogen movement through soil than through runoff water.

Kumar et al. (1996) have measured the role of five common riparian species in reducing erosion of organic carbon and a few selected cations. The loss order of different nutrients was organic C > Ca > K > Na. Nutrient conservation values of the five species ranged from 30 to 83% for organic C, 19 to 78% for Na, 13 to 72% for K, and 29 to 52% for Ca. The strongest controlling factor is the number of fine roots. Clay content selectively eroded in greater quantities carries on its surface more adsorbed nutrients (Oades, 1988). Particulate C movement is also reported to be greater in erosional runoff by many workers (Schreiber & McGregor, 1979; Lowrance & Williams, 1988). Albert and Spomer (1985) have also measured runoff for dissolved nitrogen and phosphorus from watersheds and found that low movement is due to their low concentration in the dissolved state, as also reported by Kumar et al. (1992a, 1992b). The U.S. Environmental Protection Agency (1973) has prescribed for NO_3-N an upper permissible limit of 10 mg/L in drinking water. A high concentration of nitrate is extremely harmful to human health, particularly to children, as it has a damaging effect on their blood cells. The plant cover around wetland periphery or riparian embankments, therefore, not only prevents rapid silting of rivers and reservoirs but also ensures the safe quality of drinking water by retarding the level of nitrate/nitrogen and phosphorus to safe limits. All plants need nitrogen in available form for metabolism. Therefore, a good cover of vegetation absorbs a reasonably high quantity of runoff nitrates and lets a small and regulated quantity flow down in receiving waters, where aquatic organisms utilize them.

The canopy cover of multilayered foliage has a cushioning effect on raindrop energy (Young & Wiersma, 1973). The soft throughfall raindrop, on reaching the ground tends to convert into surface runoff along the slope. Plant residues or litter reduce the shear stress of runoff by diversions and detentions (Foster & Meyer, 1977). Therefore, a strong correlation in canopy cover and litter mass with erosion retardation has been found (Kumar et al. 1992a). Winger (1986), in the U.S. Department of Interior Fish and Wildlife Service, has published studies on forested wetlands of the Southeast and their role on water quality. Cowardin et al. (1979) classify the bottomland hardwood forests of southeast America as forested wetlands. These highly productive ecotonal zones (Teskey & Hinckley, 1977) interact with the aquatic ecosystem (Wharton, 1980), and especially during hydroperiods or high floods, there is considerable deposition of silt organic matter and soil nutrients, which in turn adds to a rich biodiversity and level of primary productivity. Klopatek (1975) regards ecotonal forested wetlands as an extremely open system by virtue of rapid inflow–outflow nature. Ambasht (1962) had therefore rightly selected both inflow and outflow forces of erosion, silting and inundation, in his studies of underground parts of riverbank plants. Brinson et al. (1981) have shown that, depending on the hydroperiod and amount of suspended material, the thickness of vertical accretion ranges from a few millimeters to more than one meter per year. Water-tolerant species such as *Taxodium distichum*, *Acer* spp., *Ulmus* spp., *Fraxinus* spp., and *Quercus* spp. form hardwoods and harbor rich fauna of fishes, aves, and mammals. These tree stands greatly stabilize the habitat and reduce water-current and soil-cutting force of water. They act as nutrient buffers and filters. Furthermore, these provide a greenbelt that prevents habitat encroachment overdevelopment. Forest products are obtained as a bonus, besides the great ecological services they render in oxygenating the air, retarding the beating effect of raindrops, adding to infiltration and reducing runoff quantity, and above all, checking soil erosion and nutrient losses of ecotones and preventing water bodies from silting and eutrophication. In fact, these ecotones are unique in recreation, educational, and research values (Jahn, 1979; Reimold & Hardisky, 1979). Shaw and Fredine (1956) estimated 127 million acres of wetlands in the United States, which was reduced to only 70 million acres by 1970 (Goodwin & Niering, 1974). In India, the rate of wetland ecotone habitat loss is very high, but reliable and authentic records are not available.

Thronson (1978) regards agricultural nonpoint sources affecting water quality to be fertilizers (i.e., nutrients, pesticides, and organic and inorganic material). While doing so, the ameliorating role of ecotone vegetation comes into play. The effectiveness of their role has been described with data obtained in field and culture conditions. Forested wetlands also act as nutrient sinks, in the sense that they store nutrients in spring and early summer, and release them slowly in late summer and fall (Kibby, 1979). Another very important way to regulate excess nitrogen is through denitrification in flooded soils into nitrous oxide and molecular nitrogen (Reddy & Patrick, 1984).

There is a constant struggle between plant cover and erosion, and under natural conditions, almost always the plants are winners and habitat remains stabilized. Thornes (1989) has provided a model of vegetation and erosion struggle, and helps in evaluating the critical stage when a "push" would shift the system to either a good vegetal cover side or into a denudation side. Anwar et al. (1989) consider shrubs more effective in halting runoff and erosion. However, Ambasht's works (1970, 1985) show a very high efficiency of grasses, particularly when the herbage cover is good and biotic disturbance is least.

Frankenberg (1989) has found that in the riparian corridors of R. Murray in Australia, *Phragmites australis* is a very good soil conserver against erosive forces. Ambasht (1985) has reported that, in a watershed, the input of nitrogen and phosphorus in a 12-month period through rainfall was 8.7 kg N/ha^{-1}/yr^{-1} and 0.65 kg P/ha^{-1}/yr^{-1}, respectively. From the protected grass covered plots, erosional loss was only 4.27 kg N/ha^{-1}/yr^{-1} and 0.28 kg P/ha^{-1}/yr^{-1}. But scraping the herbage cover increased the nutrient losses to 109 kg N/ha^{-1}/yr^{-1} and 7.7 kg P/ha^{-1}/yr^{-1}. Judicious management allows enough herbage removal during winter and early summer (November to May) during the dry phase, when there is almost no rainfall to cause any erosion by water. Groeneveld and Griepentrog (1985), in their geomorphological studies of Carmel River, California, have found that although the streambank was fairly stabilized by the reinforcing nature of roots, there occurred severe events of erosion in certain years. Species of *Salix*, *Populus*,

Platanus, and *Alnus*, which are well adapted to erosion-prone riparian wetlands, constituted the dominant vegetation. Schultze and Wilcox (1985) have recorded that severe storms from 1978 through 1983 caused widespread damage to streams of California. A large number of research projects were undertaken by different workers, in which revegetation was an important component. Some important plants included in multiple projects were *Hordeum* spp., *Bromus*, *Trifolium*, *Salix*, and *Populus*.

In undisturbed riparian forest and wetlands during 3 years of studies, Rhodes et al. (1985) have found an astounding 99% value of removing nitrate nitrogen. *Salix* and *Alnus* trees, and *Equisetum*, a peridophyte, and many sedges and grasses constituted the flora of studied site in the Sierra Navada of United States. Lowrance and Shirmohammadi (1985) have developed a model for riparian ecosystem management (REM) that predicts water and nutrient input to riparian ecosystems from uplands and their fate in the context of soil, leaf, litter, and vegetation conditions. The soil submodel is complex and includes daily changes in riparian forest soil nutrient pool. Likewise, in the litter submodel, the daily changes in litter nutrient pool are calculated and in vegetation submodels, nutrients in vegetal biomass minus released through harvest and litterfall are considered on a daily-change basis.

Roseboom and Russel (1985), working in large alluvium stream valleys in the prairies of western Illinois, have found that channelization to divert flood water only temporarily gives benefit to individual landowners, but in the long run, it is extremely harmful. From aerial photograph comparisons in 1940 and 1979, they have recorded complete erosion of some of studied sites.

Plant cover is the primary mode of habitat stabilization and prevention of soil erosion. Some erosion is inevitable and may be desirable. After all, even from lush green watersheds, slow soil erosion has led to deposition elsewhere, and most of the fertile alluvial valleys and plains have formed that way. It is only in damage and degradation of vegetal cover, accelerated erosion and silt deposition, that a double-edged menace takes place. There is a heavy denudation of surface soil at the place of erosion, exposing parent rocks and siltation or upwelling of receiving lakes and rivers. Under a natural state, there is a dynamic equilibrium between formation and disruption processes, and vegetal cover plays a key role in maintaining such a balance.

To feed the burgeoning human population, it has been necessary to convert forests into croplands. Wise use of ploughing, bunding, terracing, creation of windbreaks, shelter belts, and many other appropriate techniques have been developed to reduce considerably the loss of soil and fertility. The quantity of river-borne sediments carried into oceans has increased from 10 billion tons/year before the phase of intensive agriculture to between 25 and 50 million tons in modern times. This has caused heavy desertification and input of high fertilization cost in agriculture practiced on such eroded habitats. Wind erosion gets accelerated in treeless situations, and deserts march to adjoining fertile lands. According to the United Nations Environmental Program (UNEP 1992), the Sahara desert has been continuously increasing, as found through satellite-based analysis of vegetation index. It has increased from 8,942,000 sq km in 1981 to 9,269,000 sq km in 1990. In many parts of the world, deserts are expanding, primarily due to loss of vegetal cover and loss of fertility. Globally, about 295 million ha of land (i.e., about the size of India) are already strongly degraded (i.e., original biotic functions are destroyed). Roughly 910 million ha are moderately degraded and can be restored with proper management.

The rate of conversion of wetlands to other kinds of ecosystem uses is not well documented except for the United States. The states Iowa (99%), California (91%), and

Nebraska have lost most of the original wetlands. At least half of the original wetlands have disappeared in America and Europe (UNEP, 1992).

Sharma et al. (1992) have carried out field research on the watershed management in Sikkim Himalaya and measured runoff water, soil eroded, organic carbon, and total nitrogen losses across agricultural fields, agroforestry systems, natural forests, and bare lands. Using Ambasht's paired-plot technique, they have calculated Cv for water, soil, organic carbon, and total nitrogen. Crop cover is least effective, with only 26% soil Cv, 21% water Cv, 9% organic carbon Cv, and 53% nitrogen Cv. Agroforestry practice to raise the cardamom crop with *Alnus nepalensis* as nurse trees has considerably improved the soil, with carbon and nitrogen efficiencies being 60% soil Cv, 73% carbon Cv, and 83% total nitrogen Cv, but the water Cv declined to 14%. Natural forests in this rather fragile mountainous system showed the best effect of vegetal cover and percentage Cv's were 31% for water, 93% for soil, 88% for organic carbon, and 94% for total nitrogen.

Water Cv is always less than soil Cv in almost all studies made so far. Ambasht (1998a) has made field experiments of simulated rainfall equivalent to 25 mm rainfall per showering treatment on the upland vegetation around Surha Lake (3,000 ha oxbow lake) in India. At one site, water Cv was 44% and soil was Cv 91.5% during the rainy season and 58% and 93%, respectively, in the winter season. At another site, the soil Cv was around 93.7% and water Cv around 58% in both seasons. However, total losses are much higher in the rainy season. Partly computed and partly projected data reveal that annual erosional soil loss from embankment added to lake is on the order of 1.7 $t/ha^{-1}/yr^{-1}$, whereas, with loss of vegetal cover in the ecotone (experimentally scraped plots), the soil erosion level rose to 23.3 $t/ha^{-1}/yr^{-1}$ and overall soil Cv was 93%.

Owen (1971) has described various factors that affect the rate of soil erosion. Besides topography, intensity, and duration of rainfall, the kind of vegetal cover is emphasized. Bare soil erodes 2.5 times more rapidly than soil planted with cotton crop, 4,000 times more rapidly than grass-covered soil, and almost 32,000 times more quickly than land covered with virgin forests (Bennet, 1955). These estimates appear to be on high side. With respect to agricultural practices, some of the traditionally recommended methods are (1) contour farming (i.e., plowing, seeding, cultivating and harvesting across the slope reduces erosion up to 65%), (2) terracing or steps of flat plots down the hills, (3) gully reclamation, (4) plantation of shelter belts and windbreaks. A windbreak plantation tree height of 35 ft trains the wind to take an upward course, leaving the ground surface, and wind speed is reduced both on windward and leeward sides. Some protection is extended up to 175 ft on the windward and 1,500 ft on the leeward side (Owen, 1971).

Conclusions

Wetlands, both lotic and lentic, are undergoing degradation and ecotonal vegetation is overexploited. This has resulted in silting of river beds and upwelling of lakes and reservoirs. The water-carrying capacity is thus reduced. Agrochemicals applied in fields are finding ways to reach water bodies, causing eutrophication and pollution. The vegetal cover on embankments is very important in checking soil erosion, and retarding nutrient and water runoff. The ecotone vegetation also acts as sinks and harvesters of pollutants. Different species have different conservation efficiencies. Grasses are, by and large, much more efficient than forbs and weeds.

Acknowledgments

The authors are thankful to the Council of Scientific and Industrial Research (CSIR), New Delhi, for the Emeritus award (to RSA) and the Senior Research Associate award (to NKA). Thanks are also due to the Indian National Science Academy for awarding Honorary Scientistship (to RSA).

References

Albert, E. E., & Spomer, R. G. (1985). Dissolved nitrogen and phosphorus in runoff from watersheds in conservation and conventional tillage. *Journal of Soil and Water Conservation, 40*, 153–157.

Ambasht, R. S. (1962). Root habits in response to soil erosion, silting and inundation. Doctoral thesis, Banaras Hindu University, India.

Ambasht, R. S. (1970). Conservation of soil through plant cover of certain alluvial slopes in India. *Proceedings of the IUCN XI Technical Meeting* (pp. 44–48). Gland.

Ambasht, R. S. (1985). Primary productivity and soil and nutrient stability in Indian hilly savanna lands. In J. C. Tothill & J. J. Mott (Eds.), *Ecology and management of world savannas* (pp. 217–219). Canberra: Australian Academy of Sciences.

Ambasht, R. S. (1995). Primary productivity and soil and nutrient stability in Indian hilly savanna lands. In J. C. Tothill & J. J. Mott (Eds.), *Ecology and management of world savannas* (pp. 217–219). Canberra: Australian Academy of Sciences.

Ambasht, R. S. (1998a). *Impact analysis of anthropogenic activities in productivity, soil erosion, eutrophication and sedimentation of Surhatal (Ballia) Lake and surrounding wetlands and habitat restoration.* Final Technical Report M.O.E.F Project No. J22012/9/91, pp. xxv, 79.

Ambasht, R. S. (1998b). World water and wetland resources. In R. S. Ambasht (Ed.), *Modern trends in ecology and environment* (pp. 115–130). The Netherlands: Backhuys.

Ambasht, R. S., & Ambasht, N. K. (1998). Ecology of Indian wetlands. In S. K. Majumdar, E. W. Miller, & F. J. Brenner (Eds.), *Ecology of wetlands and associated systems* (pp. 104–116). Pennsylvania Academy of Sciences.

Ambasht, R. S., & Shankar, M. (1992). Rehabilitation of degraded river corridors. In J. S. Singh (Ed.), *Restoration of degraded land: Concepts and strategies* (pp. 191–209). Meerut, India: Rastogi Publications.

Ambasht, R. S., Singh, M. P., & Sharma, E. (1984). Soil, water and nutrient conservation of certain riparian herbs. *Journal of Environmental Management, 18*, 99–104.

Ambasht, R. S., & Srivastava, N. K. (1994). Restoration strategies for the degrading Rihand River and reservoir ecosystems in India. In W. J. Mitsch (Ed.), *Global wetlands: Old World and New* (pp. 483–492). Amsterdam: Elsevier.

Anonymous. (1982). The state of India's environment—a citizen report. New Delhi: Centre for Science and Environment, Ambassador Press.

Anwar, C., Baheramsyah, K., & Hamzah, Z. (1989). Effectiveness of shrubs and agroforestry in Kadipaten village (Indonesia). (Upper citanduy sub-watershed) in trimming runoff and erosion. *Bulletin Penelititian Hutan, 551*, 1–8.

Babbit, B. (1985). Foreword to *Riparian Ecosystems and their management: Reconciling and conflicting uses.* USDA Forest Service.

Bennet, H. H. (1955). *Elements of soil conservation.* New York: McGraw-Hill.

Blackburn, W. H., & Wood, J. C. (1990). Nutrient export in stormflow following forest harvesting and site preparation in East Texas. *Journal of Environmental Quality, 19*, 402–408.

Boyt, R. L., Bayley, S. E., & Zoltek, J. (1977). Removal of nutrients from treated municipal wastewater by wetland vegetation. *Journal of Water Pollution Control, 49*, 789–799.

Brinson, M. M., Swift, B. L., Plantico, R. C., & Barclay, J. S. (1981). *Riparian ecosystems: Their ecology and status.* U.S. Fish Wildlife Service, FWS/OBS-81/17.

Byron, E. R., & Goldman, C. R. (1989). Land use and water quality in tributary streams of Lake Tahoe, California–Nevada. *Journal of Environmental Quality, 18*(1), 84–88.

Cooper, J. R., Gilliam, J. W., & Jacobs, T. C. (1986). Riparian areas as a control of non point pollutants. In D. L. Correll (Ed.), *Watershed research perspectives* (pp. 166–192). Washington, DC: Smithsonian Institution Press.

Cowardin, L. M., Carter, V., Golet, F. C., & La Roe, (1979). *Classification of wetlands and deepwater habitats of the United States*. Washington, D.C.: U.S. Fish and Wildlife Service.
Davis, T. J. (Ed.). (1994). *The Ramsar convention manual*. Ramsar Convention Bureau. Gland.
Dugan, P. J. (1990). *Wetland conservation—a review of current issues and required action*. IUCN-Gland.
Ewel, K. C. (1978). Riparian ecosystems: Conservation of their unique characteristics. In R. R. Johnson & J. F. McCormick (Eds.), *Strategies for protection and management of floodplain wetlands and other riparian ecosystems* (pp. 56–62). General Technical Report, WO-12, Forest Service. Washington, DC: USDA.
Foster, G. R., & Meyer, L. D. (1977). Soil erosion and sedimentation by water—an overview. In Soil erosion and sedimentation. *Proceedings of the National Symposium on Soil Erosion and Sedimentation by Water* (pp. 1–13). ASAE Publication No. 4-77. St Joseph, MI: ASAE.
Frankenberg, J. (1989). The role of vegetation in protecting river bank erosion. In *The MDFRC Annual Report (1988–89)* (pp. 10–13). Albury, Australia.
Goodwin, R. H., & Niering, W. A. (1974). Inland wetlands: Their ecological role and environmental status. *Bulletin of Ecological Society of America*, 55(2), 2–6.
Groeneveld, D. P., & Griepentrog, T. E. (1985). Interdependence of groundwater, riparian vegetation and streambank stability. A case study. In USDA Forest Service (Ed.), *Riparian Ecosystems and their Management* (pp. 44–48).
Jahn, L. R. (1979). Values of riparian habitats to natural ecosystems. In R. R. Johnson & J. F. McCormick (Eds.), *Strategies for protection and management of flood plain wetlands and other riparian ecosystems* (pp. 157–160). U.S. Forest Service, General Technical Report, WO-12.
Kadlec, R. H., & Kadlec, J. A. (1979). Wetlands and water quality. In P. E. Greeson, J. R. Clark, & J. E. Clark (Eds.), *Wetland functions and values: The state of our understanding* (pp. 436–456). Minneapolis: American Water Resource Association.
Kibby, H. V. (1979). Effects of wetlands on water quality. In R. R. Johnson & J. F. McCormick (Eds.), *Strategies for protection and management of flood plain wetlands and other riparian ecosystems* (pp. 289–298). U.S. Forest Service, General Technical Report, WO-12.
Klopatek, J. M. (1975). The role of emergent macrophytes in mineral cycling in fresh water marsh. In F. G. Howell, J. B. Gentry, & M. H. Smith (Eds.), *Mineral cycling in south eastern ecosystems* (pp. 367–392). Energy Research Development Adminstration, ERDA CONF-740513.
Kumar, R., Ambasht, R. S., & Srivastava, N. K. (1992a). Conservation efficiency of five common riparian weeds in movement of soil, water and phosphorus. *Journal of Applied Ecology*, 29, 737–744.
Kumar, R., Ambasht, R. S., & Srivastava, N. K. (1992b). Nitrogen conservation efficiency of five common riparian weeds, in a runoff experiment on slopes. *Journal of Environmental Management*, 34, 47–57.
Kumar, R., Ambasht, R. S., Srivastava, A. K., Srivastava, N. K., & Sinha, A. (1997). Reduction of nitrogen losses through erosion by *Leonotis nepataefolia* and *Sida acuta* in simulated rain intensities. *Ecological Engineering*, 8, 233–239.
Kumar, R., Ambasht, R. S., Srivastava, A. K., & Srivastava, N. K. (1996). The role of some riparian wetland plants in reducing erosion of organic C and selected cations. *Ecological Engineering*, 6, 227–239.
Lazarus, D. S. (1990). Save our soil. *Our Planet*, 2(4), 10–11.
Lowrance, R., Leonard, R., & Sheridan, J. (1985). Managing riparian ecosystem to control nonpoint pollution. *Journal of Soil Water Conservation*, 40(1)., 87–91.
Lowrance, R., & Shirmohammadi, A. (1985). A model for riparian ecosystem management in agricultural watersheds. In USDA Forest Service (Ed.), *Riparian ecosystems and their management* (pp. 237–240).
Lowrance, R., Todd, R., Fail, J., Hendrickson, O., Jr., Leonard, R., Jr., & Asmussen, L. (1984). Riparian forests as nutrient filters in agricultural watersheds. *Bioscience*, 34(6), 374–377.
Lowrance, R., & Williams, R. G. (1988). Carbon movement in runoff and erosion under simulated rainfall conditions. *Soil Science Society of America Journal*, 52, 1445–1448.
Lyngby, J. E., & Brix, H. (1982). Seasonal and environmental variation in concentrations of calcium, copper lead and zinc in eelgras (*Zostera mariana* L) in the Limfjord, Denmark. *Aquatic Botany*, 14, 59–74.
Maltby, E., & Turner, R. E. (1983). Wetlands of the world. *Geography Magazine*, 55, 12–17.
Mitsch, M. J., & Gosselink, J. G. (1993). Wetlands (2nd ed.). New York: Van Nostand Reinhold.
Oades, J. H. (1988). An introduction to organic matter in mineral soils. In J. B. Dixon & S. B. Weed (Eds.), *Minerals in soil environment* (2nd ed., pp. 89–159). Madison, MI: Soil Science Society.
Owen, O. S. (1971). Natural resource conservation. New York: Macmillan.
Peterjohn, W. T., & Correll, D. L. (1984). Nutrient dynamics in agricultural watershed: Observation on the role of a riparian forest. *Ecology*, 65, 1466–1475.
Ramsar. (1990). *Proceedings of the fourth meeting of contracting parties*. Gland.

Reddy, K. R., & Patrick, W. H. (1984). Nitrogen transformations and loss in flooded soils and sediments: CRC critical review. *Environment Control, 13*, 273–309.

Reimold, R. J., & Hardisky, M. A. (1979). Nonconsumptive use values of wetlands. In P. E. Greeson, J. R. Clark, & J. E. Clark (Eds.), *Wetland functions and values: The state of our standing* (pp. 558–564). American Water Resource Association, Technical Publication No. 79-2.

Rhodes, J., Skan, C. M., Greenbe, D., & Brown, D. L. (1985). Classification of nitrate uptake by riparian forets and wetlands in an indisturbed head water watershed. In USDA Forest Service (Ed.), *Riparian ecosystems and their management* (pp. 175–179).

Roseboom, D., & Russell, K. (1985). Riparian vegetation reduces stream bank and row crop flood. In USDA Forest Service (Ed.), *Riparian ecosystems and their management* (pp. 241–244).

Schreiber, J. D., & McGregor, K. C. (1979). The transport and oxygen demand of organic carbon released to runoff from crop residues. *Progress in Water Technology, 11*, 253–261.

Schultze, R. F., & Wilcox, G. I. (1985). Emergency measures for streambank stabilization: An evaluation. In USDA Forest Service (Ed.), *Riparian ecosystems and their management* (pp. 59–61).

Schwer, C. B., & Clausen, J. C. (1989). Vegetation filter treatment of dairy milkhouse waste water. *Journal of Environmental Quality, 18*, 446–451.

Sharma, E., Sundariyal, R. C., Rai, S. C., Bhatt, Y. K., Rai, L. K., Sharma, R., & Rai, Y. K. (1992). Integrated watershed management: A case study in Sikkim Himalaya. Gyanodaya Prakashan, Nanital: G. B. Pant Institute.

Shaw, S. P., & Fredine, C. G. (1956). *Wetlands of the United States*. U.S. Fish and Wildlife Service, Circle 39.

Smith, R. L. (1996). Ecology and field biology (5th ed.). New York: Harper Collins.

Srivastava, N. K., & Ambasht, R. S. (1990). Impact of industrial effluents in the limnology of Pant Sagar and Rihand River. In J. S. Singh, K. P. Singh, & M. Agrawal (Eds.), *Environmental degradation of Obra Renukoot–Singrauli area and its impact on natural and derived ecosystems* (pp. 265–284). Final MAB Technical Report, Banaras Hindu University.

Teskey, R. O., & Hinckley, T. M. (1977). *Impact of water level changes on woody riparian and wetland communities: Vol. 1. Plant and soil responses*. U.S. Fish Wildlife Service, FWS/OBS-77/58.

Thornes, J. (1989). Solutions to soil erosion. *New Scientist, 1667*, 45–49.

Thronson, R. E. (1978). *Nonpoint source control guidance, agricultural activities*. U.S Environment Protection Agency, EPA/440/3-78/001.

Triquet, A. M., McPeek, G. A., & McComb, W. C. (1990). Songbirds diversity in clearcuts with and without a riparian buffer strip. *Journal of Soil Water Conservation, 45*(4), 500–503.

U.S. Environmental Protection Agency. (1973). *Water quality criteria*. Washington, DC: U.S. Government Printing Office.

UNEP. (1992). *The world environment 1972–1992*. London: Chapman & Hall.

Van der Valk, A. G., Davis, C. B., Baker, J. L., & Beer, C. E. (1979). Natural fresh water wetlands as nitrogen and phosphorus traps for land runoff. In *Wetland functions and values: The state of our understanding* (pp. 457–467). Minneapolis: American Water Resource Association.

Verry, E. S., & Timmons, D. R. (1982). Water borne nutrient flow through an upland-peatland watershed in Minnesota. *Ecology, 63*, 1456–1467.

Wharton, C. H. (1980). Values and functions of bottom land hardwoods. *Trans North American Wildlife Natural Resource Conference, 45*, 341–353.

White, E. M., & Williams, E. J. (1977). Soil surface crust and plow layer differences which may affect run-off quality. *Proceedings of the South Dakota Academy of Sciences, 56*, 225–229.

Winger, P. V. (1986). *Forested wetlands of the Southeast: Review of major characteristics and role in maintaining water quality*. Washington, DC: U.S. Department of International Fish and Wildlife.

Young, R. A., & Wiersma, J. L. (1973). The role of rainfall impact in soil detachment and transport. *Water Resources Research, 9*, 1629–1636.

Young, R. A., Oustad, C. A., Bosch, D. D., & Anderson, W. P. (1987). *ANGPS, agricultural non-point source pollution model*. Conservation Research Report No. 35. Morris, MN: USDA-ARS.

13

Identification, Assessment, and Mitigation of Environmental Impacts of Dam Projects

Robert Zwahlen

Introduction: Dams and the Environment

For a long time, dams have been a very important means for storing water, mainly for irrigation. Originally, these were small structures made of earth, bamboo, and other material. Often, these weirs were impermanent structures that were usually destroyed during the wet season and had to be rebuilt again, as for example, as described by Gao (1998) for the early rice cultivation systems of the Dai in Xishuangbanna, Yunnan, in southern China PDR. Irrigation schemes have been of special importance in seasonally dry areas, as in Sri Lanka, where sophisticated systems of cascading tanks feeding irrigation systems have been built. Some of these systems have been operating over the centuries. Rational use of water was the main issue, and it is reported that already in 1153 A.D., a Sri Lankan king stated that "not a single drop of water received from rain should be allowed to escape into the sea without being utilized for human benefit" (Baldwin, 1991, p. 11). Irrigation systems have played a major role in the development of centrally governed states, because their construction, management, and maintenance required investments and efforts that were beyond the possibilities of an individual household or a single village (Johnson & Earle, 1997, p. 209).

In recent times, with the new possibilities offered by the development of technology, dams have become more important for power generation. Simultaneously, they have become much larger, partially regulating and using the water resources of large rivers. Building dams for energy production started around 1890; today, there are more than 45,000 large dams worldwide, most of which were built after 1950.

Robert Zwahlen • Environment and Social Development Specialist, Electrowatt Engineering Ltd. (EWE), Hardturmstr. 161, P.O. Box 8037, Zurich, Switzerland.

This chapter deals mainly with large dams built or planned for hydropower generation, following the definition of what constitutes a large dam, provided in Box 13.1. Nevertheless, some examples from experience with smaller dams, with dams built for irrigation or drinking water supply, and with multipurpose projects, are also cited.

Box 13.1. Types of Large Dams

There are various definitions of large dams. The International Commission on Large Dams (ICOLD), established in 1928, defines a large dam as a dam with a height of 15 m or more from the foundation. If dams are between 5 and 15 m high and have a reservoir volume of more than 3 million m^3, they are also classified as large dams. Using this definition, there are over 45,000 large dams around the world.

The two main categories of large dams are reservoir type storage projects and run-of-river dams that often have no storage reservoir and may have limited daily pondage. Within these general classifications there is considerable diversity in scale, design, operation, and potential for adverse impacts.

Reservoir projects impound water behind the dam for seasonal, annual, and, in some cases, multiannual storage and regulation of the river.

Run-of-river dams (weirs and barrages, and run-of-river diversion dams) create a hydraulic head in the river to divert some portion of the river flows to a canal or power station (World Commission on Dams, 2000, Box 1.2, p. 11).

The usefulness of dams has always been obvious, be it the energy generated, the increase in water availability for irrigation and drinking water supply, or additional (or in some cases, even main) benefits of flood regulation and protection, navigation, tourism and so on. Only gradually, however, has it been noted that dams, in addition to their benefits, also can have considerable adverse effects on the environment and affected human populations. Although such effects have been (and in some cases still are) considered inevitable but acceptable impacts justified by the "greater good," gradually this attitude has changed: Dams are still being built (Fig. 13.1), but it is now generally accepted that adverse impacts need to be identified, and that measures have to be sought for preventing, minimizing, or compensating such effects (the so-called "mitigation measures"). Two main developments have contributed to this change in paradigm: first, the general change in public awareness concerning environmental problems (triggered to a large degree by the 1972 United Nations Conference on the Human Environment in Stockholm; see Tolba et al., 1993, pp. 661ff., for a more detailed account of this development); in addition to the general increase in environmental awareness, this has led or contributed to the development of a coherent and comprehensive environmental legislation in an increasing number of countries. The second important development was the increasing resistance of directly affected peoples against development and mostly dam projects, the most well-known example for this being the case of Narmada in India. In this field, the role of nongovernment organizations (NGOs) in the public discussion and as pressure groups has increased considerably.

Today, there is a large and growing general opposition against dam projects (e.g., Goldsmith & Hildyard, 1984; Goldsmith & McCully, 1993). The most important and active group in this respect is probably IRN (International Rivers Network), a California

Figure 13.1. Dachaoshan hydropower plant, China PDR, construction site, Lanchang (Mekong) River. The river is diverted around the dam site by means of a coffer dam and diversion tunnels (center right). View into the future reservoir.

based NGO that is active worldwide, with a very high commitment against dam projects (see McCully, 1996).

Without any doubt, considerable environmental and socioeconomic damage has been done by many of the large dams, and especially by the largest ones. In the past, many of these impacts have not been anticipated, but mitigation measures, if any, have only been taken once such impacts became manifest. And in a considerable number of dam projects, appropriate mitigation measures have not been implemented. This is aggravated further by the fact that in a number of cases, the project achievements (in terms of energy production, economic output, duration of useful life span, etc.) were considerably below expected values. This can lead to a case in which the benefits of the project are not the anticipated ones, whereas the negative impacts might be more important than expected. Therefore, a critical attitude toward dam projects is certainly required and important. However, opposition against dams in the recent past has sometimes taken the form of overall rejection of such projects, just as they were promoted uncritically earlier. Both attitudes are obviously wrong and do not do justice to the problems at hand. In spite of all the potential negative effects, hydropower is still an important renewable and, if properly planned, implemented and managed, environmentally friendly source of energy. In addition to that, water resources and their careful use and management are of growing importance in the light of increasingly scarcer availability and higher demand (see, e.g., Cans, 1997; de Vaulx, 1997). Dams will continue to be an important contribution to the adequate management of water resources. On the other hand, water—and rivers as ecological entities and not just recipients filled with water—have other roles to play, in addition to their potential use as energy source. And not always is a dam the right—or best—solution in a given situation (see, e.g., Zwahlen, 1992, for a discussion of this topic in relation to irrigation projects, which includes the "no project" option as a valuable and sometimes better alternative).

A number of publications deal in a more or less general way with environmental impacts and impact assessment of hydropower projects (e.g., Aegerter & Messerli, 1981; International Commission on Large Dams [ICOLD], 1985; Schuh, 1986; Zwahlen, 1995; World Wide Fund for Nature [WWF], 1999). Recently, the World Commission on Dams (WCD) has undertaken the task of bringing the views of proponents and opponents in the "dam business" together in a wide discussion, initiated in 1997 by a joint effort of World Conservation Union (IUCN) and the World Bank (WB) (Dorcey et al., 1997), with the aim of establishing guidelines for the development of dam projects. This effort culminated in the recent publication of the WCD report (2000), which emphasizes two—in my view, absolutely essential—aspects for the planning of new dams:

- *Overall assessment of resource development options:* Because rivers, as has been said already, have functions other than mere energy production, and in view of the growing scarcity of water resources, these should be used in the best possible way. This can include dams but, in specific cases, other options can prove to be better and ultimately more sustainable. In this context, it is also important to point out that the unit in view should be the entire river basin and not just an already selected "ideal dam site."
- *High importance of environmental and socioeconomic impacts:* These should be taken into consideration in a very early planning stage. Of very high importance in this concept is the involvement of all stakeholders in the planning process.

This chapter deals mainly with this second point. Although in the past, environmental issues have been dealt with, if at all, once they became manifest problems, the present approach is to deal with these issues in parallel to the planning process (mainly during feasibility and detailed design studies). The new approach is now to start addressing these issues in an even earlier stage, as soon as the possibility of using a river starts to be evaluated, and before a specific project site and project type has even been identified and defined (see Box 13.2). This approach is, of course, receiving criticism from entities involved in the planning and implementing of dams (e.g., many state electric authorities), who see this as an "antidevelopment" approach that will make the construction of dams impossible, or at least much more difficult than previously. In my opinion, this is an erroneous view of the problem. With all the technological knowledge in the field of dam construction, dam projects are not being abandoned because of engineering problems. If the conditions of a site (i.e., mainly hydrology, geomorphology, and geology) are suitable for dam construction, and if it seems economically feasible, it will also be technically feasible. If, today, a dam project is contested, it is not for engineering reasons, but for environmental and social reasons. This can lead to massive delays in the construction of a project, and such a delay is very costly. However, in most cases, it would have been possible to deal with these problems long before the project was finalized and construction started. If early investigations show that the project will encounter difficulties of this kind, it is then possible to deal with these issues in time. If a solution can be found, it can be assumed that it will then be possible to build the plant without any undue delay, and if not, unnecessary costs for developing an ultimately useless project can be saved, and these resources can be used for a more promising project. In this sense, I think that this new approach is in the interest of everybody involved in any way in such development projects.

Box 13.2. Planning a Hydropower Plant Using the "New" Approach

In 1990, the Swiss utility Forces Motrices de Mauvoisin, after having heightened the Mauvoisin double arch concrete dam by 13.5 m (from 237 to 250.5 m), decided to raise the installed capacity of the plant from 350 to 900 MW, in order to enable it to produce more high-value, peak energy. The project consisted of the construction of a new powerhouse (planned as an underground cavern power plant), a headrace tunnel about 18.4 km in length, and a 1,768-m-long underground pressure shaft. The waterways would bring the water from the dam, with full supply level (FSL) at 1,975 m asL, to the power house in the valley, using a total head of 1,470 m.

Having gained experience with the previous project of dam heightening, the company decided to "turn planning around": Instead of making the project and then submitting it to an environmental impact assessment (EIA), it was decided to start entirely from the environmental point of view. With a still relatively rough project concept, EIA work was started. Even more importantly, from the very beginning, all relevant stakeholders (federal and cantonal agencies involved in the licensing process, the communities where the plant was situated, fishermen's associations and, very importantly, all NGOs active in this field, which could have opposed the project), were invited right from the start to participate in the planning process.

The following main issues were identified:

- Locations of access galleries for constructing the long tunnels.
- Sites for disposal of the excavation material from the tunnels (a total of approximately 520,000 m^3).
- Influence on the Rhone River below the power plant, because increased peaking would sharply increase the short-term variation in discharge pattern of the river, creating massive surge waves when the turbines were turned on.

For all these problems, adequate solutions could be found that satisfied all the stakeholders. One of the main measures was the integration of a large compensation pond in the project, and extensive modeling of operation and resulting discharge patterns was carried out to optimize the size of this basin and its operation. The technical project had always only been advanced as far as it was required for assessing and solving the environmental problems, and the dialogue with all parties involved had been maintained during the whole process.

The success of this approach could be seen in the fact that all the necessary licences could be obtained in a very short time, and that no objections against the project were raised at the time of the publication of the project documents (Swiss legislation provides the possibility for directly affected parties, which includes recognized NGOs active in the field of environmental protection, to deposit objections against such a project at this moment; such complaints can result in a major project delay). Therefore, the construction permit was obtained in record time.

The fact that, in spite of this, the project has not been carried out had nothing to do with environmental concerns, but with the situation on the power market: Just when construction was about to start, the European power market was liberalized, and suddenly there was a very extensive offer of cheap power available. This made the project no longer interesting from an economic point of view (Ecotec & Ecosens, 1993).

The EIA as an Instrument

Like most major projects, dam projects today usually require an EIA (environmental impact assessment). Starting in the United States, where it first had been introduced, the EIA has developed into a widely used and accepted tool for identifying environmental and socioeconomic impacts. Today, most countries have environmental legislation that prescribes the elaboration of an EIA for a series of projects, including major power generation projects. In addition to country-specific legislations and standards, and often used where the latter are not (yet) available, there are international guidelines and standards, mainly those of the World Bank (1991; Dixon et al., 1989) and other multilateral development banks such as the Asian Development Bank and others (1990, 1991; Lohani et al., 1997; Interim Mekong Committee, 1982). One of the major shortcomings of the EIA, especially in the light of the "new approach," as outlined earlier, is the fact that by its very concept and nature, it is focused on a specific project. Of course, it would be possible to carry out evaluations and comparisons of alternatives within the framework of an EIA, and it has often been postulated that this should be so. Nevertheless, in practice, the fact remains that an EIA in most cases is carried out for a specific project. This has to be kept in mind when the purpose or the outcome of such a study is being discussed.

A comprehensive EIA for a complex project such as hydropower will always cover a large range of subjects; these main topics are covered and illustrated with examples in the following sections of this chapter. Therefore, an EIA normally requires the involvement of a number of specialists that deal with the various aspects.

It is important to point out that the interdisciplinarity in EIAs for dam projects is not limited to the exchange between the various environmental and social specialists involved. Close cooperation is required with the engineers involved in planning the project itself, both for understanding technical requirements and constraints, and for making environmental requirements and constraints understood. Effective mitigation measures can only be elaborated if they are accepted by the technical specialists, and if they are technically feasible and economically acceptable. Furthermore, environmental concerns have to be seen within an institutional and legal framework. This requires discussions and exchange with representatives of the various state and other public institutions involved, with local scientists, and so on. Last, but not least, there is always a population in the project area itself, which in one way or another will be directly affected by the project. Here, as well, mutual understanding of the situation is essential for reaching satisfactory results.

This chapter does not intend to be a comprehensive sourcebook or guideline for elaborating EIAs for dam projects. In the context of the book, this chapter emphasizes aspects of aquatic ecology; these issues will be dealt with in more detail, although by no means in an exhaustive manner. Nevertheless, because an EIA is not limited to one discipline or aspect of the environment, and because there are often direct and indirect links between the various topics to be treated, the discussion will not be limited to water and effects on water. Rather, the intention is to show "the whole picture," although some aspects that are often of very high importance, such as impacts on terrestrial habitats, or the whole, complex issue of socioeconomic impacts and mainly project-induced resettlement, will be dealt with here in a somewhat more cursory way. This, however, does not mean that these aspects are of less importance!

Wherever possible, I try to illustrate the more theoretical discussions with examples from actual projects. In doing so, I rely heavily on my own experience (i.e., on projects in which I have been involved myself). This, of course, is not meant to exclude other examples. There is one main difficulty when discussing environmental impacts of real projects and the findings of EIA reports: In most cases, these reports have not been published and are therefore not easily accessible, although they usually are "public good," in the sense that they have been made available to interested and involved parties at one time in the planning cycle. Nevertheless, such reports contain a wealth of information on the projects in question, and on their specific environment. Whenever such reports are cited here, they are labeled as "unpublished report" in the references.

The Main Steps of Environmental Impact Assessment

An EIA—or every discussion of environmental impacts, whether following formal rules or not—basically consists of three main steps: (1) description of the prevailing situation, (2) identification of impacts, and (3) the formulation of mitigation measures. As simple as this may seem, each of these steps has its own problems and difficulties. These are discussed very briefly and on a more theoretical basis in the following paragraphs. Examples stemming from practical experience from a variety of projects are provided in the sections dealing with the individual environmental issues at hand.

Description of the Present Situation

It is obvious that effects of any project on the environment at a given site can only be identified if the situation of the environment at this precise location is known. For this reason, a description of the prevailing situation, or, more to the point, of "the situation without project," has to be made. This simple statement, in its implementation, meets two main practical problems:

1. *The complexity of the environment.* What "the environment" really is in a specific case has to be defined and what and how much of it should be known and described in order to have an acceptable database for the other two steps. The problems involved become immediately clear if we just consider briefly the biological components of the environment. In a temperate region (e.g., a central European site), several thousand animal species may be present, most of which are of microscopic size; in a moist tropical environment, this number is in the hundreds of thousands. It is obvious that under such conditions, a "complete" description of the environment is not possible, all the more so, if one takes into account that a mere species list is still far from being a description of this environment, because the interactions between the various species also have to be taken into account. Therefore, every such description of the prevailing situation needs to be limited to certain key factors or indicator organisms (e.g., vegetation types, with their most prominent species instead of the whole flora; certain key groups of animals, such as birds, reptiles, large mammals, and so on; or endangered and protected species). The understanding here is that if the environment is in a state in which these identified species can live, it will implicitly also allow all the other, unidentified species to survive.

2. *The instability of the environment over time.* A given environmental situation today is not necessarily the same situation it will be in some years' time, when construction or operation of the project actually takes place. This is all the more so if other (man-made) influences are manifest in a certain site, and the decision whether to implement the project or not can be decisive for a series of other potential developments. Therefore, the

"situation without project" is not just a point in time ("today") but can stretch far into the future, covering planning, construction, operation, and postoperation phases.

For actually defining "the environment" to be described in its spatial, material, and time dimensions, a good knowledge of the project in question is, of course, a prerequisite. It is obvious that, in a way, potential project impacts are already anticipated here.

Identification of Impacts

Impacts can only be identified if we know the "recipient" of the impact and the probable manner in which it will react. Therefore, the correct identification of the "relevant environment" (discussed earlier) is essential.

Some impacts are easy to identify, to understand, and to assess in their magnitude, whereas others are not. So, for example, it is easy to say what will happen with a forest located within the reservoir area of a hydropower project: at reservoir impoundment, it will disappear, be destroyed completely, and so will all the plants and animals living there. It is, however, much less straightforward to see what this actually means for this forest type in the wider area, for the populations of the affected species, or even for individual animals. This requires some knowledge of the situation that goes beyond the limits of the area under direct influence. A special challenge can be such parameters that will evolve over time, for example, water quality and fish populations in the affected water bodies. Here, besides experience from other projects, modeling can be an important, although not always easy to handle, tool.

The following pairs of expressions are normally used to characterize impacts:

- *Direct/indirect:* A direct impact is caused immediately by the project itself, whereas an indirect impact is a consequence of a change caused by the project. So, for example, submersion of a forest by the new reservoir, or trees cut for the right of way of the transmission line would be direct impacts, whereas a rise in malaria incidence caused by the fact that *Anopheles* mosquitoes find good living conditions in the drawdown area of the reservoir would be an indirect impact.
- *Total–partial:* This is used to characterize the magnitude of the impact, considering a specific item. A forest, a wetland, or a human settlement in the reservoir area, depending on its exact location and its dimensions in relation to the reservoir, can either be totally or only partially submerged (where "partially" can mean from almost 0% to almost 100%).
- *Permanent–transient:* This characterizes the duration of an impact. Whereas the interruption of a river by a dam, for all practical purposes, is a permanent impact, noise and dust emissions caused during the construction period take place only during a limited time.
- *Important–negligible:* This is an attempt to characterize the meaning of an impact; terms that are synonyms are "relevant" and "irrelevant." The submersion of Abu Simbel by Lake Nasser, the reservoir formed by the Aswan Dam, was certainly perceived as important, whereas the submersion of desert land by the same reservoir was not.
- *Positive–negative:* Sometimes also termed "desirable" and "undesirable," these expressions are used mostly to characterize impacts on human activities, or (potential or actual) human uses or perceptions of impacts and their consequences.

Submersion of villages and their fields is definitely a negative impact, whereas the generation of electric energy and the possibility for fishing in the reservoir would be positive impacts; here, note that, very often, positive and negative impacts do not affect the same part of the human population (see a later section for a short discussion of this point).
- *Acceptable–not acceptable:* All this finally ends in a dichotomy on the acceptability of an impact. "Acceptable" usually means "according to legal standards," but in the absence of such standards (e.g., when discussing the visual impact on a landscape), it can also mean that a consensus (among whomever) has been reached. The consequence of an unacceptable impact would be that the project in question cannot be implemented.

Although, of these categories, direct–indirect, total–partial, and permanent–transient might be rather clearly identifiable and therefore undisputed in most cases, this is definitely not the case for the three remaining categories. As a matter of fact, what is important, positive, and acceptable is a matter of personal preferences or conventions (whether defined legally, culturally, or by means of an *ad hoc* consensus). To give just one, rather extreme example, the author of the Kabalega Falls Hydropower Project in Uganda told me that his project, far from destroying the falls, would have actually saved them (which would have turned the main impact of this project into a positive, desirable one). As a matter of fact, these falls cause a backward erosion, cutting into the escarpment, which, in some million years, will eventually cause the disappearance of the falls. Whether this is a justification for a hydropower plant with an expected life span of maybe 100 years probably depends largely on the personal hierarchy of values.

Mitigation Measures

In spite of the difficulties involved, it is important to classify the impacts into categories as mentioned earlier, because whether mitigation measures are required will depend on the perceived importance and "direction" of the impacts and the direction in which they have to point. Here, again, the basic principle is simple and easy to understand, but application is often somewhat more complex and less clear-cut. Basically, the measures should lead to enhancing positive and reducing negative impacts, where required. This, of course, requires a consensus on what is positive and what is negative, and under which conditions anything is required or not (and, not in the least, how much cost is allowed).

When discussing the mitigation of negative impacts, there are always three categories to be considered, in the following order:

1. *Avoidance:* Measures (project modifications, choice of an alternative site, etc.) that will prevent a certain impact from actually happening, in many cases, lead to a yes-or-no type of decision, because certain impacts can only be avoided if the project is not implemented; this, of course, should be considered as an option.

2. *Minimization:* Reducing the magnitude of an impact by applying adequate measures (e.g., reducing dam height and therefore reservoir size to limit the area to be submerged, or preimpoundment reservoir clearing to reduce the impact on water quality).

3. *Compensation:* In cases where an impact cannot be avoided and adequate mitigation is not possible, compensation can be an acceptable solution. This would mean that the lost value is replaced by something of equal or higher value located

elsewhere. Such a compensation would, for example, provide agricultural land of the same size and value to a farmer who loses his land because of a project, or replace a loss in forested area by reforesting a similar area with a comparable forest type elsewhere.

Obviously, it is not sufficient just to formulate mitigation measures. The really decisive point is that they are implemented if the project is becoming reality. This requires the following additional conditions: (1) careful planning of the measures, including identification of suitable and available land in the case of compensation measures; (2) identification of the costs of the measures and their integration into the overall project budget, (3) a firm commitment of the project proponent for actually carrying out the measures, (4) firm conditions formulated by the competent authorities making the measures part of the project, and (5) a follow-up implementation control; this latter point is usually a part of the monitoring program.

Structure of an EIA Report

Most EIA reports follow more or less narrowly the general outline shown in Box 13.3.

Box 13.3. Sample Outline of a Project-Specific EA Report

Environmental Assessment (EA) reports should be concise and focused on the significant environmental issues. The detail and sophistication of the report should be commensurate with the potential impacts. The target audience should be project designers, implementing agencies, proponents, and financiers. The report should include the following:

1. *Executive summary*: Concise discussion of significant findings and recommended actions.
2. *Policy, legal, and administrative, framework* within which the EA is prepared: The environmental requirements of any cofinanciers should be explained.
3. *Project description* in a geographic, ecological, social, and temporal context: This includes any off-site investments that may be required by the project (e.g., dedicated pipelines, access roads, power plants, water supply, housing, and raw material and product storage facilities).
4. *Baseline data*: dimensions of the study area and description of relevant physical, biological, and socioeconomic conditions, including any changes anticipated before the project commences. Current and proposed development activities within the project area (but not directly connected to the project) should also be taken into account.
5. *Environmental impacts*: The positive and negative impacts that are likely to result from the proposed project should be identified and assessed. Mitigation measures, and any residual impacts that cannot be mitigated, should be identified. Opportunities for environmental enhancement should be explored. The extent and quality of available data, key data gaps, and

uncertainties associated with predictions should be identified/estimated. Topics that do not require further attention should be specified.
6. *Analysis of alternatives*: Proposed investment design, site, technology, and operational alternatives should be compared systematically in terms of their potential environmental impacts, capital and recurrent costs, suitability under local conditions, and institutional, training, and monitoring requirements. To the maximum extent possible, for each of the alternatives, the environmental costs and benefits should be quantified and economic values attached, where feasible. The basis for the selection of the alternative proposed for the project design must be stated.
7. *Mitigation plan*: Feasible and cost-effective measures that may reduce potentially significant adverse environmental impacts to acceptable levels should be proposed, and the potential environmental impacts, capital and recurrent costs, and institutional and training requirements of those measures duly estimated. The plan (sometimes known as an "action plan" or "environmental mitigation or management plan") should provide details on proposed work-programs and schedules, to ensure that the proposed environmental actions are in phase with engineering and other project activities throughout implementation. The plan should consider compensatory measures, if mitigation measures are not feasible and cost-effective.
8. *Environmental management and training*: The existence, role, and capability of environmental units at the on-site, agency, and ministry levels should be assessed and recommendations made concerning the establishment and/or expansion of such units and the training of staff, to the point that EA recommendations can be implemented.
9. *Monitoring plan* regarding environmental impacts and performance: The plan should specify the type of monitoring, who would do it, how much it would cost, and what other inputs (e.g., training) are or may become necessary.

Appendices:

List of EA preparers: individuals and organizations.

References: written or published materials used in study preparation. This is especially important in view of the large amount of unpublished documentation that is often used.

Record of interagency/forum consultation meetings: This includes lists of both invitees and attendees. The record of consultations to obtain the informed views of the affected people and local NGOs should be included (Goodland et al., 1993).

Personally, I prefer a somewhat different structure, as shown in Box 13.4, mainly for practical reasons.

Box 13.4. EIA Report for a Hydropower Project: Sample Table of Contents

Executive summary

A. Baselines
   1. Introduction
   2. Administrative and legal framework
   3. The project
   4. The study area
   5. The environment: general considerations

B. Sectoral studies

   I. The physical environment

   6. Geology and soils
   7. Climate
   8. Water

   II. The biological environment

   9. Vegetation

      9.1. Introduction
      9.2. Material and methods
      9.3. Prevailing situation
      9.4. Impacts
      9.5. Mitigation measures
      9.6. Conclusions

   10. Terrestrial fauna
   11. Aquatic fauna
   12. Protected areas

   III. The human environment

   13. Settlements and population
   14. Land use
   15. Public health
   16. Cultural sites, archaeology

C. Synopsis
   17. Main impacts
   18. Mitigation measures
   19. Resettlement and compensation planning
   20. Watershed management plan
   21. Environmental monitoring
   22. Environmental Management Plan

ANNEXES

   Annex 1. Literature cited
   Annex 2. (Any material not to be placed in the text)

The basic content of the studies is, of course, the same, regardless of the structure of the report. The first outline (Box 13.3) leads basically to a "horizontal" structure of the report: After the first general chapters, the existing environment is described (baseline data) for all the aspects. Then, in the following chapter, the impacts likely to occur are identified and described, again covering all the relevant environmental parameters. In a next chapter, mitigation measures are developed, and so on. The other approach leads to a "vertical" structure of the report. After the general chapters in Part A, Part B provides the so-called sectoral studies: Here, every environmental field to be treated is dealt with in a similar manner. The structure provided for Chapter 9 (in Box 13.4) as an example is basically used for each of these aspects. This sectoral part is then followed by a synthetic part, where the important impacts, measures, and results are described and discussed, and, very importantly, the link between the various aspects or chapters is being made. The advantage, in my view, of this structure, is a merely practical one: Because normally in the elaboration of an EIA, a specialist will deal with a certain aspect of the environment (e.g., vegetation, water, or aquatic fauna), this arrangement will make the development of the work easier and the editing of the report less cumbersome. On the other hand, most readers of such reports, and especially the specialists of the agencies involved in project licensing, are interested either in the general outcome (in which case they can read Parts A and C) or in a specific topic; in this latter case, they will find all the relevant information in one chapter and will not have to look for it in at least three different chapters (present situation, impacts, and mitigation).

However, the quality of the results and of the report will not depend on the choice of the outline, but rather on the competence of the specialists involved and on the data available. In any case, a good synthesis, which makes the necessary links between the different aspects, is essential, avoiding a situation in which mitigation measures proposed for one impact will ultimately result in a much larger impact in another field. On the other hand, with careful planning, it is often possible to formulate mitigation measures in a way that will enhance their effects by covering a range of issues.

One basic problem of every EIA, which is mentioned again later, is the fact that, here, we deal with events that will happen in the future. This means that the identified impacts (as well as the proposed mitigation measures) are not yet reality and therefore cannot be observed. So we have to make a prognosis on the outcome of effects in the future. The inevitable uncertainty that has to be accepted here is one reason for the necessity of a monitoring program. On the other hand, *ex post* analyses should be made in order to see whether the predicted impacts have actually occurred in the manner and the order of magnitude predicted (and if not, why?). Because we have to rely largely on experiences with actually implemented projects to predict the outcome of certain impacts, such a control would be very useful for a further development of the EIA capability. Unfortunately, not many such studies have been made so far, and even when a monitoring program is carried out, the data are very often not accessible. The importance of this feedback has been demonstrated, for example, in the study by Buckley (1991), which revealed a rather small accuracy of prediction of environmental impacts compared to the actual outcome. Likewise, Boothroyd et al. (1995), in an *ex post* analysis of the social impacts predicted in 75 impact assessment reports on a total of 52 large projects in Canada, found that such reports tend to overestimate positive as well as negative impacts, and that the benefits for local populations derived from the projects are usually smaller than predicted.

Legal and Administrative Framework

One point that has to be clear for every EIA is the legal and administrative framework for the study. Three points have to be mentioned specifically:

1. *The relevant legislation under which the project itself and the EIA has to be prepared.* This legal background defines the procedures to follow, the standards to be respected, and so on. In most countries, by now there is a specific environmental legislation that normally also contains specific regulations regarding the EIA (lists of projects for which an EIA is required, standards to be respected, identification of the competent authorities involved, etc.). It is also possible that in cases in which the country does not have its own comprehensive environmental legislation, especially in the absence of certain environmental standards, international standards, or those of another country, are being applied. For example, the Chilean environmental legislation (Chile, 1997) states that in cases in which there are no standards defined in the law, Swiss standards apply. In such cases, these standards form an integral part of the legal framework.

2. *The administrative framework.* It is important to know which authorities will be involved in the process of establishing and mainly approving the EIA. In most cases, this is a specific environmental authority; very often, there are such authorities at the national and at the provincial/state level. Depending on the nature of the project, one or both can be involved in such a process. Furthermore, in addition to these environmental authorities, there are often other bodies that deal with specific topics (e.g., forestry, coastal protection, water resources) and also have to be involved when a project has influence in their domain.

3. *The procedure to follow.* These procedures can be different, depending on the specific national legislation.

The Project

Every EIA requires a detailed description of the project in question. This is not necessarily a technical description, because structural details that can be essential for the construction of the plant may not be needed. However, it is essential that all the project structures and operational details that are of relevance for the environmental impacts are provided. Figures 13.2 and 13.3 provide, in a very generalized and schematic way, the main relevant parts of a hydropower plant.

The dams that have received the most interest are those perceived as being "very large." However, it is not easy to determine which dam is actually the largest one: the answer would depend largely on the criterion used for determining "large." Table 13.1 provides a series of "largest dams" according to different parameters.

Table 13.2 gives the main characteristics of some of the dams used as examples for discussing selected topic in the following sections.

The main structures of a dam project and their dimensions are very important for identifying potential impacts. This, however, is not sufficient. It is equally important to have at least a basic idea of the planned operation of the plant. This depends on a series of conditions, such as the following:

- The size of the reservoir (i.e., the storage capacity) can range between no or very small (daily) storage in run-of-river (ROR) projects, to seasonal storage in projects with larger reservoirs (usually used for storing water in the rainy or wet season, to be used during the dry season), and even to overannual storage in the case of very

Environmental Impacts of Dam Projects

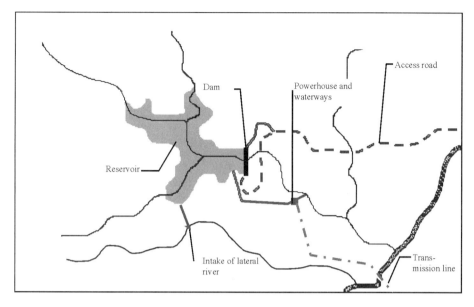

Figure 13.2. Main components of a hypothetical hydropower project.

large reservoirs, which can be used for regulating water availability in regions with unpredictable rainfall and large year-to-year variation.

- Rainfall or snow melting pattern in the catchment (i.e., water availability and its seasonal distribution).
- The main purpose of the project (hydropower, irrigation, drinking water supply, or a combination thereof).
- Intended additional benefits or uses of the reservoir (e.g., for fisheries, flood protection, tourism, navigation).
- Intended or possible method of operation of the power station, mainly whether it will be used for peak power generation (normally the case in hydropower plants with large storage, leading to short-term, often very marked and abrupt fluctuations

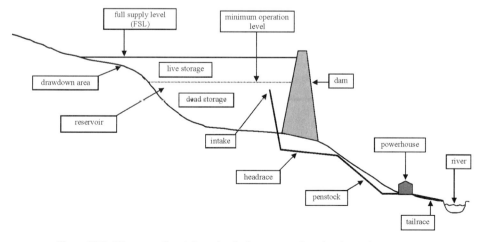

Figure 13.3. Diagrammatic cut through a hydropower project showing main components.

Table 13.1. The Largest Dams

Name	Country	Year[4]	River	Dam height m	Crest length m	Res. volume 10^6 m^3	Installed capacity MW
1. By dam height[1]							
Nurek	TJK	1980	Vakhsh	300	704	10,500	2,700
Grande Dixence	CHE	1961	Dixence	285	695	401	1,700
Inguri	GEO	1980	Inguri	272	680	1,100	1,250
Vaiont	ITA	1961	Vaiont	262	190	169	?
Chicoasén	MEX	1980	Grijalva	261	485	1,613	2,400
Mauvoisin[3]	CHE	1957	Drance	250.5	520	181	397
Guavio	COL	1989	Guavio	246	390	1,020	1,600
Sayano-Shushensk	RUS	1980	Yenisei	245	1,066	31,300	6,400
Mica	CAN	1973	Columbia	242	792	24,700	2,610
2. By reservoir volume[2]							
Kakhovskaya	UKR	1955	Dnieper	37	3,620	182,000	?
Kariba	ZMB	1959	Zambezi	128	579	180,600	1,266
Bratsk	RUS	1964	Angara	125	4,471	169,270	4,500
Aswan	EGY	1970	Nile	111	3,830	168,900	2,100
Akosombo	GHA	1965	Volta	134	671	148.000	912
Daniel Johnson	CAN	1968	Manicouagan	214	1,314	141,852	1,292
Guri	VEN	1986	Caroni	162	11,490	138,000	10,300
Krasnoyarsk	RUS	1967	Yenisei	124	1,065	73,300	6,000
Bennett	CAN	1967	Peace	183	2,042	70,309	2,416
3. By installed capacity							
Three Gorges	CHN	u.c.	Yangtse	185	2,150	39,300	17,680
Itaipú	BRA	1983	Paraná	196	7,297	29,000	12,600
Guri	VEN	1986	Caroni	162	11,490	138,000	10,300
Grande Coulée	CAN	1942	Columbia	168	1,272	11,795	6,494
Sayano-Shushensk	RUS	1980	Yenisei	245	1,066	31,300	6,400
Krasnoyarsk	RUS	1967	Yenisei	124	1,065	73,300	6,000
Churchill Falls	CAN	1971	Churchill	32	5,560	32,640	5,428
La Grande 2	CAN	1978	La Grande	168	2,826	61,715	5,328
Bratsk	RUS	1964	Angara	125	4,471	169,270	4,500
Ust-Ilim	RUS	1977	Angara	102	33,725	59,300	4,320

[1] Only plants in operation; Rogun (Tajikistan), planned to be 335 m high, construction stopped, has been omitted.
[2] Only man-made reservoirs. Therefore, the 150 MW Owen Falls Power Plant in Uganda, which uses Lake Victoria as reservoir, with a volume of 2,700,000*10^6 m^3, is not included in the list.
[3] After heightening of the dam by 12.5 m in 1990.
[4] Year of commissioning; u.c.=under construction.

in discharge pattern downstream of the power plant) or for baseload generation (normal operation pattern for ROR plants; does not involve major changes in natural downstream discharge patterns).

A number of impacts depend on these conditions. Apart from the already mentioned influence on discharge in the downstream area, there are also the questions of drawdown (duration, regularity, and, also, depending on reservoir morphology, size of the drawdown area), and the time required for first filling of the reservoir.

The Environment

In a very general manner, three groups of environmental parameters have to be taken into consideration, namely, abiotic, biotic, and human or socioeconomic aspects (Figure

Environmental Impacts of Dam Projects

Table 13.2. Main Characteristics of Some of the Dams Mentioned

Item	Unit	Nam Ngum 2 Lao PDR Nam Ngum (planned)	Inguri Georgia Inguri 1987	Houay Ho Lao PDR Houay Ho 1998	Ralco Chile Bio Bio (planned)	Murum Sarawak Murum (planned)	Egiin Mongolia Egiin Gol (planned)	Mauvoisin Switzerland Drance 1958/1990	3 Gorges China Yangtze u.c. (2009)	Aswan Egypt Nile 1969	Ilisu Turkey Tigris (planned)	Urrá Colombia Sinú 2000
Project, Country, Region												
River												
Commissioned in												
Hydrology												
Catchment area	km²	5,640	3,174	192	5,130	2,750	45,000	167	680,000	2,500,000	35,517	4,600
Mean river flow	m³/s	196	155	10.3	269.7	248	92	8.2	30,000	2,500	490	340
Annual flow	M m³	6,900			8,510	7,826	2,988	260	960,000	80,000	15,450	10,700
Reservoir												
Full supply level	m asl	375	510	883	725	540	890	1961.5/1975	175	183	525	128.5
Min. operating level	m asl	335	430	861	705/692	515	850	1800			485	107
Max. flood level	m asl	375			730	546.8			181	185	526.8	135.2
Storage volume	M m³	3,590	1,110	596	1,223	12,400		270/300	39,300	168,900	10,400	1,740
Live storage	M m³	2,970	676	508	797			180/207		90,000	7,460	1,200
Reservoir area	km²	122	13.5	38	34.7	245	125	2.06/2.26	1,100	6,750	313	74
Normal drawdown	m	40	80	22	20	25	40				40	21.5
Dam												
Dam type		CFRD	Arch	CFRFD	RCC	CFRFD	RCC	Arch	Gravity	Earth-rockfill	Rockfill	Rockfill
Dam height	m	181.5	271.5	80	155	146	90	237/250.5	185	111	135	67
Crest length	m		728	370	370			520	2,150	3,830	1,850	1,300
Crest elevation	m asl	376.5	511.5	885	730	546	894	1962.5/1976		196	530	137
Spillways												
Design flood	m³/s	0,885	2,500		5,690	2,237					18,000	14,900
Power station												
Installed capacity	MW	3×205	1,300	150	570	900	220	350 (+550)	18,200	2,100	1,200	340
Annual energy output	Gwh	1,500	4,430	617	3,380			820 (865)	84,700	10,000	3,600	1330
Number of turbines		3	5	2	3	4	4	10	26	12	6	4
Turbine type		Francis	Francis		Francis			Pelton		Francis	Francis	Francis
Transmission line	km	150			46		150			787	160	
Waterways												
Intake design flow	m³/s	221	450	56	368	976		28.5 (+46)			1,266	700
Headrace length	km	15	1	7	1.4		4.8 (+18.4)					
Penstock length	m	3*180	260	700	234	450		443 (+1,768)			470	176
Tailrace length	m		3,200	200	227			1,200				
Total head	m	146	410	776	175			1,470		77	122	54

CFRD=concrete faced rockfill dam; RCC=roller compacted concrete; (planned)=in planning stage; u.c.=under construction. Mauvoisin: xxx/yyy=before/after heightening of the dam in 1990; numbers in parenthesis: planned-for enhancement of capacity (not implemented).

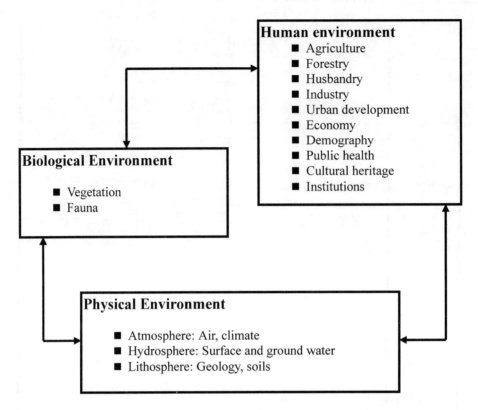

Figure 13.4. Main components of the environment. The biological (living) environment depends largely on the conditions provided by the physical environment but also has an influence on it. The human environment, again, is superimposed on the components of the natural environment, depending on them but influencing and changing them in an often quite decisive way.

13.4). The description of the environment has to cover all the aspects that can be affected in any noticeable way by the project. This means that in the choice of the parameters to be considered, in the methods used, and especially in the depth of analysis for a given parameter, the nature, extent, and potential effect of the project itself has to be considered. Therefore, and to ensure that environmental issues are duly incorporated into project planning, in order to minimize undesirable effects as far as possible, close collaboration of environmental specialists with the engineers involved in project planning and design is essential.

The Abiotic Environment

The abiotic components comprise the physical and nonliving parts of the environment, such as the atmosphere (climate, air), lithosphere (geology, soils), and hydrosphere (water, hydrology). Some of these may themselves be affected by the project, such as the water regimen in the river, or water quality, whereas others simply form the basis for the biotic and human aspects.

The Biotic Environment

The biotic components comprise the animals (fauna) and the vegetation. It is not possible to identify and list in detail all the existing animal and plant species within a given area. Rather, studies have to concentrate on those groups that can serve as biological indicators and/or that are mostly to be affected by the project. Vegetation and fauna depend largely on the abiotic conditions and are influenced by human activities. For the purpose of an EIA for a hydropower plant, in most cases, terrestrial as well as aquatic biological resources will have to be addressed. Very often, special emphasis will have to be placed on those resources that form an essential basis for the human population (e.g., forest resources and river fish).

The Human Environment

The questions in relation to the human population presently living within the study area are treated here, including the human population itself, its utilization of the natural resources (agriculture, fisheries, etc.), the available infrastructure, as well as any need to compensate the population directly affected by the project.

The Area of Influence

From the general project layout in Figure 13.2, it is evident that the definition of the area of influence depends to a very large degree on the parts or aspects of the project about which one is talking. It is important that, in an early stage of the process, the project area is well defined; normally, it will be necessary to define various subareas, because not all of them will be affected in the same manner by the project. The kind of investigation to make, however, depends very much on the expected impacts. The most important "project area types" are briefly described here. As for other aspects, area definition is highly project-specific and will have to be made according to the case presented by any individual situation.

The Immediate Project Area

This is usually defined as the area where the project itself (i.e., its main components) is located. It can be subdivided further into the following specific areas:

- *Dam site*: the place where the dam will be built. This area covers the extent of the future dam itself and its immediate surroundings, including appurtenant structures (river deviation tunnels, water intake structure, etc.), so long as these are in the vicinity of the dam. This place requires special attention from an engineering point of view, mainly for questions of dam foundation geology, and so on, but it is equally important for the EIA.
- *Reservoir area*: the land surface that will be covered by water once the project goes into operation. Its delimitation is defined usually at the full supply level (FSL; the normal maximum water level). In special cases, especially when a flood level is considerably higher and likely to be reached frequently, this additional area will have to be included in the assessment.

- *Drawdown area*: the surface between FSL and normal low water (minimum operation) level. This area is covered by water for prolonged periods (e.g., at the end of the wet season) but dry during others (e.g., at the end of the dry season). For certain aspects, this area will be dealt with as part of the reservoir area; for other aspects, however, it might need special attention (e.g., for drawdown area cultivation or visual impact assessment).
- *Immediate catchment area*: the land immediately surrounding the reservoir, which drains directly into it.
- *Powerhouse site*: area occupied by the powerhouse and appurtenant structures (administration buildings, maintenance infrastructure, switchyard, etc.). If the powerhouse is located directly at the foot of the dam, it might be considered part of the dam site; however, depending on the design of the project, the powerhouse can be several kilometers away from the dam.
- *Construction site*: area occupied during construction, including permanently used areas (i.e., sites of buildings and structures such as dam, powerhouse, waterways, intakes and outlets, permanent housing for plant staff, etc.) and temporarily used ones (construction site facilities, workers' camps, burrow and dumping areas, etc.). Where main structures are distant from one another (e.g., when there is a long headrace tunnel from the reservoir to the powerhouse), there may be more than one construction site.
- *Downstream area*: the river below the dam as far as it is affected by the project. Here, two main parts can be identified: (1) between the dam and powerhouse, where river discharge is reduced or zero in extreme cases, and (2) below the powerhouse water outlet, where the river discharge pattern is influenced by power plant operation. The downstream area is not necessarily limited to the river itself but also includes the floodplain and other areas indirectly affected by the change in discharge pattern.

Of special importance is the downstream area in transbasin projects (i.e., projects affecting more than one river). There are two main possibilities: (1) Rivers in parallel valleys are taken in and their water is redirected into a reservoir to increase the available amount of water, and (2) the water from a reservoir is redirected to a site other than the original river (Box 13.5). In the first case, the discharge in the river below the intake structure will be much reduced; it can even be zero for a certain stretch. In the second case, discharge is reduced in one river, below the dam, and increased in the other river, which receives the water from the power plant.

Box 13.5. Delimitation of the study area

The definition of the study area can sometimes be difficult. This is true especially for the downstream effects: How far do they reach, and, therefore, how far down from the project site needs the investigation to go?

The Curciusa hydroelectric project (HEP), in Switzerland, consisted of constructing a dam on a small lateral tributary of the Rhine River, and directing its water through a tunnel southwards to the Moesa, which belongs to the drainage basin of the Po River. In doing so, the project would have crossed the continental divide, redirecting water from the North Sea basin to the Mediterranean basin. The question was, how far down to investigate potential impacts. This can only be decided on a case-to-case basis. The following solution was adopted here:

- For the northern side (abstraction of water): the whole of the affected stream and the main stream after confluence as far down as a point where the abstracted water was less than 5% of average flow; this was well within natural fluctuations and therefore no longer directly observable.
- For the southern side: the whole valley down to the point where the river joined the next larger one; the main impact here was due to power plant operation and not to the difference in absolute water quantity (Electrowatt Engineering [EWE], 1987b).

The Wider Project Area

This encompasses all the other areas affected by the project that have to be considered, but where the impacts are normally less than (or different from) the ones in the immediate project area. The most important ones are as follows:

- *The access roads corridor*: If a new road is being built, its direct as well as potential indirect effects will have to be taken into consideration (mainly making new areas easily accessible).
- *The transmission line corridor*: The impact here is normally limited to certain restrictions in land use (e.g., tree crops no longer possible in the power line right of way [ROW]) and to visual impacts. However, it can be more important in cases where the transmission line runs through densely populated areas, forests, or natural, protected areas.
- *The resettlement area*: If resettlement is required for a project, the corresponding impacts on the land chosen for this purpose must be included in the considerations. This is also true for areas designated for any other compensatory measures.
- *The watershed area*: This area is normally not affected—at least not directly—by the project, but it has important effects on the project, mainly concerning reservoir siltation and eutrophication.

Environmental Aspects and Main Impacts

The environment and the entity formed by a project and its environment, including all the effects on each other, form a complex system with a lot of ramifications, interactions, and indirect effects (i.e., activity–response chains). Therefore, it is not possible to establish a linear system of environmental impacts that would place every effect in a logically well-defined order. A few examples may be sufficient to illustrate the difficulties encountered:

- Sediment and sediment transport has to do with water (hydrosphere) as the transport medium, and with soils and geological underground (lithosphere).
- Water quality (hydrosphere) is influenced not only by geology and soils (lithosphere), but also by submerged vegetation in the reservoir and by input from human activities, and itself has effects on organisms living in the water and on human use of this resource.

- Vector-borne diseases depend on climate, water availability and quality, and the local fauna (presence or absence of a specific animal vector), and have impacts on human health.

Therefore, when structuring a report or deciding where to deal with a certain subject, decisions have to be made that are not always, or not entirely, founded on logic alone. It is, for instance, not very important whether vector-borne diseases are treated under the main heading of "fauna," "public health," or as a separate chapter. The important point is that the question is addressed in every case in which it could be an issue.

The impacts of a hydropower plant are always highly project-specific, whereby the kind and design of the project itself and the site-specific environmental conditions of the project site interact. This always has to be kept in mind when attempting to draw up a standard procedure or checklist for a "typical" hydropower project or a "standard" EIA.

In spite of this, it can be said that the major direct and indirect impacts of any hydropower project, which consists of the main components illustrated in Figures 13.2 and 13.3, are always the same; their relative importance, of course, will be determined by the site- and project-specific conditions. These main impacts are listed as follows:

- *Interruption of a river continuum*. The fact that a dam is built across a river will always interrupt a system that was, up to now, an entity. Direct consequences of this interruption are a change in river flow patterns, a change in sediment transport (mainly due to sediment retention in the reservoir), an interruption of fish migration (complete for upstream migration, obstacle and risk for downstream migration), and the interruption of drift (i.e., the more or less passive movement of various organisms downriver).
- *Change in river discharge pattern downstream of the dam*. This effect is closely related to the first one. In this respect, two main parts of the river can be identified: (1) between the dam and powerhouse outlet, where discharge is reduced, in extreme cases to zero, and (2) downstream from the power house outlet, where river discharge is influenced by plant operation.
- *Change from river to lake conditions in a part of the former river at the formation of the reservoir*. Water quality will change due to this effect, and the new lake is a habitat very different from that of the former river.
- *Destruction of terrestrial habitats*. All terrestrial habitats within the reservoir area will be permanently destroyed, because they are going to be covered with water. This has effects on vegetation and fauna, as well as on the human population living in this area.
- *Access to the area provided by new access roads*. Although the direct impact of the roads (e.g., on vegetation) might be rather small, the roads can trigger a development, especially in cases when hitherto inaccessible areas are opened in this way, that can have very considerable environmental effects.
- *Social impacts*. These can be manifold. The most important in many cases is the involuntary resettlement as a consequence of a dam project, but there are also other socioeconomic effects, such as effects on the population in the downstream area (through disruption of river floodplain dynamics, groundwater table changes, etc.); immigration into the area, especially during the construction phase, as a consequence of job opportunities; and effects on the host population for the resettlers.

Environmental Impacts of Dam Projects

Most other impacts that may arise are likely to be related to these major effects, very often as secondary consequences.

Abiotic Environment

The Atmosphere

Climate

Large water bodies influence the climate of their surroundings, especially the temperature and the humidity. The most noticeable effects are a general cooling in summer, a warming in winter, and a reduction of the daily and seasonal temperature variation. This effect can be clearly seen when the climate of a place on the seashore, with maritime climate, is compared to that of a place far away from sea influence, with a continental-type climate (see Figure 13.5). Direct measurements have documented this effect. It has been demonstrated, for instance, that a small island in the Finnish Bay exhibited a January temperature that was 1.3 °C higher than that of a station in the nearby land some distance away from the coast, whereas the average May temperature was 2.5 °C lower. Similar effects have been demonstrated in the vicinity of two lakes (Lake Peipus, Lake Chelkap), although, in these cases, the differences were smaller, reaching only +0.3 ° in winter, −0.7 ° and −1.8 °, respectively, in summer (Alissow et al., 1956). This is also illustrated for Lake Aral in Figure 13.6. This is mainly due to the fact that the water is able to store a considerable amount of heat and reacts very slowly to changing temperatures. While on land, the temperature (soil and air layers close to it) can exhibit large daily fluctuations; this is not the case for water. In Irkutsk, the daily variation in

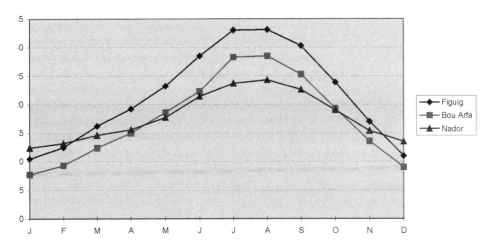

Figure 13.5. Mean monthly temperatures of three stations on a south–north gradient in Morocco showing the difference between a continental (Figuig, 400 km from the coast on the northern fringes of the Sahara desert) and a maritime climate (Nador, on the Mediterranean coast). Bou Arfa lies in between, temperatures being lower due to higher elevation (EWE, 1998).

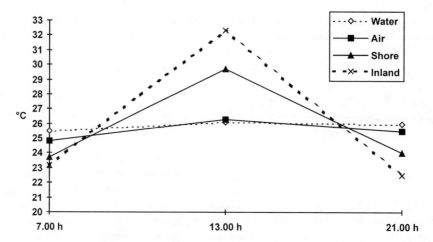

Figure 13.6. Daily temperature variations as influenced by a lake. Measurements from Lake Aral, August 1902. The amplitude of the water temperature is merely 0.6 °C, of the air above the water, 1.5 °C, at the shore, 6.9 °C, and in greater distance from the lake, 9 °C (Alissow et al., 1956).

summer (i.e., difference between daily minimum and maximum) is between 13.5 and 21.7 °C; in winter, when it is 5.7 to 14.5 °C (air 2 m aboveground), the soil surface variation is even greater (29.8 °C in summer, 6.2 °C in winter; measurements for a day with rather low air temperature variation). It will be noted that Irkutsk is exhibiting a continental climate in spite of the proximity of Lake Baikal, which is one of the largest inland water bodies of the world. The biggest daily variations in air and soil temperature are recorded from deserts, which lack the cooling effect of evapotranspiration. In African deserts, daily surface temperature differences of up to 43 °C have been recorded, and in middle Asian deserts, up to 50 °C.

In larger water bodies (seas and very large inland waters), normal daily fluctuations of temperature close to the surface are usually below 1 °C (and smaller for deeper layers), and this is also the case for the air layer approximately 1 m above the water surface, with this difference becoming more pronounced in higher air layers (Alissow et al., 1956; Geiger, 1950).

In the case of smaller lakes, it is very difficult to define clearly the effects of the proximity of the lake on ambient temperatures; the complex mixing processes due to local winds tend to blur this influence. Nevertheless, it can be said from experience that, as a general rule, lakes have a beneficial effect on the local microclimate. This is shown by the fact that in temperate regions, some plants normally limited to a warmer climate grow exclusively, or at least much better, in the vicinity of lakes (e.g., vineyards in central Europe). Furthermore, settlements tend to concentrate around lakes. Although this has certainly different reasons—historical and scenic ones, among others—the microclimate certainly contributes to it.

A few publications are concerned with the question of evaporation from lake surfaces (Kuhn, 1977; Hoy & Stephens, 1977). However, there does not seem to exist to date a thorough climatic study in relation with a man-made lake, which would compare temperature and humidity values before and after dam construction. Ladeishchikov and Obolkin (1990) provide data on climatic effects of large, man-made lakes in Siberia (see Table 13.3). The studies that at least mention potential climatic effects of artificial lakes

Table 13.3. Effect of Large Siberian Reservoirs on the Mesoclimate of Their Surroundings

Reservoir		June		Nov.		Year			
		Average temp. °C	Average precip. mm	Average temp. °C	Average precip. mm	Average temp. °C	Average ampl. °C	Absol. ampl. °C	Precip. mm
Irkutsk	b	14.8	81	−10.6	17	−0.9	12.8	73	454
	a	15.3	60	−9.0	20	−0.2	11.8	66	434
Bratsk	b	15.4	—	−12.6	—	−2.1	12.4	78	—
	a	14.0	—	−9.8	—	−1.4	9.3	70	—
Krasnojarsk	b	15.7	57	−10.3	25	0.4	—	—	385
	a	16.2	55	−7.2	32	0.7	—	—	411

Significant changes: November temperature (all stations); June temperature (Bratsk only); yearly temperature (Bratsk and Irkutsk). No significant changes in precipitation. Note that not all changes went in the "expected" direction). No indications on sites of measurements and on lengths of observation period; b=before, a=after dam construction.
From Ladeishchikov & Obolkin (1990).

attribute to them a minor or almost negligible effect (e.g., Odingo, 1979). This seems perfectly understandable given the fact that although artificial lakes certainly create considerable impacts on the environment, climatic changes can, in the light of the details given previously, be considered of minor importance and, if at all noticeable, then rather beneficial.

In temperate and cold regions, the effect of a lake on the climate takes place only while it is not covered with ice. A compact ice layer covering the water body effectively blocks temperature exchange of the lake with its surroundings. In humid, tropical areas, the effect of a reservoir on climate will most probably be negligible or almost nil, because temperature differences between water and surrounding land are not very marked, and evaporation from the open water surface is not very different from the evapotranspiration of the surrounding vegetation (Doorenbos & Pruitt, 1977). The most marked effects may be expected from large reservoirs in a dry environment. A rise in air humidity and a balancing of temperature would certainly be regarded as beneficial impacts in such a case. But even from Lake Aswan, one of the largest man-made lakes and located definitely in a hot, arid environment, no such effect is clearly documented.

Unfortunately, and in the absence of any reliable data, the effects of reservoirs on climate seem to be taken merely as an additional argument to corroborate a general attitude vis-à-vis a project. For example, Ahmad (1999) expects a beneficial change in microclimate from semiarid to subhumid after introduction of irrigation related to the Sardar (Narmada) Sarovar multipurpose project, in a publication that emphasizes the benefits of this project. On the other hand, Biro (1998, p. 224), in a study that looks very critically at Kayraktepe Dam in Turkey, states: "The reservoir lake will change the micro-climate of the Göksu Basin, adversely affecting the flora and fauna of the region, including some rare species," without any further substantiation of either effects or species affected. A thorough scientific study of this effect is urgently needed. Nevertheless, in my opinion, this is a topic where common sense should be used, considering

- that a very large water body in a cold or temperate climate can, to some limited degree, smooth temperature variations (cooling effect during very hot and warming effect during cold weather);

- that a similarly large water body in an arid climate can lead to a certain reduction of high ambient temperatures and a rise in relative humidity;
- that a large water body in a warm, humid climate will not have any noticeable effect on temperature or humidity; and finally
- that there is, all over the world, a very high preference for living next to a lake (presumably because it is considered attractive from an aesthetic point of view, and probably also because the climatic conditions are agreeable, but probably not because there is an intolerable climate in the vicinity of a lake!).

It is probably safe to conclude that any climatic effects expected from a reservoir will be felt as agreeable rather than as worsening the situation.

Obviously, when comparing living conditions around a reservoir with those around most natural lakes, one reservoir characteristic should not be forgotten, namely, the drawdown of the reservoir. This fact can eventually make living near a reservoir much less attractive than in the case of a lake, with a more or less stable water level year-round (see Box 13.6).

Box 13.6. Case Study Egiin HEP, Mongolia

Egiin HEP is situated in northern Mongolia, on the river Egiin Gol near its confluence with the Selenge; this latter river is the largest in Mongolia and the main tributary of Lake Baikal. The project area has a cold to moderate, strongly continental type of climate, characterized by marked seasonal and daily variations in temperature; night frosts may occur in every month of the year (see Figure 13.7). The Egiin HEP watershed lies within the wettest part of Mongolia, with yearly precipitation of 300 mm and more; in the highest parts of the watershed area, precipitation is above 500 mm per year. The largest part of the precipitation falls normally during the summer, July and August being the wettest months (Sodnom & Yanschin, 1990). Relative humidity is rather low, especially during the summer months, normally ranging between 40 and 50%. Basic studies for this project were made in 1992 to 1994 (EWE, 1994), but did not go beyond the feasibility level.

The effects of lakes on the local climate, as mentioned earlier, could also to some degree be attributed to the reservoir on Egiin Gol. The heat accumulated in the lake during summer would have a warming effect on its immediate surroundings, whereas its cooling down in autumn would result in a slightly later occurrence of frosts in autumn and in a delay in the onset of snow cover in the proximity of the lake; this effect would last until the lake was covered by ice. On the other hand, the melting of the ice in spring could have a cooling effect in this period. However, the changes in temperature would certainly be too small to affect the vegetation period in any noticeable way. Due to the low water level in spring, the effect would be less marked than in autumn. However, considering the relatively small surface and volume of the lake (e.g., compared with the large natural lakes Hovsgol and Baikal), the effect on climate would certainly be too small to prevent night frosts under the prevailing climatic conditions. The same can be said for an increase in air humidity, which would be too small to have any significant effect on plant growth.

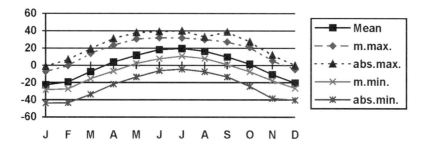

Figure 13.7. Temperature in the Egiin Gol project area, Mongolia. In this very continental climate, temperature differences (not only the annual differences shown here, but also daily variation) are very high (roughly between −40 and +40 °C). The curve for the absolute minimum shows that in every month of the year, there is a risk for night frosts (EWE, 1994).

Greenhouse Gases

Hydropower is considered an environmentally friendly form of energy also because the alternative of generating the same amount of energy by means of fossil-fueled thermal power plants invariably leads to massive CO_2 emissions. This gas is the main so-called greenhouse gas, responsible for global warming. It should be noted that, also, from a scientific point of view, it has been contested whether the well-documented rise in CO_2 concentration in the atmosphere is a cause or a consequence of global warming (e.g., Calder, 1999); however, even if the latter should be the case, a reduction in greenhouse gas emission certainly has many beneficial environmental effects. Besides, waiting to take action until the last scientific doubt has been eliminated seems not a very reasonable attitude.

This environmental plus of hydropower has been questioned. Rudd et al. (1993) and Fearnside (1995, 1997) stated that hydropower reservoirs that emit large quantities of methane might ultimately, calculated over the lifetime of the plant, have greenhouse effects of the same order of magnitude or even greater than a thermal power plant producing the same amount of energy (0.3–0.5 equivalent $TgCO_2$/TWh compared to 0.4–1.0 equivalent from a carbon-fired thermal power plant). This is mainly due to the fact that the greenhouse effect of methane is higher than that of CO_2 by a factor of 21 (Institute of Electrical Engineers, 1994); methane is produced when in a reservoir biomass is being degraded under abiotic conditions (see the section on water quality).

This view, however, has been strongly contested. Gagnon and Chamberland (1993) and Gagnon (1998) showed that extremely high emissions of greenhouse gases only occur in very special cases (when a reservoir submerges extremely high quantities of organic material; e.g., in mires, organic soils in cold climates, or rain forests in tropical climates) and when the ratio of reservoir surface per unit of energy produced is extremely high. Furthermore, they stated that the publications mentioned earlier made a series of very unfavorable and unrealistic assumptions, such as a total degradation of submerged biomass and a very high value for the conversion of methane in CO_2 equivalents. Overall, and with the potential exception of a few very special cases (e.g., Balbina in Brazil, which flooded tropical rainforest with a very high amount of biomass and has an unusually high ratio of area per energy output), it can be said that hydropower has considerably lower greenhouse gas output than fossil-fueled thermal power (see Table 13.4).

Table 13.4. Comparison of Greenhouse Gas Emissions of Different Energy Systems (Data from Québec, in Equivalent g/MJ)

Energy option	CO_2 emission during final combustion	CO_2 emission by energy system	CH_4 emissions equiv. CO_2	Total emissions equiv. CO_2	Relative ratio to hydroelectricity
Hydroelectricity	—	6.8	0.3	7.1	1
Oil-fired heating	95	7	0	102	14
Natural gas heating	68	9	17–34	94–111	13–16
Oil-fired plants	190	14	—	204	29
Natural gas plants	136	18	17–34	171–188	24–26

From Gagnon and Chamberland (1993).

Air Quality

Air contamination, in the context of a hydropower plant, will usually be restricted to the construction phase, in which construction activities (excavation and transport of dam material, etc.) can have a localized effect on air quality. However, this is definitely a transient effect limited to the construction period; although, in the case of a larger dam, this can last up to 10 years, I personally do not know of any hydropower project in which air contamination (with exhaust gases and dust) was considered a major problem. The issue has to be addressed in an EIA, and measures, if required, have to be taken (e.g., protection of workers' camps from dust), but in comparison with other issues, this one is rather marginal; the same is true for noise, which also can be caused by construction activities. In most cases, these will rather be workplace safety and work health rather than environmental issues. They are not discussed in any further detail here.

There is just one other point we might have to consider. In some cases, within an EIA, a comparison is being made, usually on a rather general level, between the impacts of the hydropower project compared to the impacts that would arise if the same amount of energy were produced by an alternative plant, usually a thermal power plant using fossil fuel. Without going into any further details here, it is obvious that air contamination is a major issue in thermal power plants. A few data are provided in Table 13.5.

The Hydrosphere: Water

General Considerations

Hydropower is primarily using water as a resource for generating energy, and its main components (dams and waterways) interfere directly and in various ways with water. For this reason, impacts on water have a very important place in every EIA dealing with any dam project. In addition to this, because aquatic ecology is the main subject of this book, there are good reasons to deal with this issue in some length. A HEP will normally influence surface water bodies in three ways:

1. By interrupting the river continuum through the construction of a dam, thus interrupting the movements of organisms (by active migration or by drift) and sediment transport.

Environmental Impacts of Dam Projects   309

Table 13.5. Annual Emissions from Typical UK 2,000 MW Fossil Fuel Power Stations (in t)

Emission	Coal-fired conventional (no FGD)	Coal-fired with flue gas treatment (FGD, LowNOx)	Oil-fired conventional	Gas-fired combined cycle gas turbine
Airborne particulates	7,000	5,000	3,000	Negligible
Sulphur dioxide	150,000	15,000	170,000	Negligible
Nitrogen oxides	45,000	30,000	32,000	10,000
Carbon monoxide	2,500	2,500	3,600	270
Hydrocarbons	750	500	260	180
Carbon dioxide	11,000,000	11,000,000	9,000,000	6,000,000
Hydrochloric acid	5,000–20,000	2,500–10,000	Negligible	Negligible
Solid waste (bottom and fly ash)	840,000	840,000	Negligible	Negligible
Radioactive emissions (bequerel)	10^{11}	10^{11}	10^9	10^{12}

Radioactive emission: dependent on source of fuel. For coal largely, for gas, exclusively radon.
Trace elements: fuel source–dependent, may include arsenic, chromium, copper, vanadium, nickel, lead, zinc, selenium, cadmium, antimony.
Source: IEE (1994).

2. By creating a lake, that is, by changing a portion of the river from its natural running water into a stagnant water condition.
3. By changing water discharge patterns downstream from the dam.

A reservoir or a lake presents very different living conditions for aquatic organisms in comparison to those that prevail in a river. Many riverine species will not be able to adapt to these new conditions, whereas others will thrive. Furthermore, reservoir water quality will depend a great deal on the preimpoundment condition of the reservoir area. Whenever a large amount of biomass is submerged, the rotting plant material will lead to water quality deterioration in the deeper layers of the reservoir. This in turn can lead to fish deaths, but it can also impair water quality downstream from the dam. This aspect is especially important when the river or the reservoir itself is used as a source of drinking water.

The dam and the reservoir, in relation with turbine operation, can change the river discharge downstream from the dam considerably. This is especially the case in strongly seasonal climates with a marked change between rainy and dry seasons. In such cases, water is stored during the wet period to be released during the dry period. The reservoir has a flood mitigation effect depending mainly on its size and storage capacity. Other effects of the HEP are short-term fluctuations caused by turbine operation. When the powerhouse is not situated directly below the dam, but farther downstream, a stretch of the river between the dam and water outlet will eventually become completely dry. Whereas attenuating effects on river flows are normally welcome by people living along the river, other effects can be less beneficial, and certain floodplain habitats that depend on river dynamics might be negatively affected.

One general problem when dealing with impacts of hydropower projects on water rivers in tropical climates is the fact that although there is a very vast limnological literature for rivers from temperate climates, mainly Europe and North America, this is not the case for tropical rivers (Wetzel & Gopal, 1999; Zauke et al., 1992). Much general research work remains to be done to improve our understanding of these important habitats.

Hydrology

Hydrological data are an important basis for an EIA for a hydropower plant and are of course an essential part of hydropower planning, because this is the resource basis on which the whole project is founded. Three parameters are of special interest:

1. *Average annual discharge*: gives a rough impression of the "size" of the river and allows calculation of the water available on average.

2. *Seasonal variation*: provides information on within-year variation of river discharge and seasonal availability of water; this variation can be rather small in climates with year-round precipitation and very high in extremely seasonal climates (very marked dry season and/or large proportion of the precipitation falling as snow).

3. *Interannual variation*: gives information about the reliability of water availability; this variation is especially high in semiarid or arid climates with a very unreliable precipitation pattern (see Figure 13.23).

Some information on annual river discharge patterns in various climates is provided in Figure 13.8 and Table 13.6.

ROR projects have a very marginal influence on river discharge patterns, because they use the water of the river "as it comes." However, especially in conditions with high seasonal variability, storage projects are built with the aim of storing water during the wet (rainy or snow melt) season to make it available for energy production (and/or irrigation, as the case may be) during the dry or cold season, when demand is usually highest. Such projects, depending on their regulating capacities, can have more or less marked effects on the discharge pattern in the downstream area, as illustrated in Figures 13.9 and 13.10. This can have effects on downstream water users; whether they are positive or negative depends mainly on the specific situation. If more water is available during the dry season, this can have very positive effects for the population living there. On the other hand, floodplain dynamics can be changed very decisively, and so can production patterns or livelihood systems depending on such floodplains and their natural dynamics. One famous example is

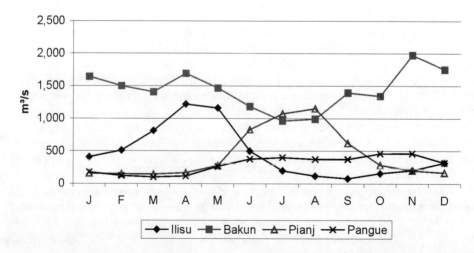

Figure 13.8. Discharge pattern at dam sites on rivers from different climates (see Table 13.6 for further explanations).

Environmental Impacts of Dam Projects

Table 13.6. Characteristics of Rivers from Different Climate Types (see Figure 13.8 for Discharge Patterns)

Project	Ilisu	Bakun	Pianj[1]	Pangue
River	Tigris	Rajang	Pianj	Bio Bio
Region	SE Anatolia	Sarawak	Gorno-Badakhshan	Concepción
Country	Turkey	Malaysia	Tajikistan	Chile
Mean annual discharge (m^3/s)	475	1,443	436	296
Total annual discharge ($10^6 \, m^3$)	14,949	45,455	13,727	9,332
Catchment area (km^2)	35,517	14,750	n.a.	5,430
Climate	Temperate subtropical mediterranean	Humid tropical	Temperate, extremely continental high mountain	Mediterranean
Main origin of water	Winter (Dec. to March) rainfall and spring (March to June) snow melt	Year-round rainfall with seasonal variation, no real dry season	Snow melt in very high mountains, no rainfall during summer	Winter (June to Aug) rainfall, spring (Oct., Nov.) snow melt

[1] This is no hydropower project site, but just results from a hydrological station on this river forming the border between Tajikistan and Afghanistan.

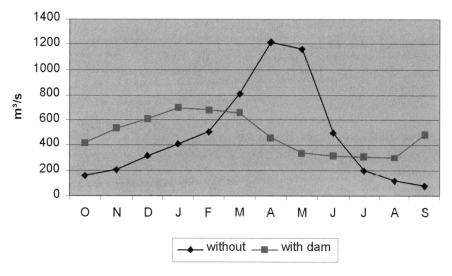

Figure 13.9. Tigris River discharge downstream of the Ilisu Dam, Turkey. Main effects of the dam are a massive reduction of the wet season flow (snow melt in upper catchment) and a considerable rise in dry season flow. Mean monthly discharge does not show the daily variations, which can be considerable in a hydropower project used for peak electricity production. Such a massive change in river flow pattern will certainly influence the biota in the river. Whether effects on water use in the downstream area are positive or negative depends on use pattern. However, higher flow during the dry season can be expected to have positive effects.

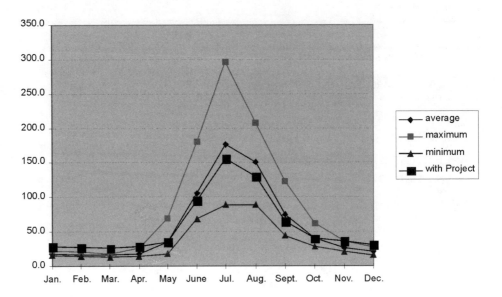

Figure 13.10. Gunt River discharge at site of Pamir 1 Hydropower Station: average, maximum, and minimum monthly discharge without, in comparison to average discharge with regulating structure at the outlet of lake Yashilkul, the natural lake serving as reservoir for this power plant. In this case, the effect is minimal and lies mostly within natural fluctuations (EWE, 1999).

the Nile Valley below the Aswan high dam, which retains up to 98% of the Nile sediments, as well as the natural seasonal floods, and prevents in this way the natural fertilization of the fields (Moatassem et al., 1993; Biswas, 1995). That such effects have to be taken seriously is also shown by the study of Barbier and Thompson (1998), which shows that the benefit which the local population drew from the natural floodplain dynamics is higher than that obtained from the project.

Such reservoirs, besides stocking water during the wet season, also have the effect of reducing floods. Flood protection can even be one of the main project objectives, as in the highly contested Three Gorges Project presently under construction on the Yangtze in China (Edmonds, 1992).

Very often, hydropower plants have the main purpose of producing peak energy. This is the case especially where they are operated in a grid together with ROR or thermal power plants; the latter have a very low flexibility in operation and are therefore normally used for producing baseload energy, whereas storage plants are very flexible in the sense that their production can literally go from 0 to 100% in a matter of seconds, which makes them highly suitable for responding to short-term demand variations. Operating a power plant under such a regimen obviously leads, as the negative part of the energetic advantage, to extremely rapid and often very high fluctuations in water output, and therefore in discharge of the river downstream of the plant.

This case is illustrated, as one example of many, with the project of raising the installed capacity of the existing Mauvoisin hydropower plant in Switzerland. The existing plant, which uses a reservoir formed by a 250-m-high arch dam, has a maximum water output of 28.5 m³/s, and is operated as a peaking plant, usually producing power for only a few hours daily and mainly in winter, often with several on–off movements daily. The project envisaged the construction of an additional power station, which would have more than doubled the installed capacity, from 350 to 900 MW; because no additional water was going to be captured, the total amount of energy produced would have remained the same (except for a small gain through a rise in efficiency of the installations), but it would have been possible to produce a narrower, higher peak (i.e., twice the amount of energy for half of the time). This, of course, would have more than doubled the water release to the river. During winter, the Rhone at this place, the river receiving the water outfall, carries about 40 m³/s (seasonal low discharge in winter). So operating Mauvoisin alone would have almost tripled discharge, and this once or twice a day for a very short period. The effect was even more marked by the fact that other power plants operate with a similar pattern on the same river.

Such short-time discharge amplitudes can have the following negative effects:

- Riverbed erosion due to the surge wave caused by the rapid increase in discharge.
- Destruction of fish breeding grounds.
- Impacts on fish and other water organisms due to the rapid changes in water temperature (water being brought down from high mountain reservoirs can be considerably colder than the water in the lowland river).
- Danger to downstream users of the river.

The measure usually applied for mitigating this impact is the construction of a regulation pond. In the case of the mentioned Mauvoisin II project, such a pond with a capacity of 470,000 m³ was planned; this would not have allowed for storing the water discharged during a supposed 2-h period of plant operation to release it, then continuously over the whole day, but it would have been sufficiently large to slow down considerably the increase speed of the discharge, thus reducing the surge wave, then somewhat filling the "trough" between two operation peaks. This basin alone is a major structure covering an area of 9.95 ha (450 × 200 m), surrounded by a 10-m-high dam.

A different way to deal with the same problem was chosen in the case of Ralco HEP on the river Bio Bio in Chile: Here, the discharge would have been released into the already existing reservoir of Pangue, which then would have acted as a regulation pond. The Pangue reservoir is too small to act as seasonal storage, so that Pangue is operated almost as an ROR plant; nevertheless, the reservoir has sufficient storage for regulating daily fluctuations. (Note that, so far, neither Ralco nor Mauvoisin II have been built.)

A special problem for the downstream discharge can arise during the filling phase, especially in the case of very large reservoirs that store the inflow of several months or even years. A complete closing of the dam would then mean that no water at all would flow below the dam for a long period, a situation that is definitely not acceptable, because it would have very severe impacts on all the organisms living in the river, on the whole surroundings of the river, and on all downstream users of river water. The potential problem can be even more serious in cases in which such a reservoir is on an international river and the owner of the dam could "close the tap" for lower lying countries. This is the case e.g., for Attatürk Dam on the Turkish part of Euphrates River or for the planned Ilisu HEP on the Tigris. In such cases, international agreements—and firm commitments of the upper lying countries—are required, something not always easily accomplished (e.g., Laki, 1998; Turan, 2000).

Groundwater

A hydropower plant can influence groundwater mainly in the following ways:

- *Around the reservoir.* Especially in rather flat areas, where surrounding land is not much higher than FSL of the reservoir, the groundwater table will rise to the level of the reservoir or somewhat higher. This can have a positive effect (e.g., in semiarid climates with a marked dry season) inasmuch as during the wet season, and as long as the reservoir water table is high, the groundwater will be supplemented. This can have the effect that wells used as drinking water sources (which usually provide a better water quality than the reservoir itself) will not dry out or, at least, not as quickly as before.
- Rather steep reservoir shores with a high amount of loose overburden (loose material covering the solid rock) will be saturated with water. If the reservoir is being drawn down very quickly, this can trigger massive landslides, because the water is not being drained quickly enough from this material.
- The backwater effect at the inflowing river, if this is not steep, will lead to a high groundwater level in its floodplain. On the other hand, drawdown of the reservoir can lead to a (seasonal) lowering of this groundwater table (see Box 13.7).
- If the powerhouse is not located immediately below the dam, but a certain distance farther downstream, or even in a different river, the flow in the river between the dam and the water outlet will be very much reduced (often to zero as far down as the nearest lateral tributary). Especially in high-pressure power plants in mountainous areas, this is a very frequent situation, because the aim here is to make use of the head to enhance power production (see Table 13.10). On the other hand, of course, very often in such cases the rivers below the dam flow in steep, narrow valleys, so that the effect on floodplain groundwater level is not really felt. This, however, is not always the case, especially not in transbasin projects. Here, again, it is important to deal with every project as a special case and consider its specific situation. To mitigate this effect, many countries have regulations concerning the release of a minimum residual flow (mandatory release, riparian release) that should be sufficient to conserve this part of the river as a habitat for river organisms (see Table 13.11); needless to say, in spite of this, it will no longer be the river habitat it was originally.
- Below the powerhouse and the water outlet, there will be the combined effect of the dam and the powerhouse: change in seasonal pattern and in daily discharge. Daily discharge variations normally do not affect the groundwater level in a very marked way, but changes in seasonal flow patterns can be very decisive (again, depending on the situation and the parameter looked at, positive or negative). The fact that seasonal floods no longer occur or are at least very much reduced can have a very far-reaching effect on the groundwater table of a floodplain, and on floodplain dynamics as a whole.

Box 13.7. Pamir HEP, Gorno-Badakhshan, Tajikistan

The completion of this project, whose first stage was inaugurated in 1985, will make use of Lake Yashilkul, a natural lake situated in the Pamir mountains at an altitude of 3,720 m asl. The aim is to build a structure at the outlet of this lake that would allow

Figure 13.11. Floodplain of the river flowing into Lake Yashilkul, Pamir mountains, Tajikistan (3,720 m asl). Drawing down the lake permanently would lead to a drying out of this area, which is an important pasture in an otherwise rather arid environment.

for retention of some additional water during summer and to release it during winter to be used for energy production. The power plant is a ROR plant located some 100 km downstream from the lake. Low winter discharge (see Figure 13.10) does not allow power production at full capacity during this time, when energy shortage is most severe and demand is highest. At the main inflows into the lake, there are two natural floodplains (see Figure 13.11) that are very important pastures for the population living near the lake (at altitudes of around 4,000 m asl, and in an arid climate, where livestock breeding is the only livelihood basis, and which will not profit from the energy generated by the project). The project as designed originally (before 1991) had planned to permanently lower the lake level by 2 m and to draw it down during winter by an additional 6 m. There is a very high probability that with the lower lake level during summer, the productivity of the floodplains would have been destroyed or at least markedly reduced. The following measures were recommended during the feasibility study:

- A change in project design allowing it to maintain summer water level at its present altitude.
- To minimise winter drawdown to the absolute minimum required.
- To ensure a rapid filling of the lake in spring (during snow melt) in order to bring it up to its natural summer level.
- To investigate the possibility to include a small hydropower station into the weir at the outlet of the lake for supplying nearby settlements with electricity.
- To carry out a thorough monitoring program which also includes the possibility of compensation payments if pasture should be lost in spite of the measures taken (EWE, 1999, 2001).

One other effect can be closely related to the groundwater situation in the reservoir and dam area: the question of reservoir area, and especially of dam-site seepage. The presence of fissures or faults in the reservoir bed and the rock forming the dam foundation can lead to major water losses. Seepage through the dam foundations can even jeopardize dam stability, whence the necessity for detailed geological investigations of dam sites.

Water Quality

Very often, the role of reservoirs is not limited to the production of energy; rather, they serve other purposes as well, such as irrigation, supply of drinking water, fisheries, recreation, and so on. For these uses, water quality is important. Standards have been established that have to be met; otherwise, the usefulness of the reservoir is limited. Expensive measures may have to be taken (e.g., for the treatment of drinking water).

A number of publications deal with water quality issues in reservoirs in a general manner (ICOLD, 1994; Zwahlen, 1998). A recent review of WB-financed hydropower projects mentions the water-quality issues briefly, however, without going into details, and finally states that the report "has not come across any case where these problems have become unmanageable or would appear to have significantly affected economic returns from the project" (Liebenthal, 1996, p. 36). Nevertheless, water quality is one of the main topics in every EIA for a hydropower project.

Basically, there are two groups of water-quality problems associated with reservoirs: (1) effects of first filling of the reservoir and its development during the first few years, and (2) the development of the reservoir during the operation phase. However, we should not completely forget the construction phase, during which water quality can also be affected, mainly due to an increased silt load stemming from the construction activities and a risk of water contamination with concrete, chemicals, and fuels used on the construction site.

Impoundment

Impacts. The water-quality issue associated directly with impoundment is oxygen depletion, especially in the deeper layers of the reservoir (see Figure 13.12), due to the oxygen consumption for the breakdown of the submerged organic material. This is mainly the vegetation covering the reservoir area. As Table 13.7 shows, this can be very large amounts. In specific cases (mainly in cool temperate climates, but exceptionally also in the tropics, namely, in the case of tropical peat forests, as in lowland areas of Kalimantan), the soils can contain very high amounts of dead biomass; peat layers can be several meters thick and contain at least 65% biomass (MacKinnon et al., 1996, p. 120). Nevertheless, most tropical soils are not very rich in organic material due to the quick breakdown under warm and moist conditions, so that the relevant parameter is plant biomass.

There is only limited information available on the actually observed water quality changes and their long-term development in reservoirs. The best known example of massive water-quality problems is probably Brokopondo in Suriname (Van der Heide, 1982; Zauke et al., 1992, p. 113), where the formation of H_2S was so severe that during a certain period, the workers had to wear gas masks. For other studies covering the impoundment period, see Visser (1973), Biney (1987), Simanov and Kantorek (1990), and Schetagne (1994). It is highly probable that there is rather a lot of information on this topic stemming from monitoring exercises of various reservoirs. However, since these data are usually used only as internal reports, they are not easily accessible.

Environmental Impacts of Dam Projects

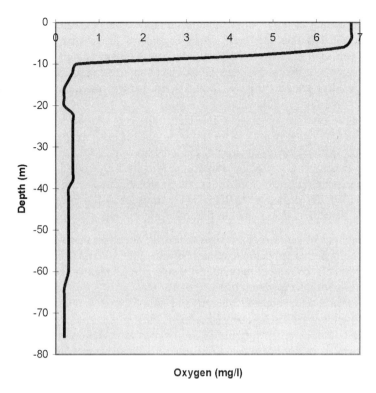

Figure 13.12. Oxygen content in Batang Ai (Sarawak, Malaysia) reservoir about a year after impoundment. A large part of the reservoir area had been covered with primary or secondary rain forest, which has not been removed prior to impoundment. The reservoir has developed a rather stable thermal stratification. Below 10 m, the water is practically anoxic.

Table 13.7. Biomass Recorded from Various Tropical Forests (t/ha Dry Weight)

	1	2	3	4	5	6	7	8	9	10	11
Leaves	10	8	7.3	9	8.2	8.4	8.9	8.6	3	7	9
Twigs, branches	101	93	99					96	78	55	98
Trunks	234	230	215					222.5	64	136	299
Lianas		23						21.9			
Roots	125	32	61	20.5		3.3	40		25	128	111
Fine litter	7.2			10.5		23	7.7	10.9		7	7
Coarse dead wood	25.8							25.8			
Total biomass	493	363	382.3	668	457	357.3	357.7	385.7	170	333	524

1 = Amazon rain forest. Estimated dry weight from original fresh weight figures: Walter and Breckle (1984, p. 8)
2 = Thailand, rain forest. Walter and Breckle (1984, p. 58)
3 = Cambodia, rain forest. Walter and Breckle (1984, p. 58)
4 = Malaya lowland evergreen rain forest. Whitmore (1985, p. 114)
5 = Malaya lowland rain forest. Whitmore (1985, p. 114)
6 = Malaya semievergreen rain forest. Whitmore (1985, p. 114)
7 = New Guinea lower montane rain forest. Whitmore (1985, p. 114)
8 = Amazon rain forest. Fearnside (1995)
9 = Zaire, Miombo forest. De Angelis et al. (1981)
10 = India, dry deciduous tropical forest. De Angelis et al. (1981)
11 = Brazil, rain forest. De Angelis et al. (1981; see there also for data on temperate zone forests)

When estimating the effect of this biomass on oxygen content in a future reservoir, it has to be considered that the "soft" biomass, mainly leaves, twigs, and fine litter, decomposes very quickly, whereas branches with a diameter of only a few centimeters take much longer, and tree trunks, especially when they are completely and permanently submerged, will last almost "forever." For this reason, it is the soft part that has to be considered and not the total biomass.

Assuming a "mass formula" of $C_6H_{12}O_6$ for the composition of biomass, 1.07 t of oxygen will be required to break down 1.0 t of biomass. If less oxygen is available, which is usually the case in deeper reservoirs with seasonal or overannual storage covering a high amount of biomass, anoxic conditions will result. The problem tends to be more serious in reservoirs in moist tropical than in dry or temperate regions (see, e.g., Leentvar, 1973; Biswas, 1973; Van der Heide, 1982; Biney, 1987; Simanov & Kantorek, 1990; Schetagne, 1994). This difference is due mainly to the following reasons:

- The amount of submerged biomass is usually higher in these cases.
- Due to the higher temperature of air and water, the decomposition process is much more rapid, according to the rule that the speed of a physiological process at least doubles with a temperature increase of 10 °C.
- Higher water temperature also means a lower total oxygen content in the water (O_2 saturation at 10 °C is 11.1 mg/L, at 30 °C 7.2 mg/L).
- Under certain conditions, tropical lakes or reservoirs can develop a stable thermal stratification in contrast to lakes in more seasonal climates, where a circulation once or twice a year is the rule. This stable stratification prevents oxygen from reaching the deeper, anoxic water layers.
- The higher temperatures in tropical lakes lead to a much higher biological productivity in the upper water layers (mainly planktonic algae), where there is light. This effect can be further increased by the high concentrations of plant nutrients in the water, stemming from the decomposing submerged soils and vegetation. This biological productivity adds to the biomass in the deeper parts of the lake and therefore to oxygen depletion.
- Floating aquatic weeds can form thick mats on the surface of a reservoir, which reduce the oxygen uptake of the water from the atmosphere and the production of oxygen in the water body itself. Here again, dead plant material contributes to the organic load in the water and to oxygen depletion.

Oxygen depletion and the formation of anoxic deep water layers can have the following adverse, undesired secondary effects: (1) formation of toxic substances, mainly hydrogen sulphide (H_2S) in the anoxic layer of the reservoir; (2) fish deaths in cases where anoxic water containing toxic substances comes to the surface (e.g., caused by wind or strong rainfalls); (3) because this water is very acid, with a low pH, it is aggressive to concrete and metals (i.e., turbines and waterways of the power plant); (4) release of such water may cause problems in the downstream area (bad odor, toxicity); (5) in the case of multipurpose reservoirs, this quality problem may make water unsuitable as drinking water or even for irrigation purposes.

Modeling. For the assessment of water-quality problems in reservoirs, computer models are being used to predict the evolution of water quality over time. The aim of these models is to detect potential problems and to obtain an idea on kind and magnitude of problems, in order to formulate adequate mitigation measures. Basically, such models are being used to predict the development of water quality in two separate and ecologically

Environmental Impacts of Dam Projects

very different, although functionally closely related, parts of the water body, namely, in the reservoir itself, and in the river downstream from the reservoir. As always in such cases, we have to deal again with events that will (or might) take place in the future, and for this reason are not accessible to direct observation. On the other hand, we deal with a very complex system of which we do not know all the relevant parameters.

For the Pangue project (upper Río Bio Bio in Chile), water-quality development in the future reservoir has been predicted with a model developed for dynamic processes related to stratification, water turnover, and nutrient cycles in natural lakes (Ulrich, 1991; EWE, 1995). As is always required in such cases, a number of modeling runs has been carried out to simulate a range of different conditions, the main variable here being the quantity of biomass left in the reservoir. One obvious shortcoming of this model for an application in such a case was the fact that it did not model the gradual filling of the reservoir; rather, it had to be assumed that on "day 0," the reservoir would already have reached its full supply level ("day 0" in Figure 13.13). In spite of this, the model provided results which seemed quite plausible; one example, limited to only a few results of a series resulting from one

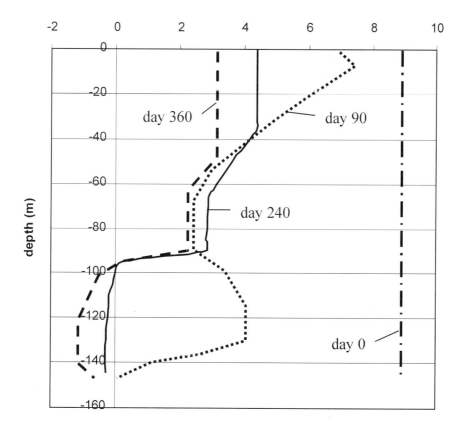

Figure 13.13. Results of model calculation of O_2 concentration in the water of Ralco reservoir (Río Bio Bio, Chile), days 0 to 360 after first filling of the reservoir. It is assumed that at first filling (day 0), O_2 concentration is high (near saturation) over the whole water column. The other curves show a progressive depletion of O_2, especially in the lower layers of the reservoir. Negative values for O_2 concentration indicate that O_2 has not only been depleted, but also that there is actually an O_2 deficit (i.e., a surplus of not yet degraded biomass). Source: EWE (1995).

specific condition, is provided in Figure 13.13. This shows that under the assumed conditions (main parameter: 37 t/ha of soft, easily degradable biomass left in the reservoir area), rather quickly (already 90 days after impoundment), oxygen concentration would drop to 0 at the bottom of the reservoir, and that after about 240 days, there would be an oxygen deficit. It is obvious that "negative" oxygen concentrations cannot really occur, because the content cannot possibly drop below zero. Nevertheless, the calculated negative values are important in indicating that, in this case, there is a complete oxygen depletion while there is still degradable biomass left. In such anoxic conditions, anaerobic biomass breakdown will take place, which can lead to the genesis of toxic substances. It has to be pointed out again that a model (or any "prediction" made in the framework of an EIA) does not have the aim of actually exactly predicting a specific situation at a specific moment in time, but that it is (or should be) used for indicating if, and under what conditions, there is probably going to be a problem, and if yes, what the expected severity of the problem will be. These results are then used to decide on the necessity and type of mitigation measures to be taken. One problem of computer modeling, however, is that people tend to "believe" the results, which often leads to discussions of the type: "Why does O_2 saturation reach 0% in a depth of 95 m at day 240 and then stays constant for the next 120 days?" It seems obvious that no model of this kind provides answers to this type of question.

Personally, I prefer a much simpler "model" for making a prediction of the expected situation in the future reservoir. As a matter of fact, the main question to be answered is as follows: "Will there be a water-quality problem, yes or no, and if yes, of what kind and approximately what magnitude?" and not "What will the situation in a depth of x m y days after first impoundment be?" For this reason, I usually use the calculation shown in Table 13.8 for getting this answer, or at least a first rough idea. It is based on the following assumptions and input data:

- Biomass in the reservoir area can be obtained in two different ways: (1) by actually determining the biomass for the main vegetation types present and multiplying it by the area covered by this type, or (2) by applying date from the literature to the vegetation types present. In both cases, a vegetation type map of the reservoir area is indispensable. Considering the difficulties of actually determining standing crop (=biomass; see, e.g., Whitmore, 1985), it is obvious that the most important source for errors probably lies here.
- Oxygen content in the inflow water can be obtained either from direct measurements (usually made in an EIA), or by deduction from water temperature. Although in a rather turbulent, not organically polluted river, oxygen content is usually at saturation, somewhat lower values should be applied to be on the safe side.
- Hydrological data (average river discharge, total annual inflow, and total reservoir volume) can be obtained from the general project information.
- The composition of the (soft, easily degradable) biomass is assumed to correspond to the overall general formula $C_6H_{12}O_6$, which is quite accurate. The breakdown would then follow the general formula for respiration:

$$C_6H_{12}O_6 + 6O_2 \rightarrow 6CO_2 + 6H_2O,$$

which in turn means that 1.07 t of oxygen are required for completely decomposing 1.0 t of biomass.

Table 13.8. Simple Calculation Form for Estimating O_2 Conditions in Future Reservoir, with Examples

Project River Country Parameter	Unit	Formula	Nam Ngum 2[4] Nam Ngum Lao PDR 100%	50%	20%	Urrá Sinú Colombia	Dachao-shan Lanchang (Meekong) China PDR	Ralco Bio Bio Chile
1 Biomass total (soft only)[1]	t	(2*3)	488,800	244,400	97,760	74,000	27,011	35,000
2 Reservoir area	ha		12,220	12,220	12,220	7,400	2,625	3,500
3 Biomass average (soft biomass only)	t/ha		40	20	8	10	10	10
4 Mean annual river discharge	m³/s		200	200	200	340	1,332	270
5 Water in reservoir, total volume	10⁶ m³		3,600	3,600	3,600	1,740	890	1,233
6 Annual inflow	10⁶ m³	(4)*31.5	6,300	6,300	6,300	10,710	42,000	8,505
7 Oxygen concentration in inflow water[2]	mg/l		8	8	8	8	8	10
8 Oxygen per m³	g/m³	(= 7)	8	8	8	8	8	10
9 Oxygen total at first filling	t	(8*5)	28,800	28,800	28,800	13,920	7,120	12,330
10 Oxygen total in annual inflow	t	(8*6)	50,400	50,400	50,400	85,680	336,000	85,050
11 Oxygen demand per t of biomass	t		1.07	1.07	1.07	1.07	1.07	1.07
12 Oxygen demand total	t	(1*11)	523,016	261,508	104,603	79,180	28,902	37,450
13 O_2 balance 1: O_2 (1st filling) −O_2 demand	t	(9−12)	−494,216	−232,708	−75,803	−65,260	−21,782	−25,120
14 % O_2 balance 1: O_2 demand in % of O_2 in 1st filling	%	(12*100/9)	1,816	908	363	569	406	304
15 O_2 balance 2: O_2 (annual inflow) −O_2 demand	t	(10−12)	−472,615	−211,108	−54,203	6,500	307,098	47,600
16 % O_2 balance 2: O_2 demand in % of O_2 in annual inflow	%	(12*100/10)	1,038	519	208	92	9	44
17 Oxygen required (1)[3]	t/ha	(3*11)	43	21	9	11	11	11
18 Oxygen required (2)[3]	t/ha	(12/2)	43	21	9	11	11	11

The shaded fields are those requiring project-specific input data.
[1] From the total biomass in the reservoir; only the soft, easily degradable part is taken into consideration.
[2] Obtained either from measurements or from an (estimated) water temperature; lower than saturation values chosen to be on the safe side.
[3] Control lines: the two values have to be identical.
[4] Three cases calculated for 100, 50, or 20% of original soft biomass remaining in reservoir at impoundment.
Results: Lines 13 to 16; explanation in the text.

With these values, it is possible to calculate the total amount of oxygen required for breaking down the biomass left to be submerged in the reservoir and to put it in relation to the oxygen available in the water. The results are then the following:

1. Oxygen balance 1 (lines 13 and 14 in Table 13.8): If the result here is positive, then this would mean that the oxygen available from first filling is sufficient for breaking down the whole biomass in the reservoir and most probably there will not be a water quality problem.
2. Oxygen balance 2 (lines 15 and 16 in Table 13.8): If the result here is positive, then this would mean that the inflow of oxygen to the reservoir is high enough to break down the biomass. In addition, in reservoirs with a high turnover rate (i.e., total reservoir volume much smaller than annual inflow), there is a high flushing effect that would contribute to solving the problem, if there should be any in the beginning.

The important point is that the interpretation of the results is being made the other way round: If the results show highly negative values, this means that most probably there will be a water-quality problem, and that measures have to be taken to reduce this risk.

It is obvious that this "model" does not take into account a number of parameters that also are of importance for the water-quality problem: (1) reservoir morphology (a deep reservoir will show conditions other than a shallow one); (2) oxygen exchange at the surface, with the atmosphere (depends on reservoir morphology, temperature, wind and wave action, etc.); (3) biological activity within the reservoir (oxygen production in the upper layers, additional biomass to be decomposed; depends on temperature, nutrient content, etc.); and (4) exchange dynamics within the reservoir (stratification, internal circulation, insertion of inflowing water into the water column, etc.). All these parameters play a role and can influence water quality positively or negatively. Dynamic water-quality models do take into account such effects. However, their application is justified only in cases in which the corresponding data are available, and this, in my experience, is very often not the case.

I want to illustrate this with a practical example. For the planned and already constructed Urrá hydropower plant in Colombia (Figure 13.14), a water-quality model had been applied to predict effects of the impoundment on the downstream water quality in the Sinú river. The results of the modeling showed an almost total oxygen depletion in the reservoir, almost oxygen-free water at the power station outlet, and a very slow recovery (i.e., reoxygenation) in the almost 120 river km down to its estuary. This part of the river flows through a densely populated area, and the water is used for irrigation and partially also for drinking water. Furthermore, there is an important inland fishery exploiting fish that show a complex seasonal migration and reproduction pattern involving the river and two large seasonal swamps formed by it. Water quality such as that shown by the model would probably have eradicated or at least heavily impacted these fish populations, and would have had negative impacts on other forms of water use. For this reason, the Colombian Environmental Authority hesitated to give the operation permit for Urrá. As an additional mitigation measure, the implementation of a massive artificial breeding program was requested for the fish species in question; however, because the reproduction biology of these fish is not yet completely understood, such a program, which involved catching the larger part of the fish stock present, could just as well have proved detrimental for these species. From the data I received, I got the impression that in this specific case, there would

Environmental Impacts of Dam Projects

Figure 13.14. Urrá hydropower plant, Sinú River, Colombia, before impoundment. Power house at the left, bottom outlet (still open) at the foot of the main dam (center); high turbulence in water flowing out leads to quick reoxygenation. Reservoir area in the background.

not be a serious water-quality problem, because no severe oxygen depletion was to be expected in the reservoir (see column "Urrá" in Table 13.8). Inspection of the site and a careful analysis of existing data (water-quality measures had been carried out during the whole construction period) provided the following results:

- During construction, a part of the reservoir had seasonally already been filled due to the backwater effect of the dam and the waterways. This had temporarily covered up to 25% of the reservoir area, eliminating the soft biomass from this area (and, of course, consuming the corresponding amount of oxygen).
- This oxygen-depleted water had been released continuously through the bottom outlet of the dam, which had been kept open the whole time.
- In spite of this, water-quality controls during the whole period in the reservoir (near the dam site) and in the river have shown O_2 concentrations that were near saturation. Even if a reduction in oxygen in the reservoir had actually taken place, this was more than compensated for through the reoxygenation due to the high turbulence of the water at the outlet. As a matter of fact, O_2 concentrations were consistently a little bit higher in the river than in the reservoir (see Figures 13.15 and 13.16).

Based on these results, the operation permit was granted. The ongoing monitoring program (which was part of the normal procedure that had also included an EIA) revealed that, in fact, there was no water-quality problem.

I do not want to say that the model applied was "wrong," and even less that such a modeling is useless. However, the results were obviously wrong in this case. This clearly shows three points that I think should always be taken into account when such models are applied:

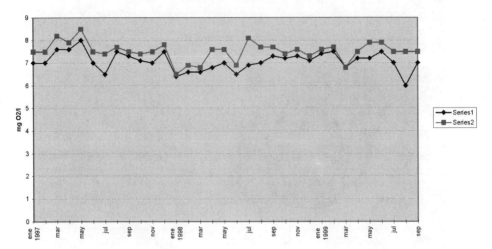

Figure 13.15. Urrá hydropower plant: Water quality expressed as oxygen content during the construction phase (January 1997 to September 1999). Series 1 is upstream from the dam; series 2 is downstream (about 1 km) from the dam. All values are near saturation, and there is no marked deterioration in periods when the water level in the "prereservoir" was high and a large amount of biomass was therefore submerged (see Figures 13.14 and 13.16).

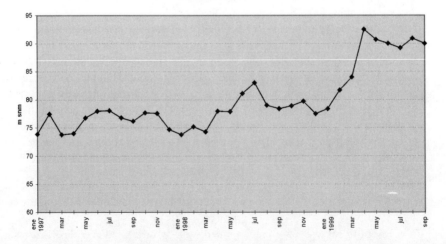

Figure 13.16. Urrá hydropower plant, Columbia: Water level within the future reservoir during the construction period. The retention effect of the dam led to a preimpoundment submersion of up to 25% of the future reservoir area already during construction (Zwahlen, 1999).

1. Models meant to make predictions on a highly complex issue, such as the development of water quality in a future reservoir, need a very accurate and site-specific database as input; now, very often, the data available are not sufficient.
2. If the results of a model do not seem plausible in light of experiences with similar cases—or just according to common sense—there is a high probability that they are actually wrong! In such a case, a thorough discussion of the situation, including probable explanations for the results and expected probable "real" outcome, might be more helpful than the model.
3. All available "real-world" data should be used, because they are always more reliable than modelling results.

Mitigation Measures. If water-quality problems caused by reservoir impoundment are likely to occur, adequate measures have to be formulated (and, of course, implemented!). One of the most important measures in the case of forest-covered reservoir areas is preimpoundment reservoir clearing. This consists in removing as much as possible of the biomass prior to impoundment, with the aim of reducing the amount left, if possible, to "below risk values."

There are at least three good reasons for clearing a reservoir prior to its filling:

1. To reduce the biomass left to decompose in the water, with the aim of creating conditions for good water quality in the reservoir.
2. To make use of the timber, which is a resource otherwise lost.
3. To remove trees that, if left standing in the reservoir, would be a hazard for any boat traffic on the future lake and an obstacle to fishing.

On the other hand, reservoir clearing can cause considerable costs, especially in the case of large, densely forested and not easily accessible reservoirs, and even more so if the trees are not very interesting commercially. For this reason, reservoir clearing cannot be considered as a measure to be taken in any case; costs and benefits have to be taken into account carefully, and the decision has to be made based on this (i.e., considering the risk potentially avoided, the benefits from taking out the timber, and the costs involved). If such a measure is actually being taken, it has to be implemented carefully (and it should be monitored), because there is always a risk that it will not produce the expected results. Especially in the case of reservoirs in moist, tropical climates, if forest is cleared, there is a very quick regrowth of all kinds of bushes, young trees, and herbs. Productivity in such conditions can be 10–40 t/ha annually, and all this new regrowth is soft, easily degradable biomass. In this way, forest clearing could actually increase the risk for bad water quality in a new reservoir. An ideal way to carry out the clearing, therefore, would be to clear-cut the area during the dry season (where such a season exists!) to take out useable timber and other wood, and to burn the remaining material before the onset of the rainy season; this, of course, would have to be done in the dry season preceding the reservoir impoundment. Leaving such cleared areas for another rainy season would invariably lead to the regrowth mentioned earlier.

One other problem is that, with regard to water quality, it is of no use to take out only the timber and to leave the remaining biomass in the reservoir area, because this is the part that creates the main problem. Therefore, it will be required to cut even small trees and bushes to let this material dry and then burn it shortly before impoundment. There is usually no other practical way to eliminate such large amounts of biomass from a reservoir. Burning, however, poses the risk of fire spreading beyond the future high-water level, and this has to be prevented in any case. But especially when reservoir clearing is being done by private timber companies, their commitment to clearing and eliminating this part of the vegetation is very limited, because only the larger trunks are interesting from a commercial point of view. For this reason, clear arrangements and a close control are required.

Two examples follow:

- *Nangbéto*: Dam located near Atakpamé, on the river Mono, in central Togo. A preimpoundment study had revealed a potentially severe water-quality problem and a high necessity for reducing the amount of biomass in the future reservoir (EWE & SOGREAH, 1984). The following main reasons were given for this: (1) high probability of water-quality problems due to decomposing biomass; (2) necessity to use the resource (possibility for producing charcoal, because there was already a fuel wood shortage, at least regionally); and (3) dead trees remaining in the water would be an obstacle to fisheries, considered an important additional benefit of the reservoir. Main problems for implementing this plan were the costs involved, the high cost for transporting charcoal, and the fact that the major part of biomass did not stem from forests but from high grass savannas, which are difficult to remove. The dam was commissioned in 1987. Unfortunately, no follow-up study has been made.
- *Houay Ho*: Dam located on the small river of the same name in the Xe San basin, southern Laos. Here again, the EIA (EWE, 1996) had shown the necessity for a preimpoundment clearing. In this case, the fact that there was a high amount of valuable timber (mainly *Pinus kesiya*) made the clearing interesting from a commercial point of view. Two timber companies (one from Thailand, one from Viet Nam) were commissioned with this task. Reservoir clearing was carried out during the construction period of the dam; the contracts with the firms clearly stated that they also had to remove—by means of cutting and controlled burning—the nontimber biomass. During the implementation, there were discussions of whether or not logging of trees had also taken place at sites above the planned full supply level. A site visit in February 2001 did not reveal any evidence for this; on the other hand, the high amount of small trees still standing in the drawdown area (see Figure 13.17) was at least an indication of a possible shortcoming in the elimination of the soft biomass; this could have been the reason for the water-quality problems reported for the downstream area (personal communication, Mr Gary Oughton, February, 2001).

Another possibility to deal with this problem is to build an intake structure that allows abstraction of water from different levels within the reservoir. Usually, there is one intake at a level several meters below the minimum operation level of the reservoir (depending on reservoir morphology, hydrology, and reservoir operation), and this can be at a very considerable depth relative to the full supply level. In such a case, if a water-quality problem exists, water would be taken from deep, oxygen-free layers (possibly containing H_2S, which is not only toxic but also, due to its acidity, highly corrosive, and can severely damage concrete waterways and the metallic structures such as the turbines). A multiple-story intake would allow regulation of the water intake in a way that water from a rather superficial level, having an acceptable quality, would be used. For example, this solution has been proposed for the (not yet implemented) Nam Ngum 2 project (EWE, 1998b; see also Table 13.8).

Figure 13.17. Drawdown area of Houay Ho Reservoir, Lao PDR, three years after impoundment. The trees have been removed during the preimpoundment clearing, but from the high number of dead bushes present it can be concluded that the amount of soft biomass submerged was rather high.

Other measures sometimes recommended, such as active oxygenation of a reservoir by pumping air into deeper layers, are normally not feasible, especially not in large reservoirs, because this requires a complicated, expensive structure and a high amount of energy.

Finally, there is the possibility of doing nothing at all. As a matter of fact, the problem will resolve itself over the years, once the biomass has been eliminated from the reservoir through the biological decomposition process. Of course, this approach will not be an acceptable solution in cases where the conditions will lead to a long transition period, with very bad water quality having adverse effects downstream from the dam (e.g., in cases where there are settlements using the river as source of drinking water, or if there are important fishing or fish breeding grounds that might be affected). Such a condition is the rule rather than the exception. However, the development of the reservoir can lead to water-quality problems similar to those described here for the impoundment period (see section on eutrophication).

Reservoir Development

Once the reservoir has been filled and normal operation has started, the situation is likely to stabilize. However, this does not mean that the reservoir from then on will be a stable entity in the sense that it will not undergo any changes over a long period. On the contrary, a reservoir is a dynamic system that will evolve and change over time. This evolution has been illustrated by Balon (1974) in his classical graph on the probable evolution of Lake Kariba (see Figure 13.18). To the best of my knowledge, no better (long-term) study is available that actually followed the evolution of a reservoir from the preimpoundment situation over the period of first impoundment, then onwards, until such

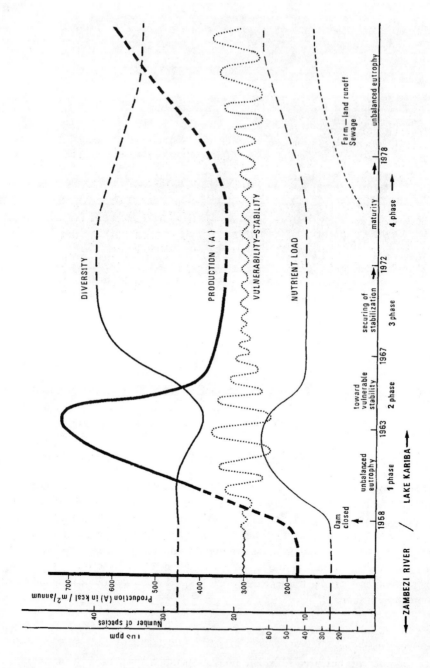

Figure 13.18. Development of Lake Kariba after closing the dam on the Zambezi River in 1958. The first dramatic rise in productivity was followed by a sharp decline. The development beyond 1972 is hypothetical (Balon, 1974, p. 137).

a period when actual stabilization of the reservoir would be reached. Such studies, under different conditions, would be very useful and are urgently needed.

There are mainly two effects that greatly influence the reservoir: siltation (or sedimentation) and eutrophication.

Siltation. Every river carries a certain sediment load, depending on its capacity for transporting such material. In a fast-flowing mountain stream, there will be a high sediment load with generally coarser material (stones, gravel), whereas farther downstream, with a more gentle gradient and consequently a slower flow, only finer material will still be transported (sand, silt). On its way, and depending on the flow regimen, a river will therefore erode, transport, and accumulate riverbed material. Floods play an important role in this process. A reservoir changes this dynamic completely. Because the water is stagnant in a reservoir, as in a natural lake, the sediment will be deposited there. By this process, the reservoir will gradually be filled. As long as only the deeper layers of the reservoir are being filled (the so-called dead storage, i.e., the water below the intake level), this does not affect reservoir operation. However, when live storage is starting to silt up, the useful reservoir storage is being reduced, which diminishes the capacity for energy production (or for storage of irrigation water, as the case may be). Ultimately, the reservoir will be completely filled with sediment and will then have lost its entire storage capacity. In such a case, it could still be used as an ROR plant (the total amount of water would still be the same, although only available according to the natural flow regimen, and the total head could still be used). Nevertheless, the main purpose, that of storage, would be lost.

Siltation is a natural process and also occurs in natural lakes, which undergo the same development. The rapidity depends on the erosion processes in the catchment area and the carrying strength of the inflowing rivers. In many cases, however, erosion is influenced (and in most cases, greatly enhanced) by human activities, for example, deforestations in the catchment area, which can lead to a very massive increase in sediment load. Such a development can actually reduce the useful life span of a reservoir very considerably.

In most cases, there is not much that can be done about siltation. In some cases, and when the material is fine enough, it can be flushed through a bottom outlet (see Box 13.8). In other cases, it is possible to exploit the sediment, because it consists of valuable sand and gravel. In most cases, however, the material does not present any value, the quantities are far too large to be eliminated, and transport distances are too long, leading to prohibitive costs. The only feasible way is to implement a careful watershed management program that aims at reducing erosion and in this way the sediment load.

Box 13.8. Sediment Flushing at Mauvoisin HEP

Mauvoisin HEP, in the high mountains of southern Switzerland, has been completed since 1958. It is alimented mainly with water from melting snow and glaciers; a high percentage of its catchment area of 167 km^2 is covered with glaciers. Sediment reaching the reservoir is mainly very fine silt stemming from rock abrasions by the glaciers. Over the years, this material accumulates in the reservoir, and if nothing is done, it could block the bottom outlet, reducing the capacity for flood evacuation. In order to keep this important safety measure intact, two things are being done:

- Every few years, an artificial flood is created in the river downstream of the dam, by releasing a high quantity of water from the dam. This measure aims

at keeping the river (which normally has a much reduced flow) open and capable of evacuating a flood peak.
- About once every 25 years the reservoir is being drawn down completely, and at this moment, large quantities of the fine sediment are evacuated through the bottom outlet. This aims mainly to keep the ground release functional and to prevent its blocking by sediments. At the same time, the lower part of the dam is inspected.

This release regularly leads to a "catastrophe" in the Dranse, the affected river, below the dam. The high quantity of very fine silt being transported during a very short period leads to congestions of all pores in the normal river sediments, thus killing all organisms living there. Furthermore, the concentration of silt is so high and so fine that it leads to congestions of the gills of fish and other water organisms, which then suffocate. In this way, substantial damage is caused to fish stocks.

Reservoir Eutrophication and Contamination During Operation. The postimpoundment phenomena are specific to reservoirs and do not occur in natural lakes. Of course, natural damming of rivers can occur, such as the prehistoric volcanic eruptions that led to forming the Lago Todos los Santos in southern Chile (Villagrán, 1980, p. 15; EWE, 1984), or the massive landslide in the Pamir mountains of Tajikistan in 1911, which, by creating a natural dam about 700 m high (by far the highest "dam" in the world) led to the formation of Lake Sarez (Alford et al., 2000). But these events are too rare to be of any practical importance.

However, once a reservoir has formed and reached its state of stability, the further dynamic is very much the same as that of a natural lake. One difference can be the fact that through a deep water intake, a drawdown is possible and occurs regularly, which leads to lake-level fluctuations that are much larger than normally possible in a natural lake. Furthermore, dams often have a bottom outlet that functions for sediment flushing, something that obviously does not occur in a natural lake. Nevertheless, older reservoirs can be considered as lakes, and the experience gained from lakes can be applied to reservoirs as well (see Box 13.9).

During the operation phase, water-quality problems in reservoirs are caused primarily by the inflow of organic material and nutrients (eutrophication, mainly from domestic wastewater and runoff from agricultural areas) and/or toxic substances (mainly from industries and agriculture), both groups of pollutants stemming from the watershed area of the reservoir. Activities in the reservoir itself, such as aquaculture (fish cage farming) can contribute to the problem, because feeding the fish directly adds to the organic load. Eutrophication can make the water unsuitable for recreational purposes or as drinking water (hygienic and toxicity problems). Eutrophication processes and measures to be taken have been dealt with extensively by Ryding and Rast (1989). In the following paragraphs, only a few specific issues arising in this context are addressed very briefly.

Box 13.9. Case Study of Saguling and Cirata Reservoirs, Indonesia

Three dams forming a cascading system have been built on the Citarum River in West Java, Indonesia: Saguling (1986), Cirata (1988), and Jatiluhur (1964). Whereas the two upper reservoirs in the cascade, Saguling and Cirata, serve mainly for energy production, the lowermost, Jatiluhur, is a multipurpose project used mainly for irrigation and drinking water supply for Jakarta.

Mainly in Saguling, but also in Cirata, severe water-quality problems arose and gradually worsened over time. One main symptom was the periodical, mass fish deaths recorded for the lakes, affecting aquaculture (floating cage fish breeding), presumably caused by upwelling of bad quality (anoxic and even toxic) water from the deeper layers of the reservoir. In a World Bank–financed study aimed at identifying the causes and finding solutions for these problems, the following main problems were identified:

- The whole catchment area is densely populated, the largest settlement being Bandung, with about 4 million inhabitants. All domestic sewage reaches Saguling reservoir without any treatment.
- The catchment is intensively used for agriculture. Like the sewage, runoff from agricultural areas reaches the reservoirs immediately, resulting in a high input of agrochemicals.
- Sewage from industries, mainly in Bandung (metal and textile industry), also reaches Saguling reservoir without any treatment.
- Cage fish breeding had been identified as one possibility for compensating people who lost their land due to the impoundment of the two upper reservoirs, and it was a part of the resettlement package. Originally, a carrying capacity for about 6,000 fish cages had been identified, and these had been allocated to project-affected people. Over the years, however, this rather lucrative business had largely been taken over by others, and the number of fish cages had risen to over 20,000 in the mid-1990s. Because the fish in these cages have to be fed, this leads to an additional very massive input of organic matter into the reservoirs.

Interestingly, water quality is worst in Saguling (the topmost reservoir in the cascade) and best in Jatiluhur (the lowermost; Whitten et al., 1996, p. 457). It can be said that Saguling (which directly receives most of the pollutants) serves as a sewage treatment pond for the two other reservoirs.

Main recommendations for improving the water-quality situation were the following:

- Sewage treatment plants, especially for industries and major settlements.
- Improved agricultural practices in the catchment area to reduce runoff and agrochemical input into the lakes.
- Reforestations, mainly in the steeper parts of the catchment area, to reduce surface runoff and erosion.
- Better control of the number and better management of fish cages.

All this, finally, would lead to a comprehensive watershed management, which is ultimately the only solution or (in the case of new reservoirs) prevention of this type of problems (see section on watershed management for an additional discussion of this issue) (EWE & AMYTHAS, 1998).

- *Organic material*: The main source of organic material input into a lake or a reservoir is untreated domestic sewage. High biological oxygen demand (BOD), that is, high concentrations of organic substances in the water, leads to high oxygen consumption. This can have the same effect as described earlier, namely, depletion of oxygen in the deeper water layers of the lake, below the thermocline, where anoxic conditions result.
- *Nutrients*: Nutrients are inorganic substances that are essential for primary production, that is, for plant growth (mainly nitrates and phosphates). Main sources of these nutrients are domestic sewage and runoff from agricultural areas. These nutrients serve as fertilizers for plant (mainly algae) growth. In the upper layer of the lake or reservoir, where light penetrates, very high plant productivity is possible in the presence of high nutrient concentrations, which can even lead to an oversaturation with oxygen. Dead organisms sink to the bottom of the lake, where their bacterial breakdown adds to oxygen depletion; in a highly eutrophicated lake, this can lead to anoxic conditions.

    Here, the principle of the "minimum factor" is important: Plant growth is limited by the nutrient available in lowest concentrations. In temperate lakes, this is normally phosphate, but it can be nitrate (i.e., nitrogen in a form that makes it available for plants) in other conditions. This effect has been observed very clearly in Swiss lakes. Due to the massive inflow of untreated domestic sewage, almost all Swiss lakes exhibited very bad water quality in the late 1950s and early 1960s. In spite of massive investments in sewage treatment, water quality improvement was not as expected. Although BOD input dropped drastically due to mechanical and biological sewage treatment, nutrient input remained high and even continued to rise. This led to a massive in-lake biological productivity, with frequent algal blooms, and oxygen depletion continued to be a problem. The situation improved drastically after a further step was added to sewage treatment, namely, the elimination of phosphates, together with prohibiting the use of phosphates in detergents, which had become the major source of phosphate input into Swiss waters. Today, most larger Swiss lakes have returned to an almost oligotrophic state, with clear, oxygen-rich water that can again be used for bathing, swimming, and even for drinking water supply.
- *Aquatic weeds*: A specific and often very serious problem of tropical reservoirs is the mass development of aquatic weeds. Floating plants, mainly *Eichhornia crassipes*, *Salvinia molesta* (Figure 13.19), and a few others, can form thick mats that cover the surface of the water body completely (Bechara & Andreani, 1989). By shading it out, and by yet another massive input of organic material, they add to oxygen depletion once again. Besides, they block waterways. On the other hand, some of these floating plants can be used, such as *Echinochloa staginata*, a floating grass covering large parts of Lake Kainji, in Nigeria, which can be harvested and used as fodder for cattle during the dry season (Morton & Obot, 1984; Obot & Ayeni, 1987).

    There is no easy way to get rid of these plants. Some promising experiments have been made with the beetle *Cyrtobagus*, a natural enemy of *Salvinia* (Room & Thomas, 1985; Thomas & Room, 1986; Julin et al., 1987). Limiting nutrient input into the lakes might help to reduce the growth of aquatic weeds as well.
- *Health*: Domestic sewage input into water bodies also means input of germs. High concentrations of coliform bacteria are regularly found in water contaminated with

Figure 13.19. Two of the most important aquatic weeds, *Eichhornia crassipes* and *Salvinia molesta*, which can form dense floating mats on reservoirs.

human waste, and germs causing salmonellosis and cholera can seriously threaten the health of the human population depending on contaminated water as source of drinking water. Furthermore, reservoirs can be breeding grounds for disease vectors and intermediate hosts, mainly malaria and schistosomiasis.
- *Toxic substances*: Toxic substances in the water of lakes and reservoirs stem mainly from industrial or mining activities (heavy metals) and from agriculture (pesticides) in the catchment of the reservoir. Such substances can become a threat to human health when they are accumulated in the food chain, as has been experienced in various instances with mercury (see, e.g., Isaacson & Jensen, 1992, Vol. 2, p. 31, 153ff). In the Cirata reservoir, Java, Indonesia, mercury concentrations of 0.16 µg/L have been recorded from the water, whereas mercury content in fish tissue ranged between 0.05 and 0.5 mg/kg; this corresponds to a 300- to 3,000-fold concentration of mercury in fish tissue.

Solving the Eutrophication Problem. It is evident that these problems are not caused by the reservoir or the hydropower development, but are a consequence of all the activities, including settlements, industries, and use of land and water within the catchment area. And it is equally evident that the solution to the problems as well lies mainly in the catchment area, and to a much lesser extent in the reservoir itself.

The main solutions to these problems are known and can be considered as well-established technology: sewage treatment plants for domestic sewage, which include a mechanical (elimination of waste), a biological (breakdown of organic compounds, i.e., reduction of biological oxygen demand [BOD] and chemical oxygen demand [COD]), and a chemical (elimination of phosphates and possibly nitrates) part, and special waste-water treatment facilities for the treatment of industrial waste. The major difficulty in the application of these technologies is the massive costs involved; especially for the domestic sewage treatment, not only treatment plants are required but also, and often mainly, efficient sewerage and drainage systems.

Even more difficult to control are the inflows from nonpoint sources, mainly agriculture. Fertilizers and pesticides are applied on enormous surfaces, and runoff and seepage water from these areas are contaminated. Here, efficient technologies have to be applied that, most of all, prevent the application of too-high doses of these products. This, again, requires appropriate training of the farmers. In this context, it is certainly worth mentioning that a general rise in environmental public awareness is urgently needed in many places.

The Lithosphere

The lithosphere is the third component of the abiotic environment. Three aspects are dealt with briefly here.

Geology

In the planning of a high dam, geology is important mainly for the following reasons:

- *Dam foundation*: For the type of dam to be chosen, and for its stability, the geology of the dam site has to be thoroughly investigated.
- *Reservoir area*: Geological conditions are important mainly for knowing whether the reservoir will be sufficiently impermeable, or whether there is the risk of a major loss of water (e.g., through tectonic faults).
- *Waterways*: Especially if waterways (headrace and pressure shafts) are underground structures, as is usually the case in high-pressure plants in mountainous areas, the geology of the rocks in which these tunnels will be constructed is important to know.
- *Construction material*: All high dams, but especially earth and rockfill dams, require a large amount of material, which has to be found through geological investigations. For economic reasons, this material should be found as close as possible to the construction site.

These geological investigations are usually not considered part of the EIA; they form a part of the "technical" project planning. The EIA will use the information obtained from these investigations to provide a short description of the geological situation. The project itself will not change the geology of the site, and for this reason, measures will not have to be taken.

There is, however, at least one point with which the EIA has to deal: the question of material moved during construction. There are two kinds:

- *Construction material*: Used, for example, for dam construction, either as concrete aggregates or as filling material in the case of an earth or rockfill dam. The main concern here from an environmental point of view is the place of origin of this material (excavation pits, quarries). If possible, this material should be taken from within the reservoir area, because this considerably reduces the impact of the project on the landscape. If this is not possible (mainly because of lack of availability or accessibility of suitable material), measures will have to be taken

to rehabilitate these sites after the end of construction. Many examples from dams around the world show that no such measures have been taken.
- *Excavation material*: For the dam foundations and for waterways in the form of tunnels, massive excavations need to be made. Very often, it is not possible to use this material, either because it is not suitable, or there is no demand for it, or because transport costs would be prohibitive. In such cases, it has to be deposited, which, again, can have very adverse impacts on the landscape. Landscaping measures have to be sought and implemented for integrating such deposits into the landscape.

Reservoir-Induced Seismicity

Dams have to be built in such a way that they can withstand an earthquake in the project area. For assessing this risk, an analysis of the site-specific seismic situation has to be made.

A part of this analysis is the question of reservoir-induced seismicity. A high dam and a large reservoir create an additional considerable mass that has to be carried by the local earth's crust. If this is not stable, there is a certain risk of additional seismic activity induced by the project, especially during the impoundment phase of the reservoir.

Soils

Soils are lost by impoundment of the reservoir. This, however, is mainly of concern in relation to prevailing land use (i.e., by the population living in the project area). This, therefore, has to be taken into account in the framework of a resettlement and compensation plan (see section on resettlement and land use).

The Biotic Environment

Terrestrial Vegetation, Fauna, and Habitats

Aims and Scope

Strictly speaking, impact of hydropower plants on terrestrial habitats could be ignored in the framework of a book on applied aquatic ecology. However, the interactions of water, aquatic habitats, and their surroundings are numerous, and a line is not easy to draw. In addition, in an attempt to provide at least a sketch of the "complete picture," terrestrial biota cannot be left out altogether. Therefore, these aspects are briefly included here.

The potential loss of valuable habitats from a biological or biodiversity point of view, caused by a hydroelectric project, has to be addressed in the EIA. The importance of this point will depend largely on the site-specific conditions. Three examples can illustrate this point:

- *Nangbéto, Togo*: The reservoir area was rather densely populated, and natural habitats had been very much reduced due to human activities, mainly slash-and-burn cultivation. Since only small rests of natural habitat, and therefore a very restricted associated fauna, lived in this area, this point was of rather marginal importance in comparison to human use aspects (EWE & (SOGREA), 1984).

- *Egiin Gol, Mongolia*: The reservoir area was inhabited mainly by seminomadic herders and used as pasture for their cattle. However, use intensity in this case was well below carrying capacity, so that it certainly influenced the local vegetation, without, however, damaging it. In addition, the area contained a considerable amount of valuable habitats, mainly highly dynamic river floodplains. In this case, vegetation and (to a somewhat lesser extent) fauna were considered to be of high value, together with aspects of human use of the area (EWE, 1994a).
- *Ayago Nile, Uganda*: This project was located in the center of Kabalega Falls National Park, the largest national park in Uganda. There was no human population and no direct utilization of vegetation and fauna (at the time of the feasibility study, tourism was completely down in Uganda, and the only "use" of the fauna was poaching). The park provided habitats for a rich variety of wildlife, among which were a number of endangered species. Therefore, in this case, impact on vegetation and fauna was considered the most important of all potential impacts (EWE, 1980).

As is the case for a number of other potential impacts, the one caused by the reservoir is usually the most important when dealing with biodiversity and habitats, mainly because this affects the largest area. Nevertheless, potential impacts of other project components have to be considered as well, for example, dam, burrow, and dumping areas for construction and excavation material; and construction site infrastructure (workers' camp, temporary material deposits, etc.). In addition, a number of potential indirect impacts have to be considered:

- Access roads can provide easy access to large areas that were previously not accessible; this can also include areas in the catchment of the dam. Access can trigger all kinds of activities (immigration of colonizers, logging, poaching, etc.). Such effects have to be taken into consideration, and measures have to be taken if there is a serious threat to valuable habitats or species.
- The presence of a high number of workers, often coming from other areas, during the construction phase can lead to heavy poaching during this time in areas where interesting wildlife is available. Here again, measures may be required.
- One effect that is often neglected is the impact on floodplain systems in the downstream area. If a very superficial approach is chosen, it can be concluded that, because the same total amount of water will still flow in the river below the powerhouse, there will be no change. However, floodplain habitats are not mainly regulated by the amount of water but rather by its seasonal distribution, which consists of (rather predictable) seasonal changes between high and low flow, but includes (unpredictable) flood events. Both are important for maintaining floodplain dynamics, and both are usually changed considerably with the construction of dams in the upstream area (Dister, 1984; Hughes, 1990). As has already been observed, this effect can be most serious in the case of transbasin projects that affect at least two rivers in a very different way.

Baseline Data

Three topics are usually distinguished in EIAs when dealing with the biological aspects: fauna, flora, and vegetation. *Fauna* deals with the animals, and especially with identifying animal species present—or potentially living—in the project area. *Flora* does the same with plants, with identification of plant species living in the project area or parts thereof, and establishment of species lists. *Vegetation*, however, does not deal with individual species but rather with typical assemblies of plant species forming the habitat for other (animal) species.

When dealing with flora and fauna, narrow limits are set, especially in tropical areas and mainly with regard to animals but often also with plant species, given the sheer number of species present. For various reasons, it is never possible to make a complete inventory. First, the time required to carry out such an investigation is never available in an EIA. Second, if such a study were actually carried out, it would produce a wealth of information that, although very interesting from a scientific point of view, would be almost useless within the framework of the EIA. Why this is so is briefly discussed in Box 10.

Box 13.10. Plant Species in an EIA

In an EIA for a limestone quarry in Switzerland, an inventory of the vegetation at the prospective quarry site had to be made. The main interest of the botanist carrying out the fieldwork for this aspect was mosses, so special attention was paid to this (otherwise usually ignored or neglected) group. The investigation revealed the presence of two species of mosses that previously had been found only once in Switzerland.

This discovery, which might have had some importance from a scientific point of view, was of no use for the EIA in question: Because nobody had been looking for these mosses in Switzerland before, it was not known whether this now meant the discovery of rare species, or whether they were widespread and rather common. Answering this question may be important from a floristic point of view and also for defining adequate protection measures in case these actually were rare species, but this cannot be done within the framework of an EIA.

Because, therefore, a complete inventory is not feasible, the adequate methodology and depth of investigation has to be found. Most important is the accumulation of already existing knowledge of a specific area (from publications and other sources). Discussions with local people can provide much information (e.g., on game animals and useful plant species). Normally, some specific groups of plants and animals are chosen as indicators because their presence or absence from a specific area provides information on its status and potential as a habitat for other species. Fieldwork, at least to some extent, is always required to gain direct, firsthand knowledge of the area in question, but it will have to concentrate on these key or indicator groups.

A vegetation map, often derived from aerial photographs or satellite imagery and then verified on the ground, is an important and useful basis for decisions on measures. Such a map will usually be made for the core areas of the study (e.g., for the reservoir area). It is, among other things, a basis for estimating the biomass present in the reservoir area, it serves to identify the presence or lack of valuable habitats, and it is at least a basis, but not necessarily sufficiently detailed, for identifying land use types and the amount of land presently used. In this way, it is a good starting point for identifying land requirements for compensation and resettlement programs.

One point will usually be investigated: the presence or absence of rare, endangered and protected species, and habitats suitable for them. There are a number of reasons why these species get special attention. First, there are "red lists," and so on, mainly those prepared by IUCN, but also some by other organizations (often country-specific); these lists make the species in question rather easily identifiable, and their limited number makes them an entity that can be handled. Second, in spite of being rare (and some because they are rare), their biology is usually rather well known. Many have specific habitat requirements, one (main) reason why they are rare. This makes them good indicator species. If elephants, tigers, orangutans, or Nile crocodiles are found, it can be assumed that the habitat in question is in a good-enough state to allow the survival of a large number of other, less "prominent" species as well. Third, because they are so prominent, they get a great deal of attention, and if a project threatens a large population of such a prominent species this can lead to considerable pressure against the project (see Box 13.11). And finally, in many cases, these species are legally protected, and as has been stressed, an EIA serves, first of all, to verify whether a project is in compliance with the applicable legislation.

Box 13.11. The Case of Kabalega Falls HEP

In 1968, a plan was set forth to construct a 600 MW hydropower plant utilizing the head of the Kabalega Falls (formerly called Murchison Falls), located on the Victoria Nile about 35 km upstream of Lake Albert, in western Uganda. Because the Kabalega Falls represented the main attraction of the Kabalega (Murchison) Falls National Park, the largest of its kind in Uganda, and the power plant site was in the very center of the park, the project received very strong opposition, mainly from conservationist organizations all over the world. The park was one of the main tourist attractions of the country and was therefore a very important foreign exchange earner and the habitat of, among other species, elephants, the two African rhinoceros species, and the Nile crocodile; of the latter, it probably contained the largest remaining population on the whole continent. Finally, the project was abandoned. It might have been the first major hydropower project in the world that was dropped (among other reasons) for environmental causes.

A sequel to this project, Ayago Nile, some 30 km upstream from the Kabalega Falls, was studied in 1978–1980. It would not have had any impact on the falls, but its adverse effects on the National Park would have been just as important, mainly due to the fact that the construction of two access roads, about 40 km each, cutting through the park would have been required. Like its predecessor, this project has not been implemented either (Anonymous, 1968, 1970; Katete, 1968; Cott, 1968; Seers et al., 1979; EWE, 1980).

Impacts

Investigation of flora and fauna provides the necessary information for an evaluation of the area from a nature conservation point of view, which can span from trivial (already strongly degraded, only very widespread and common habitats with very common species present) to very valuable or unique (valuable and rare habitat types in very good condition, with no or limited human influence and presence of very rare, endangered, or endemic species in viable populations). Here, one important point has been mentioned before, namely, the fact that the environment—with or without the project—is not a stable entity but will evolve over time. So, for example, it could be found that the project will negatively affect a population of endangered species and, on closer inspection, it might be possible to see that without the project, given the increasing pressure on land resources, there is a very high probability that this population would be eliminated anyway. Such a situation is not necessarily an argument in favor of a project, but it at least shows that very often, in order to maintain a situation (e.g., to save endangered wildlife), just not building a dam is not a sufficient solution.

Something that is less easy to identify but can be of much more importance than the direct and certainly destructive effect of reservoir impoundment on habitats, flora and fauna are the possible indirect impacts of a project. One such potential indirect impact is created by the construction of access roads to the project site. In cases where this project site lies in a rather remote, inaccessible area, these roads open a path and make the area much more accessible. A consequence of this can be the immigration of settlers into such an area and their encroachment on forests or other habitats, first along these roads, and then spreading farther. In this way, it is possible that the whole catchment is being made accessible, which in the long run can have adverse effects on the reservoir itself (by increasing siltation). If such impacts are likely to occur, specific measures have to be formulated for their mitigation as well.

Mitigation

Based on the identification and evaluation of potential impacts, recommendations will be made to mitigate undesirable effects or avoid them altogether, if they are considered unacceptable from a conservation point of view. Such measures can consist, among other possibilities, in project changes, in capturing valuable species prior to impoundment and bringing them to suitable alternative habitats, in the protection of habitat types similar to the ones to be destroyed in another area, or in the active restoration and rehabilitation of areas affected during construction.

Box 13.12. Mitigation Measures: Examples

- Heightening the Mauvoisin Dam in Switzerland by 13.5 m led to the submersion of approximately 0.5 km^2 of floodplain habitat upstream from the existing reservoir; this is a legally protected habitat type in Switzerland. Two separate sets of measures were adopted in this case. The first, which consisted of rehabilitating an area immediately downstream from the dam, had three purposes: (1) rehabilitating the construction site after finishing the construction activities for this project; (2) rehabilitating the construction site originally used for dam construction in the 1950s (and which 50 years later still looked like a construction site!); and (3) creation of wetland habitats for various species that had become rare in the area. Second, at a site further downstream along the river, an additional wetland area was created, mainly using water from tributaries to the main stream, with the additional aim of

> creating habitats adjacent to the river that offer a refuge for aquatic organisms during sediment flushing from the dam (EWE, 1987a).
> - On the Nokai Plateau in Laos, there is a population of about 150 elephants (*Elephas maximus*) left. Due to an increasing human population, and therefore increasing human–elephant conflicts, the long-term survival of this population seems doubtful. A part of its habitat will be submerged by the Nam Theun 2 hydropower project. For this reason, an extensive research program has been started by the developer, with the aim of establishing and implementing a management plan for these elephants that would guarantee their survival (SEATEC, 1997).
> - The Three Gorges Project affects the habitat of the highly endangered Chinese river dolphin (*Lipotes vexillifer*). Investigations have shown that, given the situation, even if the dam were not built, there is a high risk that this species would become extinct rather soon. As a part of the construction of this highly contested dam, a program has been started to investigate the species and to create better conditions for its survival.
> - Ralco HEP, in Chile, would have further affected fish species in this river, already affected by the Pangue Dam, by fragmenting their habitat and splitting the population into several isolated subpopulations. The main concern was for *Diplomystes nahuelbutaensis*, a catfish endemic to Chile. Investigations have shown clearly that this species was endangered mainly through the introduction of exotic species of trout and salmon. Nevertheless, in the framework of the project, a program was started to investigate the endemic species, especially its reproductive biology and artificial breeding (EWE, 1997).
>
> In these last three examples, there remains obviously one problem: Mitigation consists of a program that aims at conservation of these threatened species; however, whether the program is successful will only be evident much later, when the project has been already implemented. On the other hand, it has to be pointed out that in each of these cases, the main threat did not come from the project, and that without the measures taken in the framework of the project, prospects for these populations would not have been very good.

The aim is always to avoid impacts if possible, to mitigate their effect, or to provide compensation for impacts that cannot be avoided or mitigated. One difficulty involved, which should not be forgotten, is the fact that here, as well, we are making predictions on a future, possible situation, which, however, is influenced by various external factors, among others, and perhaps most importantly, by the commitment of the parties involved for actually implementing the proposed measures. A few examples are provided in Box 13.12.

Aquatic Habitats

Aims and Scope

The construction of a dam and the subsequent creation of a reservoir affect aquatic habitats mainly in the following ways:

- Interruption of the river continuum. Apart from influences on sediment transport, this mainly affects the movements of water organisms (active up- and downstream migrations and drift).
- Creation of a lake, or a lake-like water body (which can have some features, mainly due to reservoir operation, not normally observed in natural lakes). Changing a part of the river, with its riverine conditions, to lacustrine conditions also deeply affects the organisms living there.
- Reduction of river discharge between the dam and the powerhouse outlet (or, in the case of transbasin projects, in the whole downstream river area).
- Changes in discharge pattern downstream from the powerhouse outlet.

I have already dealt with impacts of these project-induced effects on hydrology and water quality. Here, I discuss the impacts on aquatic habitats and on aquatic organisms briefly. The focus here is on fish, not only because these species are usually used as indicator organisms for aquatic habitats, but also because of their importance as a resource for human use. Obviously, other aquatic organisms are also affected by the construction of dams (e.g., river dolphins; Reeves & Leatherwood, 1994). But again, this topic cannot be treated exhaustively here. However, there is a very extensive literature on the subject, some of which are mentioned in the reference section. The main aim of this section is to point out the problem, provide some additional information or examples, and indicate potential solutions, their inherent difficulties, and remaining open questions.

Baseline Data

As has been described for terrestrial habitats, information has to be collected on affected aquatic habitats and species, and, as in terrestrial habitats, a choice needs to be made about what groups of organisms to investigate. Usually, the emphasis lies on fish (see Box 13.13), and some other groups such as benthic organisms (either for their importance as indicators for habitat and water quality, or because of their importance as fish food), or plankton are included in the investigation. Again, there are three main sources of information, all of which need to be tapped: publications or other records on previous investigations, the knowledge of local people, and fieldwork. Usually, the focus will be on the following questions, which have to be answered for a proper identification of impacts and adequate measures:

- Species present (most frequent; most important from a fisheries point of view; rare, endangered, or endemic species).
- Reproductive behavior (migrations, seasonality of migrating behavior, spawning grounds).
- Prevailing conditions for the fish fauna (mainly, general state of the habitat, water quality, fishing pressure, and other projects that can have an impact on the populations, etc.).

Box 13.13. Fish Species in Lao Rivers

The case of fish species in the rivers of Lao (PDR) is a good illustration of the problems encountered in many cases when addressing biodiversity conservation. The investigations made for Nam Theun 2 recorded, from only a short period of fieldwork, a total of 185 species, among which 25 species were recorded for the first time in Laos and 11 new species not known from anywhere else. In earlier publications, a total number of 205 species of fish has been recorded from the territory of Lao PDR (SEATEC, 1997).

From this, it could be concluded that the rivers investigated (Nam Theun and Xe Bangfai) are highly diverse fish habitats with an extraordinarily high proportion of endemic species. But this is not necessarily so. It is absolutely possible that the new species found there are actually quite widely distributed in the Mekong basin, or that, at the very least, they occur in other tributaries of the Mekong as well. However, because the fish fauna of the water courses of Laos is not yet well investigated, there is no way of knowing it for the time being. This also means that it is almost impossible, with the present knowledge, to establish a reliable concept for maintaining the fish diversity of the country. Again, an EIA for a specific project cannot produce this kind of information.

Impacts and Mitigation Measures

Interruption of Rivers

The fact that dams interrupt fish migrations is probably one of the most widely known environmental impacts of dam projects, not in the least because especially migrations of salmonid fish in northern temperate regions have been affected. In Basel, Switzerland, located on the Rhine River, in late 19th century, there was a law prohibiting city households to feed servants salmon more than twice a week; at that time, salmon were abundant and correspondingly very cheap. Half a century later, there were no salmon left in the river (due not only to extensive straightening of the river and water pollution, but also due to dams, which made migrations to upstream spawning grounds impossible), and the salmon had become a highly sought after and very expensive delicacy.

The obvious solution for this impact are fish passes and fish ladders, installations that allow migrating fish to pass this artificial obstacle and to reach, in spite of it, their spawning grounds in the headwaters of the rivers (such fish ladders are mainly for anadromous fish species, i.e., species that migrate upstream for reproduction like salmon, and only of limited usefulness for katadromous species such as the eel, which migrate downstream for reproduction). Of course, fish ladders are not only of use for those species that migrate from the sea to rivers, but also potentially for a great number of species that migrate much more limited distances within the rivers themselves. For the design of a fish pass, it is important to know what species are targeted by the measure, because not all species are able to use the same type of structure. There is an extensive literature on fish passes (real fish ladders being only one possibility among several), namely, a vast Russian experience with such structures (Pavlov, 1989). This is not the place to deal with details of fish pass design, which, however, is essential for the success of the measure.

In spite of their usefulness in many places, there are limitations to fish passes. First, they are applicable only for passing obstacles of not much more than 10-m height. A large dam, according to the definition provided in the introduction to this chapter, therefore, cannot be crossed by means of such a structure. Here, other ways have to be found, such as fish lifts, or the capture of fish and their transport (by road) around the obstacle, to release them upstream from the dam.

Another point that has to be considered is the fact that most experience gained so far with fish ladders and similar structures stems from temperate (northern) climates. Little is known so far about the effectiveness and limitations of such structures in tropical waters; very often, knowledge of relevant aspects of fish biology is lacking as well for these rivers.

Finally, it has to be recognized that fish ladders are useful only for upstream migrations, not for downstream movements (neither for juvenile nor adult fish). The main problem involved is that, in the reservoir, the water movement caused by the fish ladder in comparison to that caused by the water intake for turbines is too weak, which makes it impossible for the fish to find the way. Therefore, the downstream migration will go through the turbines, which implies a very high risk of injuries and a high mortality. There are mainly two reasons for death or injuries among fish passing through the turbines. One is the physical contact with the turbine blades; here, incidence of injuries and deaths depends largely on the turbine type and the size of the fish (see Table 13.9). The second problem consists in the very high pressures, and mainly in the differences of pressure, the fish undergo when passing through the turbines (Montén, 1985; Berner & Geiger, 1988). The only mitigation in this case consists in providing structures and measures (fish screens) that will prevent fish from entering the waterways leading to the turbines.

Measures of a different kind altogether consist in artificially breeding fish species for restocking purposes. This measure has proved to be quite efficient—although rather expensive—in cases where natural reproduction was no longer granted, at least not to an extent that would have been sufficient for maintaining the fish stock at the desired high level. This has been, for instance, the case in Switzerland since the 1950s, mainly due to water pollution, and it was done only for "interesting" species (i.e., for those commercially attractive or valued for sports fishing). At least in the long run, it is not a real substitute for intact aquatic habitats that allow the natural reproduction of fish.

One additional impact on fish related to powerhouse operation has to be mentioned briefly. Very frequently, water restitution from the turbines, or spilling from dams, results in a very high turbulence. This can have the positive effect of reoxygenating water from the reservoir, as has been mentioned for the case of Urrá, but it can also lead to an oversaturation of the water with gases (mainly O_2 and N_2). This, in turn, can adversely affect fish by causing the gas bubble disease (too high partial pressure of these gases in the blood, leading to the formation of air bubbles in the circulatory system). This can be lethal for fish (Zhong & Power, 1996; Reichenbach-Klinke, 1980, p. 358).

Table 13.9. Probability for Fish to Be Injured or Killed When Passing through Turbines of Different Types

Turbine type	Fish size cm	Injured or killed %
Francis	10–15	> 50
	30–50	100
Kaplan	10–15	5–10
	30–50	15–30
Improved turbine*	10–15	< 5
	30–50	5–10

*Newly developed turbine type, to be installed in the upgraded Rheinfelden hydropower plant (Rhine River, Swiss–German border).
From Monten (1985) and Berner and Geiger (1988).

Creation of a Reservoir

The transition from a river to a lake environment has consequences for the fish fauna (and, of course, for other species of plants and animals as well; see Zhong & Power, 1996, for an overview). Of the species living in the river and therefore adapted to this habitat, not all will be able to live in the reservoir. Although a few species might actually profit from this change and become much more numerous than they were before, others will decrease in numbers or even disappear altogether. This is not necessarily the case for migrating species only, and to make species-specific predictions in this respect is impossible if the biology—and especially the reaction to such changes in habitat quality—is not known. This knowledge can only be gained from investigations of actual cases. One example of such a change is illustrated in Figure 13.20. Unfortunately, there are only very few such longitudinal studies (e.g., Antwi & Ofori-Dansin, 1993), and I personally do not know any that cover the situation in the river before impoundment (for a sufficiently long period to be aware of natural fluctuations), and during the construction, and operation periods of the reservoir, until such a time when stability was reached.

It is a well-known—but apparently not a very well-documented—fact that fisheries yields in reservoirs, after a peak period following impoundment, drop sharply to stabilize at a far lower level, as suggested by the graph provided by Balon (1974; see Figure 13.18). This presumably is due to the high amount of nutrients stemming from submerged biomass and soils, available shortly after first filling of the reservoir. This leads to a high primary productivity in the new lake and therefore to a high supply of food for fish. So in spite of the water-quality problems (anoxic conditions, and sometimes even the presence of highly toxic hydrogen sulphide) in the hypolimnion of the reservoir, there are good conditions for fish near the surface. After a number of years, however, once the nutrients have been washed out, conditions become more oligotrophic. This usually means an improvement of water quality in the lake, but also, because of a shortage of food, a diminishing fish stock. The best known reservoirs in this respect are the large African ones (see, e.g., Lelek, 1973; Lelek & El-Zarka, 1973; Lewis, 1974; Petr, 1974; Blake, 1976; Kaptesky & Petr, 1984). However, considering the problems in actually monitoring fish stocks in view of all the

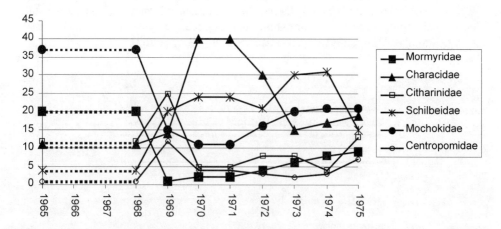

Figure 13.20. Change in the composition of the fish fauna in the Niger and the Kainji Reservoir after impoundment in 1968. Data on fish fauna in the preimpoundment river from Visser (1973; % specimen numbers), evolution in the reservoir (% of weight of catches) from Lewis (1974) and Blake (1977). From these figures, no information on the absolute abundance of fish can be obtained.

intervening factors—such as catch effort and reliability of reporting—the situation is not easy to analyze. Fishing in reservoirs, of course, is not limited to large reservoirs; small ones can also be very important, mainly for supplementing the diet of the local population or for supplying local markets (Baijot et al., 1994).

One important and not always unproblematic way to increase fisheries yield in reservoirs consists of introducing exotic species. The aim here is to have species that will thrive in a lake, especially in cases where there were no species present in the original river system that easily adapt to this situation; Balon (1974, p. 136) describes the history—which, in this case, was a success—of the introduction of *Limnothrissa miodon*, a pelagic sardine from Lake Tanganyika, to Lake Kariba, whereas the introduction of *Tilapia macrochir* failed. However, there is always a risk involved with the introduction of exotic species, mainly because this can have very adverse effects on indigenous species, either because the new species actively preys on them or competes for food, or because diseases are introduced with the exotic species (Dumont, 1986; Worthington, 1989; Vila Pinto, 1989; Worthington & Lowe-McConnell, 1994). Finally, as had been mentioned already for the Cirata and Saguling reservoirs in Indonesia, aquaculture by means of floating fish cages is another way of increasing productivity. All these activities, of course, interfere with the lake ecosystem and can have negative impacts.

Reduction in River Discharge

Two parameters are important for the production of energy: the amount of water and total head available. The main purpose of reservoirs is to stock water to make it available at the time when it is needed for energy generation. In addition to this, however, the dam also increases the available head. On the other hand, additional head can also be created by arranging the powerhouse not directly at the foot of the dam, but further downstream, adducting the water by means of headrace waterways, normally tunnels, often several km in length. This, then, means that in the stretch of the river between dam and powerhouse, outlet discharge is permanently reduced and can be virtually zero, at least as far down as the next tributary (Table 13.10). It is obvious that this also has consequences for the function of the river as a habitat for aquatic species, and on any form of human use of the river water.

The mitigation measure usually applied for this impact is to designate a certain amount of water that has to be released from the dam and left flowing in the river (Orbig, 1989; Schmidtke & Ottl, 1988). This amount is normally defined in relation to the size of

Table 13.10. Impacts on Rivers below Dams

Plant	Country	FSL m asL	Length of headrace structure km	Total head m	Length of affected river stretch km
Enguri	Georgia	510	15	410	80[2]
Ralco[1]	Chile	725	7	175	10
Houay Ho	Lao PDR	883	1	776	10[3]
Mauvoisin	Switzerland	1971	14.7	1490	33[3]
Mattmark	Switzerland	2197	3.7	1024	??

[1] Project not implemented to date.
[2] Transbasin project, water restitution to different river.
[3] Water restitution at a different place within the same main river basin; figure is length of affected part of used river (till inflow into next larger one).

Table 13.11. Definition of Residual Flow to Be Released from Water Intakes or Dams According to Swiss Legislation (Law on Water Protection)

$Q_{347}(L/s)$	Residual flow (L/s)	For every additional flow of x L/s	Additional release of x L/s
≤ 60	50	10	8
160	130	10	4.4
500	280	100	31
2,500	900	100	21.3
10,000	2,500	1,000	150
60,000	10,000	—	—

The reference point is Q_{347}, the discharge reached or exceeded during 347 days per year.
From the Swiss Confederation (1991).

the river and especially in relation to its natural minimum flow. In Switzerland, for example, this residual flow is defined in function of the Q_{347} (i.e., the discharge reached or exceeded during 347 days per year (Table 13.11). Besides this rather rigid regulation defining the amount of water to be released, this law also specifies some conditions under which higher releases can be defined, or lower ones can be tolerated.

Such a regulation aims at maintaining some basic functions of the affected part of the river more or less intact, for example, to maintain a continuous flow of water, providing a habitat for fish and other aquatic species, and maintaining a diluting effect for inflowing (treated) wastewater. Although these aims in general will be reached by this measure, it has to be pointed out quite clearly that this does not mean, from an ecological point of view, that the river habitat as such is kept intact. The fact remains that the amount of water is considerably reduced and the natural discharge pattern is changed (although it is also possible to imitate this pattern, if the release is not interpreted as a rigid "all the time" release, but rather as a total amount to be released, which can vary seasonally). In order to maintain intact river habitats, with their whole set of species, it might therefore be preferable to introduce measures at a different level. It could be determined for instance, that for a certain number of rivers used for energy production, one would be set aside and left intact. This sounds like a very simple measure, but in fact it is very complex and certainly beyond the scope and possibility of an EIA for a hydropower project, because it requires coordination—and legal power—at a considerably higher level.

Changes in River Regimen

The changes in river regimen have been described earlier. When discussing their impact on river habitats, we have to distinguish between three quite different effects, namely, (1) daily fluctuations, (2) seasonal fluctuations, and (3) sediment transport.

Daily Fluctuations. Power plants operated for peak power energy production often release high amounts of water during very short times only (a few hours or less), and on–off movements can take place several times a day. This leads to massive daily amplitudes in river discharge, with several negative effects on river habitats. One severe impact is the surge wave created when the turbines are turned on, because this is usually "from 0 to 100" in a matter of seconds. This surge wave has a flushing effect on organisms and fine sediment, and can in this way destroy certain populations or especially fish breeding grounds. A second point is the difference in temperature between the water released from the reservoir and the receiving river. In the case of a tropical lowland reservoir, water temperature can be considerably higher than that in the river, whereas in the case of reservoirs in mountainous areas, especially in high-pressure schemes using a high head, reservoir water temperature will be considerably lower than that in the lowland river receiving the water. A third factor, finally, is the fact that changes in discharge result in a considerable change of submerged area. A part of the riparian zone of the river will be flooded and dry out again, sometimes several times a day. This makes survival of (sessile) organisms in the riparian zone, which is usually the most productive one, very difficult and can in this way reduce productivity of the river system, an impact that will have repercussions, for example, on fishing communities in the downstream area.

Mitigation for these impacts can consist of an adaptation of the operation scheme of the power plant (e.g., by a stepwise, gradual starting of the turbines to reduce the surge wave effect). This, however, has technical limitations and reduces efficiency (and therefore usefulness) of the powerplant, because it will not be able to respond in the desired way to short time fluctuations in power demand. Therefore, the solution applied more frequently consists of constructing regulation ponds, which, if used properly, have the effect of reducing the surge wave, of smoothing, at least to some extent, the discharge fluctuations, and can also reduce the temperature difference.

Seasonal Fluctuations. Many floodplain habitats depend not only on the availability of water as such, but also on its seasonal distribution. This distribution includes seasonal floods that occur more or less regularly, although with very large interannual differences. Absence of a normal seasonal fluctuation in river discharge caused by a dam can therefore change the floodplain dynamics radically. In western Europe, floodplains count among the rarest and most threatened habitat types as a result of more than 100 years of very extensive human interference with river dynamics, for flood protection as well as for energy production (Walter & Breckle, 1989, p. 126; Figures 13.21 and 13.22). On the other hand, floodplains can be highly productive ecosystems, and changes in their dynamics can profoundly affect them, as well as human populations depending on them, which makes clear that in terms of revenue for the affected populations, irrigation projects interfering with this dynamic are not necessarily the better solution (Zwahlen, 1992; Barbier & Thompson, 1998). Mitigation for this type of impact is usually not easy, especially in the case of large and complex floodplains. If such a system is being affected seriously by a dam project, every possibility will have to be considered very carefully, and this will have to include the no-dam option as one possibility. Again, it is not possible to give a generally acceptable answer, because every project presents a specific case that has to be analyzed carefully. For smaller systems, it is sometimes possible more or less to maintain—or even restore, in cases where it has been lost already—floodplain dynamics by releasing artificial floods from the dam at appropriate times of the year. Needless to say, this is a costly measure, because it can considerably reduce the amount of water available for energy generation. Therefore, the better option might often be to use the resources—including monetary resources—available for compensation measures in the sense that legal protection is granted or improved to still existing similar systems at a different place.

Changes in Sediment Transport. It has been already said that sediments are trapped in reservoirs. Reducing the sediment load can have effects on the erosive capacity of the river below the dam, an effect that often leads to increased riverbed erosion. On the other hand, especially the absence of more or less regular flood events can lead to an increased accumulation of sediments, mainly of those brought by lateral tributaries, in the riverbed. In the event of a large flood that has to be released through the flood discharge structures of

Figure 13.21. Egiin Gol floodplain, Mongolia. In the river systems of northern Mongolia, this highly dynamic habitat type is still very widespread and not influenced by human activities.

the dam, the riverbed is then blocked by these sediments and is no longer capable of eliminating this high flood peak. This can considerably increase flood damage.

Another topic is the role of fine sediment that is retained especially in very large reservoirs. Depending on its composition, fine sediment can play an important role in the

Figure 13.22. Densely populated and intensively used landscape in Valais, Switzerland. As in most of western Europe, virtually nothing is left of the floodplain habitats. The rivers flow in narrow, straightened beds between flood dikes.

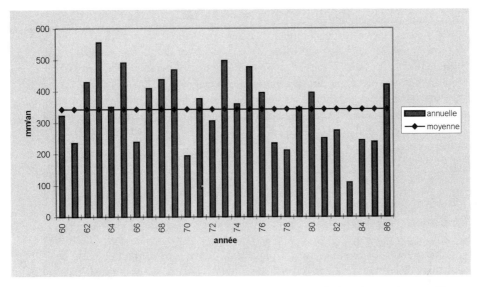

Figure 13.23. Rainfall during the period 1960–1986, Oujda meteorological station, northeastern Morocco. Interannual differences are very marked, and a draught period from 1981 to 1985 is clearly visible (EWE, 1998).

floodplain, where it acts as a fertilizer. The best-known case in which sediment retention has a wide-ranging effect on the downstream area is undoubtedly presented by the Aswan high dam on the Nile. This multipurpose dam project has been controversial from the very beginning, first of all because of political issues, namely, the question of whether the United States or Russia would finance construction of this dam. One main point among others such as water losses due to evaporation, however, was always the question about potential effects of sediment retention in the reservoir. This sediment, which has always been deposited by floods on the agricultural lands of Egypt, had a very important fertilizing effect that now has to be replaced by artificial fertilizers. Other effects of sediment retention are a change in the Nile delta dynamics, with increased erosion and saltwater intrusions, and impacts on fisheries in the Mediterranean sea around this delta (see Box 13.14). On the other hand, thanks to water stored in Lake Nasser, Egypt suffered much less from the severe drought that badly affected the whole of northern Africa in the late 1980s (Afifi, 1993; Figure 13.23).

Box 13.14. Aswan and Mediterranean Fish Stocks

One well-known, major impact of the Aswan Dam is the fact that it triggered a decline in sardine fisheries in the Mediterranean Sea. This part of the sea was fertilized by the input of Nile sediment, which was the reason for a high productivity and, ultimately, the high fish stocks. After construction of the dam, fisheries yields dropped dramatically, and this effect was explained by the lack of the fertilizing effect of Nile sediments, which were now retained in Lake Nasser (Goldsmith & Hildyard, 1986).

That the situation might not be that simple and clear-cut was shown by Bilio (1986). Fisheries in the Egyptian part of the Mediterranean started at about 6,000 t in

> 1929, and rose to an annual yield of 35,000–38,000 t in the late 1950s. Since about 1960, yields started to decline, and this was attributed to overfishing. In the mid-1960s, in parallel with the start of the impoundment of the Aswan reservoir, a steep decline in captures was observed, and landings reached a mere 8,000 t in 1970. This latter effect was attributed to the retention of fertilizing sediments in the reservoir. On the other hand, fishing in the reservoir started in 1963, and yields rose from 4,000–17,000 t in 1977, with a reduction in the following years due to the lack of water. The potential for fisheries in the reservoir is estimated at 20,000–25,000 t yearly.

The Human Environment

Ultimately, the most important impacts of dams are certainly those on human populations. The main impact, and the one intended, is certainly the fact that energy is being produced and put at the disposal of the society—or at least of a part of the society—of the countries in question. Doubtlessly, an economic development that goes beyond a subsistence economy is not possible without energy input. Electricity is an important form of energy, with still growing importance, and hydropower is a "clean," environmentally friendly, renewable source of energy (within the limitations that have been discussed).

In spite of this, there are social and socioeconomic problems involved in the use of hydropower that have long been largely ignored, or at least not properly addressed. I raise these issues here, although only briefly. Available space does not allow a complete discussion, but I think that these problems are too important—and in many respects, too narrowly linked to applied aquatic ecology—to be ignored here altogether. That they are dealt here in less detail than a few other issues is justified by the overall subject of the book and does not mean that they are less important.

One complex issue related to energy generation, namely, in developing countries, is not touched here at all: the question of access to electricity by different groups of the population. As a matter of fact, it is difficult to see why in Southeast Asia and in other regions, huge amounts of electricity should be used for cooling offices so far down as to constitute a health hazard, whereas rural electrification is often neglected. However, here we deal only with the more or less direct effects of dams and related structures, setting this question aside.

Socioeconomic Impacts: Affected Populations

With very few exceptions, wherever a dam is being built to create a reservoir, there is a human population that will be affected by this project (see Box 13.15). *Affected*, as it is used here, means that this population will suffer more or less severe losses because of the project, without getting any direct benefits from it (Cernea, 1996, 1997, 1999). Just to mention one frequent observation, it is very often the case that a population that is displaced because of a hydropower project has no electricity, neither before nor after implementation of the project.

There are mainly two ways a human population can be directly affected by a hydroelectric project (and also by a number of other development projects):

- *Completely affected*: people losing their houses, their land and other resources, or their entire livelihood basis. This is usually the case for settlements in river valleys, which very often concentrate in such areas because there are good, mainly alluvial, soils cultivated by their inhabitants.
- *Partially affected*: cases in which the villages themselves are situated above FSL and will therefore not be submerged, but inhabitants lose a part of their resources, again, mostly land.

These direct effects are caused mainly by the reservoir and to a lesser extent by other structures such as the dam, access roads, powerhouse and appurtenant structures, transmission line ROW, and so on.

Box 13.15. Project-Affected Peoples (PAPs)

It is not always easy to identify clearly all the people directly affected by a project. There are different reasons for this, as shown in the following examples:

- In a socioeconomic study of the Nam Ngum watershed in Lao PDR, English (1998) concluded that for several hydropower projects, the number of affected people had been underestimated, mainly because access to the area had not been possible due to safety reasons.
- The Resettlement Action Plan for Ilisu hydropower project in Turkey came to the conclusion that the number of people entitled to compensation was considerably higher than estimated at first, mainly because many families had left the area earlier (for reasons other than the project), but still possessed land in the reservoir area (Semor, 2000).
- For the Three Gorges Project in China PDR, the reported number of resettled people varied between 500,000 and 1,400,000 (Ryder, 1990). This depended mainly on how the water level requiring resettlement was defined, namely, whether people above normal FSL, but below maximum flood level, would also have to be resettled.

In addition to the usually obvious, direct impacts, there are a number of indirect and often less obvious negative effects of such projects on human populations. One often neglected aspect is the potential effects in the downstream area (Scudder, 1996; Barbier & Thompson, 1998; Zwahlen, 1992):

- *Reduced river discharge in the downstream area*: Settlements between the dam and the powerhouse outlet might suffer from the reduced water flow in the river either directly (more difficult access to water, less water available for drinking or irrigation, river no longer manageable by boat, etc.) or indirectly (e.g., a drop in groundwater level as a consequence of the reduced river discharge).
- *Increased river discharge*: In the case of a transbasin project, where the water from the reservoir is being released to a different river, this river can be changed fundamentally by the often very substantial increase in water discharge: increased flood risk or flood intensities; changes in general river dynamics (meandering pattern and river bank erosion; see Figure 13.24); and so on.

Figure 13.24. Nam Himboun River, Lao PDR, downstream of the water outfall from Theun Himboun power plant. Through diversion of the water from Nam Theun into Nam Himboun, discharge in this latter river has increased markedly, and the resulting erosion threatens some villages.

- *Changes in downstream river discharge*: Danger from surge waves (for people using the river for washing, fishing, etc.); seasonal changes affecting water availability (an effect that can be adverse or beneficial); and so on.

Finally, there are a number of other indirect effects. The following list is not necessarily complete:

- Water-related diseases.
- Other public health problems.
- Cultural impacts.

These issues are briefly addressed in a later section of this chapter.

Resettlement

In larger projects, the reservoirs can be of considerable size, 100 to several hundred square kilometers being the rule rather than the exception. Because river valleys often contain good soils for agricultural production and the rivers themselves often provide drinking water and are the main transport routes, human populations tend to concentrate in such valleys. This means that, in most HEPs, a considerable human population loses its land and homes, and will therefore have to be resettled (see Table 13.12 for a few examples from African dam projects). Undoubtedly, the largest of those dam-induced resettlements is required for the Three Gorges project on the Yangze, in China, which requires the resettlement of about 1.2 million people (Chau, 1995; Ryder, 1990, p. 36; Yao, 1991, p. 64).

In the past, people affected in this way often have suffered greatly because of such developments, because their losses have not been duly compensated, leading to impover-

Table 13.12. Resettlement in Some African Dam Projects

Reservoir	River	Country	Completed in	Displaced population
Assouan	Nile	Egypt	1969	108,000
Akosombo	Volta	Ghana	1964	80,000
Kossou	Bandama	Ivory Coast	1970	75,000
Kariba	Zambesi	Zambia/Zimbabwe	1958	57,000
Kainji	Niger	Nigeria	1968	44,000
Cabora Bassa	Zambesi	Mozambique	1974	25,000
Buyo	Sessandra	Ivory Coast	1980	16,000
Selingue	Sankarani	Mali	1980	15,000
Manantali	Bafing	Mali/Senegal	1982	12,700
Nangbeto	Mono	Togo	1987	10,000

From Lassally-Jacob (1996).

ishment of this part of the population. Even when the problem was actually recognized, it was very often dealt with inappropriately. In some cases (e.g., subsistence farmers given compensation in the form of money, which they did not know how to handle), the result was usually a complete failure of such a procedure, because nothing was left of this compensation after a short period of seeming luxury. This has been described in detail by Hong (1987) for the case of Batang Ai resettlement, Sarawak. Other examples can be found in McCully, 1998, p. 65ff.). Such experiences have triggered a growing resistance against dam projects worldwide (Charest, 1995; Oliver-Smith, 1996; Gray, 1996).

Today, it is generally recognized and accepted "best practice" that people affected by a development project—be it a dam, a road, or anything else—should not only not suffer from the project economically, but they should also get their fair share of the development benefits. In the very least, after the resettlement, they should not be worse off and, if possible, should be better off than before. However, this generally accepted standard in itself is not yet a guarantee of adequate implementation of project-induced resettlements.

Resettlement has become one of the most important aspects in the planning of a hydropower scheme, and guidelines have been established by the leading development banks as to how such resettlement should be carried out (e.g., Asian Development Bank, 1998; World Bank, 1999; World Bank OD 4-30, 1990).

Proper resettlement planning mainly requires the following steps:

- *Assessment and inventory of the prevailing situation*: the number of people living in a given area, socioeconomic condition, livelihood, infrastructure, social system, land tenure, ethnic groups, and so on. This also needs to include sociocultural questions.
- *Assessment of the impact of the project*: number of settlements affected, number of people involved, amount of land to be submerged, infrastructure to be destroyed, routes to be interrupted or newly established, and so on.
- *Potential effect of the construction phase*: job opportunities, social effects of workers in large numbers coming to the area from outside, potential effect of job and income losses at the end of the construction period, and so on.
- *Requirements for properly reestablishing the affected population*: People should, at the very least, not be worse off after resettlement than before; if possible, their situation should improve.

- *Views of the affected population itself on how resettlement should be done*: participation in resettlement site selection, resettlement options, compensation mechanisms, and so on.
- *Assessment of the potential of the sites chosen as resettlement areas*: space, soil and water resources, access to markets and medical services, and so on.
- *Identification of the institutional framework required for handling the resettlement issue*: existing agencies, NGOs, and so on. In many cases, a project-specific resettlement agency will have to be established. It is important that the population to be resettled have adequate representation within this organization, so that the people will be heard at any stage of the process.

Cases in which ethnic minorities are affected by a project can have especially far-reaching consequences. In addition to social and economic issues, there are also cultural and ethnic issues involved, such as the risk of losing ethnic identity, culture, and language (see, e.g., Manser, 1992). Specific guidelines have been set up for such cases (e.g., World Bank OD 4-20, 1990; Jalal et al., 1997). Resettlement guidelines always point out the necessity for special emphasis on vulnerable groups such as ethnic minorities, poor and landless rural people, households headed by women, and so on. In places that have been inhabited by human populations for a long time, the question of archaeological sites can be very important. In the case of Ilisu HEP, this led to very extensive and well-documented excavation campaigns (Tuna et al., 2001).

The resettlement has to be planned very carefully; socioeconomic, cultural, spatial, as well as institutional and financial aspects need to be included. The elaboration of a Resettlement Action Plan is a complex endeavor (e.g., Nam Theun 2 Electricity Consortium, 1997; SEMOR, 2000), but it is a very important and indispensable tool for the implementation of an adequate resettlement.

As for other measures, the best mitigation would be avoidance of resettlement; if this is not feasible, then it should be minimized as far as possible (see Table 13.13). Regardless of the number of people that have to be resettled, fair compensation must be provided.

The work carried out for the resettlement itself needs to be integrated into the framework of the HEP construction process. It has to be kept in mind that site preparation for resettlement (e.g., in cases where tree crops are a part of the livelihood of the population) might require several years. Attention has to be paid to prevent other people from encroaching on the resettlement area during the process. Participation of the affected population is crucial. During the planning phase, they should be integrated into the process, and their views and opinions should be respected whenever possible. Official agencies' views of how these people might benefit from development associated with resettlement can differ substantially from views of the affected population itself. Especially for ethnic minorities that differ culturally from the society at large, the changes imposed by

Table 13.13. Alternatives Studied and Consequences for Resettlement

Alternative		High dam	Low dam	ROR
Full supply level	(m asL)	242	230	206
Reservoir area	(ha)	2,100	990	100
Families to be resettled	(N)	2,291	714	27

Source: Kukule Gang Joint Venture (1992) and Zwahlen (1995).
Mainly because of the resettlement issue, the run-of-river (ROR) option was finally chosen.

resettlement should not lead to the disruption of their ethnic integrity and their culture (Chang et al., 1995). A forced integration of such groups into mainstream society is very likely to end in social disaster. On the other hand, unrealistic expectations of the affected people (e.g., concerning money to be obtained in the act) have to be counterbalanced.

A proper resettlement is possible only with adequate funding. The necessary funds must be included in the overall project budget. As with other aspects related to an HEP, resettlement needs an *ex post* follow-up (monitoring), so that corrections can be made, if necessary.

Land Use

One of the most important and critical issues in resettlement planning and implementation is land availability. In most cases, river valleys that contain fertile alluvial soils are densely populated. This invariably leads to conflicting interests in every hydropower project; very often, large amounts of such valley soils are submerged by the formation of the reservoir. People living there or cultivating these soils have to be resettled.

Land and suitable soils are the most important resource for most rural populations, whether for subsistence cultivation or for cash crop production. Unfortunately, the best soils are normally found in the floodplains of the rivers. Much of this soil is lost when a river is dammed for energy production or irrigation. Sloping land above the reservoir level is normally less suitable for cultivation and presents a higher erosion risk. Other areas with fertile soils are, in most cases, occupied and not available for resettlement.

The following points have to be addressed:

- Prevailing land use in the reservoir area: amount of land used for agricultural or other purposes by the local population; proportion of this land that will be lost due to the project; its suitability, appropriateness of the prevailing form of cultivation, and subsistence crops and revenue generated by this land. Very often, in addition to individually used agricultural land, "commons" (e.g., pastures, forests for timber, fuel wood, and NTFP i.e. nontimber forest products), accessible to everybody, play a very important role in the subsistence economy of an affected population and are especially important for landless, poor people who do not possess any assets that yield compensation. For example, effects of not compensating these commons have been discussed by Ulluwishewa (1997).
- The present land tenure situation, amount of land available, form of use, and generated revenue have to be known. Depending on the social system, this has to be known on a household level (in the case of individual land ownership) or on a village level (in the case of corporate ownership by the community); the question of whether land cultivated belongs to the farmer cultivating it or to somebody else is important.
- Identification of land requirements of the population that will have to be resettled (including the aforementioned commons).
- Prevailing land use and soil suitability in the surroundings of the reservoir, in the catchment area, and, if required, in other areas.
- Identification of characteristics of potential resettlement areas, including prevailing land use, soil suitability, available soils, and suitable cultivation methods for these soils.

- Available land resources and land ownership in the future resettlement areas have to be known. It must be ascertained whether enough land of suitable quality will be available, and that there will be no conflicts over questions of ownership of such land. Again, the often conflicting official and traditional land tenure systems have to be taken into account.
- If necessary, an agricultural extension program must be put in place, in order to train and prepare the resettlees for their new conditions. This is required, for example, when people who practiced shifting cultivation have to adapt to a sedentary agriculture type, or when people who have grown irrigated paddies in the valley floors must adapt to hill paddies or other rain-fed cultivations.

Land use and soil suitability maps are normally established as a tool for further planning. Prevailing unsustainable forms of land use (e.g., shifting cultivation, with an insufficient fallow period) have to be identified, and the specific habits and capabilities of the local population have to be taken into account.

The aims of these specific studies are to provide a sound basis for resettlement planning and other aspects of the ongoing project elaboration.

Vectors, Water-borne Diseases, and Other Public Health Issues

In tropical climates, several diseases are strongly linked with water, because their agents are either transmitted by contaminated water, or their intermediate hosts or vectors live or develop in water (see Table 13.14).

The most important of these diseases are as follows:

- *Malaria*: disease caused by protozoan blood parasites transmitted by mosquitoes of the genus *Anopheles*. The larvae of these mosquitoes develop in stagnant water, with different species inhabiting different types of water bodies (small reservoirs, ponds, drainage ditches, etc.). Malaria is very widespread, and one of the four forms (the tertian or cerebral malaria caused by *Plasmodium falciparum*) is a very dangerous illness with a high mortality rate. Resistance against insecticides and drugs have developed in the mosquito and in the parasite, respectively, which makes malaria more difficult to control.
- *Schistosomiasis*: disease caused by several species of parasitic flukes that develop in the bloodstream of man. The first stages develop in aquatic snails, and

Table 13.14. Categories of Diseases Related to Water, as Defined by WHO

WATER-BORNE (water acts as a passive vehicle for infective agent): bacterial [*Salmonella* (typhoid), *Enterobacteria*, *Escherichia coli*, *Campylobacter*, cholera, Leptospirosis, etc.]; viral (hepatitis A, poliomyelitis, rotaviruses, enteroviruses); parasitic (amoebiasis, giardiasis, intestinal protozoa, *Balantidium coli*.)

WATER-WASHED (infections that decrease as a result of increasing the volume of available water): enteric (some diarreas and gastroenteritis); skin (scabies, ringworm, ulcers, pyrodermitis); louse-borne (typhus and related fevers); treponematoses (yaws, bejel, pinta); eye and ear (otitis, conjunctivitis, trachoma).

WATER-BASED (a necessary part of the life cycle of the infective agent takes place in an aquatic organism): crustaceans (Guinea worm, paragonimiasis); fish (Diphyllobotriasis, Anisakiasis, flukes); molluscs (flukes, schistosomiasis).

WATER-RELATED (infections spread by insects that breed in water or bite near it): mosquitoes (malaria, filariasis, yellow fever, dengue, hemorrhagic fever); tsetse flies (tripanosomiasis); blackflies (onchocerciasis).

From Tolba and El-Kholy (1992, p. 549) and WHO (1992, p. 118).

infestation of humans takes place when they come into contact with water that contains the free-swimming larvae, which then penetrate actively through the skin. Irrigation schemes (reservoirs, irrigation canals, drainage ditches) provide suitable habitats for these snails.

- *River blindness*: caused by a parasitic nematode whose vector is the blackfly (*Simulium* sp.). The parasite is transmitted by the biting blackfly and develops under the human skin. It can lead to destruction of the eye and therefore blindness. Blackflies develop in rapidly moving, well-oxygenated water such as rapids or water outlets of power stations. The disease was widespread in central and western Africa. Through a concerted effort, it has been virtually eradicated (World Health Organization, 1999, p. 25).
- *Dengue*: A viral infection transmitted by *Aedes* mosquitoes, which develop in the smallest water bodies (empty cans, water jars, etc.). The disease occurs mainly in densely populated areas (cities).
- *Diarrheal diseases (cholera, among others)*: caused by various agents transmitted mainly by drinking water of insufficient quality (e.g., contaminated by domestic wastewater). The hygienic conditions are of crucial importance in both the spread and prevention of these diseases.

Other diseases that are less relevant from an overall public health point of view and/or have a restricted range are equally linked with water [e.g., filariasis (elephantiasis), dracunculiasis (Guinea worm), and others]; however, locally they can be of very high importance and should not be neglected. Likewise, the prevalence of many intestinal parasites depends on hygienic conditions such as water quality, among other factors (Figure 13.25).

Figure 13.25. Although this drainage channel in Burkina Faso is a highly welcomed source of water, it is also the breeding place of vectors and intermediate hosts for diseases, and its use for drinking water is unsafe.

The creation of a reservoir can enhance the conditions for the spreading of these diseases, mainly those depending on a vector or an intermediate host, by creating favorable living conditions for these organisms (Oomen, 1983; Madsen et al., 1988; Lai, 1992; Tyagi & Chaudhary, 1997; Gheberyesus et al., 1999). The situation has to be analyzed carefully, with attention to the conditions that are specific for the project area in question. Specific guidelines can be applied here (Tiffen, 1991; Asian Development Bank, 1992; Hunter et al., 1993). The following points are important:

- *Location*: Many vectors and therefore the diseases transmitted by them have specific geographic distribution ranges.
- *Climate*: Climate type (perhumid, seasonally arid) plays an important role in the development potential of vectors.
- *Prevailing conditions*: Vegetation, soils, water bodies, and so on, are habitat conditions for vectors and parasites.
- *Local public health and socioeconomic situation*: prevailing diseases (frequency, importance), access to medical infrastructure, nutrition, and so on.
- *Exposure*: the local population's contact with the infecting agents.

The effects of the project have to be analyzed. The most important questions are whether the living conditions for vectors are changed considerably (e.g., creation of new habitats for aquatic snails through artificial formation of stagnant water bodies), and whether the exposure of the local population to infecting agents will be changed. If there is a risk of increased occurrence of such diseases, or if the analysis of the prevailing situation has revealed unacceptable conditions concerning such diseases, mitigation measures must be proposed. These measures can be grouped in three main categories:

- *Public health measures*: direct improvement of the health situation of the local population by improvement of the medical infrastructure in the project area (or in the resettlement areas, if applicable). This also includes training of the population, especially of mothers, in personal and food hygiene.
- *Sanitation measures*: supply of good-quality drinking water; elimination of wastewater and solid waste; drainage of settlements.
- *Environmental management measures for vector control*: draining of ditches or temporary pools forming in the drawdown area of a reservoir; landfills in areas where seasonally stagnant water accumulates; proper maintenance of irrigation canals, and so on. The aim of these measures is to eliminate or at least reduce breeding places for vectors and intermediate hosts (Rafatjah & Kuo, 1982; World Health Organization, 1993).

Shortcomings in implementation, but especially in operation and maintenance, can lead to a situation in which the expected benefits of a project are not achieved. Unfortunately, in such cases, very often the negative aspects (e.g., proliferation of intermediate hosts for schistosomiasis and malaria vectors) persist or are even aggravated. This was found to be the case in an assessment of the need and possibilities for the rehabilitation of 35 small irrigation schemes in Burkina Faso (EWE, 1989). Most of the schemes had more or less broken down, but schistosomiasis and malaria prevalence were very high. The relation between planned and implemented achievements is illustrated in Figure 13.26.

An additional series of specific public health issues has to be addressed in relation to the construction phase of large dams, especially given that a large number of workers, up

Environmental Impacts of Dam Projects

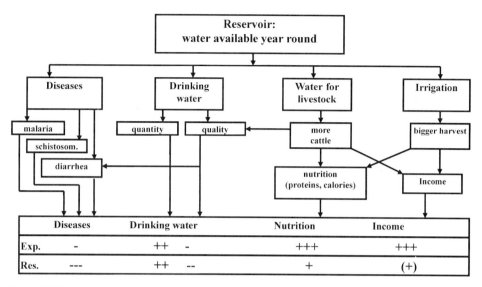

Figure 13.26. Expected (Exp.) outcome of an irrigation program compared to actual results (Res.) in the absence of proper operation and maintenance. The positive effects on nutrition and income are not achieved, whereas the adverse effects on health are worse than anticipated (Zwahlen, 1992).

to several thousand in the case of large dams, mostly come from outside to the project area. The main risks here are exposure of workers to locally endemic diseases to which they have not been exposed before, and therefore have not developed resistance on the one hand, and the risk that diseases brought by the workforce from outside to the project area then spread to the local population on the other. The only way to prevent and mitigate such effects is to have good, on-site health services for the workforce, which should also be accessible to local population.

Environmental Monitoring

Every EIA is a prediction of the future behavior and development of a highly complex system. It tries to anticipate undesirable future effects and intends to counterbalance them by means of mitigation measures. It is obvious that the development that takes place will depend on a large variety of variables, the project itself being only one of them. Other intervening variables include demographic development (through intrinsic growth or migrations), land use development in the wider area, or economic development in the region or in the country as a whole. Such developments can either take pressure off an area or increase pressure. Undesirable environmental effects can in this way be compensated or increased, or new effects, which are completely independent of the project can be triggered. However, especially when the development is in the catchment area, the project itself may be affected.

On the other hand, measures have been proposed to counterbalance impacts. The suitability of these methods for reaching the preset aims will have to be proved. In order to achieve these aims, the measures will have to be implemented properly. But even the best measures can fail if the conditions change in an unforeseen and unpredictable manner.

The EIA identifies the sensitive aspects of the specific environment in a given situation for a specific project. It makes statements about the development of these aspects in the future, with and without the project, identifies potentially adverse developments, and establishes measures either to enhance desirable or minimize or even avoid undesirable effects.

For the important aspects likely to undergo changes in the future, a monitoring program has to be established. Points to be monitored include the following, among others:

- Water quality
- Groundwater around the reservoir and in the downstream area
- Hydrology in the downstream area
- Development of the fish fauna in the reservoir and in the downstream area
- Siltation of the reservoir
- Land use in the surrounding of the reservoir and in the catchment area
- Public health

This list is not exhaustive; the aspects to be monitored depend on the specific conditions. A separate monitoring program will usually be needed (or should be set up) for all aspects related to resettlement and compensation of the affected human population.

At the same time, an institutional setting has to be defined that will be able to carry out the monitoring program. It is important that this be started before the construction period and maintained throughout this period and beyond. Usually, the owner of the plant will be responsible for the monitoring and might have to create a special environmental unit for this purpose. Official organizations (e.g., specific agencies of the Ministry of Environment or corresponding entities) should then be responsible for supervising the monitoring. If required, a specific authority will have to be created. It is important that appropriate equipment and funding be made available. The monitoring program has to answer the following main questions:

- Does the development of the various parameters take place in the predicted way, or are there any important deviations to be seen?
- Have the mitigation and compensation measures been implemented in an appropriate manner?
- Do they produce the expected results? If not, what is the reason?

The answers to these questions allow corrective measures to be taken, should they be required. At the same time, they also provide evidence for the problem of changes provoked by the project itself versus changes caused by other factors. This can, in certain cases, be important for questions of liability.

Watershed Management

Effects of the prevailing situation and of human activities in the catchment area on reservoirs have been briefly mentioned. This in itself already explains the necessity for a watershed management plan, which has to be part of a comprehensive EIA. The aims of such a plan are to identify existing or potential situations in the catchment area that could negatively affect the reservoir, through siltation and eutrophication, and to propose measures to prevent such a situation from developing or to reduce already existing adverse effects.

Hydroelectric or irrigation projects normally do not have any direct impacts on their catchment area. However, they can have indirect effects in mainly two ways:

- *Through resettlement*: People currently living in the reservoir area, when forced to move out, may have no alternative other than moving upstream into the catchment area of the reservoir. At the same time, they may be forced to change from irrigation agriculture in the river floodplain to upland, rainfed agriculture on hill slopes, a condition to which they are not adapted. This can lead to a rapid increase in vegetation destruction and consequent erosion damage in the watershed area. This is especially the case in projects where resettlement is insufficiently planned or inadequately implemented.
- *Through access*: In some cases, the catchment area of a project, which was almost inaccessible, is made easily accessible through the construction of access roads required for the construction of the project. This can lead to an uncontrolled immigration of squatters into hitherto uninhabited areas, with consequences for vegetation and soils as just described.

The EIA must identify problems that are likely to arise or already manifest in the watershed area, whether caused by the project itself or by independent conditions (see Figure 13.27 for the various parameters likely to have an influence). The watershed management plan lists the required measures for the short-, medium-, and long-term management of the area, with the aim of maintaining it in a condition that will allow for a maximum life span of the project. This, at the same time, will lead to a sustainable use of the natural resources (forest, water, soils, etc.) within the catchment area itself. The watershed management plan must address the following points:

- Protection of intact vegetation (e.g., remaining natural forests or other vegetation types or habitats).
- Maintenance of properly managed land use systems, be they agricultural, forestry, or other.
- Rehabilitation of degraded systems, for example, through reforestation, conversion of unsustainable land use types to sustainable ones (replacement of slash-and-burn cultivation with agroforestry) or other measures to reduce or halt erosion and soil destruction.

Development trends or potentially harmful future developments must be anticipated in order to formulate appropriate conservation or prevention measures. It is often necessary to establish an appropriate institutional framework to deal with the issues of planning and implementing land use in the watershed area. Close contacts between different agencies that often act independently are essential (forestry, agriculture, nature conservation, mining, and others) in order to achieve good results on a long-term basis.

When addressing the watershed management plan for an HEP, created in the framework of project planning, one fundamental difference to most of the other issues addressed in this chapter has to be pointed out. Problems in the (future) reservoir originate from places where the owner of the plant has very limited, if any, possibilities for direct intervention. Very often, a power company is not even the legal owner of the reservoir; it has only the right to use the water. And, of course, this company does not have the jurisdiction to force communities or industries in the catchment of their reservoirs to erect and operate sewage treatment stations.

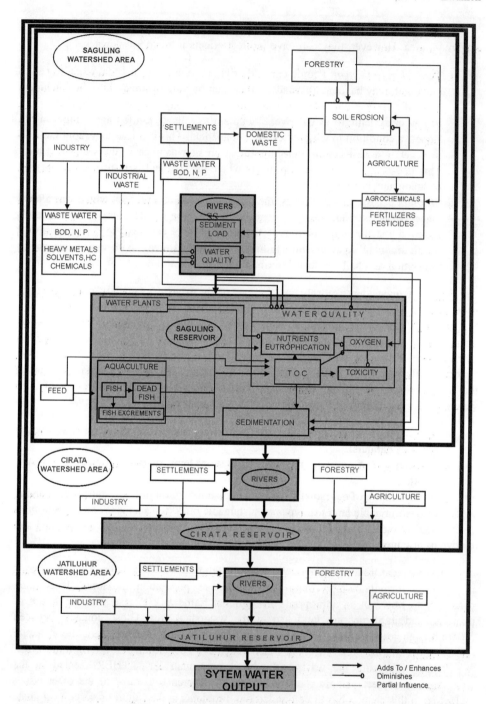

Figure 13.27. Diagrammatic presentation of the watershed in a cascading system of three reservoirs on Citarum River, Java, Indonesia. Some of the potential effects are illustrated only for the topmost part of the watershed (draining directly into Saguling Reservoir). Watershed management is made more complex by the fact that here, as in most cases of larger river basins, more than one political entity is involved (EWE and AMYTHAS, 1998).

For these reasons, watershed management or, to speak in the terms of Ayres et al. (1996), an integrated Water Resources Management (WRM) has to be implemented on a different level, and the EIA for a specific project can only provide input and assistance for this process. For the implementation of a watershed management, the following points have to be considered:

- A comprehensive approach includes all uses of water, and all structures and activities in the watershed.
- The unit is the watershed, not the political unit; all political units concerned must be integrated.
- Stakeholder participation to develop WRM at the community level, involving all concerned parties.
- Strengthen policies on land use in watershed to minimize erosion and chemical input.
- Information is required on water quality and quantity, land use changes, and trends in water demand.
- Establish mechanisms for solving conflicts.
- Improve water allocation and use efficiency (by proper pricing or other mechanisms).
- Ensure pollution prevention and abatement by regulation/enforcement and economic instruments.

Because it is in the common interest of all potential users of the water—within the catchment of the reservoir as well as in the downstream area—the situation also has to be addressed by all stakeholders in common. It goes without saying that such a comprehensive approach contains many problems in itself. For example, stakeholder involvement can be difficult due to the fact that impacts become manifest further downstream. This means that polluters are not harmed by the pollution they cause and are therefore not motivated to pay for mitigation measures from which they will not benefit.

An important contribution of the owner of the reservoir consists in carrying out careful water-quality monitoring, making the data available for the relevant political bodies, and actively participating in watershed management implementation. Simulation models can be useful for identifying the role of the various pollution sources and their relative importance in generating the water-quality problems.

Watershed management is crucial for the long-term operation of a hydropower project. Although water-quality problems caused by impoundment, even in the absence of mitigation measures, will be solved after a few years, reservoir eutrophication and siltation stemming from the watershed area will increase over the life span of the reservoir, unless the appropriate watershed management measures are being taken. It is evident that appropriate water-quality protection and watershed management measures have to be taken in any case, independent of whether there is a reservoir or not. Careful development and management of the water resources can contribute to an improvement of the overall situation.

Beyond EIA: The River Basin as Unit

It has been pointed out repeatedly that an EIA can only focus on an individual project and does not solve all the problems involved. This is quite obvious for the question of watershed management (Doolette & Magrath, 1990). On the other hand, the problem of overall optimization of resource allocation and use, considering all parties involved and all potential uses—as opposed to the maximization of one specific use—cannot be solved by an EIA. This has to be addressed by means of a much more comprehensive planning approach that focuses on the entire river basin as the planning unit.

References

Aegerter, S., & Messerli, P. (1981). The impact of hydroelectric power plants on a mountainous environment: A technique for assessing environmental impact. *Fachbeitr. Schweiz. MAB-Inf. Nr*, 9, 1–38.

Afifi, A. K. (1993). Role of Aswan High Dam in safeguarding Egypt from floods and droughts. In *High Aswan Dam, vital achievement fully controlled*. ICOLD 61st Executive Meeting and Symposium, Cairo, 1–6 November, pp. 139–150.

Ahmad, A. (1999). The Narmada Water Resources Project, India: Implementing sustainable development. *Ambio*, 28(5), 398–403.

Alford, D., Cunha, S. F., & Ives, J. D. (2000). Lake Sarez, Pamir Mountains, Tajikistan: Mountain hazards and development assistance. *Mountain Research and Development*, 20(1), 20–23.

Alissow, B. P., Drosdow, O. A., & Rubinstein, E. S. (1956). *Lehrbuch der Klimatologie*. Berlin: Deutscher Verlag der Wissenschaften.

Anonymous. (1968). Hydro-electric power development threatens Uganda's Murchison Falls National Park. Kampala, 8pp. (unpublished report).

Anonymous. (1970). Crocodiles going in Uganda. *Oryx*, 10, 207.

Antwi, L. A. K., & Ofori-Dansin, P. K. (1993). Limnology of a tropical reservoir (the Kpong Reservoir in Ghana). *Tropical Ecology*, 34(1), 75–87.

Asian Development Bank. (1990). *Environmental guidelines for selected infrastructure projects*. Manila: Office of the Environment, Author.

Asian Development Bank. (1991). *Environmental guidelines for selected agricultural and natural resources development projects*. Manila: Office of the Environment, Author.

Asian Development Bank. (1992). *Guidelines for the health impact assessment of development projects*. ADB Environment Paper No. 11, 45 pp. + App. Manila: Author.

Asian Development Bank. (1998). *Handbook on resettlement: A guide to good practice*. Manila: Author.

Ayres, W. S., Busia, A., Dinar, A., Hirji, R., Lintner, S. F., McCalla, A. F., & Fobelus, R. (1996). *Integrated lake and reservoir management: World Bank approach and experience*. World Bank Technical Paper No. 358.

Baijot, E., Moreau, J., & Bouda, S. (1994). *Aspects hydrobiologiques et piscicoles des retenues d'eau en zone Soudano-Sahélienne*. Wageningen: CTA.

Baldwin, M. F. (Ed.). (1991). Natural resources of Sri Lanka: Conditions and trends. Report prepared for the Natural Resources, Energy and Science Authority of Sri Lanka. Colombo 280pp.

Balon, E. K. (1974). Fishes of lake Kariba, Africa. Hong Kong: T.F.H. Publications.

Barbier, E. B., & Thompson, J. R. (1998). The value of water: Floodplain versus large-scale irrigation benefits in northern Nigeria. *Ambio*, 27(6), 434–440.

Bechara, J. A., & Andreani, N. L. (1989). El macrobentos de una laguna cubierta por Eichhornia crassipes en el valle de inundación del Río Paraná (Argentina). *Tropical Ecology*, 30(1), 142–155.

Berner, P., & Geiger, W. (1988). Fischereigutachten Neubau Kraftwerk Rheinfelden. KWR, EWI, 106pp. (unpublished report).

Bilio, M. (1986). Fischerei. In: *Der Assuan-Staudamm und seine Folgen*, pp. 116–118. Frankfurt: Kreditanstalt für Wiederaufbau.

Biney, C. A. (1987). Changes in the chemistry of a tropical man-made lake, the Densu reservoir, during five years of impoundment. *Tropical Ecology*, 28(2), 222–231.

Biro, Y. E. K. (1998). Valuation of the environmental impacts of the Kayraktepe Dam/Hydroelectric Project, Turkey: An exercise in contingent valuation. *Ambio*, 27(3), 224–229.

Biswas, A. K. (1995). Environmental sustainability of Egyptian agriculture: Problems and perspectives. *Ambio*, 24(1), 16–20.

Biswas, S. (1973). Limnological observations during the early formation of Volta lake in Ghana. In W. C. Ackermann, G. F. White, E. B. Worthington, & J. L. Ivens (Eds.), *Man-made lakes: Their problems and environmental effects. Geophysical Monograph, 17,* 121–128.

Blake, B. F. (1976). Lake Kainji, Nigeria summary of the changes within the fish population since the impoundment of the Niger in 1968. *Hydrobiologia, 53*(2), 131–137.

Blake, B. F. (1977). The effect of the impoundment of Lake Kainji, Nigeria, on the indigenous species of mormyrid fishes. *Freshwater Biology, 7,* 37–42.

Boothroyd, P., Knight, N., Eberle, M., Kawaguchi, J., & Gagnon, C. (1995). The need for retrospective impact assessment: The megaprojects example. *Impact Assessment, 13*(3), 253–271.

Buckley, R. C. (1991). How accurate are environmental impact predictions? *Ambio, 20*(3–4), 161–162.

Calder, N. (1999). The carbon dioxide thermometer and the cause of global warming. *Energy and Environment, 10*(1), 1–18.

Cans, R. (1997). *La bataille de l'eau.* Paris: Le Monde Editions.

Cernea, M. (1997). The risks and reconstruction model for resettling displaced populations. *World Development, 25*(10), 1569–1588.

Cernea, M. M. (1996). Understanding and preventing impoverishment from displacement: Reflections on the state of knowledge. In C. McDowell (Ed.), *Understanding impoverishment: The consequences of development-induced displacement:* Vol. 2. Refugee and Forced Migration Studies (pp. 13–32). Oxford, UK: Berghahn Books.

Cernea, M. M. (Ed.). (1999). The economics of involuntary resettlement: Questions and challenges. Washington, DC: World Bank.

Chang, Y. T., Khoo, K. J., Loo, A. C., Piggott, G. A., & Zwahlen, R. (1995). Human resettlement in dam projects— an overview of issues and sensitivities. *Proceedings of the MWA International Conference on Dam Engineering,* 1–2 August 1995, pp. 549–553. Kuala Lumpur, Malaysia.

Charest, P. (1995). Aboriginal alternatives to megaprojects and their environmental and social impacts. *Impact Assessment, 13*(4), 371–386.

Chau, K.-C. (1995). The three-gorges project of China: Resettlement prospects and problems. *Ambio, 24*(2), 98–102.

Chile. (1997). Reglamento del Sistema de Evaluación de Impacto Ambiental. Núm. 3, Santiago, 27 de marzo de 1997, Art. 7. Ministerio Secretaría General de la Presidencia de la República.

Cott, H. B. (1968). The status of the Nile crocodile below Murchison Falls. *IUCN Bulletin, 2*(8), 62–64.

DeAngelis, D. L., Gardner, R. H., & Shugart, H. H. (1981). Productivity of forest ecosystems studied during the IBP: The woodlands data set. In D. E. Reichle (Ed.), *Dynamic properties of forest ecosystems* (pp. 567–673). International Biological Programme 23. Cambridge, UK: Cambridge University Press.

de Vaulx, M. (1997). Vers une guerre de l'eau en méditerranée au XXIème Siècle? In H. El Malki (Ed.), *L'Annuaire de la Méditerrannée 1997* (pp. 112–120). Rabat/Paris: GERM-PUBLISUD.

Dister, E. (1984). Zur ökologischen Problematik der geplanten Donau-Staustufe bei Hainburg/Niederösterreich. *Natur und Landschaft, 59*(5), 190–194.

Dixon, J. A., Talbot, L. M., & LeMoigne, G. J. M. (1989). *Dams and the environment: Considerations in World Bank projects.* World Bank Technical Paper No. 110, 34pp. Washington, DC: World Bank.

Doolette, J. B., & Magrath, W. B. (Eds.). (1990). *Watershed development in Asia: Strategies and technologies.* World Bank Technical Paper No. 127, 227pp. Washington, DC: World Bank.

Doorenbos, J., & Pruitt, W. O. (1977). *Guidelines for predicting crop water requirements.* FAO Irrigation and Drainage Paper No. 24, 144pp. Rome. Source: IEE, 1994: The Environmental Effects of Electricity Generation.

Dorcey, T., Steiner, A., Acreman, M., & Orlando, B. (Eds.). (1997). *Large dams: Learning from the past, looking at the future.* Workshop Proceedings, Gland, Switzerland, April 11–12, 1997. Gland and Washington, DC: IUCN/World Bank.

Duckworth, J. W., Salter, R. E., & Khounboline, K. (1999). Wildlife in Lao PDR: 1999 Status Report. Vientiane: IUCN, WCS, CPAWM.

Dumont, H. J. (1986). The Tanganyika sardine in Lake Kivu: Another ecodisaster for Africa? *Environm. Conserv., 13*(2), 143–148.

Ecotec, Ecosens. (1993). Projet d'aménagement hydroélectrique Mauvoisin II. Augmentation de la puissance 550 MW. Rapport d'impact sur l'environnement. Forces Motrices de Mauvoisin S.A., Sion (unpublished report).

Edmonds, R. L. (1992). The Sanxia (Three Gorges) project: The environmental argument surrounding China's super dam. *Global Ecology and Biogeography Letters, 2*(4), 105–125.

English, R. (1998). *Socio-economic profile of the Nam Ngum Watershed*. ADB, TA 2734-LAO: Nam Ngum Watershed Management.
EWE and AMYTHAS. (1998). Cirata and Saguling: Environmental studies and training. Final report (unpublished). Jakarta: PLN, Environmental Division.
EWE and SOGREAH. (1984). Aménagement Hydroélectrique de Nangbéto: Déboisement de la cuvette du barrage, projet de développement de la pêche. Faisabilité technique et économique. Communauté Electrique du Bénin, République Togolaise (unpublished report).
EWE. (1980). Ayago-Nile Hydroelectric Project, Uganda: Ecology. Electrowatt Engineering, 75 pp. (unpublished report).
EWE. (1984). Estudio Ambiental y ecológico de la centrales Petrohué y Canutillar: Vol. I. Central Petrohué, Vol. II. Central Canutillar, Vol. III. Centrales Petrohué y Canutillar. ENDESA, Santiago, Chile (unpublished report).
EWE. (1987a). Projet de surélévation du barrage de Mauvoisin. Analyse des effets sur l'environnement. Forces Motrices de Mauvoisin, Sion, Switzerland (unpublished report).
EWE. (1987b). Saisonspeicherwerk Curcuisa. Bericht zur Umweltverträglichkeitsprüfung (UVP). Misoxer Kraftwerke, Zurich, Switzerland (unpublished report).
EWE. (1989). Aménagement hydroagricole de 35 barrages: Santé. 45 pp. Environnement. 16 pp. Ministère de l'Eau, Burkina Faso (unpublished report).
EWE. (1994a). Egiin Hydroelectric Project. Environmental Impact Study. Asian Development Bank; Central Energy System, Mongolia (unpublished report).
EWE. (1994b). Murum Hydroelectric Project. Feasibility Study. Final Report, Vol. 3A: Environmental Impact Assessment (EIA). Sarawak Electricity Supply Corporation, Sarawak, Malaysia (unpublished report).
EWE. (1995). Proyecto Central Hidroeléctrico Ralco. Evaluación Preliminar de Impacto Ambiental. Empresa Nacional de Electricidad S.A. (ENDESA). Santiago, Chile (unpublished report).
EWE. (1995). Proyecto Central Hidroeléctrico Ralco. Evaluación de Impacto Ambiental. Empresa Nacionalde Electricidad SA (ENDESA), Santiago, Chile (unpublished report).
EWE. (1996). Houai Ho Hydroelectric Project, Lao PDR: EIA, Daewoo Corporation, Seoul, Korea, 211 pp. (unpublished report).
EWE. (1997). Central Pangue: Estudio condiciones agues abajo. Pangue S.A., Santiago de Chile (unpublished report).
EWE. (1998). Monographie régionale de l'environnement. Région Economique de l'Oriental. Rapport de synthèse. Royaume du Maroc, Ministère de l'Environnement (unpublished report).
EWE. (1998). Nam Ngum 2 Hydroelectric Project, Lao P.D.R.: Environmental Impact Assessment. The Shlapak Group (unpublished report).
EWE. (1999). Gorno-Badakhshan Energy Development Project, Tajikistan. Electrowatt Engineering for IFC, Washington, and AKFED, Dushanbe.
EWE. (2001). Evaluation of the GBAO electric networks and assistance in project preparation: Vol. 3. Environmental and socio-economic measures. International Finance Corporation and Government of Tajikistan (unpublished report).
Fearnside, P. M. (1995). Hydroelectric dams in the Brazilian Amazon as sources of "greenhouse" gases. *Environmental Conservation*, *22*(1), 7–19.
Fearnside, P. M. (1997). Greenhouse-gas emissions from Amazonian hydroelectric reservoir: The example of Brazil's Tucuruí dam as compared to fossil fuel alternatives. *Environmental Conservation*, *24*(1), 64–75.
Gagnon, L. (1998). State of research in 1998 confirms that hydropower is a tool to reduce greenhouse gas emissions. Modelling, Testing and Monitoring for Hydro Powerplants—III, Aix-en-Provence, 1998; Proceedings: 95–105.
Gagnon, L., & Chamberland, A. (1993). Emissions from hydroelectric reservoirs and comparison of hydroelectricity, natural gas and oil. *Ambio*, *22*(8), 568–569.
Gao, L. S. (1998). On the Dais traditional irrigation system and environmental protection in Xishuangbanna. Kunming: Yunnan Nationality Press.
Geiger, R. (1950). Das Klima der bodennahen Luftschicht. Braunschweig: Vieweg.
Gheberyesus, T. A., Haile, M., Witten, K., et al. (1999). Incidence of malaria among children living near dams in northern Ethiopia: Community based incidence survey. *British Medical Journal*, *319*, 663–666.
Goldsmith, E., & Hildyard, N. (1984). The social and environmental effects of large dams: Vol. 1. Overview, Vol. 2. Cast studies, Vol. 3. A review of the literature. Wadebridge, UK: Wadebridge Ecological Centre.
Goldsmith, E., & McCully, P. (1993). Warum Staudämme gebaut werden. *E+D Entwicklung Déveleppement*, *41*, 21–23.

Goodland, R. J. A., Daly, H. E., & Serafy, S. E. (1993). The urgent need for rapid transition to global environmental sustainability. *Environmental Conservation, 20*(4), 297–309.

Gray, A. (1996). Indigenous resistance to involuntary relocation. In C. McDowell (Ed.), *Understanding impoverishment: The consequences of development-induced displacement* (pp. 99–122). Refugee and Forced Migration Studies, Vol. 2. Oxford, UK: Berghahn Books.

Hong, E. (1987). Natives of Sarawak: Survival in Borneo's vanishing forests. Malaysia: Institut Masyarakat.

Hoy, R. D., & Stephens, S. K. (1979). Field study of lake evaporation. Canberra: Australian Government Publishing Service.

Hughes, F. M. (1990). The influence of flooding regimens on forest distribution and composition in the Tana river floodplain. *Kenya Journal of Applied Ecology, 27*(2), 475–491.

Hunter, J. M., Rey, L., Chu, K. Y., Adekolu-John, E. O., & Mott, K. E. (1993). *Parasitic diseases in water resources development: The need for intersectoral negotiation.* Geneva: World Health Organization.

ICOLD. 1. (1985). *Dams and the environment: Notes on regional influences.* International Commission on Large Dams, Bulletin No. 50, 91pp.

ICOLD. (1994). Dams and environment: Water quality and climate. Barrages et environnement: Qualité de l'eau et climat. ICOLD, Commission Internationale des Grands Barrages, Bulletin 96, 87pp.

Institute of Electrical Engineers. (1994). *The environmental effects of electricity generation.* London: Author.

Interim Mekong Committee. (1982). *Environmental impact assessment: Guidelines for application to tropical river basin development.* Bangkok: Mekong Secretariat.

Isaacson, R. L., & Jensen, K. F. (Eds.). (1992). *The vulnerable brain and environmental risks*: Vol. 1. Malnutrition and hazard assessment, 268pp., Vol. 2. Toxins in food, 332pp., Vol. 3. Toxins in air and water, 358pp. New York/London: Plenum Press.

Jalal, K. F., Kelles-Viitanen, A., & Wilkinson, G. (1997). *The Bank's policy on indigenous peoples.* Asian Development Bank Working Paper 4-97.

Johnson, A. W., & Earle, T. (1997). *The evolution of human societies: From foraging group to agrarian state.* Stanford, CA: Stanford University Press.

Julin, H. M., Bourne, A. S., & Chan, R. R. (1987). Effects of adult and larval *Cyrtobagus salviniae* on the floating weed *Salvinia molesta. Journal of Applied Ecology, 24*(3), 935–944.

Kapetsky, J. M., & Petr, T. (Eds.). (1984). *Status of African reservoir fisheries.* CIFA Technical Paper No. 10. Rome: FAO.

Katete, F. X. (1968). The mistake at Murchison Falls. *Africana, 3*(8), 18–21.

Kuhn, W. (1977). *Berechnung der Temperatur und Verdunstung alpiner Seen auf klimatologisch-thermodynamischer Grundlage.* Arbeitsbericht der Schweizerischen Meteorologischen Zentralanstalt Nr. 70. Zürich.

Kukule Ganga Joint Venture. (1992). Kukule Ganga Hydropower Project. Feasibility Report, Vol. 3, Annex I: Environmental Impact Assessment. Ceylon Electricity Board, Colombo, Sri Lanka (unpublished report).

Ladeishchikov, N. P., & Obolkin, V. A. (1990). Climate and climatic resources of reservoirs. *Arch. Hydrobiol. Beih., 33*, 689–693.

Lai, P. F. (1992). Batang Ai HEP: Post-impoundment medical study. Sarawak Electricity Supply Corporation, 60pp. (unpublished report).

Lai, P. F. (1992). The Bakun Hydroelectric Project: Medical ecological baseline study and health impact assessment. Sarawak Electricity Supply Corporation, 70pp. (unpublished report).

Laki, S. L. (1998). Management of water resources of the Nile basin. *International Journal of Sustainable Development and World Ecology, t*(4), 288–296.

Lassally-Jacob, V. (1996). Land-based strategies in dam-related resettlement programmes in Africa. In C. McDowell (Ed.), *Understanding impoverishment: The consequences of development-induced displacement* (pp. 187–199). Refugee and Forced Migration Studies, Vol. 2. Oxford, UK: Berghahn Books.

Leentvaar, P. (1973). Lake Brokopondo. In W. C. Ackermann, G. F. White, E. B. Worthington, & J. L. Ivens (Eds.), *Man-made lakes: Their problems and environmental effects.* Geophysical Monograph, *17*, 186–196.

Lelek, A. (1973). Sequence of changes in fish populations of the new tropical man-made lake, Kainji, Nigeria, West Africa. *Archives of Hydrobiology, 71*(3), 381–420.

Lelek, A., & El-Zarka, S. (1973). Ecological comparison of the preimpoundment and postimpoundment fish faunas of the River Niger and Kainji. In W. C. Ackermann, G. F. White, E. B. Worthington, & J. L. Ivens (Eds.), *Man-made lakes: Their problems and environmental effects.* Geophysical Monographs, *17*, 655–660.

Liao, W.-L., Bhargava, D. S., & Das, J. (1988). Some effects of dams on wildlife. *Environmental Conservation, 15*(1), 68–70.

Liebenthal, A. (Ed.). (1996). *The World Bank's experience with large dams: A preliminary review of impacts.* The World Bank, Operations Evaluation Department.

Lohani, B. N., Evans, W. N., Everitt, R. R., Ludwig, H., Carpenter, R. A., & Tu, S. L. (1997). *Environmental impact assessment for developing countries in Asia*: Vol. I. Overview, 364pp., Vol. II. Selected case studies, 314pp. ADB, Mainla.

MacKinnon, K., Hatta, G., Halim, H., & Mangalik, A. (1996). *The Ecology of Kalimantan: Indonesian Borneo*. The Ecology of Indonesia Series, Vol. III. Singapore: Periplus Editions (HK).

Madsen, H., Daffalla, A., Karoum, K., & Frandsen, F. (1988). Distribution of freshwater snails in irrigation schemes in the Sudan. *Journal of Applied Ecology, 25*(3), 853–866.

Manser, B. (1992). *Stimmen aus dem Regenwald: Zeugnisse eines bedrohten Volkes*. Bern: WWF/Zytglogge.

McCully, P. (1996). *Silenced rivers: The ecology and politics of large dams*. London: Zed Books.

Moattassem, M., Hassan, S. K. A. G., Makary, A. Z., & Sherbini, A. (1993). River Nile water quality status (July 1991/April 1992). In *High Aswan Dam, vital achievement fully controlled* (pp. 411–432). ICOLD, 61st Executive Meeting and Symposium, Cairo, 1–6 November.

Montén, E. (1985). *Fish and turbines: Fish injuries during passage through power station turbines*. Vattenfall.

Morton, A. J., & Obot, E. A. (1984). The control of *Echinochloa staginata* (Retz.) P. Beauv. by harvesting for dry season livestock fodder in Lake Kainji, Nigeria—a modelling approach. *Journal of Applied Ecology, 21*(2), 687–694.

NTEC. (1997). Nam Theun 2 Hydroelectric Project: Resettlement Action Plan. Draft, Nam Theun 2 Electricity Consortium, Vientiane, Lao PDR (unpublished report).

Obot, E. A., & Ayeni, J. S. O. (1987). Conservation and utilization of aquatic macrophytes in Lake Kainji, Nigeria. *Environmental Conservation, 14*(2), 168–170.

Odingo, R. S. (1979). *An African dam: Ecological survey of the Kamburu/Gtaru hydroelectric dam area, Kenya*. Ecological Bulletin 29. Stockholm: Swedish Natural Science Research Council.

Oliver-Smith, A. (1996). Fighting for a place: The policy implications of resistance to development-induced resettlement. In C. McDowell (Ed.), *Understanding impoverishment: The consequences of development-induced displacement* (pp. 77–97). Refugee and Forced Migration Studies, Vol. 2. Oxford, UK: Berghahn Books.

Oomen, J. M. V. (1983). *Rapport de la mission de courte durée "Aspects sanitaires du projet 40 petits barrages" en Haute Volta*. Rapport polycopié, 82 pp., Euroconsult, Arnhem, NL.

Orbig, K. E. (1989). Wasserwirtschaftliche Grundsatzfragen bei Ausleitungsstrecken. Informationsber. Bayr. Landesamt Wasserwirtsch., München, pp. 9–18.

Pavlov, D. S. (1989). *Structures assisting the migration of non-salmonid fish: USSR*. FAO Fisheries Technical Paper No. 308, 97pp.

Petr, T. (1974). Distribution, abundance and food of commercial fish in the Black Volta and the Volta man-made lake in Ghana. *Hydrobiologia, 45*(2–3), 303 337.

Rafatjah, H. A., & Kuo, C. (Eds.). (1982). *Manual on environmental management for mosquito control*. WHO Offset Publications, No. 66 283pp. Geneva: WHO.

Reeves, R. R., & Leatherwood, S. (1994). Dams and river dolphins: Can they coexist? *Ambio, 23*(3), 172–175.

Reichenbach-Klinke, H. H. (1980). *Krankheiten und Schädigungen der Fische*. Stuttgart: Fischer.

Room, P. M., & Thomas, P. A. (1985). Nitrogen and establishment of a beetle for biological control of the floating weed Salvinia in Papua New Guinea. *Journal of Applied Ecology, 22*(1), 139–156.

Rudd, J. W. M., et al. (1993). Are hydroelectrical reservoirs significant sources of greenhouse gases? *Ambio, 22*(4), 246–248.

Ryder, G. (Ed.). (1990). *Damming the three gorges: What dam-builders don't want you to know*. Toronto: Probe International.

Ryding, S. O., & Rast, W. (Eds.). (1989). *The control of eutrophication of lakes and reservoirs*. Man and the Biosphere Series, Vol. 1. Paris: UNESCO, and Park Ridge, IL: Parthenon.

Schetagne, R. (1994). Water quality modifications after impoundment of some large northern reservoirs, *Arch. Hydrobiol. Beih., 40*, 223–229.

Schmidtke, R. F., & Ottl, A. (1988). Bibliographie zum Thema Dotation/Mindestabfluss Restwasserführung in wasserkraftbedingten Ausleitungsstrecken. *Wasser, Energie, Luft, 80*(11/12), 304–306.

Schuh, G. E. (1986). Environmental problems of irrigation projects. *Environmental Conservation, 13*(1), 74.

Scudder, T. (1996). Development-induced impoverishment, resistance and river-basin development. In C. McDowell (Ed.), *Understanding impoverishment: The consequences of development-induced displacement* (pp. 49–74). Refugee and Forced Migration Studies, Vol. 2. Oxford, UK: Berghahn Books.

SEATEC (1997). Nam Theun 2 Hydroelectric Project. Environmental Assessment and Management Plan (EAMP). Draft Final Report. SEATEC International in association with Sinclair Knight Merz, ECI, Inc., and EDAW, Australia (unpublished report).

Seers, D., et al. (1979). The rehabilitation of the economy of Uganda: Vol. II. Sector papers (pp. 25–359). London: Commonwealth Fund for Technical Co-operation, Commonwealth Secretariat.
SEMOR. (2000). Ilisu hydropower project. Draft Resettlement Action Plan. DSI, Ankara, Turkey (unpublished report).
Simanov, L., & Kantorek, J. (1990). Water quality in the Sance reservoir (North Moravia). *Arch. Hydrobiol. Beih.*, *33*, 869–874.
Sodnom, N., & Yanschin, A. L. (Eds.). (1990). *National atlas*. Ulan Bator and Moscow: Peoples' Republic of Mongolia, Academy of Science MPR, Academy of Science USSR. (in Russian)
Swiss Confederation. (1991). *Bundesgesetz über den Schutz der Gewässer*. (Gewässerschutzgesetz, GSchG).
Thomas, P. A., & Room, P. M. (1986). Successful control of the floating weed *Salvinia molesta* in Papua New Guinea: Useful biological invasion neutralizes a disastrous one. *Environmental Conservation*, *13*(3), 242–248.
Tiffen, M. (1991). Guidelines for the incorporation of health safeguards into irrigation projects. PEEM Guidelines Series 1, 2nd ed. Geneva: WHO.
Tolba, M. K., El-Kholy, O. A., El-Hinnawi, E., Holdgate, M. W., McMichael, D. F., & Munn, R. E. (Eds.). (1993). *The world environment 1972–1992: Two decades of challenge*. London: Chapman & Hall for UNDP.
Tuna, N., Öztürk, J., & Velibeyoglu, J. (Eds.). (2001). Salvage project of the archaeological heritage of the Ilisu and Carchemish dam reservoirs: Activities in 1999. Ankara: METU Centre for Research and Assessment of the Historic Environment.
Turan, I. (2000). International aspects of water issues. In *Water and development in Southeastern Anatolia: Essays on the Ilisu dam and GAP* (pp. 38–45). Proceedings of a Seminar, Turkish Embassy, London.
Tyagi, B. K., & Chaudhary, R. C. (1997). Outbreak of falciparum malaria in the Thar Desert (India), with particular emphasis on physiographic changes brought about by extensive canalization and their impact on vector density and dissemination. *Journal of Arid Environments*, *36*(3), 541–555.
Ulluwishewa, R. (1997). Women, development and the emergence of household energy crisis in the dry zone of Sri Lanka: Implications for sustainable development. *Sri Lanka Journal of Social Sciences*, *20*(1–2), 89–115.
Ulrich, M. (1991). Modelling of chemicals in lakes—development and application of user-friendly software (MASAS & CHEMSEE) on personal computers. Diss. ETH/EAWAG No. 9632. Zurich.
van der Heide, J. (1982). Lake Brokopondo: Filling phase limnology of a man-made lake in the humid tropics. Amsterdam: Diss. Univ. Amsterdam.
Vila Pinto, I. (1989). Diagnóstico y evaluación de la fauna íctica del Rio Maule. Universidad de Chile; Pehuenche S.A.: Informe Policopiado.
Villagrán, M. C. (1980). Vegetationsgeschichtliche und pflanzensoziologische Untersuchungen im Vicente Pérez Rosales Nationalpark (Chile). *Dissertationes Botanicae*, *54*, 165pp.
Visser, S. A. (1973). Preimpoundment features of the Kainji area and their possible influence on the ecology of the newly formed lake. In W. C. Ackermann, G. F. White, E. B. Worthington, & J. L. Ivens (Eds.), *Man-made lakes: Their problems and environmental effects*. Geophysical Monographs, *17*, 590–595.
Walter, H., & Breckle, S. W. (1984). *Oekologie der Erde: Vol. 2. Spezielle Oekologie der tropischen und subtropischen Zonen*. Stuttgart: UTB, Fischer.
Walter, H., & Breckle, S. W. (1986). *Oekologie der Erde; Band 3: Spezielle Oekologie der gemässigten und arktischen Zonen Euro-Nordasiens*. Stuttgart: Fischer.
WCD. (2000). *Dams and development: A new framework for decision-making*. The Report of the World Commission on Dams. London: Earthscan.
World Bank. (1991). *Environmental Assessment Sourcebook: Vol. I. Policies, procedures, and cross-sectoral issues*. World Bank Technical Paper No. 139, 227pp. Washington, DC: Environment Department, Author.
World Bank. (1991). *Environmental Assessment Sourcebook: Vol. II. Sectoral guidelines*. World Bank Technical Paper No. 140, 282pp. Washington, DC: Environment Department, Author.
World Bank. (1991). *Environmental Assessment Sourcebook: Vol. III. Guidelines for environmental assessment of energy and industry projects*. World Bank Technical Paper No. 154, 237pp. Washington, DC: Environment Department, Author.
World Bank. (1999). *Resettlement and rehabilitation guidebook*. CD-Rom. Washington, DC: Author.
World Bank Operational Directive 4-30. (1990). *Involuntary resettlement*. Washington, DC: Author.
World Bank Operational Directive 4-20. (1991). *Indigenous peoples*. Washington, DC: Author.
Wetzel, R. G., & Gopal, B. (Eds.). (1999). Limnology in developing countries (Vol. 2). New Delhi: International Association for Limnology, International Scientific Publications.
Whitmore, T. C. (1985). *Tropical rain forests of the Far East* (2nd ed.). Oxford, UK: Oxford University Press.

Whitten, T., Soeriaatmadja, R. E., & Afiff, S. A. (1996). *The ecology of Java and Bali*. Ecology of Indonesia Series, Vol. II: 1028pp. Singapore: Periplus.
World Health Organization. (1992). *Our planet, our health*. Report of the WHO Commission on Health and Environment. Geneva: Author.
World Health Organization. (1993). *The control of schistosomiasis*. Second Report of the WHO Expert Committee. WHO Technical Report Series No. 830, 86pp. Geneva: Author.
World Health Organization. (1999). *Removing obstacles to healthy development: Report on infectious diseases*. Geneva: Author.
World Wide Fund for Nature. (1999). *A place for dams in the 21st century: A WWF International Discussion Paper*. Gland: WWF International.
Worley & Lahmeyer International. (2000). Hydropower strategy development for Lao PDR. Final Report, Volume A: Hydropower strategy, 195pp. (unpublished report).
Worthington, E. B. (1989). The lake Victoria Lates saga. *Environmental Conservation*, *16*(3), 266–267.
Worthington, E. B., & Lowe-McConnell, R. (1994). African lakes reviewed: Creation and destruction of biodiversity. *Environmental Conservation*, *21*(3), 199–213.
Yao, J. (Ed.). (1991). *Three gorges: What future benefits for China*? Beijing: China Today Press.
Zauke, G.-P., Niemeyer, R.-G., & Gilles, K.-P. (1992). *Limnologie der Tropen und Subtropen: Grundlagen und Prognoseverfahren der limnologischen Entwicklung von Stauseen*. Landsberg/Lech: Ecomed.
Zhong, Y., & Power, G. (1996). Some environmental impacts of hydroelectric projects on fish in Canada. *Impact Assessment*, *14*(3), 285–308.
Zwahlen, R. (1992a). Failure of irrigation projects and consequences for a different approach: A case study. *Ecological Economics*, *5*(2), 163–178.
Zwahlen, R. (1992b). The ecology of Rawa Aopa, a peat-swamp in Sulawesi, Indonesia. *Environmental Conservation*, *19*(3), 226–234.
Zwahlen, R. (1995). How obstructive is environmentalism to hydropower development? Proceedings, MWA International Conference on Dam Engineering (pp. 541–548), 1–2 August, Kuala Lumpur, Malaysia.
Zwahlen, R. (1998). Research needs for water quality issues relating to hydro reservoirs. *Hydropower and Dams*, *5*(4), 71–75.
Zwahlen, R. (1999). Proyecto Multipropósito Urrá I. Calidad del agua. Conclusiones y recomendaciones. Gómez, Cajiao y Asociados, Bogotá (unpublished report).

INDEX

Abiotic environment, 303
Above ground biomass, 35
Acetyl cholinesterase, 239
Achhanthes minutissima, 12, 58
Acid drainage (AD), 133
Acid precipitation, 69
Acid sensitive, 68
Acid sensitive species, 69
Acidification, 180
Acidophilous, 67
Acropora, 182
Action of pollutants, 228
Aerosols, 154
Ageratum conyzoides, 273
Air quality, 308
Algal blooms, 250, 261
Alhagi camelorum, 272
Allozyme loci, 89
Alluvial floodplains, 65
Alnus, 276
Alnus nepalensis, 277
Alpha-mesosaprobic, 3
Aluminium, 134
Aluminium competition, 48
Alyssum bertolonii, 134
Amblystegium riparium, 81, 82, 87
Amblystegium tenax, 67, 86
Amelioration, 185
Amphora, 14
Amphora pediculus, 58
Anabaena, 203
Anacystis nidulans, 211
Anopheles, 288
Anoxic water layers, 318
Antarctic, 163

Antarctic algae, 160
Antecedent soil moisture, 273
Anthopleura, 192
Aquaculture pond, 137
Aquatic bryophytes, 55, 65, 66, 75, 81
Aquatic system health, 1
Aquatic weeds, 332
Aquifer, 249
Arctic algae, 160
Arctic aquatic ecosystem, 164
Arrested succession, 103
Artesian water, 249
Ascidians, 182
Automated biomonitors, 239

Bacillaria, 14
Bacterioplankton, 156
Baetis, 180
Benthic interface, 202
Beta-mesosaprobic, 3
Bioaccumulation of toxicants, 232
Biodiversity, 231
Biofilms, 57, 135
Bioindication methods, 231
Biological diversity, 21
Biological early warning systems, 239
Biological fertilizers, 157
Biological indicators, 2, 299
Biological monitoring, 2, 229
Biomass of tropical forests, 317
Biomonitoring, 66
Biomonitors, 2, 17, 67
Bioremediation, 134
Biosensors, 220, 240
Biotic environment, 335

Biotop, 234
Biotypological classification, 226
Blidingia, 103
Boeckella, 176
Boeckella gracilipes, 182, 183
Bog, 270
Bosmina, 181
Bottom dwelling organisms, 178
Brachinus caliciflorus, 177, 187
Brachythecium plumosum, 69
Brachythecium rivulare, 47
Bromus, 276
Bryophytes, 52
  as biomonitors of water quality, 72
Buffering capacity, 84
Bunding, 276
Buoyancy regulation, 201

Cadmium, 48
Cage fish breeding, 331
Calanus finmarchicus, 182
Calanus propinquus, 181
Calcifuge, 67
Calcium, 48
Calliarthon articulatum, 105
Calliergonella cupsidata, 84
Calmagrostis, 36
Calothrix, 209
Canalized succession, 120
Canonical Correspondence Analysis (CCA), 6
Canopy algae, 121
Capel wetlands, 258
Carassius auratus, 184
Cassia tora, 273
Cataracts, 175
Catchment area, 300
Caulerpa, 160
Ceratophyllum, 50
Chara, 137
Charophytes, 137
Chemical monitoring, 229
Chiloscyphus pallescens, 80
Chiloscyphus polyanthos, 69, 85
Chironomid larve, 178
Chlorocarbons, 202
Chlorofluorocarbons (CFCs), 202
Chlorogloepsis, 209

Chlorophyll florescence, 57
Chromatic adaptation, 202
Chromophore, 211
Chromosomal damage, 239
Chronosequences, 101, 114
Chydorus, 181
Cichlasoma nigrofasciatum, 185
Cinclidotus, 67, 70
Cinclidotus danubicus, 69
Cinclidotus fontinaloides, 54
Cinclidotus riparius, 87
Cladophora, 43, 51, 52, 160
Cladophora glomerata, 45, 236
Climacium americanum, 89
Climate, 303
Climate shifts, 113
Colonization, 105
Colonizer species, 258
Competitive dominance, 111
Conservation, 252, 272
Conservation value percentage (Cv), 272
Constructed wetland, 38
Continuum concept, 225
Contour farming, 277
Control function, 220
Copper, 49, 57
Corallina, 112
Cratoneum filicinum, 67, 68, 70, 80
Creation of reservoirs, 344
Crotalaria medicaginea, 273
Crown land, 253
Crustacea plankton, 235
Crytobagus, 332
Ctenophora, 14
Cyanobacteria, 157, 201
Cyclobutane pyrimidine dimers, 175, 210
Cyclops abyssorum, 181
Cyclotella, 14
Cymbella, 14
Cymbella minuta, 12
Cynodon dactylon, 272

Dam foundation, 334
Dam project affected peoples, 351
Dam project area types, 299
Dam projects, 281
Daphnia, 176, 181

Daphnia magna, 185, 187, 189
Daphnia monitor, 240
Deforestation, 133
Dengue, 357
Descriptive model for flooding, 25
Deterministic succession, 109, 120
Detoxifying enzymes, 81
Diaphanosoma, 176
Diaptomus minutus, 184
Diarrah, 270
Diarrheal disease, 357
Diatom index, 58
Diatoms, 1
  as biomonitors, 3
Dictyota dichotoma, 160
Digitaria adscendens, 273
Diplomystes nahulbutaensis, 340
Discharge pattern at dam sites, 310
Dissolved organic carbon (DOC), 156
Dissolved organic matter (DOM), 174
Diversity index, 7
Diversity of diatoms, 3
DNA damage, 210
Dobson units (DU), 150
Drawdown area, 300
Dreissena basket, 237
Dreissena polymorpha, 233, 241
Drepanocladus fluitans, 86
Dynamic biomonitors, 238

Echinochloa staginata, 332
Ecological polymorphism, 65
Ecological services, 275
Ecology of wetlands, 247
Ecosub, 266
Ecosystem integrity, 150
Ecotonal vegetation, 270
Ecotops, 226
Ecotoxicology, 221
Edge-to-area ratio, 271
Effluent water, 141
Eichhornia, 52
Eichhornia crassipes, 52, 56, 146
Elephas maximus, 340
Elodea, 50
Emergent plants, 256
Emissions from power stations, 309
Endangered species, 338

Engraulis mordex, 175
Enteromorpha, 103, 160
Environmental impact assessment (EIA), 285, 286
Environmental stresses, 201
Epipelic, 3
Epiphytic, 3
Episamic, 3
Epizoic, 3
Equisetum, 276
Equisetum arvense, 51
Erosion, 13
Erosive forces, 270
Erratic colonization, 104, 117
Erythema in human skin, 183
Escherichia coli, 211
Eucalyptus, 13
Euhalobous, 10
Eulittoral zone, 159
Eunotia, 14
Euphorbia hirta, 272
Euphotic zone, 158
European rivers, 65
Euryhalobous, 10
Eutrophic, 69
Eutrophication, 65, 133, 273, 333
Exotic weeds, 13
Extratrerrestrial UVR, 178

Facilitation model, 97, 113
Fenitrothion, 81
Fiddler crab, 238
Fish ladders, 343
Fish monitor, 240
Fish species, 341
Fish stocks, 349
Fissidens, 49
Fissidens polyphyllus, 47
Fissidens pulillus, 67, 70
Flood plain, 58
Floristic gradient, 69
Fluorimetry, 57
Fontinalis antipyretica, 47, 53, 67, 69, 81, 82
Food webs, 256
Forest products, 275
Forested wetlands, 274
Fragmentation of a river, 225

Fraxinus, 275
Freshwater ecosystems, 163
Frog hollows, 259
Frustulia, 14
Fucus, 103
Fungia, 182
Furbish lousewort, 22

Gadus morthua, 183
Ganga river, 273
Genetic changes due to metals, 50
Geographical information systems, 265
Gigartina, 115
Gloeocapsa, 208
Gomati river, 273
Gomphonema parvulum, 57, 58
Greenhouse gases, 307
Greenbelt, 275
Ground water, 262, 314
Gully reclamation, 277

Halocarbons, 154
Hard waters, 67
Heat shock response, 212
Heavy metals, 134
  in freshwaters, 43
Herbivory, 115
Hierarchical concept, 230
Homeostasis, 231
Hordeum, 276
Houay Ho, 326
Human environment, 299
Humpbacked model, 32
Hydrochemical determination, 67
Hydroelectricity, 65
Hydrographic, 87
Hydrology, 310
Hydroperiods, 274
Hydropower generation, 282
Hydrosphere, 308
Hygrohypnum duriusculum, 69
Hygrohypnum luridum, 70
Hyperaccumulator of Ca, Mg, Mn, Fe, Cu, and Zn, 141
Hyperaccumulator of minerals, 134
Hypertrophic status, 67
Impacts of dam construction, 288

Impoundment, 316
Indicator value of species, 7
Indicators, 257
  of bioavailability, 49
  of water quality, 80
Inhibition model, 97, 113
Intercellular metals, 82
Interruption of rivers, 342
Isoetes, 35

Jungermannia, 81, 85
Jungermannia vulcanicola, 55

Klebormidium, 50

Labophora variegata, 160
Lacustrine, 270
Lake acidification, 150
Lake invertebrates, 143
Lakes from quarrying, 250
Laminaria, 134
Land use, 355
Largest dams, 296
Lead, 48
Lemna, 52
Lemna trisulca, 52
Length polymorphism, 89
Lentic, 277
Leonotis nepataefolia, 273
Leskea polycarpa, 82
Lethal concentrations, 84
Life history traits, 113
Light repair, 183
Linker polypeptides, 204
Lipotes vexillifer, 340
Lithosphere, 334
Liththrix aspergillum, 105
Littoral systems, 99
Lobellia, 35
Longitudinal gradients, 225
Lotic, 277
Luciferase, 233
Luciferin, 233
Luminescence, 233

Macroalgae, 159
Macrocystis, 134
Macrocystis pyrifera, 105, 111, 123

Macroinvertebrates, 232, 237
Macrovertebrates, 17
Magnesium, 50
Malaria, 365
Management of the large rivers, 65
Manganese deposition, 54
Maps of metal contamination, 45
Marine, 270
Marsh, 270
Marsupella emarginata, 67
Mastogloia, 14
Mathematical model for species richness, 30
Mature assemblage, 119
Mauvoisin dam, 339
Mayfly, 178
Measure of similarity, 110
Melanins, 181
Membrane permeability, 204
Mercury, 2
Mercury pollution, 47
Mesohalobous, 10
Metal
  accumulation, 44, 48
  binding, 44
  contents of charophytes, 143
  loss, 49
  pollution, 236
  uptake, 49
Metals, 82
Methyl mercury, 47, 48
Microbial loop hypothesis, 156
Microcystis, 212
Microspora, 146
Microthamnion, 50
Microtox, 233
Midaltitudes, 152
Mine effluent, 54
Mine-void wetlands, 133
Minnow, 239
Mitigation measures, 325, 339
  for dams, 288
Modeling, 318
Models of succession, 113
Monitoring metal contamination, 51
Moss, 46
Mougeotia, 135, 146
Multicollinear predictors, 72

Multidimensional scaling, 13
Multisensor monitors, 241
Mussel monitor, 240
Mussels, 238
Mussel-watch, 233
Mustelus canis, 175
Mycosporine-like amino acid (MAAs), 150, 180, 207
Myriophyllum, 50
Mytilus edulis, 233

Nangbeto, 326
Natural biofertilizer, 203
Nature reserve, 13
Niches, 87
Nitella, 137, 141
Nitrogen conservation, 271
Nitrogen fixing cyanobacteria, 203
Nitrogenase, 205
Nitzschia palea, 58
Nonsuccessional colonization, 104
Nostoc, 203, 210
Nostoc spongiaforme, 157
Nuphar lutea, 81
Nutrient conservation, 269
  enrichment, 66
  filter, 271
  sinks, 275
Nutrients, 332

O_2 concentration in the water, 319
Oedogonium, 146
Oligohalobous, 10
Oligosaprobic, 3
Oligotrophic, 67, 69
Onchorynchus mykiss, 174, 175
Oncorhynchus clarki, 175
Ordination, 13
Organic pollutants, 82
Organic pollution, 3
Oryzias laptipes, 181
Oxygen balance, 322
Oxygen depletion, 318
Ozone hole, 192, 202
Ozone layer, 150

Paired plot technique, 277
Palustrine, 270

Papyrus, 36
Parthenium hysterophorus, 273
Passive monitoring, 234
Pathogens, 189
Perca flavescens, 174
Periphyton, 3, 7, 232
  sampler, 8
Petroleum spills, 187
Phalaris, 36
Phenomenology of succession, 99
Phosphorus, 271
  conservation, 271
  runoff, 273
Photoactivation, 185
Photobacterium phosphoreum, 233
Photochemical smog, 149
Photodeactivation, 185
Photoinduced tolerance, 184
Photoinduced toxicity, 187
Photoinhibition, 160
Photokinesis, 203
Photon yield, 57
Photoorientation, 149
Photophobia, 203
Photoprotection, 149, 206
Photoreactivation, 156
Photorecovery, 184
Photorepair mechanisms, 184
Photosynthetic active radiation (PAR), 149
Photosynthetic electron pathway, 202
Photosynthetic quantum yield, 161
Phototaxis, 203
Phragmites, 35
Phragmites australis, 275
Phycobiliprotein, 157, 204
Phycobilisomes, 205
Phycocyanin, 204
Phycoremediation, 133, 135
Physicochemical balance, 68
Phytochelatins, 56
Phytoextraction, 134
Phytomining, 134
Phytoplankton, 158
  blooms, 137, 262
  productivity, 149
Phytoremediation, 134
Picoplankton, 156

Pigment ratios, 55
Pimephales pomelas, 187
Pinnularia, 12, 14
Pistia stratiotes, 52
Pit wetlands, 133
Plant succession, 97
Plasmodium falciparum, 356
Platanus, 276
Pleurocarpous mosses, 89
Plume effect, 226
Pollutants in the aquatic ecosystem, 224
Polluted environments, 66
Pollution tolerant species, 87
Polychlorobiphenyls, 81
Polygonum amphibium, 272
Polymorphic, 86
Polyodon spathula, 175
Polysaprobic, 3
Populus, 275
Porella, 84
Posidonia, 265
Postmining environments, 249
Potamogeton pectinatus, 52
Potentiation, 185
Preconcentrator of heavy metals, 237
Predators, 189
Predicted error, 75
Prediction function, 220
Predictive model for shoreline wetlands, 23
Predictive models, 6
Principle axis correlation, 14
Pristine sites, 14
Productivity models, 156
Prokaryotes, 201
Protected species, 338
Pseudocyclic replacements, 104
Public health issues, 356
Public health measures, 358

Quercus, 275

Racomitrium aciculare, 67
Radiation amplification factor (RAF), 152
Radionuclides, 2
Ralfsia, 118
Ramsar convention, 269

Rana cascadae, 175
Rare species, 338
Reactive oxygen species, 189
Red lists, 338
Redundancy analysis, 73
Remediation, 134
Reoxygenation, 322, 323
Repair of DNA, 210
Reservoir
  development, 327
  eutrophication, 330
  induced seismicity, 335
Resettlement, 361
Resettlement area, 301
Resource development options, 284
Retina, 190
Rhine action program, 65
Rhizobium, 212
Rhopalodia, 14
Rhynchospora, 36
Rhynchostegium, 52, 53
Rhynchostegium ripariodes, 45, 68, 69, 70
Riccia fluitans, 82
Ricciocarpos natans, 82
Riparian, 271
Riparian and aquatic plants, 255, 261
Riparian degradation, 13
Riparian ecosystem, 271
Riparian ecosystem management (REM), 276
Rissoella verruculosa, 109, 119
River
  basin, 363
  blindness, 357
  characteristics, 311
  continuum, 302
  Danube, 236
  discharge, 302
  ecosystems, 224
  regimen, 346
Riverbed erosion, 313
Riverine, 270
Roads corridor, 301
Rubisco, 205
Ruellia tuberosa, 273
Run-of-river (ROR), 294
Rutilus rutilus, 190

Saccharum benghalensis, 272
Saccostrea commercialis, 105
Sahara desert, 276
Salinity, 9
Salinization, 133
Salix, 275
Salmo gairdneri, 241
Salmo salar, 174, 181
Salmo truta, 175
Salvinia, 52
Salvinia molesta, 332
Sand mine wetlands, 140
Sanitation measures, 358
Saprolegnia, 189
Sardar (Narmada) Sarovar, 305
Sargassum, 134
Scapania undulata, 47, 55
Schistidium alpicola, 82
Schistosomiasis, 356
Scoparia dulcis, 272
Scytonema hofmanii, 157
Scytonemin, 209
Sea urchins, 182
Seagrass, 159
Seagrass meadows, 264
Seagrasses, 265
Sediment
  flushing, 329
  load, 329
  transport, 348
Sediments, 54
Seeders, 255
Sentinel organisms, 232, 233
Shannon–Weaver index, 12, 58
Shelterbelts, 276
Shorelines, 134
Siberian reservoirs, 305
Sida acuta, 273
Signal function, 220
Sikkim Himalaya, 277
Silent spring, 219
Siltation, 13, 133, 329
Simulium, 357
Slimes, 261
Slow release fertilizer, 262
Smoltification, 174
Social impacts of dams, 302
Socioeconomic impacts, 284

Soil
    conservation, 269
    conserver, 275
    submodel, 276
Solar constant, 173
Species
    indicator values, 73, 75
    indicator variables, 75
    richness in wetlands, 33
    richness, 34
Spectral sensitivity, 182
Sphagnum, 270
Sphagnum auriculatum, 54
Sphagnum cupsidatum, 83
Sporebank, 145
Sporopollenin, 163
Sprouters, 255
Static biomonitors, 235
Stigeoclonium, 47, 48
Stoneworts, 137
Stratospheric ozone, 152
Stratospheric ozone layer, 15
Submerged biomass, 318
Submerged macrophytes, 256
Submerged moss, 53
Succession
    determinants, 114
    in rockey intertidal habitats, 102
    in salt marshes, 115
    patterns, 101, 102
Successional theory, 257
Sunburn in fishes, 175
Supralittoral zone, 159
Syechocystis, 212
Synergism, 185, 227

Tabellaria, 14
Tailings pond, 251, 252
Talitrus saltator, 192
Taxodium dischum, 275
Temperature variations, 304
Terracing, 276, 277
Terrestrial wetlands, 264
Testable predictions, 121
Thamnobryum alopecurum, 68, 69
Thermotolerance, 213
Tigriopus californianus, 191
Tilapia macrochir, 345

Tolerance model, 97, 113
Total mapping spectrometer, 202
Total protein profiles, 205
Toxic metal ions, 234
Toxicity bioassays, 229
Toxicometer, 241
Trajectory of succession, 103
Transmission line corridor, 301
Transverse gradients, 225
Trichopteras, 236
Trifolium, 276
Trimers, 205
Troposphere, 149
Trout hatcheries, 69
Turf forming algae, 121
Turnover time, 100
Tussock-forming, 272
Types of large dams, 282
Typha 35, 261

Ulmus, 275
Ulva, 115, 160
Unio pictorum, 241
Unionidae, 233
Upslope, 35
Uranium mine, 252
UV
    index, 154
    measurements, 152
    penetration into water, 154
    protective compounds, 180
    sensitivity hypothesis, 174
    vision in aquatic animals, 190
UV-A, 191
UV-A vision, 191
UV-B induced mortality, 182
UV-B radiation, 149
UVR effect on aquatic animals, 173
UVR in the fresh water, 179
UV-screening compounds, 180

Varzea, 270
Vector control, 358
Vectors, 356
Vertical gradients, 225

Water based disease, 356
Water borne diseases, 356

Water
  chemistry, 2
  level fluctuations, 22
  quality, 254, 65
  quality assessment, 229
Water related disease, 356
Water washed disease, 356
Waterbirds, 253
Watershed area, 301
Watershed management, 360
Watersheds, 271, 276
Waterways, 334
Wet meadow, 24
Wetland, 269
  creation, 249, 254
  definitions, 270
  for contaminant removal, 253
  in mine voids, 11

Wetland (*cont.*)
  losses, 276
  management for water resources, 38
  margins, 269
  for education, 253
  in gravel pits, 251
  on mined landscapes, 251
  for recreations, 253
Windbreaks, 276
World heritage area, 251

Xanthophyll cycle, 160
Xenobiotic factors, 65
Xenobiotic level, 233
Xenobiotics, 81
Xenopus laevis, 175
Xyrauchen texanus, 175

Zn concentration, 47